SMALL-SCALE TURBULENCE
AND MIXING IN THE OCEAN

FURTHER TITLES IN THIS SERIES

Elsevier Oceanography Series, 46

SMALL-SCALE TURBULENCE AND MIXING IN THE OCEAN

PROCEEDINGS OF THE 19TH INTERNATIONAL LIEGE COLLOQUIUM ON OCEAN HYDRODYNAMICS

Edited by

J.C.J. NIHOUL
University of Liège, B5 Sart Tilman, B-4000 Liège, Belgium

and

B.M. JAMART
MUMM, Institute of Mathematics, 15 Avenue des Tilleuls, B-4000 Liège, Belgium

ELSEVIER

Amsterdam — Oxford — New York — Tokyo   1988

ELSEVIER SCIENCE PUBLISHERS B.V.
Sara Burgerhartstraat 25
P.O. Box 211, 1000 AE Amsterdam, The Netherlands

Distributors for the United States and Canada:

ELSEVIER SCIENCE PUBLISHING COMPANY INC.
52, Vanderbilt Avenue
New York, NY 10017, U.S.A.

ISBN 0-444-42987-5 (Vol. 46)
ISBN 0-444-41623-4 (Series)

FOREWORD

The International Liège Colloquium on Ocean Hydrodynamics is organized annually. The topic differs from one year to another in an attempt to address, as much as possible, recent problems and incentive new subjects in physical oceanography.

Assembling a group of active and eminent scientists from various countries and often different disciplines, the Colloquia provide a forum for discussion and foster a mutually beneficial exchange of information opening on to a survey of major recent discoveries, essential mechanisms, impelling question-marks and valuable recommendations for future research.

The Scientific Organizing Committee and the participants wish to express their gratitude to the Belgian Minister of Education, the National Science Foundation of Belgium, the University of Liège, the Scientific Committee on Oceanographic Research (SCOR), the Intergovernmental Oceanographic Commission and the Division of Marine Sciences (UNESCO), and the Office of Naval Research for their most valuable support.

We thank the members of SCOR Working Group 69 for reviewing the papers published in these proceedings.

Jacques C. J. Nihoul   Bruno M. Jamart

TABLE OF CONTENTS

VIII

X

LIST OF PARTICIPANTS

ADAM Y., Dr., Management Unit of the Mathematical Models of the North Sea and the Scheldt Estuary (MUMM), Liège, Belgium.

ALLEN K.R., Dr., The Johns Hopkins University, Laurel, Maryland, USA.

BARKMANN W., Dr., University of Southampton, Southampton, UK.

BERGER A., Prof. Dr., Université Catholique de Louvain, Louvain-la Neuve, Belgium.

BLOSS S., Dr., Universität Hannover, Hannover 1, FRG.

BOISSIER Chr., Eng., Service Hydrographique et Océanographique de la Marine, Toulouse, France.

BRUMMELL N., Mr., Imperial College, London, UK.

CARR M.-E., Miss, Dalhousie University, Halifax, Nova Scotia, Canada.

CHABERT D'HIERES G., Eng., Institut de Mécanique de Grenoble, Saint-Martin-D'Heres, France.

CLEMENT F., Mr., University of Liège, Liège, Belgium.

DAHL F.-E., Eng., Veritas Offshore Technology and Services, Veritec, Høvik, Norway.

DAUBY P., Dr., University of Liège, Liège, Belgium.

DAVIES A.M., Dr., Proudman Oceanographic Laboratory, Birkenhead, UK.

DELEERSNIJDER E., Eng., University of Liège, Liège, Belgium.

DESAUBIES Y., Dr., IFREMER, Brest, France.

DJENIDI S., Dr., University of Liège, Liège, Belgium.

DJURFELDT L., Mr., Göteborg University, Göteborg, Sweden.

EIFLER W., Dr., Commission of the European Communities, Ispra, Italy.

EVERBECQ E., Eng., University of Liège, Liège, Belgium.

FEDOROV K.N., Prof. Dr., Acad. Sci. USSR, Moscow, USSR.

FERER K.M., Mr., Naval Ocean Research and Development Activity, NSTL Station, USA.

FLEBUS C., Mr., Commission of the European Communities, Ispra, Italy.

FOREMAN S.J., Dr., Meteorological Office, Bracknell, Berkshire, UK.

FRANKIGNOUL Cl., Prof. Dr., Université Pierre et Marie Curie, Paris Cedex, France.

GARGETT A.E., Dr., Institute of Ocean Sciences, Sidney, British Columbia, Canada.

GARRETT C.J.R., Prof. Dr., Dalhousie University, Halifax, Nova Scotia, Canada.

GASPAR Ph., Dr., Centre National de la Recherche Météorologique, Toulouse, France.

GIBSON C., Prof. Dr., University of California at San Diego, La Jolla, USA.

GILBERT D., Mr., Dalhousie University, Halifax, Nova Scotia, Canada.

GMITROWICZ E., Mr., Ministry of Agriculture, Food and Fisheries, Lowestoft, Suffolk, UK.

GOFFART A., Miss, University of Liège, Liège, Belgium.

GREGG M., Prof. Dr., University of Washington, Seattle, Washington, USA.

GREGORIS Y., Eng., Centre National de la Recherche Météorologique, Toulouse, France.

HACKETT B., Mr., Veritas Offshore Technology and Services, Veritec, Høvik, Norway.

HAPPEL J.J., Eng., University of Liège, Liège, Belgium.

HEBERT D., Mr., Dalhousie University, Halifax, Nova Scotia, Canada.

HECQ J.H., Dr., University of Liège, Liège, Belgium.

HENDERSON-SELLERS B., Dr., University of Salford, Salford, UK.

HOLLOWAY G., Dr., Institute of Ocean Sciences, Sidney, British Columbia, Canada.

ITSWEIRE E., Dr., The Johns Hopkins University, Baltimore, Maryland, USA.

JAMART B.M., Dr., Management Unit of the Mathematical Models of the North Sea and the Scheldt Estuary (MUMM), Liège, Belgium.

JENKINS A.D., Dr., Oceanographic Center, Trondheim, Norway.

JONES I.S.F., Dr., University of Sydney, Sydney, Australia.

JOSEPH R.I., Prof. Dr., The Johns Hopkins University, Baltimore, Maryland, USA.

KARAFISTAN-DENIS A., Dr., Middle East Technical University, Ankara, Turkey.

KARELSE M., Dr., Delft Hydraulics Laboratory, Delft, The Netherlands.

KELLEY D., Dr., Woods Hole Oceanographic Institution, Woods Hole, Massachusetts, USA.

KRAUS E.B., Prof. Dr., Cooperative Institute for Research in Environmental Sciences, Boulder, Colorado, USA.

KROHN J., Dr., GKSS Forschungszentrum, Geesthacht, FRG.

LEBON G., Prof. Dr., University of Liège, Liège, Belgium.

LEDWELL J.R., Dr., Lamont-Doherty Observatory, Palisades, New York, USA.

LOHRMANN A., Mr., Veritas Offshore Technology and Services, Veritec, Høvik, Norway.

LUECK R.G., Dr., The Johns Hopkins University, Baltimore, Maryland, USA.

MC DOUGALL T.J., Dr., CSIRO, Hobart, Australia.

MOLCARD, Dr., Unesco, Paris, France.

MOUCHET A., Miss, University of Liège, Liège, Belgium.

MULLER P., Dr., University of Hawaii at Manoa, Honolulu, Hawaii, USA.

MURTHY C.R., Dr., Canada Centre for Inland Waters, Burlington, Ontario, Canada.

NIHOUL J.C.J., Prof. Dr., University of Liège, Liège, Belgium.

OAKEY N., Dr., Bedford Institute of Oceanography, Dartmouth, Nova Scotia, Canada.

OLBERS D., Prof. Dr., Alfred Wegener Institut for Polar and Marine Research, Bremerhaven, FRG.

OZMIDOV R.V., Prof. Dr., Acad. Sci. USSR, Moscow, USSR.

PEGGION G., Dr., Saclant Centre, La Spezia, Italy.

PETERS H., Dr., University of Washington, Seattle, Wahington, USA.

PIACSEK S.A., Dr., Saclant Centre, La Spezia, Italy.

PINGREE R.D., Dr., Marine Biological Association of the UK, Plymouth, UK.

RAJKOVIC B., Dr., Institute of Meteorology, Belgrade, Yugoslavia.

RHINES P.B., Prof. Dr., University of Washington, Seattle, Wahington, USA.

RICHARDS K.J., Dr., The University Southampton, Southampton, UK.

RŒD L.P., Dr., Veritas Offshore Technology and Services, Veritec, Høvik, Norway.

RONDAY F.C., Dr., University of Liège, Liège, Belgium.

RUDDICK B.R., Dr., Dalhousie University, Halifax, Nova Scotia, Canada.

SALUSTI S.E., Prof. Dr., Università La Sapienza, Roma, Italy.

SANDSTROM H., Dr., Bedford Institute of Oceanography, Dartmouth, Nova Scotia, Canada.

SARMIENTO J.L., Prof. Dr., Princeton University, Princeton, New Jersey, USA.

SCHMITT R.W., Dr., Woods Hole Oceanographic Institution, Woods Hole, Massachusetts, USA.

SHAFFER G., Dr., Göteborg University, Göteborg, Sweden.

SHERWIN T., Mr., University College of North Wales, Menai Bridge, UK.

SIMONOT J.-Y., Eng., Université Pierre et Marie Curie, Paris, France.

SIWECKI R., Dr., Polish Academy of Sciences, Sopot, Poland.

SJOBERG B., Mr., Göteborg University, Göteborg, Sweden.

SMETS E., Dr., Ministerie Van Openbare Werken, Borgerhout, Belgium.

SMITZ J., Eng., University of Liège, Liège, Belgium.

SPITZ Y., Miss, Management Unit of the Mathematical Models of the North Sea and the Scheldt Estuary (MUMM), Liège, Belgium.

STANEV E., Dr., Sofia University, Sofia, Bulgaria.

STIGEBRANDT A., Prof. Dr., Göteborg University, Göteborg, Sweden.

TAYLOR N., Mr., University College of North Wales, Menai Bridge, UK.

THORPE S.A., Prof. Dr., The University Southampton, Southampton, UK.

VAN HAREN J.J.M., Mr., Netherlands Institute for Sea Research, Den Burg, Texel, The Netherlands.

VETH C., Mr., Netherlands Institute for Sea Research, Den Burg, Texel, The Netherlands.

WEBER S.L., Mrs., KNMI, De Bilt, The Netherlands.

WESSON J.C., Mr., University of Washington, Seattle, Wahington, USA.

ZUUR E.A.M., Eng., University of Neuchâtel, Neuchâtel, Switzerland.

INTRODUCTION

This volume contains the proceedings of the nineteenth International Liège Colloquium on Ocean Hydrodynamics (4-8 May 1987, Liège, Belgium) on the subject of **Small-Scale Turbulence and Mixing on the Ocean**. The Colloquium has been organized jointly by the University of Liège and SCOR Working Group 69. For this purpose the Group created a Programme Committee which included K. Fedorov, C. Garrett, M. Gregg, S. Thorpe and J. Woods. The aim of the Committee was to develop with Prof. J. Nihoul of the University of Liège the scientific programme of the Colloquium, to issue the announcement, and to invite contributions. It was decided to focus the Colloquium programme on the relationships between small-scale mixing and large-scale features, transports and processes. In this connection, the presentation of papers on various methods of parameterization of small-scale turbulent mixing for numerical ocean models was particularly encouraged by the Programme Committee. This resulted in more than 1/3 of the 54 papers presented at the Colloquium dealing in one way or another with the parameterization problems; many of these papers demonstrate the direct results of modelling. These proportions are well reflected in this volume of proceedings and thus emphasize once more the importance of small-scale turbulence research for such vital practical applications as ocean modelling and forecasting.

The Colloquium demonstrated the considerable progress achieved in the studies of oceanic turbulence and mixing over the period of 8 years which has passed since the discussion of the same topics in Liège in 1979. This progress was analyzed in a summary presented at the closure of the Colloquium by Prof. S. Thorpe. His summary concludes this volume.

The SCOR Working Group 69 has also considered it desirable to include in this volume a Glossary of terms which are used in dealing with oceanic turbulence and mixing. This Glossary has been prepared by the Group in a series of very thorough discussions both before and after the nineteenth Liège Colloquium. The Group felt that there should be a more uniform and correct usage of terms aimed at preventing as much as possible any designation of different numbers, scales, characteristics and processes with the same term or, to the contrary, the use by different scientists of different terms for the same numbers, scales, characteristics or processes. The preparation of the Glossary proved to be a difficult task. A small panel of group members consisting of T. McDougall (Chairman), S. Thorpe and C. Gibson must be given special credit for the success of this work. The Glossary was circulated to the Colloquium participants in the hope of reaching a consistent usage of terms in the proceedings.

It would only be appropriate to finish this introduction with the expression of thanks to SCOR for its support of the Group which led us all to this very interesting Colloquium, to Prof. J.C.J. Nihoul and all his assistants at the University of Liège, and to all the members of the SCOR Working Group 69 on Small-Scale Turbulence and Mixing in the Ocean for their efforts in bringing this work to conclusion.

Prof. K.N. FEDOROV
Chairman, Working Group 69
of SCOR

SMALL-SCALE TURBULENCE AND MIXING IN THE OCEAN: A GLOSSARY

This glossary was prepared by a subpanel of SCOR working group 69, consisting of Dr Trevor McDougall (chairman), Professor Steve Thorpe and Professor Carl Gibson. It was reviewed and ammended by the working group at its meeting immediately following the Liège Symposium.

CONTENTS

1 TURBULENCE

Turbulence is a condition of fluid flow in which (i) each of the components of velocity and vorticity is irregularly and aperiodically distributed in both space and time, (ii) energy is transferred between large and small scales where it is dissipated, and (iii) there is diffusion of properties at a rate much in excess of the molecular rates that would occur in a laminar flow with the same average distribution of flow and scalar properties.

Small-scale, active turbulence is defined as a nearly isotropic, eddy-like state of fluid motion where the inertial forces in the eddies are larger than the buoyancy and viscous forces. It consists of random motions, with Reynolds and Froude numbers that exceed critical values. The length scales of such three-dimensional turbulent motion are smaller than about $0.6\,L_R$ and larger than about $11\,L_K$ (e.g. Stillinger et al. 1983; Gibson, 1987), where $L_R = (\varepsilon/N^3)^{0.5}$ is the Ozmidov length scale (see section 4), and $L_K = (\nu^3/\varepsilon)^{0.25}$ is the Kolmogoroff length scale (see section 4).

There are small-scale fluctuations of velocity and scalar properties in the ocean that fulfil criteria (i), (ii) and (iii) above, but which do not satisfy the definition of active turbulence. Some of these fluctuations may be the remnants of previously active turbulence; they have been described as "fossil turbulence" (Woods et al. 1969, Gibson 1986). Two-dimensional turbulence can exist at larger scales with motion constrained by buoyancy forces to nearly horizontal planes (Monin and Ozmidov 1985).

2 DISSIPATION RATES OF KINETIC ENERGY AND OF THERMAL VARIANCE

Due to the dissipative and irreversible nature of turbulence, kinetic energy is lost by molecular viscosity acting on the very smallest length scales and appears as thermal energy. The rate at which turbulent kinetic energy is lost is an important property of turbulence: it is usually labelled ε and is given by $\varepsilon = 2 \, \nu \, (e_{ij})^2$, where $e_{ij} = \frac{1}{2}(u_{i,j} + u_{j,i})$ is the rate of strain tensor (e.g. Lamb 1974, section 369) and ν is the kinematic viscosity of sea-water. The preferred units of ε are $m^2 s^{-3}$ or W /kg.

ε is usually measured in the ocean by instruments that measure one velocity component in a direction perpendicular to the direction of motion of the measuring vehicle. It is often assumed that the turbulence is isotropic on the microstructure length scales on which the molecular viscosity is effective. Under this assumption, ε can be estimated by the formula,

$$\varepsilon = \frac{15}{2} \, \nu \, \overline{\left[\frac{\partial v'}{\partial x} \right]^2}$$

(Batchelor 1953), where x is the spatial direction along which the instrument measures the turbulent fluctuations v' of the velocity component v. The overbar is generally taken as an average over a short section of the data corresponding to an instrument path length of about 1 m.

The dissipation rate of thermal variance, χ, is given by

$$\chi = 2 \, \kappa_T \, \overline{(\nabla T)^2},$$

where κ_T is the molecular diffusivity of heat. The corresponding dissipation rates of the variances of other scalars can be defined similarly.

Note that ε is defined in terms of the rate of dissipation of kinetic energy (half the variance of speed), whereas χ is derived from the variance of temperature gradient (no factor of half).

3 BUOYANCY FREQUENCY, N

The buoyancy frequency, N, is given by

$$N^2 = g \left(\alpha \frac{\partial \theta}{\partial z} - \beta \frac{\partial S}{\partial z} \right),$$

where g is the gravitational acceleration, θ is potential temperature, S is salinity and z is defined positive upwards (Gill 1982). The thermal expansion coefficient, α, and the saline contraction coefficient, β, are here defined by

$$\alpha = -\frac{1}{\rho} \frac{\partial \rho}{\partial \theta} \bigg|_{S,p} = \left[-\frac{1}{\rho} \frac{\partial \rho}{\partial T} \bigg|_{S,p} \right] \left[\frac{\partial \theta}{\partial T} \bigg|_{S,p} \right]^{-1}$$

$$\beta = \frac{1}{\rho} \frac{\partial \rho}{\partial S} \bigg|_{\theta,p} = \frac{1}{\rho} \frac{\partial \rho}{\partial S} \bigg|_{T,p} + \alpha \frac{\partial \theta}{\partial S} \bigg|_{T,p} .$$

These differ (by less than 4% at a depth as large as 4000 m) from common usage by including the terms $\left.\frac{\partial\theta}{\partial T}\right|_{S,p}$ and $\left.\frac{\partial\theta}{\partial S}\right|_{T,p}$. The partial derivatives of *in situ* density, ρ, with respect to T and S on the right-hand sides of these equations can be obtained from the International Equation of State of seawater, and the partial derivatives of θ with respect to T and S can be found from Bryden (1973). In practice, the equivalent formula

$$N^2 = -\frac{g}{\rho}\frac{\partial\rho}{\partial z} - \frac{g^2}{c^2}$$

is frequently used, where c is the velocity of sound (also derived from the International Equation of State of sea water), but care has to be taken to evaluate ρ and c consistently (Gill 1982).

4 LENGTH SCALES

Several length scales relevant to turbulence in stratified flows are in common usage. Two independent length scales are derived from N, ε and ν. The *Kolmogorov length scale*, at which viscous and inertial forces are of the same order of magnitude, is defined as $L_K = (\nu^3/\varepsilon)^{0.25}$. The vertical length scale at which the buoyancy force is of the same order of magnitude as the inertial forces is given by the *Ozmidov length scale* for which the symbol L_R has been used by Gibson et al. (1974). This is defined by $L_R = (\varepsilon/N^3)^{0.5}$. (Other symbols are also in standard usage.) Here ε is the rate of dissipation of kinetic energy by the turbulent motions and N is the buoyancy frequency. In practice, different values of both L_K and L_R will result from different procedures and averaging space and/or time scales. A third length scale, not involving ε, is $(L_R L_K^2)^{\frac{1}{3}} = (\nu/N)^{0.5}$.

The *Batchelor scale* $L_B = (\nu\kappa^2/\varepsilon)^{0.25}$, is the scale at which the steepening of scalar concentration gradients by the rate-of-strain is balanced by diffusive smoothing, where κ is the molecular diffusivity. Further vertical scales can be derived from the dissipation rate of scalar variance, χ (Gibson 1987).

A vertical profile may contain regions of static instability (negative N^2). Vertical displacements, which may indicate vertical overturning, are formed by reordering the profile so as to achieve static stability. An rms value of these displacements within some specific depth range is often called the *Thorpe scale, L_T*, (Thorpe 1977). Its empirical relationship to the *Ozmidov scale, L_R*, has been examined by Dillon (1982), Crawford (1986) and Gibson (1987).

"Overturning potential energy" is the locally averaged change in potential energy produced by vertically rearranging the water column to achieve static stability.

5 INTRUSIONS, TONGUES, INTERLEAVING MOTION

Quasi-horizontal layers of water that, from their motion, or from their potential temperature, θ, salinity, S, or chemical species characteristics, can be identified as having come from a neighbouring mass of water, are called "intrusions"(or, when their three-dimensional form is to be stressed, "tongues"). Their movement is referred to as an "interleaving motion".

6 STABILITY RATIO AND TURNER ANGLE

The stability ratio (or density ratio) has been defined by

$$R_\rho = \left[\alpha \frac{\partial \theta}{\partial z} \Big/ \beta \frac{\partial S}{\partial z} \right]^{\pm 1},$$

with α and β defined as in section 3. It is most often used in conjunction with studies of double-diffusive convection. In the "finger" regime of double-diffusive convection, the positive exponent is chosen in the above expression, while in the "diffusive" regime the negative exponent is used, thus ensuring that $R_\rho > 1$ in both double-diffusive regimes. The Turner angle, in degrees, is defined as the four quadrant arctangent by

$$Tu \ (\text{deg}) = \tan^{-1} \left(\alpha \frac{\partial \theta}{\partial z} - \beta \frac{\partial S}{\partial z}, \ \alpha \frac{\partial \theta}{\partial z} + \beta \frac{\partial S}{\partial z} \right),$$

as shown in the figure. Here z is defined to be positive upwards. Tu removes the ambiguity of the sign of the exponent in the definition of R_ρ. Note that $R_\rho = - \tan (Tu + 45)$.

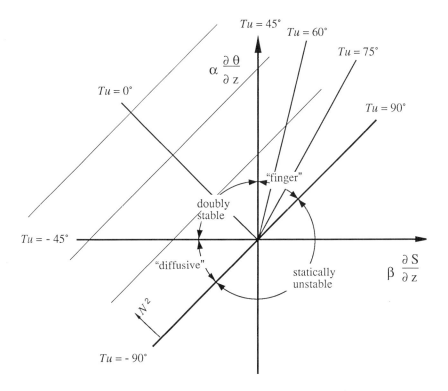

Figure 1. The axes on this figure are the contributions of the vertical gradients of potential temperature and salinity to $g^{-1}N^2$, and the figure illustrates the definition of the Turner angle, Tu.

7 NEUTRAL SURFACES

The appropriate way to parameterize transport processes in the ocean is still not entirely clear and remains an area of active research. However it is generally assumed that in the quasi-horizontal stirring by meso-scale eddies, fluid parcels move so as to do no work against gravity. The surfaces in which this is possible are termed "neutral surfaces", defined such that $\alpha \nabla_n \theta = \beta \nabla_n S$, with ∇_n referring to the two-dimensional gradient in the surface, and with α and β defined as in section 3.

If the potential temperature and salinity are split into their mean (over many eddy scales) and fluctuating parts, (i.e $\theta = \bar{\theta} + \theta'$, $S = \bar{S} + S'$), neutral surfaces may be defined for the mean field, $\bar{\theta}$ and $\bar{S}$, rather than the instantaneous, or local, θ and S. Fluid parcels, even in the absence of vertical mixing processes, do not move exactly in this surface, due to the nonlinearities in the equation of state (see section 9 below on thermobaricity), but it is assumed that the tensor describing eddy diffusion by mesoscale eddies and by small-scale processes is diagonal if axes are chosen in this surface and normal to it.

It should be noted that a neutral surface corresponds to an envelope of isopycnal surfaces (surfaces of constant potential density) when the reference pressure for the potential density is continually changed to the *in situ* pressure (McDougall, 1987a). A neutral surface differs from a potential density surface that is referenced to any fixed pressure because of the variation of α/β with pressure.

8 CABBELING

If two water parcels of the same density but different potential temperature and salinity are mixed, the resulting mixture is more dense, due to the variations of α and β with θ and S. This process is termed "densification on mixing" in Russian, but is known in English as "cabbeling". In ocean circulation models that treat θ and S separately this process is included automatically, but in models that deal only with density, cabbeling must be represented explicitly in the density equation as a source term. Cabbeling can also be interpreted as a contribution to the mean velocity across neutral surfaces (McDougall, 1987b).

9 THERMOBARICITY

The dependence of the compressibility of sea water on potential temperature and salinity means that water parcels displaced laterally without doing any work against gravity will not follow the neutral surface defined in terms of the spatially averaged potential temperature and salinity but will move off this surface. Stirring by meso-scale eddies leads to a net motion of fluid across neutral surfaces (McDougall 1987b). This process, termed "thermobaricity", arises from stirring, whereas cabbeling requires mixing at the molecular level.

8

10 INTERMITTENCE AND PATCHINESS

Turbulence is inherently intermittent. Even in unstratified environments, variations of ε and χ are large, characterized by lognormal probability distributions (Kolmogorov, 1962, Gurvich and Yaglom, 1965). In the ocean this variability is extreme because the intermittence within regions of active turbulence is compounded with a spatially patchy distribution of these regions and an apparent variety of production mechanisms and intensities.

11 MICROSTRUCTURE AND FINESTRUCTURE

Fluctuations on scales at which entropy is generated by molecular viscosity and molecular diffusion are referred to as "microstructure". These scales typically lie between 1 mm and 100 mm in the ocean. Finestructure, generally at larger scale, may represent velocity or scalar signatures of straining, mixing and/or intrusive interleaving, these mechanisms often working in combination. It is often difficult to distinguish between their respective contributions. Some of these processes may be reversible (such as that caused by internal gravity wave straining), while layered (or irreversible) finestructure may be associated with intrusions, double-diffusive convection or previous turbulent mixing events on vertical scales from tens of centimeters to hundreds of meters.

12 COX NUMBER

The Cox number of a scalar field is defined as the ratio of the mean square gradient to the square of the mean gradient, i.e. for a scalar φ,

$$C = \overline{(\nabla\varphi)^2} / (\overline{\nabla\varphi})^2 .$$

The use of the Cox number for obtaining diffusivities of scalars is described by Osborn and Cox (1972). A Cox number is usually calculated with data obtained in only one spatial direction and is defined as the ratio of the mean squared gradient in this direction divided by the mean vertical gradient squared. The distinction between these Cox numbers should always be clearly stated, with a different symbol, perhaps C_1 being used for the one-dimensional Cox number. If isotropy is assumed, $C = 3C_1$.

REFERENCES

Batchelor, G. K., 1953: *The Theory of Homogeneous Turbulence*. Cambridge University Press. Cambridge, 197pp.
Bryden, H.L., 1973: New polynomials for thermal expansion, adiabatic temperature gradient and potential temperature of seawater. *Deep-Sea Res.*, **20**, 401-408.
Dillon, T.M., 1982: Vertical overturns: a comparison of Thorpe and Ozmidov length scales, *J. Geophys. Res.*, **87**, 9601-9613.
Crawford, W.R., 1986: A comparison of length scales and decay times in stably stratified flows. *J. Phys. Oceanogr.*, **16**, 1847-1854.
Gibson, C.H., 1980: Fossil temperature, salinity and vorticity turbulence in the ocean. in *Marine Turbulence*, edited by J.C.J. Nihoul, pp 221-258, Elsevier, Amsterdam.

Gibson, C.H., 1986: Internal waves, fossil-turbulence, and composite ocean microstructure spectra. *J. Fluid Mech.*, **168**, 89-117.

Gibson, C.H., 1987: Fossil turbulence and intermittency in sampling oceanic mixing processes. *J. Geophys. Res.*, **92**, 5383-5404.

Gibson, C.H. L.A. Vega and R.B.Williams, 1974: Turbulent diffusion of heat and momentum in the ocean. In: *Advances in Geophys.*, **18A**, New York, 357-370.

Gill, A.E., 1982: *Atmosphere-Ocean Dynamics*. Academic Press. New York, 662pp.

Gurvich, A.S.and A.M.Yaglom, 1967: Breakdown of eddies and probability distributions for small-scale turbulence. *Phys. Fluids Suppl.*, **10**, 59-65.

Lamb H., 1974: *Hydrodynamics*. 6th ed. Cambridge University Press, Cambridge, 738pp.

Kolmogorov, A.N., 1962: A refinement of previous hypotheses concerning the local structure of turbulence in a viscous incompressible fluid at high Reynolds number. *J. Fluid Mech.*, **13**, 82-85.

McDougall, T.J., 1987a: Neutral surfaces. *J. Phys. Oceanogr.*, **17**, 1950-1964.

McDougall, T.J., 1987b: Thermobaricity, cabbeling, and water-mass conversion. *J. Geophys. Res.*, **92**, 5448-5464.

Monin A.S. and R.V. Ozmidov, 1985: *Turbulence in the Ocean*, Reidel, 247pp.

Osborn, T.R. and C.S. Cox, 1972: Oceanic fine structure. *Geophys. Fluid Dyn.*, **3**, 321-345.

Stillinger, D.C. K.N. Helland, and C.W. Van Atta, 1983: Experiments on the transition of homogeneous turbulence to internal waves in a stratified fluid. *J. Fluid Mech.*, **131**, 91-122.

Thorpe, S.A., 1977: Turbulence and mixing in a Scottish loch. *Phil. Trans. Roy. Soc. London*, **A286**, 125-181.

Woods, J.D. (cd.), V.Hogstrom, P.Misme, H. Ottersten and O.M.Phillips., 1969: Report of working group: fossil turbulence. *Radio Sci.*, **4**, 1365-1367.

.

THE USE OF DELIBERATELY INJECTED TRACERS FOR THE STUDY OF DIAPYCNAL MIXING IN THE OCEAN

JAMES R. LEDWELL
Lamont-Doherty Geological Observatory of Columbia University, Palisades, New York, 10964 (U. S. A.)

ANDREW J. WATSON
Marine Biological Association of the United kingdom, The Laboratory, Citadel Hill, Plymouth PL1 2PB (United Kingdom)

ABSTRACT

A plan for a series of tracer release experiments to measure diapycnal mixing in the open ocean during the next two decades is presented. These experiments can be performed on scales of 500 km X 12 months using a few hundred kg of fluorinated tracers such as sulfur hexafluoride. Various hypotheses for the processes controlling diapycnal mixing, and formulas based on these hypotheses, are briefly reviewed. The physical measurements that must be included in the tracer release experiments to test these hypotheses are discussed.

1 INTRODUCTION

We are developing a direct method of measuring diapycnal mixing in the open ocean by releasing a tracer as close as possible to an isopycnal surface, and measuring the evolution of its distribution in density space. The tracers to be used are fluorinated compounds, detectable in concentrations as low as 10^{-17} mol/L using gas chromatography with an electron capture detector. This sensitivity is such that a 12 month experiment, with a lateral length scale of 500 km, can be performed with just 100 kg of tracer. A prototype experiment, using sulfur hexafluoride and perfluorodecalin, has been performed in Santa Monica Basin, 50 km west of Los Angeles (Ledwell et al., 1986). Plans are being made to perform ocean scale experiments in the upcoming World Ocean Circulation Experiment because of the importance of diapycnal mixing in understanding ocean circulation and its interaction with the earth's climate (U. S. Scientific Steering Committee for WOCE, 1986).

We intend to perform a series of tracer release experiments at a variety of oceanic sites over the next decade to determine how diapycnal mixing depends on the hydrodynamics of the environment. Our purpose is to stimulate discussion of the sites to be chosen, the parameters to be measured, and the hypotheses to be tested. The ultimate goals are to understand the processes controlling diapycnal mixing, and to establish a means of estimating the climatology of the diapycnal diffusivity from methods inexpensive enough to be applied on a routine basis.

2 OPEN OCEAN EXPERIMENTS

The first open ocean tracer experiments will involve the injection of about 100 kg of tracer in a 100 km-scale patch in a series of long streaks as near as possible to an isopycnal surface. Experience so far has shown that the initial rms vertical dispersion of the tracer plumes can be kept to less than 10 m. Around 30 isopycnal floats would be released with the tracer to track it, and to study lateral stirring and mixing. These floats would be instrumented to monitor the ambient internal wave field, and a few of them could be instrumented to measure fine structure and microstructure characteristics . Prior to the experiment, one or more moorings will be set in the region to monitor the mean hydrography, eddy statistics, and the internal wave field. These moorings might also be instrumented to measure fine and microstructure characteristics.

A sampling ship, guided by the float positions, will determine the initial diapycnal distribution of the tracer by towing a vertical array of integrating samplers across the injection streaks. This array will consist of 20 samplers which take in water over a period of 3 hours. A CTD at the center of the array will be held on the target isopycnal surface of the injection. A CTD will also be placed at the bottom of the array so that a large number of high quality CTD down traces will be obtained during the experiment. The samplers will be accompanied by thermistors, so each tow will double as a thermistor array tow as well.

An intermediate tracer survey, again guided by the float positions, will be performed at the time estimated for the lateral scale of the patch to grow to around 250 km (of order 6 months, but depending on the experiment). The vertical array of integrating samplers described above will be used for these surveys, as it is anticipated that the tracer distribution may still be quite streaky. Each 3-hour tow will sample a track length of order 15 km, so a continuous star shaped pattern of 250 km diameter could be traced out with a month of ship time to sample the distribution well even in the face of extreme streakiness.

A final tracer survey will be performed at the time estimated for the lateral scale of the patch to be 500 km (of order 12 months). A 500 km star, with 15 km gaps interspersed with 15 km tows, can be traced out in a month to measure the distribution. This pattern will detect large scale inhomogeneities in the diapycnal distribution of the patch, and will still probably overcome the streakiness, especially since by the time of this survey the streakiness may have largely disappeared (Garrett, 1983). This final patch scale of 500 km is large compared to the eddy scale, and small compared with scale of an ocean basin; the corresponding duration is long enough in all cases considered so far to accurately measure diapycnal spreading, even for relatively small diffusivity (<0.1 cm^2/s).

2 HYPOTHESES

Diapycnal mixing in the ocean is constrained by the gravitational stability associated with the vertical gradients of temperature and salinity, and is driven by the processes feeding energy to 3-dimensional turbulence. If the potential temperature decreases with depth, and the salinity increases with depth, then the fluid is stable with respect to double diffusive effects, and the buoyancy frequency, N, is sufficient to characterize the stability. In many parts of the ocean, however, particularly in much of the Atlantic, salinity increases upward, and double diffusive effects may play a role in diapycnal mixing. Then the density ratio, R_ρ, the ratio of the stabilizing contribution of the temperature gradient to the destabilizing contribution of the salinity gradient to the density gradient, becomes an important characteristic of the environment along with N. Let us first focus on the diffusively stable case, and consider double diffusive effects later.

2.1 Buoyancy frequency formulations

It is tempting to ignore variations in the source of energy for mixing altogether, and simply hypothesize, either implicitly or explicitly, that the diapycnal diffusivity, K, depends only on N. This approach has the virtue of being simple, because it introduces just one very accessible parameter into the analysis. It has the vice, however, of being simplistic, in that it ignores the possibility of temporal and spatial variations in the energy source at fixed N. The idea draws credibility from the scaling of the internal wave spectrum with N (e.g., Munk, 1981), and from a possible scaling of turbulence energy dissipation rates with N (Gargett and Osborn, 1982; Lueck et al., 1983; Gargett and Holloway, 1984; and Moum and Osborn, 1986). However, if the internal wave spectrum is universal, it must be because it is in some sense saturated, and saturation implies that the energy dissipation rate is a very sensitive function of the amplitude. Furthermore, the statistical significance of the correlation of dissipation rates with N appears to be low.

The hypothesis that K varies with N was put forth by Sarmiento et al. (1976), who proposed an N^2 power law dependence based on Radium-226 profiles in the bottom km of the ocean and Radon-222 profiles in the bottom 100 m. More recently, Gargett (1984), compiling data from lakes, fjords, and the ocean, argued for a dependence of the form:

$$K = aN^{-b} \tag{1}$$

with b = 1, for systems not susceptible to double diffusion and for which the kinetic energy is dominated by internal waves. The parameter, a, in this relation is proposed to vary by no more than a factor of 2 or 3 for a given system, but may vary by more from one system to another. Gargett's best estimate for its value for the open ocean was 10^{-3} cm^2/s^2, based on correlations of the kinetic energy dissipation rate with N.

Independent supporting evidence for this relation must be considered weak thus far. Gargett cites the value of K inferred from measurements by Gregg (1977) of temperature variance in the North Pacific at 1000 m which fit the relation. She also points out that at 2.5 km, where N is typically 10^{-3} s^{-1}, the relation gives the historic value for K of 1 cm^2/s (Munk, 1966), and that at 4 or 5 km where N may be as low as 0.4×10^{-3} s^{-1}, the relation predicts K = 2.5 cm^2/s, in rough agreement with the value of between 3 and 4 cm^2/s found by Hogg et al. (1982) in the Brazil Basin at 4000 m. It turns out, though, that the stratification at 4000 m and below in the Brazil Basin is anomolously strong, because of the relatively high density of the Antarctic bottom water. We have estimated the buoyancy frequency at 4000 m to be around 1.4×10^{-3} s^{-1}

from the data of Fuglister (1957). Thus the value for K found by Hogg et al. at this depth is actually a factor of 5 or so higher than would be estimated by eqn. (1).

We must conclude that, although it is plausible that K, averaged laterally over the ocean, increases with decreasing N, variations from the mean may be quite large, and the precise dependence of K on N is quite unknown. Yet, Gargett (1984) and others have shown that the qualitative nature of the interior ocean circulation may depend sensitively on the variation of K with N. For example, the vertical profile of the diapycnal velocity, and thus the contribution of diapycnal mixing to vortex stretching and the interior circulation, depends drastically on whether the exponent, b, in eqn. (1) assumes the value 0, 1, or 2.

The buoyancy frequency varies by only a factor of twelve or so from the abyss (below 5000 m), where it is typically between 0.3 and 0.5×10^{-3} s^{-1}, to the main pycnocline (around 500 m), where it is typically between 4 and 6×10^{-3} s^{-1} (see Fig. 1; and Levitus, 1982). Thus to understand the interior circulation, the dependence of the average value of K on N must be known accurately, which means that variations from the mean due to variations in characteristics other than N, such as those discussed later, must be quantified, and a climatology of these characteristics must be archived. It is clear, then, that although tracer release experiments should be planned to explore the possibility of a correlation between K and N, the experiments must be accompanied by measurements of the finestructure and microstructure characteristics discussed below to try to account for residual variations in K, which are expected to be large.

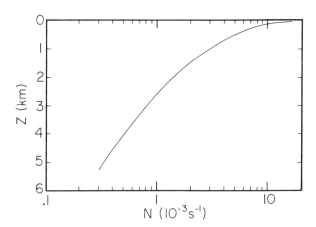

Fig. 1. Smoothed vertical profile of buoyancy frequency estimated from hydrographic data from GEOSECS Station 212 (30° 0' N, 159° 50' W, 18 Sept.1973), about 10° north of Hawaii (Broecker et al., 1982).

2.2 Microstructure characteristics

(i) Energy dissipation rate. Osborn (1980) derived an expression for diapycnal eddy diffusion of density from the turbulent kinetic energy equation for a stratified system. Assumptions of homogeneity and stationarity must be made, however, along with an estimate of the efficiency, γ, of conversion of turbulent kinetic energy to potential energy. Furthermore, measurements of the kinetic energy dissipation, ε, are usually confined to one or two components of the velocity field, so some assumption about isotropy must be made. All of this is discussed in the review by Gregg (1987), and elsewhere. The final relatioship is:

$$K = \gamma \, \varepsilon/N^2 \tag{2}$$

The value of γ is usually taken to be about 0.2. This relationship is actually involved in Gargett's hypothesis, discussed above.

The data gathered so far for ε at various values of N suggest, rather weakly in our opinion, that ε varies with some power of N between 1.0 and 1.5. If such a relation holds, and if γ is relatively constant, then eqn. (2) implies that K depends only on N. The scatter of values for ε at any one value of N covers an order of magnitude, however, so Gargett's estimate of a factor of 2 or 3 for variations in K at fixed N is probably not excessive.

The energy dissipation rate is an obvious parameter to measure during tracer release experiments to obtain a direct test of eqn. (2). It may well be that this equation, through measurements of ε, can give much better estimates of K than can use of a simple N dependence. The test is difficult, however, because K obtained from a tracer release experiment applies to an averaging time of several months, while microstructure surveys to measure ε are performed from a ship over a time scale of a few weeks at most. Even within a microstructure survey, the sampling problem is difficult because of the large range over which local dissipation rates vary in space and time. Thus there is a need for technical progress before valid comparisons of results for K from tracer release experiments can be made with results from microstructure measurements. The possibilities of improving towed devices and of developing drifting and moored devices to measure microstructure should be explored.

(ii) Thermal Cox Number. Gregg (1987) has summarized the technique of Osborn and Cox (1972) to estimate K for heat from measurements of the thermal dissipation, χ. The equation used is derived from the equation for turbulent temperature variance, again with assumptions of stationarity and homogeneity. Unless all components of the temparature gradient are measured, it is also necessary to assume something about isotropy. There is one less assumption, though, than for the kinetic energy dissipation technique, as there is no analog to the efficiency factor. The resulting equation is:

$$K = (\chi/2) \, (\partial T/\partial z + \Gamma)^{-2} \tag{3}$$

where T is the mean temperature, z is the vertical coordinate, and Γ is the adiabatic lapse rate.

Clearly, χ is another parameter that should be measured with tracer release experiments. The same sampling problems will be encountered as with measurements of ε however, and again the potential of towed, drifting and moored instruments should be explored.

2.3 Fine structure characteristics

(i) Internal wave characterisitcs. Since much of the mixing in the interior of the ocean is driven by the internal wave field, it is reasonable to seek a relationship between internal wave parameters and diapycnal diffusivity. Theoretical progress, based on analytic models of the internal wave spectrum, has been slow, however, because new complexities, which undermine the assumptions of the theories, are constantly being uncovered by observations (Gregg, 1987). Perhaps the most serious instance is the finding in recent years of 10 km scale vortices divorced from the internal wave field which may contribute strongly to the kinetic energy and to the shear at finestructure scales (Muller, 1984).

These difficulties notwithstanding, the quest for a diffusivity predictor from internal wave characteristics is irresistable. Internal wave measurements can be made on large time and space scales from moorings and floats. Thus they could be performed on the same scales as a tracer release experiment, and later they could be used on a widespread basis to establish a climatology of the diffusivity.

The most accessible projections of the internal wave spectrum are the frequency spectra of the horizontal velocity components and of the temperature perturbations, which can be measured from a single moored instrument. However, the connection between these spectra and diapycnal mixing is very dependent on presuppositions about the overall internal wave spectrum. A closer connection may be possible with the vertical wavenumber spectrum of the horizontal velocity components, or, more directly, with the frequency spectrum of the vertical shear (e.g., Munk, 1981; McComas and Muller, 1981; Gargett and Holloway, 1984; Henyey et al., 1986). Perhaps even less model dependent would be the frequency and vertical wavenumber spectra of the Richardson number. Toole and Schmitt (1987), for example, found at least qualitative correlations of Richardson number, defined for scales of a few m, with energy dissipation rates in the upper pycnocline in the northwest Atlantic subtropical front, using a profiling instrument.

In the absence of reliable theories, the best strategy is to carefully measure all of these spectra, with particular attention to the small scales, i.e., to vertical scales down to a few meters. The measurements could be accomplished with moored instruments down to vertical scales of 10 m or so, and perhaps with drifting arrays such as the Richardson number float being developed by Williams et al. (1987) to finer scales. Although technical developments may be necessary to increase the sensitivity and vertical resolution of the moored instruments, as well as to make Richardson number floats into a standard tool, the problems seem less severe than those associated with moored or drifting microstructure measurements.

(ii) Thorpe scale. Mixing events sometimes create density overturns, i.e., patches of gravitationally unstable water which can last for a time of the order of a buoyancy period. Thorpe (1977) has defined a length scale characterizing these overturns. Dillon (1982) has demonstrated for some cases that this scale is well correlated with the Ozmidov scale, defined as $(\varepsilon/N^3)^{1/2}$, and hence with ε within turbulent patches. Thus the Thorpe scale may prove to be a useful estimator for K. One advantage of this parameter is that it soon may be measurable throughout the water column with CTD casts employing fast sensors and a motion-compensating winch. Over many years a climatology of Thorpe scales might be built up from hydrographic research cruises, and, using experience from a set of tracer experiments, one could derive a climatology for K. Of course, the same might be possible with the microstructure parameters already discussed, but Thorpe scale measurements may be more routinely accessible.

2.4 Double Diffusive Enhancement of Mixing

So far, we have discussed the ocean as though double diffusion played no role in mixing. However, vast regions of the ocean may be susceptible to the salt-fingering variety of double diffusion. These regions include a great deal of the North Atlantic, the most extensively studied ocean basin. They also include much of the South Atlantic, and a good deal of the Southern Ocean, the Indian Ocean, and the South Pacific. There is strong indirect evidence that double diffusion controls the hydrographic structure of the so called Central Waters throughout the world (Schmitt, 1981). Furthermore, Mack (1985) presented observations of enhanced microstructure in patches with density ratio less than 4, and argued that enhanced mixing by double diffusive processes is prevalent in the oceans.

Eqn. (2) is invalid in the presence of active double diffusion. That equation was derived from the assumption that shear production of turbulent energy is balanced by losses to dissipation and potential energy production, and the implicit assumption that the diapycnal diffusivity for heat is the same as that for salt. In salt fingering, however, turbulent energy is produced at the expense of the potential energy, with shear production not necessarily playing an important role, and with the loss all going to dissipation. The ratio, γ', of the contributions of heat and salt fluxes to density flux must be less than 1 to satisfy energy requirements, and the ratio of the salt diffusivity, K_s, to heat diffusivity, K_t, must be greater than the density ratio by the factor R_ρ/γ'. Estimates are that γ' is around 0.7 to 0.85. If shear production, transport, and temporal changes are neglected, the turbulent energy balance equation gives the following equation for the salt diffusivity, K_s:

$$K_s = (R_\rho - 1)/(1 - \gamma') \, \varepsilon/N^2 \qquad (4)$$

As Schmitt (1988) and McDougall (1988) point out, K_s and K_t can be much larger than implied by eqn. (2), and in fact can be quite substantial even if the dissipation is immeasurably low.

It is important that tracer experiments be performed in environments susceptible to double diffusion as well as in diffusively stable regimes. An experiment could be performed in the C-SALT region (see Schmitt, 1987) to obtain a very low density ratio with clearly defined steps due presumbly to double diffusion to obtain a strong contrast with the diffusively stable case. Another experiment could be performed at R_ρ of about 2 somewhere in the Central Waters to examine whether the mixing is enhanced over the stable case. Tracer experiments will give a good estimate of the salt diffusivity, since the molecular diffusivity of the tracer is much closer to that of salt than to that of heat. Microstructure measurements should be performed during these experiments to try to measure χ and ε to estimate K_t, and to examine the relationship between the diffusivities, the density ratio, and the dissipation.

2.5 <u>Special regions</u>

(i) <u>Boundaries</u>. There are several sources of extra energy for mixing near the boundaries of the ocean basins. Turbulent boundary layers associated with boundary currents, and with internal wave currents impinging on the boundaries must enhance diapycnal mixing (e.g., Armi, 1978). Also, it has been proposed that large amounts of energy are trapped from the internal wave field, and dissipated, near a boundary whose slope is greater than a critical angle (Eriksen, 1985). Although the volume of water involved in boundary processes is only a small fraction of the total volume of the ocean, it is possible that enhanced mixing at the boundaries contributes a significant amount of the diapycnal mixing for the ocean as a whole.

Tracer experiments near boundaries to examine the possibility of enhanced mixing should be planned. One plan presently being considered is to measure the effective diapycnal diffusivity at some level for a large nearly enclosed basin, using a heat, salt or nutrient budget, as done by Hogg et al., 1982 for the Brazil Basin, and compare the result with a direct measurement of K using a tracer release experiment. The apparent diffusivity measured by the budget method will be affected by boundary mixing as well as interior mixing, while K measured by the tracer release experiment will be affected only by interior mixing, until the tracer reaches the boundaries. The difference between the two estimates of K will tell whether boundary mixing is important, and if the difference is large it will give an estimate of the boundary mixing.

(ii) <u>High kinetic energy regimes</u>. It has been proposed that regions of high eddy kinetic energy may also be regions of enhanced diapycnal mixing. One mechanism would of course be enhanced generation of internal waves. Another might be the trapping and focusing of near-inertial internal waves by the larger scale vorticity field associated with anticyclonic eddies, as suggested by Kunze (1985) and Kunze and Lueck (1986). One approach to examining the significance of this effect to diapycnal mixing in the overall ocean would be to perform a 500 km scale tracer release experiment in a region rich in eddy energy. Another approach would be to perform special smaller scale experiments within well defined eddies, such as warm core rings. Probably both approaches will be needed if further study shows that the diapycnal mixing in high eddy energy regions contributes a disproportionate share of the global diapycnal mixing.

3 SUMMARY

Consideration of the various parameters governing diapycnal mixing, and the various processes which might enhance that mixing, leads to a strategy for a suite of tracer release experiments to be performed over the coming years. A review of the fine structure and microstructure parameters believed related to diapycnal mixing suggests the measurements that should be performed along with tracer release experiments to gain an understanding and a predictive ability for the mixing. If a semiempirical theory can be developed for the relationship between a subset of these various parameters and diapycnal diffusivity, then it may be possible, as high quality data is gathered over time, to compile a climatology of the diffusivity.

A few experiments, in a diffusively stable regime, at a variety of values of N, should be planned to serve as a baseline with which to compare experiments in exceptional regimes. A good site would be the North Pacific, at about 30°N, a few days steam from Hawaii. Figure 1 shows the buoyancy frequency profile for GEOSECS station 212, at 30°N, 160°W. The region of diffusive stability starts at 600 m depth and extends to the bottom. A selection of depths to perform experiments might be 600, 2000, and 4500 m to obtain values of N of 5×10^{-3}, 1.4×10^{-3}, and 0.4×10^{-3} s^{-1} respectively.

An experiment to determine whether double diffusive effects enhance mixing in the Central Waters should be performed. One such site, proposed as the first tracer release experiment in WOCE, is at 650 m depth in the beta-triangle, centered at 27°N, 32°30'W (see Armi and Stommel, 1983). The buoyancy frequency there is around 3×10^{-3} s^{-1}, and thus it would be possible to compare the results of the experiment with experiments in a diffusively stable regime mentioned above. An experiment to really measure the enhancement of mixing in a salt-fingering regime could be performed in the C-SALT region mentioned earlier.

Experiments in boundary regions and in high kinetic energy regimes should be planned as the results from these first experiments come in.

All of these experiments should include measurements of fine structure and microstructure with coverage comparable to that of the tracer experiments. These are necessary to make progress in understanding the mechanisms driving diapycnal mixing, and to develop estimation schemes for diapycnal diffusivity The more accessible these parameters are, the more quickly will a climatology of estimated eddy diffusivity be compiled.

Acknowledgements. Several helpful comments on the original manuscript were made by T. McDougall and R. Schmitt. This work was made possible by National Science Foundation Grant OCE 86-14635. This is Lamont-Doherty Geological Observatory contribution number 4261.

4 REFERENCES

Armi, L., 1978. Some evidence for boundary mixing in the deep ocean. J. Geophys. Res., 83: 1971-1979.
Armi, L. and Stommel, H., 1983. Four views of a portion of the North Atlantic Subtropical Gyre. J. Phys. Oceanogr., 13: 828-857.
Broecker, W. S., Spencer, D. W., and Craig, H., 1982. GEOSECS Pacific Expedition, 3, Hydrographic data, 1973-1974. Superint. of Doc., U. S. Gov. Print. Off., Washington, D. C., 137 pp.
Dillon, T. M., 1982. Vertical overturns: A comparison of Thorpe and Ozmidov length scales. J. Phys. Oceanogr., 87: 9601-9613.
Eriksen, C. C., 1985. Implications of ocean bottom reflection for internal wave spectra and mixing. J. Phys. Oceanogr., 15: 1145-1156.
Fuglister, F. C., 1957. Oceanographic data from *Crawford* cruise ten obtained for the International Geophysical Year 1957-58. Woods Hole Oceanographic Institution Technical Report 57-54, Woods Hole, Mass.
Gargett, A. E., 1984. Vertical eddy diffusivity in the ocean interior. J. Mar. Res., 42: 359-393.
Gargett, A. E. and Holloway, G., 1984. Dissipation and diffusion by internal wave breaking. J. Mar. Res., 42: 15-27.
Gargett, A. E. and Osborn, T. R., 1981. Small-scale measurements during the Fine and Microstructure Experiment (FAME). J. Geophys. Res., 86: 1929-1944.

Garrett, C., 1983. On the initial streakiness of a dispersing tracer in two- and three-dimensional turbulence. Dyn. Atmos. Oceans, 7: 265-277.

Gregg, M. C., 1977. Variations on the intensity of small-scale mixing in the main thermocline. J. Phys. Oceanogr., 7: 436-454.

Gregg, M. C., 1987. Diapycnal mixing in the thermocline: a review. J. Geophys. Res., 92: 5249-5286.

Henyey, F. S., Wright, J., and Flatte', S. M., 1986. Energy and action flow through the internal wave field: an eikonal approach. J. Phys. Oceanogr., 91: 8487-8495.

Hogg, N., Biscaye, P., Gardner, W., and Schmitz, W. J., Jr., 1982. On transport and modification of Antarctic bottom water in the Vema Channel. J. Mar. Res., 40(Suppl.): 231-263.

Kunze, E., 1985. Near-inertial wave propagation in geostrophic shear. J. Phys. Oceanogr. , 15: 544-565.

Kunze, E. and Lueck, R., 1986. Velocity profiles in a warm-core ring. J. Phys. Oceanogr., 16: 991-995.

Ledwell, J.R., Watson, A. J., and Broecker, W.S., 1986. A deliberate tracer experiment in Santa Monica Basin. Nature, 323: 322-324.

Levitus, S., 1982. Climatological Atlas of the World Ocean. NOAA Professional Paper 13, Superint. of Doc., U.S. Gov. Print. Off., Washington DC., 173 pp.

Lueck, R. G., Crawford, W. R., and Osborn, T. R., 1983. Turbulent dissipation over the continental slope off Vancouver Island. J. Phys. Oceanogr., 13: 1809-1818.

Mack, S. A., 1985. Two-dimensional measurements of ocean microstructure: The role of double diffusion. J. Phys. Oceanogr., 15: 1581-1604.

McComas, C. H. and Muller, P., 1981. The dynamic balance of internal waves. J. Phys. Oceanogr, 11: 970-986.

McDougall, T. J., 1988. Some implications of ocean mixing for ocean modelling. This volume.

Moum, J. N. and Osborn, T. R., 1986. Mixing in the main thermocline. J. Phys. Oceanogr., 16: 1250-1259.

Muller, P., 1984. Small-scale vortical motions. In: P. Muller and R. Pujalet (Editors), Internal Gravity Waves and Small Scale Turbulence, Proceedings, Hawaiian Winter Workshop, Hawaii Institute of Geophysics, Honolulu.

Munk, W. H., 1966. Abyssal recipes. Deep Sea Res., 13: 707-730.

Munk, W. H., 1981. Internal waves and small-scale processes. In: B. A. Warren and C. Wunsch (Editors), Evolution of Physical Oceanography, Scientific Papers in Honor of Henry Stommel, MIT Press, Cambridge, Mass., 264-291.

Osborn, T. R., 1980. Estimates of the local rate of vertical diffusion from dissipation measurements. J. Phys. Oceanogr., 10: 83-89.

Osborn, T. R., and Cox, C. S., 1972. Oceanic fine structure. Geophys. Fluid Dyn., 3: 321-345.

Sarmiento, J. L., Feely, H. W., Moore, W. S., Bainbridge, A. E., and Broecker, W. S., 1976. The relationship between vertical eddy diffusion and buoyancy gradient in the deep sea. Earth Planet. Sci. Lett., 32: 357-370.

Schmitt, R. W., 1981. Form of the temperature-salinity relationship in the Central Water: Evidence for double diffusive mixing. J. Phys. Oceanogr., 11: 1015-1026.

Schmitt, R. W., 1987. The Caribbean Sheets and Layers Transects (C-SALT) Program. Eos, Trans. Am. Geophys. Union, 68: 57-60.

Schmitt, R. W., 1988. Mixing in a thermohaline staircase. This volume.

Thorpe, S. A., 1977. Turbulence and mixing in a Scottish Loch. Philos. Trans. R. Soc. London Ser. A, 286: 125-181.

Toole, J. M. and Schmitt, R. W., 1987. Small-scale structures in the north-west Atlantic sub-tropical front. Nature, 327: 47-49.

U.S. Scientific Steering Committee for WOCE, 1986. WOCE discussions of physical processes: reports of U.S. subject meetings, U.S. WOCE Planning Report Number 5, U.S. Planning Office for WOCE., College Station, Texas, 143 pp.

Williams, H. A., III, Converse, C. H., and Nicholson, J. W., 1987. Richardson number float. In: K. Wolfe (Editor), Current practices and new technology in ocean engineering-1987-OED, Vol. 12, Am. Soc. Mech. Eng., New York, pp 25-29.

SOME IMPLICATIONS OF OCEAN MIXING FOR OCEAN MODELLING

Trevor J. McDougall

CSIRO, Division of Oceanography, GPO Box 1538, Hobart, TAS 7001, Australia.

ABSTRACT

The importance of vertical mixing in ocean circulation models is briefly reviewed and several methods of estimating vertical mixing activity are discussed. Several aspects of the interpretation of dissipation measurements are discussed. It is pointed out that the two dianeutral advection processes -- thermobaricity and cabbeling -- are invisible to microstructure dissipation measurements. Formulae are developed for calculating vertical diffusivities of scalars in thermohaline staircases, where the diffusivity of salt can be up to 10 times as large as that given by the normal expression $0.2\ \varepsilon/N^2$. In deducing the upwelling velocity in the ocean from microstructure measurements, the non-linear terms of the equation of state of sea-water must be included consistently. A simple numerical example shows how large these effects can be. Finally, the water-mass conversion equation written in a neutral-surface reference-frame is used to examine the relative importance of salt-fingering and small-scale turbulence in the Central Water of the World's oceans. While salt-fingering is required to induce and maintain the observed curvature in the S-θ curve of these water masses, small-scale isotropic turbulent mixing is shown to probably be responsible for larger vertical fluxes of salt, heat and buoyancy.

1 INTRODUCTION

Vertical mixing processes in the ocean cause both vertical diffusion of scalars and vertical advection of water through neutral surfaces. Since Bottom Water is continually formed at the poles and is thought to spread out into all the ocean basins, dianeutral vertical advection is required to upwell Bottom Water through the main thermocline. Tziperman (1986) has presented a very interesting, yet simple, concept that demonstrates the importance of vertical mixing. He has considered a control volume between two neutral surfaces and has noted that the flux of freshwater into this control volume at the sea surface by Ekman pumping is nonzero. This implies that there is a corresponding net flux of water out of the control volume, which in turn implies the existence of vertical mixing processes. Vertical velocities are also the key to understanding the intensity of the mean ocean circulation at depth in the ocean; indeed the only theory we have for the circulation of the deep ocean (Stommel & Arons, 1960a,b) is driven by an imposed vertical velocity. The linear vorticity equation on which this theory is based ($\beta v = f w_z$), involves the vertical velocity *past geopotentials* (*w*), whereas vertical mixing achieves vertical advection *through neutral surfaces.*

Maps of conservative tracers in the ocean demonstrate that mixing of some kind must be occurring. However it is not often apparent that **vertical** mixing must be invoked. It has thus been very difficult to distinguish the relative importance of vertical and lateral mixing processes in the ocean. The relatively recent advent of anthropogenic tracers has held out the promise of answering this question through the use of several tracers whose lateral and vertical distributions are not linearly dependent. To a large extent this promise remains unrealized. I firmly believe that these tracer fields do hold the key to unravelling the relative strengths of vertical and lateral mixing processes, but that in order to extract this information, we need to be very careful about how mixing processes are modelled in diagnostic models. For example, it is common to read in the literature of diagnostic models that include the diffusive flux term of vertical mixing but not the concomitant vertical advective term in the model equations. Some modellers include the advective effect of vertical mixing, but not the diffusive flux divergence. For this type of tracer study it is often very important to distinguish between a potential density surface and the dynamically relevant neutral surface (see McDougall, 1987b). To deduce the relative importance of lateral and vertical mixing processes from the three-dimensional distribution of oceanic tracers it will be necessary to perform inverse models in which mixing processes are carefully included.

A possible sleeping giant in the field of oceanic mixing is the vertical mixing that occurs at or near the ocean boundaries. Armi (1978) stresses that mixed layers at seamounts or continental boundaries are periodically injected into the ocean interior by unsteady meso-scale eddies, thereby increasing the efficiency of boundary mixing above the level that would apply if the ocean were steady. He estimates that such boundary mixing can account for a large fraction of the upwelling of Bottom Water. Eriksen's (1985) theory of "near boundary" mixing is based on the relaxation of the internal wave field to the canonical Garrett & Munk spectrum after the internal waves are reflected off the sloping ocean floor. This theory can also give quite large vertical diffusivities and the elevated mixing activity would occur within 100 m to 200 m of the ocean floor, which is quite a large distance compared with the typical bottom mixed-layer depth of about 10 m. The implications of either type of boundary mixing for the ocean circulation have not been explored, but it is obvious that if most of the upwelling in the ocean occurs at or near boundaries, the mean ocean circulation would be dramatically different to that found by Stommel & Arons (1960b).

Section 2 will review how mixing processes are currently included in prognostic ocean circulation models, both of the eddy-resolving type and the lower resolution models that do not resolve mesoscale features. The "method of exponential tracers" of Rooth & Ostlund (1972) has been widely quoted as providing an upper bound on the vertical diffusivity in a certain depth range of the Sargasso sea, but it will be shown in section 3 that this is not the case. Section 4 briefly discusses the use of budget models to deduce vertical diffusivities from large-scale property variations. Section 5 discusses the use of the rate of dissipation of mechanical energy to estimate the vertical diffusivity of scalars, while section 6 uses the constant R_ρ nature of the Central Water of the world's oceans (Schmitt, 1981), together with the correct water-mass transformation equation, to deduce the relative importance of double-diffusive convection and ordinary small-scale turbulent mixing.

2 MIXING IN PROGNOSTIC OCEAN CIRCULATION MODELS

Woods (1985) has argued eloquently that ocean models that will be useful for the purposes of decadal climate prediction must include subsurface vertical-mixing processes in a realistic fashion. It may seem odd that the very smallest scales of motion should be so important in understanding how the ocean circulation, on the very largest spatial scales, responds to transient changes in the surface boundary conditions such as may be caused by the "greenhouse effect". In this section the ways in which vertical mixing is included in various prognostic large-scale circulation models are briefly reviewed.

Bryan (1969) pioneered the primitive equation approach to large-scale ocean modelling which has become quite popular in recent years as supercomputers have become more powerful and more readily available. The most recent example of this approach (Cox, 1985) has a horizontal resolution of $\frac{1}{3}°$ of latitude and longitude and so resolves most of the mesoscale eddy activity in the ocean. In these numerical models, diffusion occurs via strictly horizontal and vertical diffusivities. Since neutral surfaces are inclined to the horizontal, the large lateral diffusivity has the effect of fluxing density across neutral surfaces. This effect was originally pointed out by Veronis (1975), who showed that this "fictitious flux" of density is often equivalent to an unwanted vertical diffusion coefficient of 50×10^{-4} m^2 s^{-1} and that the vertical velocity in the interior of a model ocean can even have the wrong sign! More recently, Holland & Batteen (1986) also used this unphysical horizontal diffusivity of density in an eddy-resolved two-layer quasigeostrophic model. They found that even quite small horizontal diffusivities of density (compared to the horizontal eddy viscosity) substantially modify both the mean baroclinic circulation and the eddy kinetic energy of baroclinic motions. Holland and Batteen (1986) speculate that the baroclinic circulation in coarse resolution, noneddying models will also be similarly affected by these fluxes of density across sloping isopycnals.

Strong lateral mixing in the ocean occurs along neutral surfaces, or for the purpose of the present discussion, along isopycnals (McDougall & Church, 1986), rather than along geopotentials. In order to move and mix fluid parcels along geopotentials requires work to be done against gravitational restoring forces, and in the ocean this kind of work is relegated to the small-scale mixing processes which are parameterized by vertical diffusivities of order 10^{-4} m^2 s^{-1} or smaller. Several modelling groups are now experimenting with rotated diffusivity tensors so as to avoid these unphysical fluxes of density. Preliminary results indicate that the depth of the thermocline is significantly affected by this change, as is the penetration depth of tracers like tritium, (which Sarmiento (1983) predicted). As numerical models achieve finer horizontal resolution and so can afford to have smaller diffusivities before becoming numerically unstable, possibly the effects of these unphysical density fluxes will decrease, although the paper by Holland & Batteen (1986) suggests that this may not necessarily the case for baroclinic motions.

Bryan's (1986) analysis of the output of Cox's (1985) eddy-resolved circulation model of an ocean basin shows that the lateral transport of density by mesoscale motions is almost exactly compensated by lateral volume - and hence density fluxes - of the mean circulation, so that the strength of the thermohaline circulation is not increased by the inclusion of mesoscale eddy motions.

If one changes one's frame of reference from the Eulerian frame at a fixed x,y,z point to the Lagrangian perspective following a neutral surface (or, say, an isopycnal), this result becomes quite obvious. Only the flux of density normal to neutral surfaces can contribute to the thermohaline circulation; as the lateral eddy motions in an eddy-resolved model (like Cox, 1985) occur along the instantaneous neutral surfaces, they do not transport mass across isopycnals. Eddy-resolving circulation models tend to use smaller horizontal mixing coefficients and so may be less prone to errors caused by mixing horizontally rather than along the local neutral surface. The extra volume flux along neutral surfaces due to eddy motions is simply compensated by a different mean flow in the eddy-resolved model. The thermohaline circulation is driven by small-scale mixing processes and buoyancy exchange with the atmosphere, since both these processes can cause dianeutral buoyancy fluxes.

I have argued that a strictly horizontal diffusivity (which is normally used for numerical convenience rather than physical reality) can actually have quite significant effects on the baroclinic features of a model. This argues for rotating the diffusivity tensor and including the small-scale mixing processes as diapycnal diffusivities. But how should this diapycnal diffusivity be chosen? Numerical circulation models to date have simply employed a constant vertical diffusivity throughout the entire model space. In other ocean models the vertical diffusivity is parameterized as a function of the local Richardson number

$$Ri = N^2 / \left(u_z^2 + v_z^2 \right) . \tag{1}$$

Taking the thermal wind equations, $fu_z = \dfrac{g}{\rho} \rho_y$ and $-fv_z = \dfrac{g}{\rho} \rho_x$, the definition of the buoyancy frequency, N, $N^2 = g(\alpha\theta_z - \beta S_z)$, and recognising that the square of the slope of isopycnals to the horizontal, s^2, is given by $s^2 = \left(\rho_x^2 + \rho_y^2 \right) / \rho_z^2$, we find that the geostrophic Richardson number is given by

$$Ri = \frac{f^2}{N^2} s^{-2} . \tag{2}$$

The horizontal flux of density caused by the horizontal diffusivity, K, in these numerical models is $- K \left(\rho_x \underline{i} + \rho_y \underline{j} \right)$. Projecting this vector so that it becomes a dianeutral flux, (rather than a flux through a vertical plane), this is equivalent to a dianeutral flux of $- K \left(\rho_x^2 + \rho_y^2 \right) / \rho_z$. Dividing by (minus) the vertical density gradient and using (2), the exactly horizontal density flux is found to be equivalent to a dianeutral diffusivity of

$$D_{equiv} = K \frac{f^2}{N^2} \frac{1}{Ri} = Ks^2. \tag{3}$$

This effective vertical diffusivity has the right flavour in that it increases as Ri decreases, but it has several undesirable features. Firstly, there is no reason to expect the Coriolis parameter to affect small-scale turbulent mixing. Secondly, if we insert typical values of K (1000 m^2 s^{-1}) and of $(f/N)^2$ (10^{-3}), a value of Ri of order 10^2, as occurs in the model Western Boundary Currents, yields very large values of the vertical diffusivity, of order 100×10^{-4} m^2 s^{-1} Following Pacinowski & Philander

(1981) one would also expect to see a more sensitive power law dependence on the Richardson number than that contained in (3).

Prognostic numerical models of the circulation also include some deliberate vertical mixing that is parameterized with a vertical diffusivity. Bryan (1987) varied this diffusivity from $0.1 \times 10^{-4} \, \text{m}^2 \, \text{s}^{-1}$ to $2.5 \times 10^{-4} \, \text{m}^2 \, \text{s}^{-1}$ and found that the depth of the main thermocline changed by a factor of two and the meridional heat flux changed by a factor of 7. This work highlights the importance of vertical mixing processes on the climatologically important ocean circulation. It would be very valuable to see the results of an ocean circulation model that included some kind of depth variability in the vertical mixing intensity. As the real ocean must also have substantial lateral variations in the strength of vertical mixing processes, it would be very enlightening to impose some kind of arbitrary variation in the vertical diffusivity, perhaps based on the lateral eddy kinetic energy distribution.

Dianeutral advection is a vitally important process in driving the mean ocean circulation. Unfortunately, our ignorance of the magnitude of the vertical diffusivity and the way this mixing intensity varies in three-dimensional space is embarrassing. Estimates of the vertical diffusivity in the seasonal thermocline vary by more than an order of magnitude, and the spatial variations are even more controversial but no less important in terms of understanding the dianeutral velocities. A very recent numerical modelling experiment has used a vertical diffusivity proportional to N^{-1}. It would also be interesting to try a diffusivity proportional to N^{-2} because then the vertical turbulent buoyancy flux is independent of depth and so the dianeutral velocity (upwelling) has contributions from only the non-linear nature of the equation of state of sea-water (see equation 14 below). The circulation of the deep ocean would then be dramatically different, since the upwelling that drives the Stommel and Arons circulation would be substantially reduced.

3 BUDGET METHODS WITH SPECIAL CONTROL VOLUMES

The budget method of determining the vertical diffusivity of scalars is summarized in Gargett (1984). This method has been used by Shaffer (1979), Whitehead & Worthington (1982), and by Hogg et al. (1982) to determine the vertical diffusivity across an almost enclosed potential isotherm in the deep ocean. To determine the required heat flux across the potential isotherm, the lateral volume flux from the ocean floor to the height of the potential isotherm in question is measured at a deep strait. This method has yielded perhaps our most accurate estimates of the vertical diffusivity in the ocean, albeit only in restricted areas. It is widely believed that the vertical diffusivities obtained in these studies ($1\text{-}4 \times 10^{-4} \, \text{m}^2 \, \text{s}^{-1}$) are typical of the deep ocean environment, but a caveat must remain because the particular potential isotherms used in these studies are not, on average, very far off the ocean floor, and so have an unrepresentative amount of bottom contact (or near-bottom contact) compared with more typical deep and intermediate surfaces.

Niiler & Stevenson (1982) used a similar budget method with a control volume bounded by (i), a subsurface potential isotherm and (ii), the sea surface near the equator, thereby enclosing the so-called heat pool. Using the surface heat flux from bulk formulae, they deduced the turbulent heat flux across the subsurface potential isotherm and compared this with the heat flux measured by microstructure instrumentation. An important aspect of this method is the careful choice of control volume. For example, Niiler & Stevenson (1982) avoided both lateral and vertical (meaning through the surface θ = constant) advective heat fluxes by choosing a potential isotherm as their subsurface bounding surface. They also showed that the lateral heat flux in the mixed layer due to seasonal variability of the Ekman flux was quite small. A simple scaling analysis shows that this same conclusion applies to the magnitude of the lateral fluxes caused by mesoscale eddies. Perhaps this budget method will be used in reverse in the future so as to improve our estimates of the bulk formulae by using measured subsurface turbulent heat and salt fluxes to deduce the surface fluxes of heat and fresh water.

4 METHOD OF EXPONENTIAL TRACERS: ROOTH & OSTLUND (1972)

A method of analysing tritium and potential temperature data that purports to place an upper bound on the strength of vertical mixing in a restricted depth range of the Sargasso Sea has been proposed by Rooth and Ostlund (1972). These authers observed a linear relation between the vertical variations of the natural logarithm of the two tracers. However, they also assumed that this linear relation applied to the three-dimensional spatial variations of these logarithmic variables. But there is no reason to expect the ratio of the lateral variations of tracers along neutral surfaces to be similar to the corresponding ratio in the vertical. For example, the lateral changes of the two tracers, potential temperature and salinity, occur in the ratio β/α, whereas the corresponding vertical changes exhibit the ratio $[\beta/\alpha]R_\rho$.

Rooth & Ostlund (1972) began their analysis with conservation equations for tritium concentration, C, and potential temperature, which I shall label θ. They actually used three-dimensional velocity and gradient vectors, but here I shall decompose these into components along and across neutral surfaces, so that $\mathbf{V} \cdot \nabla C$ becomes $\mathbf{V}_n \cdot \nabla_n C + eC_z$. Changing variables to $\phi = ln[C]$ and $\varphi = ln[\theta - \theta_c]$, where θ_c is some constant value, these conservation equations become

$$\phi_t + \lambda + \left[\mathbf{V}_n - \frac{1}{h}\nabla_n (hK)\right] \cdot \nabla_n \phi + [e - D_z]\phi_z = K\left[\nabla_n^2 \phi + \nabla_n \phi \cdot \nabla_n \phi\right] + D\left[\phi_{zz} + (\phi_z)^2\right]$$

$$\tag{4}$$

$$\varphi_t + \left[\mathbf{V}_n - \frac{1}{h}\nabla_n (hK)\right] \cdot \nabla_n \varphi + [e - D_z]\varphi_z = K\left[\nabla_n^2 \varphi + \nabla_n \varphi \cdot \nabla_n \varphi\right] + D\left[\varphi_{zz} + (\varphi_z)^2\right]. \tag{5}$$

The notation here is the same as used by McDougall (1987a); K is the lateral diffusivity acting along neutral surfaces, D is the vertical difusivity, e is the vertical velocity through neutral surfaces (i.e. the dianeutral velocity) and h is the vertical distance between adjacent neutral surfaces. These equations are the same as equations 3 and 4 of Rooth & Ostlund (1972), except that here I have included the terms due to the lateral variations of hK. Over a restricted depth range, corresponding to a potential temperature range from 9°C to 17°C, Rooth & Ostlund (1972) found that the vertical variations of ϕ and φ could be accurately related by the simple linear relation $\phi = \mu\varphi + \phi_o$ where ϕ_o was taken to be a constant. Using (4) and (5), this relation leads to the simple and elegant equation, $\phi_t + \lambda = (\mu^2 - \mu) [D\varphi_z^2 + K\nabla_n \varphi \cdot \nabla_n \varphi]$, assuming that the temporal change of potential temperature (i.e. the φ_t term) is negligible. The left-hand side of this equation was estimated from the tritium data, and since μ was greater than one, both terms on the right are positive definite, so providing the upper bound on the diapycnal diffusivity, D.

The observational data available to Rooth & Ostlund (1972) were insufficient to estimate the lateral variations of ϕ, and so the observed vertical fit of the potential temperature and tritium data should have been interpreted as implying only

$$\phi(x,y,z,t) - \mu\varphi(x,y,z) + \phi_o(x,y,t), \tag{6}$$

where ϕ_o is not a constant but a function of lateral position on a neutral surface. The observed relationship between the *vertical* variations of ϕ and φ is retained in (6). Examples of distributions of variables in physical space that satisfy (6), are the separable forms

$$C(x,y,z,t) = C_1[t] \, f[x,y] \, \exp[z/H], \tag{7}$$

and
$$\theta(x,y,z) - \theta_c = \theta_1 f[x,y] \, \exp[z/\mu H], \tag{8}$$

where $f[x,y]$ is an arbitrary function of latitude and longitude, and θ_1 is a constant. These expressions have the property that the lateral variations of θ and C along a neutral surface are proportional. While this is only one of many possible assumptions about the lateral distributions of the two tracers, it is a very simple assumption compared with the much more special one of the lateral variations of $ln[C]$ being proportional to those of $ln[\theta - \theta_c]$ with the same constant of proportionality, μ, as applies to the vertical variations. Using (7) and (8), $\phi_o(x,y,t)$ is actually $ln[C_1(t)] - \mu \, ln[\theta_1] + (1-\mu)ln[f(x,y)]$, so that $\nabla_n\phi_o = (1-\mu) \, \nabla_n (ln[f])$.

Subtracting μ times equation 5 from equation 4, and using the more general (although possibly not the most general) relationship between ϕ and φ given by 6, we find

$$\phi_t + \lambda + \left[\mathbf{V}_n - \frac{1}{h}\mathbf{V}_n(hK)\right] \cdot \nabla_n \phi_o = (\mu^2 - \mu)\left[D\varphi_z^2 + K\nabla_n \varphi \cdot \nabla_n \varphi\right]$$
$$+ K\left[\nabla_n^2\phi_o + \nabla_n \phi_o \cdot \nabla_n \phi_o + 2\mu \nabla_n \varphi \cdot \nabla_n \phi_o\right]. \tag{9}$$

Since there are many extra terms that depend on $\nabla_n\phi_o$, and since these can be either positive or negative, Rooth & Ostlund's paper can only be interpreted as providing a value for the vertical diffusivity $(0.2 \times 10^{-4} \, m^2 \, s^{-1})$ if one *ab initio*, assumes a simple one-dimensional model, since this is the only consistent way of eliminating all the lateral derivatives of ϕ_o. This being the case, the transformation to logarithmic variables does not in fact assist with the intepretation of the data.

5 ON THE INTERPRETATION OF MICROSTRUCTURE DISSIPATION MEASUREMENTS

5.1 Thermobaricity and cabbeling

Thermobaricity and cabbeling are *vertical advection* processes that are caused by lateral mixing along neutral surfaces (McDougall, 1987a). It is traditional to imagine mesoscale eddy motions or the energetic range of two-dimensional turbulence as causing this lateral mixing (or stirring) although it is obvious that small-scale processes of some kind must ultimately be responsible for intimate mixing to the molecular scale. At present we have only a very basic understanding of how scalar property gradient variance is transferred from the mesoscale (10^5 m) to the microscale (10^{-3} m). The traditional thought experiment involves releasing some red dye in the ocean and thereafter taking snapshots of its evolution. Initially the dye is deformed into long thin streaks of red fluid, but at some stage, the spatial gradients become so large that the diffusion of red material begins to be important and then eventually becomes the dominant process. The lateral processes available to cause this lateral mixing are legion, and here we mention just two: lateral interleaving and shear flow dispersion. Both of these processes can flux scalar properties along neutral surfaces for any vanishingly small (but non-zero) amount of microscale turbulence. The argument here is that the fluxes along neutral surfaces are governed by meso-scale processes and that the small-scale turbulence diffuses the fluid's properties at the rate set by the meso-scale. If the micro-scale turbulence is particularly weak in a certain area, the stirring by intermediate-scale processes must proceed for a little longer before the spatial gradients are sufficiently strong for the weak microstructure to be effective.

It is clear from the above that neither thermobaricity nor cabbeling has any detectable signature in the dissipation rate of mechanical energy, ε. Nevertheless, both of these processes produce significant downwelling (or upwelling) and also significant water-mass transformation in the permanent thermocline. Since microstructure instrumentation is not capable of measuring all the vertical processes at work in the ocean, we must be prepared in the future to use information on meso-scale lateral turbulence in order to infer some of the vertical processes.

5.2 Vertical diffusivities in a thermohaline staircase

Consider now the use of the dissipation rate of mechanical energy, ε, to infer the vertical diffusivities of salinity, S, potential temperature, θ, and buoyancy in a thermohaline staircase. For simplicity, consider only the salt-finger stratification where relatively hot salty water overlies cooler, fresher water. Imagine a regular series of steps and layers with salt fingers being present at the steps, causing fluxes of heat and salt down the water column. The buoyancy flux ratio, γ_f, is defined as $\alpha F^\theta / \beta F^S$, where F^θ and F^S are the downward fluxes of potential temperature and salt. The vertical buoyancy flux is defined as $g[\alpha F^\theta - \beta F^S]$. In the absence of other mixing processes, it is approximately equal to the dissipation rate of mechanical energy, ε (Larson and Gregg, 1983). The buoyancy frequency, N, is defined by $N^2 = g[\alpha\theta_z - \beta S_z]$, where the vertical gradients of potential temperature and salinity are defined as averages over several steps and layers. The vertical eddy diffusivity of salt, D^S, can now be expressed as

$$D^S = -\frac{\beta F^S}{\beta S_z} = \frac{[R_\rho - 1]}{[1 - \gamma_f]} \frac{\varepsilon}{N^2},$$

(10)

while the vertical diffusivity of θ is,

$$D^\theta = -\frac{\alpha F^\theta}{\alpha\theta_z} = \frac{\gamma_f}{R_\rho} \frac{[R_\rho - 1]}{[1 - \gamma_f]} \frac{\varepsilon}{N^2},$$

(11)

and the diffusivity of density (or buoyancy) is

$$D^\rho = -\frac{g\left[\alpha F^\theta - \beta F^S\right]}{N^2} = -\frac{\varepsilon}{N^2}. \tag{12}$$

Taking values of $\gamma_f \cong 0.7$ and $R_\rho \cong 1.6$, appropriate to the recent CSALT measurements, equations (10)-(12) above give $D^S = 10 \times [0.2\ \varepsilon/N^2]$, $D^\theta = 4.4 \times [0.2\ \varepsilon/N^2]$, and $D^\rho = -5 \times [0.2\ \varepsilon/N^2]$. It is not surprising that the normal expression, $0.2\ \varepsilon/N^2$ (Osborn, 1980), does not apply in a thermohaline staircase, because the dissipation mainly occurs in the quasi-well-mixed layers where N^2 is small, and also because the energy supply for the motion is the potential energy stored in the mean salt stratification rather than being the kinetic energy of the flow field (as in the case of breaking internal waves). It is however important to note that a thermohaline staircase is so much more effective (by a factor of up to 10) at transporting salt than the simple expression involving ε and N^2 would suggest. Generally a thermohaline staircase receives special attention from an investigator, but as the analysis of microstructure data becomes more automated, it is very likely that thermohaline staircases will go unnoticed. In this case, salt fingering will transport salt much more effectively than would be suggested by an automatic data handling system that used the standard formula.

In the CSALT experiment, measurements of ε were between one and two orders of magnitude less than predicted on the basis of double-diffusive laboratory experiments (Gregg and Sanford, 1987), with an average value in the layers of only 1.4×10^{-10} m^2 s^{-3}, which is at the noise level of the measurement technique. However, using this low value of dissipation in our formula for the vertical diffusivity for salt, $D^S = 10 \times [0.2\ \varepsilon/N^2]$, together with the appropriate value of N^2 of 1.2×10^{-5} s^{-2}, we find that D^S is 0.25×10^{-4} m^2 s^{-1}. This is regarded as a medium to large value of a vertical diffusivity at a depth of 500 m, and is consistent with an estimate based on the diffusion of freon in this region (Schmitt, 1987). The factor of $5[R_\rho - 1]/[1 - \gamma_f]$ that multiplies the normal expression for a diffusivity in (10) means that, even though the ε measurements are much smaller than expected for a thermohaline staircase, the implied value of the vertical diffusivity is not small.

5.3 Deducing dianeutral velocities from dissipation measurements

The vertical velocity, e, through surfaces of neutral static stability, caused by small-scale turbulence with a vertical diffusivity, D, is given by

$$[e - D_z]\frac{N^2}{g} = D\left[\alpha\theta_{zz} - \beta S_{zz}\right], \tag{13}$$

(see McDougall 1984). This can be expressed in terms of the vertical derivative of N^2 as

$$eN^2 = [DN^2]_z - [DN^2]\frac{R_\rho}{R_\rho - 1}\left[\frac{\alpha_z}{\alpha} - \frac{\beta_z}{\beta}\frac{1}{R_\rho}\right]. \tag{14}$$

Using the expression, $D = 0.1 \; \varepsilon/N^2$, from the previous section of the paper, (14) can be rearranged to be a differential equation relating the vertical variations of ε to the dianeutral velocity, namely,

$$10 \; eN^2 = \varepsilon_z - \varepsilon \frac{R_\rho}{[R_\rho - 1]} \left[\frac{\alpha_z}{\alpha} - \frac{\beta_z}{\beta} \frac{1}{R_\rho} \right],$$

(15)

or, equivalently,

$$\frac{e}{D} = \frac{\varepsilon_z}{\varepsilon} - \frac{R_\rho}{[R_\rho - 1]} \left[\frac{\alpha_z}{\alpha} - \frac{\beta_z}{\beta} \frac{1}{R_\rho} \right].$$

(16)

The second term on the right-hand sides of (14-16) is due to the equation of state of sea-water being non-linear. Since β varies little in the oceanographic parameter range in comparison with α, $\left[\frac{\alpha_z}{\alpha} - \frac{\beta_z}{\beta} \frac{1}{R_\rho} \right]$ may be approximated by $\left[\frac{1}{\alpha} \frac{\partial \alpha}{\partial \theta} \theta_z + \frac{1}{\alpha} \frac{\partial \alpha}{\partial p} p_z \right]$, where p_z is simply -1 db/m. The terms $\frac{\partial \alpha}{\partial \theta}$ and $\frac{\partial \alpha}{\partial p}$ are about $1.1 \times 10^{-5} \; K^{-2}$ and $2.5 \times 10^{-8} \; K^{-1}$ (db)$^{-1}$ respectively (McDougall, 1987a), so that the nonlinear term in the above equations is zero when $\theta_z \cong 2.3 \times 10^{-3} \; K \; m^{-1}$; a typical gradient of potential temperature at a depth of about 1500 m in the ocean.

As a demonstration of the importance of the nonlinear-equation-of-state term in calculating the dianeutral velocity, a simple numerical integration of the differential equation for ε, (15), is shown in Figure 1. The square of the buoyancy frequency, N^2, was assumed to vary exponentially in the vertical with a length scale of 1000 m and a value of $5 \times 10^{-6} \; s^{-2}$ at a pressure of 1000 db. The full equation of state was used to find θ_z from N^2 at each depth, assuming for the sake of argument that S_z was zero. The nonlinear term $\left[\frac{\alpha_z}{\alpha} - \frac{\beta_z}{\beta} \frac{1}{R_\rho} \right]$ is zero at a pressure of 1463 db, where the potential temperature was arbitrarily put equal to 4°C. A constant upwelling velocity through neutral surfaces of $10^{-7} \; m \; s^{-1}$ was assumed and (15) was integrated by the Runge-Kutta technique, beginning at a pressure of 1463 db and proceeding first shallower to 1000 db and then deeper to 4000 db. The initial value of ε was chosen so that, in the absence of the non-linear terms in the equation of state, the vertical diffusivity would be $10^{-4} \; m^2 \; s^{-1}$. It can be seen from the above equations that in the absence of the non-linear terms, this same value of the diffusivity would apply at each depth.

From Figure 1 it can be seen that the non-linear terms require the vertical diffusivity, D, to more than double to over $2 \times 10^{-4} \; m^2 \; s^{-1}$ at a depth of 4000 m in order to maintain a constant upwelling velocity throughout the water column. Another way of quantifying the influence of the non-linear terms in the equation of state is by using the calculated ε (z) profile to evaluate a dianeutral velocity from a truncated form of (15). This velocity, which is labelled e_L (L for a "linear" equation of state), is simply, $0.1 \; \varepsilon_z/N^2$. The figure shows that the difference between the true dianeutral velocity and that calculated assuming a linear equation of state is as large as $0.2 \times 10^{-7} \; m \; s^{-1}$ at a depth of 4000 m. This contribution to the dianeutral velocity is "caused" by the non-linear terms in the equation of state in the sense that, while it is still proportional to ε, it occurs even when the turbulent buoyancy flux, $-DN^2$, is independent of depth.

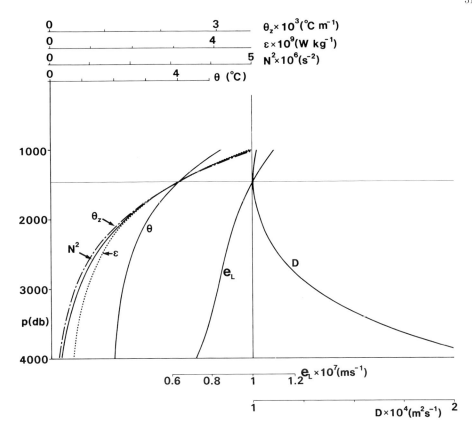

Figure 1. An illustration of the importance of the non-linear terms in the equation of state for deducing the dianeutral velocity from a measured vertical profile of the dissipation rate of mechanical energy, . The assumed profiles of N^2, θ_z and θ are described in the text. The value of ε at each depth is obtained by integrating the differential equation (15) and D is the value of the vertical diffusivity required to support an upwelling velocity of 10^{-7} m s^{-1} at every depth. If the equation of state were linear, D would be constant at 10^{-4} m^{2}s^{-1} throughout the water column. e_L is the dianeutral velocity that one would obtain from the same $\varepsilon(z)$ profile if one had ignored the non-linear terms in the equation of state.

Graphs of the nonlinear terms that appear in (16) have been shown for a part of the North Atlantic in Figures 2f, 3f and 4f of McDougall (1987a). From these it can be seen that the contribution to e/D from the nonlinear terms varies from about -10^{-3} m^{-1} at a depth of 500 m to about -0.5×10^{-3} m^{-1} at 1000 m. For a vertical diffusivity, D, of 10^{-4} m^2 s^{-1}, the difference between e and e_L varies from 10^{-7} m s^{-1} at 500 m to half this value at a depth of 1000 m. This change of e with depth, caused by the vertical variation of the non-linear term in the above equations, can also be used in the linear vorticity equation, $\beta v = f w_z$ (here β is the meridional derivative of the Coriolis frequency, f), to examine their influence on the meridional velocity, v. The contribution of the non-linear terms to e_z is -10^{-10} s^{-1} so that these terms can lead to a misinterpretation of the meridional velocity of 0.4 mm/s, a significant lateral velocity in the deep ocean.

6 THE RELATIVE ROLES OF DOUBLE-DIFFUSION AND BREAKING INTERNAL WAVES IN THE CENTRAL WATER

Schmitt (1981) has provided observational and theoretical evidence for the Central Water of the World's oceans being better fitted by a line of constant R_ρ than by a straight line on a S-θ plot. This is now taken as evidence for the importance of salt-fingering in the Central Water, since small-scale turbulent mixing with a diffusivity, D, that is the same for both heat and salt would lead to a straight line on the S-θ curve of the water masses. This issue is by no means settled, since one can imagine that lateral advection and diffusion along sloping neutral surfaces from the ocean surface would also often be important in determining the properties of the sub-surface water masses in the Central Water. The fact that the signature of anomalous winter mixing events is quickly erased in the Central Water so that a very tight S-θ curve is achieved is certainly strong evidence in favour of some kind of effective vertical mixing agent. In this section of the paper I pursue Schmitt's (1981) proposition that salt-fingering is an important process in forming the Central Water characteristics, but rather than finding that double-diffusion is much stronger than normal small-scale mixing processes, I conclude that both types of mixing perform equal and opposite amounts of water-mass conversion, and that normal vertical mixing transports significantly more buoyancy than salt-fingering, but in the opposite direction.

The starting point here is the sub-surface water-mass conversion equation in which the dianeutral advection is associated with each of the vertical processes that cause this dianeutral velocity to occur. From McDougall (1987a) the following equation for the variation of salinity on a neutral surface can be derived

$$S_t + \left[V_n - \frac{1}{h}\nabla_n(hK) \right] \cdot \nabla_n S = K \nabla_n^2 S + DgN^{-2}\theta_z^3\alpha\frac{d^2S}{d\theta^2} - F_z^s\frac{[R_\rho - r]}{[R_\rho - 1]}$$
$$+ KgN^{-2}S_z\left[\frac{\partial\alpha}{\partial\theta}\nabla_n\theta \cdot \nabla_n\theta + \frac{\partial\alpha}{\partial p}\nabla_n\theta \cdot \nabla_n p \right]. \tag{17}$$

Lateral gradients are denoted by ∇_n, $\dfrac{d^2S}{d\theta^2}$ is proportional to the curvature of the S-θ diagram formed from a vertical CTD cast, and r is the ratio of the double-diffusive flux divergences of heat and salt, $r = \alpha F_z^\theta / \beta F_z^s$. In a one-dimensional steady ocean that is considered here, S_t and all lateral gradients are zero, leaving only two terms in (17); that due to the vertical diffusivity, D, and that due to the vertical divergence of the flux of salt by salt-fingers, F_z^s. Following Schmitt (1981), let the flux of salt by salt fingers be given by $-A(R_\rho)S_z$ where the effective diffusivity of salt is a function of the vertical stability ratio, R_ρ. In the Central Water, R_ρ is taken to be constant and so A is also constant. The relevant balance in (17) can be written as

$$DgN^{-2}\theta_z^3\alpha\frac{d^2S}{d\theta^2} = -AS_{zz}\frac{[R_\rho-r]}{[R_\rho-1]}.$$ (18)

Vertical mixing in the ocean causes concomitant dianeutral advection. The above equation has the total effect (i.e. the diffusive and the advective contributions) of the vertical diffusivity on water-mass conversion on the left-hand side, and the total effect of salt-fingering on the right-hand side. Since there are only two terms, it is apparent that small-scale turbulence and salt-fingering perform equal and opposite amounts of water-mass conversion in our one-dimensional ocean.

An estimate of r can be found by using $\gamma_f = 0.7$ and a vertical length scale for the variation of the salt-finger salt flux, F^s/F^s_z, of 400 m in the relation (McDougall, 1987a)

$$r = \gamma_f - \gamma_f\frac{\beta}{\alpha}\frac{\partial[\alpha/\beta]}{\partial z}\frac{F^s}{F^s_z},$$ (19)

giving $r \cong 0.5$. The ratio of $DgN^{-2}\theta_z^3\alpha\frac{d^2S}{d\theta^2}$ to DS_{zz} is plotted in Figure 2d of McDougall (1987a) on a neutral surface in the North Atlantic, and in the region of the North Atlantic Central Water of the subtropical gyre (deeper than 500 m) this ratio is about -1. The map of potential temperature on this neutral surface in Figure 11(b) of McDougall (1987b) shows that $\nabla_n\theta \cong \mathbf{0}$, supporting the one-dimensional balance taken in (18). Using this observation, together with (18) gives the following estimate for the ratio A/D,

$$\frac{A}{D} = \frac{[R_\rho-1]}{[R_\rho-r]},$$ (20)

which is about $\frac{2}{3}$ in this region where $R_\rho \cong 2$ and using $r \cong 0.5$. The vertical flux of salt due to small-scale turbulence is then 50% greater than that due to salt fingers although the numerical uncertanties in the above quantities means that it can only be concluded that these vertical salt fluxes are similar and are both downward fluxes.

Small-scale turbulent processes transport buoyancy downwards (i.e. density upwards) in a stably statified water column, while the reverse is true of double-diffusve processes. Using (20), the relative contributions to the dianeutral buoyancy flux is given by the ratio

$$\frac{Bouyancy\ flux\ due\ to\ A}{Bouyancy\ flux\ due\ to\ D} = \frac{g[\gamma_f-1]\beta F^s}{-DN^2} = -\frac{A}{D}\frac{[1-\gamma_f]}{[R_\rho-1]} \cong -\frac{1}{5},$$ (21)

where γ_f is the buoyancy flux ratio of salt fingers and is about 0.7. This shows that double-diffusion produces only one fifth as much buoyancy flux as small-scale turbulent mixing, and that it is directed in the opposite direction.

The contribution of salt-fingering to the dianeutral velocity, e, can be found by recognizing that in the Central Water, $d\left[R_\rho\right]/dz \cong 0$, and so

$$\frac{\theta_z^2}{S_z}\frac{d^2S}{d\theta^2} = \left[\frac{\alpha_z}{\alpha} - \frac{\beta_z}{\beta}\right], \tag{22}$$

(see McDougall, 1984, equation 12). Using (18), the dianeutral velocity caused by salt-fingering in the Central water can be found as

$$D\frac{[1-r]}{[R_\rho - r]}\frac{R_\rho}{[R_\rho - 1]}\left[\frac{\alpha_z}{\alpha} - \frac{\beta_z}{\beta}\right], \tag{23}$$

which is positive, and about one third ($[1 - r]/[R_\rho - r] \cong \frac{1}{3}$) as large as the negative dianeutral velocity that arises from the non-linear terms acting with the vertical diffusivity, D, in (14).

This section of the paper has explored the implications of the water-mass conversion equation, (17), for the relative strengths of salt-fingering and small-scale turbulence in the Central Water, where, based on the work of Schmitt (1981), the vertical stability ratio, R_ρ is believed to be independent of depth. The conclusions above should be regarded as preliminary and need to be tested against a much more extensive data set in other oceans as well as in the North Atlantic. Also it may well be that the ocean is never sufficiently one-dimensional for the above balances to apply, even in the Central Water. One indication of the need to include lateral mixing processes would be if the ratio A/D of (20) was negative. Another complication would arise if the bulk of dianeutral double-diffusive fluxes were carried by quasi-lateral intrusions. The effective vertical diffusivity of salt and heat may then be negative and the ratios r and γ_f may be substantially different to 0.5 and 0.7 respectively (Garrett, 1982; McDougall, 1985). This is another example that shows that the parameterization of the effects of mixing in the ocean is in its infancy, with much work still to be done.

Acknowledgements

The author wishes to acknowledge helpful correspondence with Dr Kirk Bryan about the parameterization of mixing processes in prognostic ocean models. Encouragement from Dr Barry Ruddick led to the inclusion of section 6.

References

Armi, L., 1978: Some evidence for boundary mixing in the deep ocean. *J. Geophys. Res.*, **83**, 1971-1979.

Bryan, F., 1987: Parameter sensitivity of primitive equation ocean circulation General Circulation Models. *J.Phys.Oceanogr.*, **17**, 970-985..

Bryan, K., 1969: A numerical method for the study of the circulation ofthe World Ocean. *J. Comput. Phys.*, **4**, 347-376.

Bryan, K., 1986: Poleward buoyancy transport in the ocean and mesoscale eddies. *J. Phys. Oceanogr.*, **16**, 927-933.

Cox, M.D., 1985: An eddy resolving numerical model of the ventilated thermocline. *J. Phys. Oceanogr.*, **15**, 1312-1324.

Eriksen, C.C., 1985: Implications of ocean bottom reflection for internal wave spectra and mixing. *J. Phys. Oceanogr.*, **15**, 1145-1156.

Gargett, A.E., 1984: Vertical eddy diffusivity in the ocean interior. *J. Marine Res.*, **42**, 359-393.

Garrett, C., 1982: On the parameterization of diapycnal fluxes due to double- diffusive intrusions. *J. Phys. Oceanogr.*, **12**, 952-959.

Gregg, M.C. and T.B. Sanford, 1987: Shear and turbulence in thermohaline staircases. *Deep-Sea Res.*, in press.

Hogg, N., P. Biscaye, W. Gardner and W.J. Schmitz Jr., 1982: On the transport and modification of Antarctic Bottom water in the Vema Channel. *J. Mar. Res.*, **40**,(suppl), 231-263.

Holland, W.R. and M.L. Batteen, 1986: The parameterization of subgrid-scale heat diffusion in eddy-resolved ocean circulation models. *J. Phys. Oceanogr.*, **16**, 200-206.

Larson, N.G., and M.C. Gregg, 1983: Turbulent dissipation and shear in thermohaline intrusions. *Nature*, **306**, 26-32.

McDougall, T.J., 1984: The relative roles of diapycnal and isopycnal mixing on subsurface water-mass conversion. *J. Phys. Oceanogr.*, **14**, 1577-1589.

McDougall, T.J., 1985: Double-diffusive interleaving. Part II: Finite amplitude, steady-state interleaving. *J. Phys. Oceanogr.*, **15**, 1542-1555.

McDougall, T.J., 1987a: Thermobaricity, cabbeling and water-mass conversion. *J. Geophys. Res.*, **92**, 5448-5464.

McDougall, T.J., 1987b: Neutral surfaces. *J. Phys. Oceanogr.*, **17**, 1950-1964.

McDougall, T.J. and J.A. Church, 1986: Pitfalls with the numerical representation of isopycnal and diapycnal mixing. *J. Phys. Oceanogr.*, **16**, 196-199.

Niiler, P. and J. Stevenson, 1982: The heat budget of tropical ocean warm-water pools. *J. Mar. Res.*, **40**(suppl.) 465-480.

Osborn, T.R., 1980: Estimates of the local rate of vertical diffusion from dissipation measurements. *J. Phys. Oceanogr.*, **10**, 83-89.

Pacanowski, R.C. and S.G.H. Philander, 1981: Parameterization of vertical mixing in numerical models of tropical oceans. *J. Phys. Oceanogr.*, **11**, 1443-1451.

Rooth, C.G. and H.G. Ostlund, 1972: Penetration of tritium into the Atlantic thermocline. *Deep-Sea Res.*, **19**, 481-492.

Sarmiento, J.L., 1983: A simulation of bomb tritium entry into the Atlantic Ocean. J. *Phys. Oceanogr.*, **13**, 1924-1939.

Schmitt, R.W., 1981: Form of the temperature-salinity relationship in the Central Water: Evidence of double-diffusive mixing. *J. Phys. Oceanogr.*, **11**, 1015-1026.

Schmitt, R.W., 1987: The Caribbean Sheets and Layers Transects (CSALT) Program. *EOS Transactions, American Geophysical Union*, **68**(5)H, 57-60.

Shaffer, G., 1979: Conservation calculations in natural coordinates (with an example from the Baltic). *J. Phys. Oceanogr.*, **9**, 847-855.

Stommel, H. and A.B. Arons, 1960a: On the abyssal circulation of the world ocean - I. Stationary planetary flow patterns on a sphere. *Deep-Sea Res.*, **6**, 140-154.

Stommel, H. and A.B. Arons, 1960b: On the abyssal circulation of the world ocean - II. An idealized model of the circulation pattern and amplitude in oceanic basins. *Deep-Sea Res.*, **6**, 217-233.

Tziperman, E., 1986: On the role of interior mixing and air-sea fluxes in determining the stratification and circulation of the oceans. *J. Phys. Oceanogr.*, **16**, 680-693.

Veronis, G., 1975: The role of models in tracer studies. *Numerical Models of Ocean Circulation*, National Academy of Science, 133-146.

Whitehead, J.A. and L.V. Worthington, 1982: The flux and mixing rates of Antarctic Bottom Water within the North Atlantic, *J. Geophys. Res.*, **87**, 7903-7924.

Woods, J.D., 1985: The World Ocean Circulation Experiment. *Nature*, 314, 505-511.

MERITS AND DEFECTS OF DIFFERENT APPROACHES TO MIXED LAYER MODELLING

ERIC B. KRAUS

Cooperative Institute for Research in Environmental Sciences
University of Colorado/NOAA, Boulder CO 80309-0449 USA

ABSTRACT

Ocean mixed layer modelling has involved three different approaches. These can be listed under the headings of a) Bulk or integral models that do not resolve vertical differences within the layer, b) Diffusive models including K theories and higher closure schemes and c) Eddy-resolving models that either simulate or parameterize the larger eddies individually. Each of these three approaches is described and their relative advantages and domains of applicability are then discussed.

1 INTRODUCTION

The layers immediately below the ocean surface are almost always in a state of turbulence. They are stirred mainly by wind, waves and surface cooling. This leads to a uniform or quasi-uniform vertical distribution of temperature, salinity, and often also momentum and other conservative properties. Dissipation tends to be distributed as well almost uniformly along the vertical through the mixed layer, as demonstrated by Crawford and Osborn (1979), Shay and Gregg (1984) or Osborn and Lueck (1985), though relatively small maxima in the surface wave region below the interface and — intermittently — at the outer mixed layer boundaries are not uncommon. The turbulence kinetic energy tends to decrease downward rapidly through the stable strata that separate the mixed layers from the ocean interior.

The monthly or seasonally-averaged depth of oceanic mixed-layers, as defined by various criteria, has been mapped by Levitus (1982) and other authors. It is affected locally by annual and diurnal changes, by transient variations associated with meso-scale eddies and by internal waves at the layer boundaries. Because of these features it is necessary to use a plurality of observations for an unbiased assessment of the mean layer depth (Large et al., 1986).

The annual march of the layer depth is not harmonic and its character is quite different from the annual surface or layer temperature change. The greatest depth is reached at the end of the cooling season, that is, in March or April in the Northern

Hemisphere. After that, the layer becomes rapidly shallower to reach minimum depth only three months later in June. After the summer solstice it deepens again, though the mixed layer temperature keeps on increasing until the end of the heating season in August or September. An analogous development characterizes the diurnal surface mixed layer. The content of physics and the verisimilitude of different mixed-layer models can be judged, to some extent, by the efficacy with which they represent the asymmetries and the hysteresis in this development.

Mixed-layers contain only a small fraction of the total oceanic mass. Their direct influence on the large-scale dynamics is not very great. Many oceanic general circulation models achieve a reasonable level of verisimilitude without any explicit consideration of mixed-layers, by forcing the interior ocean directly through specified surface boundary conditions or parameterized surface fluxes. However, the dynamics of the air/sea boundary layers becomes a dominant process in the determination and prediction of the sea surface temperature, which is a crucial parameter in the working and the simulation of the joint atmosphere-ocean system.

The equitable terrestrial climate is due mainly to the storage of heat during late spring and summer by the ocean mixed-layer in temperate latitudes. This heat is released again to the atmosphere in autumn and winter. The process exerts a critical influence upon the annual/interannual evolution of the climate and upon the global circulation of heat or chemical substances.

The heat storage capacity or thermal inertia of a 50 m deep ocean mixed-layer is about 80 times smaller than that of the entire ocean, but it is more than twenty times larger than that of the whole atmosphere above. Similar ratios apply to the chemical storage capacity of the layer. The exchange of energy and matter across the air-sea interface is often more vigorous than the exchange with the interior ocean through the lower layer boundary. The critical thermal and chemical buffering role of the ocean mixed-layer is a consequence of these conditions. Consideration of this role is particularly important in the study of interannual, decadal and perhaps secular climate fluctuations, including those that are expected to be caused by changes in the system's CO_2 content.

Simulations of the mixed layer and its evolution require, as input, information about the surface fluxes, about conditions in the strata below, and about advection and antecedent conditions in the layer itself. They are expected to provide a prognostic output of the surface temperature and of the mean temperature and velocity structure. The models that can do that fall into three categories: 1) Integral models, 2) Diffusive models and 3) Eddy simulating models. I shall describe briefly each type of model in the following part of this paper. Their relative merits will then be compared in the conclusion.

2 INTEGRAL MIXED LAYER MODELS

Bulk or integral mixed layer models are based on the *a priori* consideration, that there is a mixed layer just below the surface in which the vertical gradients of conservative properties are negligibly small.

The simplest type of this kind of model involves a mixed layer of uniform constant depth, which is not subject to advection and is isolated from the interior ocean. The effect of such a layer upon the atmosphere can be compared to that of a swamp, which has a fixed capacity for storing heat but cannot move it around. Swamp models of the ocean have been used by Manabe and Wetherald (1980), Washington and Meehl (1984) and Meehl and Washington (1985) to study the atmospheric climate and the CO_2 problem.

In reality, the integral amount of thermal energy that can be stored in a mixed layer depends usually more on changes in layer depth than on changes in temperature. The depth to which an ocean layer can be thoroughly stirred depends in turn on the amount of turbulence kinetic energy that is available for this purpose. A mixed layer theory, based on this concept, was formulated originally by K. Ball (1960). It was applied to the ocean by Kraus and Turner (1967) and many other investigators.

In the absence of internal sources or sinks, the rate of change of some conservative property S in a mixed layer of instantaneous depth h is represented simply by the flux $F_0(S)$ through the air-sea interface, plus the change caused by the "entrainment" of fluid with a different concentration S_h from the interior medium. One has hence

$$h \, dS/dt = F_0(S) + (S_h - S)w_E . \tag{1}$$

The entrainment velocity w_E represents the depth of the fluid stratum that is being incorporated into the mixed layer per unit time. It is generally not equal to the change in layer depth, which is given by

$$dh/dt = w_E - w_h , \tag{2}$$

where w_h, the bulk vertical velocity at the outer mixed layer boundary, is commonly derived from the wind stress curl.

The entrainment velocity w_E can be computed from the integral of the Turbulence Kinetic Energy (TKE) equation over the layer depth h. With zero vertical variations within the mixed layer, this integral can be expressed in the form:

$$\frac{1}{2} w_E(q^2 + C^2) = Au_*^3 + \frac{\alpha g}{\rho c}(\frac{1}{2}hH_0 - \lambda J_0 + (\Delta \mathbf{V})^2(\frac{1}{2}w_E - D|\Delta \mathbf{V}|) + \int_{-h}^{0} \epsilon dz. \tag{3}$$

A detailed derivation of this equation has been published by Niiler and Kraus (1977). The terms on the left hand side represent the increases per unit time of Turbulence Kinetic Energy $(\frac{1}{2}q^2)$ and of potential energy which are caused by the entrainment of quiescent, denser fluid from below. The potential energy increase is equal to the product of the entrainment velocity w_E with the square of the long internal wave velocity C at the mixed layer base. In the ocean $C^2 = (b - b_h)h$ where b, b_h are the specific buoyancies on both sides of the mixed layer base. They are specified by $b = g(1 - \frac{\rho}{\rho_r})$ where ρ_r is a reference density. It should be noticed that b must be discontinuous across the lower mixed layer boundary in this formulation.

On the right hand side, the first term denotes the kinetic energy input by the wind, assumed to be proportional to the third power of the wind velocity or the friction velocity

u_*. The following term represents the potential energy change produced by the turbulent and radiant surface heat flux H_0 and by the heating in depth, that is caused by the penetrating part J_0 of solar radiation. J_0 is negative definite $(J_0 < 0)$ while H_0 can have either sign. (In Eq. 3, $\alpha = -\dfrac{1}{\rho}\dfrac{\partial \rho}{\partial t}$ is the temperature expansion coefficient, c is the specific heat and λ is the e-folding depth of J_0.)

The third, right-hand term of (3) indicates the rate of turbulence kinetic energy increase, that is due to the mixing of a bulk velocity shear $\Delta \mathbf{V}$ across the layer base, minus the decrease associated with the emission of internal waves, which is assumed equal to the cube of $\Delta \mathbf{V}$ times a proportionality constant D. The last term represents the integral of the dissipation ϵ.

The transport of turbulence by the turbulent eddies, that is the third order correlation term in the TKE equation, is contained implicitly in the equation (3), which has been parameterized and simplified by various authors in many different ways. The Niiler and Kraus (loc. cit.) version neglected the terms containing q^2 and $|\Delta \mathbf{V}|^3$. It assumed further that both the generation and the dissipation of wind-generated turbulence was proportional to the cube of the friction velocity — an assumption that finds some support in a paper by Oakey and Elliot (1982) — and that the dissipation of the convectively generated turbulence was proportional to the upward buoyancy flux. After minor rearrangement this yields the following simplified form of (3)

$$w_E\, C^2 - w_E (\Delta \mathbf{V})^2 = 2m_1 u_*^3 + m_2 \frac{\alpha g}{\rho c}\left(\frac{1}{2} h H_0 - \lambda J_0\right). \qquad (4)$$

The ratio of the terms on the left hand side of (4) is equal to the Richardson Number at the layer base. On the right hand side, the proportionality factors m_1 and m_2 specify the non-dissipated fractions of the mechanically and convectively generated TKE. The smaller the value of these fractions, the less energy is available for entrainment.

The factor $m_2 \equiv 1$ during sea surface heating $(H_0 < 0)$ when no thermal eddies are being generated and when the mixing depends entirely on an input of mechanical energy. When the resulting tendency towards stabilization becomes stronger than the wind stirring, the sum of the terms on the right hand side of (3) or (4) becomes negative. There is then not enough TKE available to stir the water down to the depth h. The entrainment velocity w_E vanishes in this case and water is "detrained" from the mixed layer, which reforms at a shallower depth that can be determined diagnostically by setting the right side of (4) equal to zero. This yields

$$h = -\frac{2m_1 \rho c u_*^3 - \alpha \lambda g J_0}{\alpha g H_0} \propto L, \qquad (5)$$

where $H_0 < 0$ and $J_0 < 0$. The symbol L denotes the Monin-Obukhov length. As both the wind stress and the penetrative radiation contribute to the generation of turbulence kinetic energy, it seems logical to include the J_0 term in a generalized formulation of L. Omission of the radiation term J_0 has probably caused several investigators to underestimate the depth of the mixed layer during the heating season.

The mixed layer model described above involves a measure of arbitrariness. In particular, the values of the factors m_1 and m_2 are not well known and the resulting parameterization cannot be very accurate. Niiler and Kraus quoted laboratory and oceanographic observations that suggest values for m_1 between 1.25 and 8.0. It so happens that 1.25 is the reciprocal of the von Karman constant $\kappa=0.4$, which is usually included in the denominator of the Monin-Obukhov length L. A choice of $m_1=1.25$ simply makes the summer (detrainment) mixed layer depth $h = 2L$. The same value of m_1 also seems to produce the best results in the hybrid, large scale, numerical circulation model that is the subject of another report in this volume. Unfortunately, this agreement is almost certainly fortuitous. Gaspar (1987), using the same mixed layer model, found that $m_1 = 0.63 \pm 0.08$ best fitted observations at station P in the Pacific. An even slightly smaller number was derived by Davis et $al.$ (1981) from a continuous, detailed, 20-day measurement series.

It was argued above that $m_2 = 1$ during detrainment. Ball ($loc.$ $cit.$) assumed in his original paper that convective eddies are too large to be affected by dissipation and that, therefore, $m_2=1$ also in unstable conditions with a positive (upward) buoyancy flux. Since then, authors have observed or stipulated values that range all the way from $m_2 = 1$ down to $m_2 = 0.036$ for a deepening mixed layer under the ice of a frozen lake (Farmer, 1975). Gaspar (1987) publishes a useful table, summarizing many different assessments of m_2 mainly in the atmosphere, which cluster around a value $m_2 = 0.2$.

In reality, the indicated parameterization is rather too simple and neither coefficient is a true constant. The numerical value of the factors m_1 and possibly also m_2 is likely to be affected by atmospheric stability, by characteristics of the surface wave field, by shear currents, and perhaps by other non-local influences. A decrease of m_1 and m_2 with the layer depth h has been stipulated by some authors including Resnyanskiy (1975), Garwood (1977) and Wells (1979). It seems obvious that there is no single, clear-cut answer to this problem. Fortunately, the particular form of parameterization exerts more influence upon the rate of mixed-layer development than upon the final state.

Nothing has been said so far about the model temperature change during detrainment. One can get another expression for h if one interprets Eq. (1) as a thermal energy equation, with $S=\rho cT$ (T = surface or mixed layer temperature) and $F_0(T) = H_0$ With $w_E = 0$ during detrainment, elimination of h between the resulting equation and (5) yields

$$H_0^2 = \mid \frac{2m_1\rho c}{\alpha g} \, u_*^3 - \lambda J_0 \mid \; \rho c \frac{\partial T}{\partial t} \; . \tag{6}$$

Equation (6) can be used in a model to predict temperature changes during detrainment. Alternatively, it can be considered an inversion of the integral mixed layer model, which permits us to assess the surface heat flux from observations of the surface stress, the solar surface irradiance and the surface temperature change. All these variables can be obtained from consecutive, routine satellite observations.

Apart from this possibility of inversion, the integral models have the merit of simplicity, computational economy and physical plausibility. The governing equations are very transparent, in the sense that each of their terms can be identified with some particular physical process. The reverse side of this coin is a rather arbitrary parameterization of dissipative processes and an oversimplification of the real upper ocean structure. Though some authors (see, e.g., Woods and Strass, 1986) have made allowances for shear in a thermally well mixed layer, a consistent application of the integral method involves the assumption of a uniform distribution of all conservative properties, including momentum, throughout the layer. The resulting model replaces the transition zone at the lower mixed layer boundary by a discontinuity. It also cannot be used to predict non-adiabatic changes in the stable strata below.

3 DIFFUSIVE MODELS

Integral mixed layer models involve *a priori* assumptions about the vertical structure of the upper ocean. In diffusive models, such assumptions are avoided. Diffusion also does not permit the development or the persistence of discontinuities. In numerical models, diffusion always involves in the first instance an interaction between adjacent grid points. All diffusive models are therefore fundamentally local in character.

The oldest, most orthodox and still most widely used approach to boundary layer modelling is linear. It is based upon the concept of eddy diffusivity, which arose directly from the presumed analogy between viscous and turbulent transports. Eddy coefficients, usually denoted by the symbol K, play the same role in this approach as do the molecular coefficients of viscosity, conductivity and diffusivity in kinetic theory. For example, the turbulent, vertical temperature flux is simply assumed to be proportional to K times the vertical temperature gradient:

$$\overline{w'T'} = -K\frac{\partial T}{\partial z} \ . \tag{7}$$

This approach breaks down when an up-gradient transport of one property is driven by the down-gradient transport of another property. It can become meaningless also when the gradient of the mean of a transported property becomes zero, as is the case usually in truly well mixed layers.

The eddy coefficients are determined empirically. Unlike the molecular coefficients, they cannot be considered constant. Following Prandtl's and G.I. Taylor's pioneering studies early this century, the variations of K have been parameterized frequently by the product of a mixing length l_m with a characteristic turbulent velocity. The mixing length represents the distance which a fluid parcel travels before it mixes completely with its new environment. Observations indicate that it is necessary to consider K or l_m dependent not only upon the distance from the boundary, but also the level of turbulence intensity, the stability and other factors. Many investigators have proposed therefore various parameterizations of K or l_m as a function of these processes and conditions.

In practically all of these schemes, the functional form of K in unstable conditions when the buoyancy flux is upwards differs from that in stable condition when it is downwards. Within each of these domains K is then represented as a function of the temperature and velocity distribution, or of the surface heat flux and stress, or of a combination of these quantities. The specific form of this function is based on dimensional arguments and observational data. Good reviews of these efforts with full lists of references have been published by Bush (1977) and by Sarachik (1978).

Classical K theories are restricted to the prediction of changes in the bulk or mean variables. The turbulent momentum and buoyancy transports can interact only through their effect upon the mean shear and the stratification. Changes in the mean products or squares of the fluctuating quantities cannot be considered in the framework of these linear or first-order theories. To predict stochastic variables, such as the turbulence kinetic energy or the temperature variance, one has to formulate higher-order closure schemes which include the turbulence kinetic energy equation and other non-linear predictive equations of the same type.

In ocean turbulence models, temperature is often assumed to be a unique function of the density, and pressure fluctuations are almost always neglected. This leaves the three velocity components and the temperature as the only fluctuating variables. From these four quantities one can form 10 different products and squares, which can be identified as the six components of the Reynolds stress tensor, three components of the temperature flux vector and the scalar temperature variance. The rates, at which these second-order quantities change, can be expressed formally by a set of predictive equations. To form a closed system, the triple correlations that occur in these equations have to be parameterized in terms of lower order quantities. Alternatively they have to be expressed by additional equations. To complete the system, it is of course necessary to include also the first-order equations for the changes of the bulk variables.

The system of parameterized second-order equations can be further simplified and truncated in various ways. The level of simplification has been assessed systematically by Mellor and Yamada (1974) in the framework of a hierarchy of turbulence closure models. They ordered the various terms in these equations by scaling them with a parameter that measured the deviation from isotropy. The progressive neglect of terms that were considered small on this scale gave them four different levels of simplification.

Their highest "level 4" involves all ten second-order, predictive differential equations, with triple correlations expressed in simplified or parameterized form. On level 3, the three velocity variance equations collapse into a single predictive equation for the turbulence kinetic energy. The only other predictive equation is that for the temperature variance. The turbulent momentum and temperature transports are now derived from diagnostic equations. Levels 2 and 1 involve only diagnostic equations for the stochastic variables. They are therefore related to the various parameterization schemes for the eddy viscosity or the mixing length.

A somewhat different hierarchy of models can be based on the order of the predicted quantities. First-order models predict only changes in the mean or bulk variables. Mellor and Yamada's level 1 or 2 models and all large-scale ocean circulation models fall into this category. Second-order models involve at least some transients of variances and covariances. They include levels 3 and 4 in the Mellor and Yamada scheme and have been applied to the ocean by Worthem and Mellor (1979), Klein and Coantic (1981) and others.

The second-order closure schemes have to parameterize triple correlations. They make no allowance directly for processes such as the turbulent flux of TKE from the surface to the entrainment region. Layer deepening can occur in second-order models only if a "local" TKE source is available from shear flows at the layer boundary. To consider triple correlations explicitly, one has to use higher-order closure schemes.

A third-order, level 4 model, which was designed specifically to assess the turbulent transport of turbulence kinetic energy down to the base of the mixed layer, has been described by André and Lacarrère (1985). Their approach involved only two bulk variables: a (one-dimensional) horizontal velocity u and the temperature T. That reduces the scheme to two first-order predictive equations for these variables and to six second-order equations for the products and squares of u', w' and T'. To this had to be added nine third-order predictive equations for triple correlation terms such a $\overline{w'u'^2}$ or $\overline{T'^3}$.

The model has been applied only to a hypothetical forcing situation, but the results look realistic. Such realism is perhaps not surprising as the scheme involves 14 different empirical and partly adjustable, non-dimensional constants; though the problem of tuning cannot be trivial in a system of 17 prognostic equations that have to be integrated at each grid level at each time step. The resulting total integration time is relatively large, even without consideration of two horizontal velocity components or salinity .

Unlike the integral models, the diffusive models can represent gradients and other departures from a well-mixed state within the layer. They can indicate the different rates at which different bulk properties are being mixed into the layers from the sea surface and from the outer mixed-layer boundaries. In this way they can provide a rationale for the frequently observed differences between the velocity, salinity and temperature distribution in the ocean surface layers. Third and higher-order closure models can account also for a turbulent transport of TKE. The unrealistic simulation of the outer mixed-layer boundary as a discontinuity, which is characteristic for the integral models, is avoided in the diffusive approach.

Non-dimensional proportionality constants are involved in all diffusive models. Their number increases with the order and complexity of the modelling scheme. Several of these constants have been assessed from experiments or from scale analyses. However, such an assessment cannot be exact and some of the constants used in the higher-order closure schemes have never been observed or derived independently. This gives the investigator some freedom in the fine tuning of these models. It permits very realistic simulations of the observed vertical structure, but impairs the physical significance and the universality of the results.

The number of simultaneous, prognostic equations and the large number of terms that they each contain, limits the transparency and the direct physical insight, which the higher-order closure schemes can provide. They also require considerable amounts of computer time and that affects their efficacy in the framework of general circulation models. For the same reason, no attempt has been made so far, to use them for the simulation of seasonal developments.

4 EDDY SIMULATING MODELS

Diffusive models of all levels of complexity are limited essentially to numerical interactions between variables at neighboring grid points. They are based on the assumption of a continuous structure in the limit of a sufficiently fine grid scale. Eddy-simulating models, by contrast, involve advective transports over different distances by a set of differently-sized eddies. Sub-grid-scale diffusion remains an essential process in these models and has to be stipulated or implied.

Perhaps the most ambitious simulation of a fluid boundary layer can be derived from a relatively rigorous, numerical integration of the equations of motion (Deardorff 1983, Moeng 1984). Models of this type do not parameterize the larger eddies and they do not tamper with the dynamics of the phenomenon. They cannot be exact, because small-scale motions remain unresolved even in a model with a small distance between grid points. The larger eddies have to be resolved on a relatively fine time and grid scale in three, or at least in two, dimensions. The integrations require, therefore, the use of very powerful computers for considerable periods of time. At present, this makes it practically impossible to incorporate this type of direct eddy simulation into global or large-scale circulation models.

Another, somewhat more economic approach involves the parameterization of a spectrum of turbulent eddies, which can transport fluid elements across a range of distances. The method has been formalized by Fiedler (1984), and by Stull and his collaborators (1984, 1985, 1986), though some of the basic concepts go back again to G.I. Taylor's original insights. Stull called this type of spectral parameterization "Transilient Turbulence". The name comes from the Latin "transilire" which translates into "leap across". Application of the transilient model are presented in this book elsewhere by Gaspar and by myself.

To illustrate the concept consider a column of fluid composed of N discrete cells or boxes stacked on top of each other. Each cell $j = 1,2,...N$ contains a specific amount S_j of some conservative property S. Turbulent eddies of various sizes can carry fluid mass between the different cells with resulting transport and mixing of S. The scheme is reminiscent of classical mixing length theory, except that it involves an array of fixed lengths, corresponding to the distances between the different boxes.

Turbulent mixing causes the concentration of S in cell number i at time $t + \Delta t$ to be a function of its distribution in all the other cells at an earlier time t. In the transilient scheme this is described by the relation:

$$S_i(t + \Delta t) = c_{i1}S_1(t) + c_{i2}S_2(t) + ... c_{ii}S_i(t) ... + c_{iN}S_N(t) . \tag{8}$$

The dimensionless "transilient coefficient" c_{ij} represents the portion of fluid which is mixed into box i from box j during the time interval Δt. In the absence of mixing $c_{ii} = 1$ and all the other coefficients are zero.

The distribution of S in the N cells of the domain can be represented by a vector $[S]$ of N components and the transilient coefficients c_{ij} compose a square matrix $[C]$ of N^2 elements. The preceding argument can then be generalized to represent the simultaneous turbulent exchange between all the cells or grid-points by the linear matrix equation

$$[S(t+\Delta t)]=[C(\Delta t)]\cdot[S(t)] \ . \tag{9}$$

An extension of this approach to cover continuous distributions, with an infinite number of infinitesimal cells, has been described in Stull's papers.

In contrast to the local, higher-order closure schemes, Eq.(9) symbolizes a non-local, first-order or linear closure representation of the turbulent exchange process. It contains classical K theory as a special case, that is represented by a transilient matrix $[C]$ in which only the diagonal elements c_{ii} and the elements next to the diagonal $c_{i,\,i+1}$, $c_{i,\,i-1}$ are allowed to be different from zero.

The components of $[C]$ must always satisfy conservation constraints. The total mass of fluid cannot increase or decrease during the mixing. This requires that

$$\sum_{j=1}^{N} c_{ij} = 1, \text{ for each } i \ . \tag{10}$$

Conservation of the total amount of S requires also that

$$\sum_{i=1}^{N} c_{ij} = 1, \text{ for each } j \ . \tag{11}$$

If one assumes further that the portion of fluid which is mixed from box j into box i is equal to that mixed from i into j, one has in addition

$$c_{ij} = c_{ji} \qquad \text{for } i \quad j \ . \tag{12}$$

This last "exchange hypothesis" is not essential. Its realization makes the transilient array symmetric and merges the two conservation restraints into a single one.

The amount of fluid that can be moved between distant grid boxes varies with the length of the time interval Δt. When there is no turbulence or when $\Delta t = 0$, each of the diagonal elements $c_{ii} = 1$ and each of the other elements $c_{ij} = 0$. For higher turbulence intensities and longer time intervals, the diagonal elements are less than unity and the off-diagonal elements tend to be larger than zero. In the original paper it is shown that the change in the transilient coefficients which is being produced by an increase of the time step from Δt to $n\Delta t$, can be represented by the non-linear matrix equation

$$[C\,(n\Delta t)]=[C\,(\Delta t)]^n \ . \tag{13}$$

This relationship permits the investigator to explore the use of relatively long time steps in numerical models.

The value of the transilient coefficients has to be determined as a function of the bulk or mean variables. This is probably the most arbitrary feature of the transilient method.

In general, the parameterization is based on the concept that some fluid will be exchanged ultimately even in a stably stratified environment between two vertically separated grid-points i and j, if the vector velocity shear between them is strong enough to create and maintain turbulence of sufficient intensity. In the original papers, the transilient coefficients for exchange between grid points i and j were assumed, therefore, to be functions of a bulk Richardson number

$$R_{ij} = \frac{g(\rho_i - \rho_j)(z_i - z_j)}{(\mathbf{V}_i - \mathbf{V}_j)^2} . \tag{14}$$

where $(z_i - z_j)$ is the vertical distance, ρ is the density and $\mathbf{V}$ the velocity vector.

In some later papers, including the one presented in this volume, parameterization was carried out in terms of available energy. Mixing is always expected to stop when the bulk Richardson number becomes larger than some stipulated critical value, or when the available energy becomes negative. This cessation can occur before complete mixing has been achieved.

The transilient method can be categorized as a first-order turbulence closure parameterization. Statistically it includes the effects of any desired eddy sizes. Being first-order, it cannot model directly the transport of turbulence energy by the turbulent eddies, which is a higher order effect. It makes no *a priori* assumptions about the sources of turbulence. Shear induced mixing both at the surface as well as at the thermocline and convective mixing over deeper layers are all included in the transilient approach because of the instabilities that are associated with these processes across the fluid domain.

Like the diffusive models, the transilient method can represent double diffusion phenomena, different gradients of different properties within the layer and also continuous transitions between the surface mixed-layer and the interior fluid. It is not quite as simple as the integral model, but it does provide physical insights. Its main weakness is associated with the parameterization of sub-grid-scale turbulence and with the considerable arbitrariness that is involved in the parameterization of the transilient coefficients.

The detailed vertical resolution makes considerable demands on computer time. This requirement is mitigated to some extent by the possibility of using relatively large time steps and standard matrix algebra. Though this helps, it is probably not sufficient to allow for the economic incorporation of the transilient scheme into experiments with general circulation simulations.

5 SUMMARY

Each of the three model categories — integral, diffusive and transilient or eddy simulating — has advantages and drawbacks. They all involve a measure of arbitrariness. None of them is suitable for all occasions that may require mixed-layer models.

The different models make very different demands on computer resources. I have estimated the computer times needed for the integration of one time step in an integral, a transilient and a third-order closure model to have approximate ratios of about 1:100:1000.

The superiority of the integral models in this respect is further enhanced by the fact that they can use relatively large time steps without causing computational instabilities.

The principle advantage of the integral models is their generality and simplicity. They simulate the different rates of sea surface temperature change during heating and cooling periods reasonably well, particularly when penetrating radiation is not neglected. The governing equations of integral mixed layer models can be inverted easily and that opens the possibility of their use in conjunction with satellite observations. With minimal demands on computer time, they can be incorporated in large- scale circulation models of the atmosphere or the oceans without undue costs — although the actual programming of such a combination is not a trivial matter.

The integral approach treats all conservative properties as being homogeneous throughout the layer depth. This implies a vertically uniform distribution not only of enthalpy and concentrations, but also a horizontal displacement of the whole layer like a solid slab without mean internal shears. Though this can be a rather gross oversimplification of conditions in the real ocean, it conforms with the structure of most general circulation models as a finite set of equipotential or isobaric or isopycnic layers, which are assumed to be homogeneous internally.

Transilient models can produce simulations that are more realistic in their details than can be produced by integral models. In particular, they can deal with surface layers that are not mixed completely and with transients in the stable strata below. They can be adapted to reproduce differential rates of transport and dissipation for different properties, including the effect of double diffusion processes.

The transilient parameterization requires enough computer time to make its incorporation into general circulation models rather uneconomic at this stage. At present, the method seems suitable for regional simulations, particularly in the equatorial region where the near surface vertical profiles of momentum and density are often quite different. They may also be found useful for the study of frontal phenomena and for the investigation of acoustic and biological transport processes in the ocean.

The diffusive approach is in the direct line of traditional turbulent transport studies. The parameterizations involved in this approach incorporate the conclusions and results from many more laboratory experiments, field tests, and theoretical investigations than were used in either of the two other approaches. The concept of eddy diffusivity or eddy viscosity appears to be particularly useful for the study of small-scale, highly localized features such as the "constant stress" boundary layer, for example. It tends to become arbitrary when applied to larger scales such as the mixed layer, where complex variations of K have to be stipulated. Both K theories and even more so the higher-closure models become then very complicated and involve many empirical parameters. This limits their predictive value and the physical insight that they can provide. The higher-order closure models tend to be also too costly in computer time to make their inclusion in general circulation models attractive. They may well provide standards of realism that could be helpful in the calibration of simpler models and they should be particularly useful in the modelling of special situations where a realistic, local representation of the vertical profiles is of primary importance.

Acknowledgments: The research leading to this paper has been supported by the U.S. Office of Naval Research under contract N00014-86-C-0363 and by the National Science Foundation under Grant OCE 85-00860. I would also like to thank Ms. Ann Brennan for production assistance.

REFERENCES

André, J.C. and Lacarrère, P., 1985. Mean and turbulent structures of the oceanic surface layer as determined from one-dimensional, third-order simulations. *J. Phys. Oceanogr., 15*: 121-132.

Ball, F.K., 1960. Control of inversion height by surface heating. *Quart. J. Roy. Meteor. Soc., 86*: 483-494.

Bush, N.E., 1977. Fluxes in the surface boundary layer over the sea. In: E.B. Kraus (Editor), *Modelling and Prediction of the Upper Layers of the Ocean*, Pergamon Press, pp. 72-91.

Crawford, W.R. and Osborn, T.R., 1979. Microstructure measurements in the Atlantic equatorial undercurrent. *Deep-Sea Res., GATE Suppl. II-V*, 26: 285-308.

Deardorff, J.W., 1983. A multi-limit mixed-layer entrainment formulation. *J. Phys. Oceanogr., 13*: 988-1002.

Fiedler, B.H., 1984. An integral closure model for the vertical turbulent flux of a scalar in a mixed layer. *J. Atmos Sci., 41*: 674-680.

Garwood, R.W., Jr., 1977. An oceanic mixed layer model capable of simulating cyclic states. *J. Phys. Oceanogr., 7*: 455-468.

Gaspar, P., 1987. Modelling the seasonal cycle of the upper ocean. *J. Phys. Oceanogr.* (submitted)

Klein, P. and Coantic, M., 1981. A numerical study of turbulent processes in the marine upper layers. *J. Phys. Oceanogr., 11*: 849-863.

Kraus, E.B. and Turner, J.S., 1967. A one-dimensional model of the seasonal thermocline. *Tellus, 19*: 98-106.

Large, W.G., McWilliams, J.C. and Niiler, P.P., 1986. Upper ocean thermal response to strong autumnal forcing of the Northeast Pacific. *J. Phys. Oceanogr., 16*: 1524-1550.

Levitus, S., 1982. *Climatological Atlas of the World Ocean*. NOAA Professional Paper 13., 173 pp.

Manabe, S. and Wetherals, R.T., 1980. On the distribution of climate change resulting from an increase in carbon dioxide content of the atmosphere. *J. Atoms. Sci. 37*: 99-118.

Meehl, G.A. and Washington, W.M., 1985. Sea surface temperatures computed by a simple ocean mixed layer coupled to an atmospheric GCM. *J. Phys. Oceanogr., 15*: 92-104.

Mellor, G. and Yamada, T., 1974. A hierarchy of turbulence closure models for planetary boundary layers. *J. Atoms. Sci., 31*: 1791-1805.

Moeng, C.H., 1984. A large-eddy-simulation model for the study of planetary boundary layer turbulence. *J. Atmos. Sci., 41*: 2052-2062.

Niiler, P.P. and Kraus, E.B., 1977:. One-dimensional models of the upper ocean. In: E.B. Kraus, (Editor) *Modelling and Prediction of the Upper Layers of the Ocean.* Pergamon Press, pp. 143-172.

Oakey, N.S., 1985. Statistics of mixing parameters in the upper ocean during JASIN phase 2. *J. Phys. Oceanogr., 15*: 1662-1675.

Oakey, N.S. and Elliott, J.A., 1982. Dissipation within the surface mixed layer. *J. Phys. Oceanogr., 12*: 171-185.

Osborn, T.R. and Luek, R.G., 1985. Turbulence measurements with a submarine. *J. Phys. Oceanogr., 15*: 1502-1520.

Resnyanskiy, Yu.D., 1975. Parameterization of the integral turbulent energy dissipation in the upper quasihomogeneous layer of the ocean. *Isv. Atm. Ocean. Phys., 11*: 453-457.

Sarachik, E.S., 1978. Boundary layers on both sides of the tropical ocean surface. Review papers on equatorial oceanography. Proc. FINE Workshop, Ft Lauderdale, Nova/N.Y.I.T. Univ. Press.

Shay, J.J. and Gregg, M.C., 1984. Turbulence in an oceanic convective mixed layer. *Nature, 310*: 282-285. (Corrigendum: *Nature, 311*, 84)

Stull, R.B., 1984. Transilient turbulence theory, Part 1: The concept of eddy mixing across finite distances. *J. Atmos. Sci., 41*: 3351-3367.

Stull, R.B., 1986. Transilient turbulence theory. Part 3: Bulk dispersion rate and numerical stability. *J. Atmos. Sci., 43*: 50-57.

Stull, R.B. and Kraus, E.B., 1987. The transilient model of the upper ocean. *J. Geophys. Res.* (in press)

Washington, M.W. and Meehl, G.A., 1984. Seasonal cycle experiment on the climate sensitivity due to a doubling of carbon dioxide with an atmospheric general circulation model coupled to a simple mixed layer ocean model. *J. Geophys. Res., 89*: 9475-9503.

Wells, N.C., 1979. A coupled ocean-atmosphere experiment: The ocean response. *Quart. J. Roy. Meteor. Soc., 105*: 355-370.

Woods, J.D. and Strauss, V., 1986. The response of the upper ocean to solar heating II. The wind-driven current. *Quart. J. Roy. Meteor. Soc. 112*:

Worthem, S. and Mellor, G. 1979. Turbulence closure model applied to the upper tropical ocean. *Deep-Sea Res., GATE Suppl., 26*: 237-272.

THE INCLUSION OF A SURFACE MIXED LAYER IN A LARGE-SCALE CIRCULATION MODEL

ERIC B. KRAUS†, RAINER BLECK‡ and HOWARD P. HANSON†
† Cooperative Institute for Research in Environmental Sciences, University of Colorado/NOAA, Boulder CO 80309-0449 USA
‡ Division of Meteorology and Physical Oceanography, Rosenstiel School of Marine and Atmospheric Science, University of Miami, Miami FL 33149 USA

ABSTRACT

This paper is concerned with a numerical model that combines an adiabatic, baroclinic interior ocean with a thermodynamically active mixed layer of variable depth. The interior ocean is specified in terms of isopycnic coordinates. An adjustment procedure that permits matching of these two domains during periods of mixed layer shallowing is described. Discussion of the model's formulation is followed by a presentation of some preliminary model results.

1 INTRODUCTION

Turbulent diffusion in the interior ocean, below the surface mixing zone, occurs mainly within isopycnal layers (see e.g. Sarmiento, 1983). Diapycnal (vertical) mixing is relatively small. Various investigators have made explicit allowances for this asymmetry either by a rotation of the diffusion tensor (Redi, 1982; Cox, 1987) or by formulating ocean models in isopycnal coordinates (Bleck and Boudra, 1981, 1986). Both of these approaches suffer from the absence of representative, quantitative information about the complete turbulent diffusion tensor and its oceanic distribution in space and time.

Present knowledge about diapycnal mixing in the ocean and its parameterization has been summarized by Gargett (1986). The rate of isopycnal mixing is generally assumed to be a function of the local mesoscale eddy field. In the future, it is to be hoped that statistics derived from the integration of eddy-resolving ocean models will provide information about this phenomenon. It is conceivable that some parameterization could be based on the estimated relative transport rates over various distances, which may be associated with some bulk criterion for baroclinic instability between horizontally (isopycnally) separated grid points. To date, no actual work has been carried out along these lines. Conceptually, the approach would be analogous Stull's (1984) specification of transilient mixing along the vertical as a function of bulk Richardson Numbers. (See also papers on this topic by Gaspar and by Kraus in this volume.)

The following brief account deals with ongoing work on one aspect of these problems, the interface between the highly turbulent upper ocean and the more laminar interior ocean. We have not progressed far enough to report final conclusions or to show realistic simulations of the real ocean. However, the problems and various difficulties that we have encountered so far may be of some interest to those who now contemplate a similar endeavor.

We have developed a model that simulates an adiabatic interior ocean, composed of a number of isopycnic layers, that is capped by a vertically homogeneous mixed layer. It is forced by sea-surface fluxes of heat, kinetic energy and of momentum that vary with latitude and with time. The local depth and density of the mixed layer is determined by the value of these fluxes, as well as by advection and by the entrainment of water from below.

Though salinity changes will be allowed for in future versions of the model, we have assumed in this first instance that the salinity has a constant value $S=35$ ‰. This makes for a unique relation between the potential temperatures and the densities or specific volumes. The various isopycnal layers are characterized in the model by the parameter

$$\varsigma = 1 - \rho_r/\rho . \tag{1.1}$$

A reference density $\rho_r = 1000$ kg m^{-3}, makes ς almost — but not quite — identical with the commonly used parameter σ_T. The values of ς and of the associated potential temperatures in the interior ocean are invariant model parameters. By contrast, the local density and temperature of the mixed layer can assume a continuous range of values, as determined bv air-sea interactions, mixing with the underlying strata and advection.

The simulation of the mixed layer in this paper follows the scheme described by Niiler and Kraus (1977) and that of the interior ocean is based upon the Bleck and Boudra (*loc. cit.*) approach. That approach does not allow for any diapycnal mixing but can be generalized to simulate a thermohaline circulation. In its present form, the model is too coarsely meshed (222 km at the equator; 111 km at 60° N) to resolve mesoscale eddies. Isopycnal mixing, that is turbulent transports within isopycnal surfaces, is simulated by an eddy exchange coefficient and by the effects of "numerical diffusion" associated with finite differencing. The absence of explicitly-resolved eddies is not critical in the present project, which is concerned primarily with the ultimate effects of a temporally and spatially variable injection of water mass and of conservative properties from the surface mixed layer into the interior ocean.

This injection occurs during periods when the mixed layer becomes warmer and shallower. The locally retreating mixed layer then leaves water masses below that have to be added to the appropriate isopycnal layer in the interior ocean. The mixed layer itself is detached and cannot be affected during this phase by the conditions below.

The opposite holds during the deepening phase of the evolution, when water is being entrained from below. The resulting transfer of mass then affects the local composition of the surface mixed layer, but it cannot affect the characteristics of the water in the interior.

It follows from the preceding argument that the entrainment and detrainment processes require quite different numerical modelling procedures. During layer deepening one has to compute the contribution of the entrained water to the change in mixed layer temperature or specific buoyancy. During periods of strong heating, on the other hand, one has the more complicated problem of apportioning the detrained water among the various layers in the model. The following two sections deal with the algorithms required for these two cases. The results of our first model integration are then described in a final section.

2 THE ENTRAINMENT ALGORITHM

A complete equation for the entrainment velocity w_E can be found in the Niiler and Kraus (*loc. cit.*) paper and also in another contribution by one of us (Kraus) in the present volume. In the investigation described here, we used a highly simplified version of this equation of the form

$$w_E = \frac{2m_1 u_*^3/h + m_2 B}{b - b_l} \equiv \frac{E}{b - b_l} , \qquad (2.1)$$

where b and h are the mixed layer buoyancy and depth, b_l is the buoyancy immediately below the layer base, u_* is the friction velocity and B is the surface buoyancy flux; m_1 is a proportionality factor assumed here to have a value equal to 1.25 and m_2 is equal to 1.0 for $B < 0$ or equal to 0.2 for $B > 0$. The terms involving penetrative radiation, velocity shears at the mixed layer base and changes in the mean square turbulence velocity were neglected in the present case though they do occur in the complete w_E equation. We may include some of these terms in future versions of our model.

As the mixed layer deepens, it may swallow up the water contained earlier in some of the model isopycnic layers. This must happen whenever the assigned buoyancy b_k of layer k exceeds the local mixed layer value b. Layer k is then carried on in the model locally in a 'massless' state with zero thickness, until it is 'inflated' again during the following heating season. In other words these layers are needed as a receptacle for water forming the seasonal thermocline.

Omitting the effect of a mean upwelling velocity for the purpose of the present discussion, one can use (2.1) to specify the amount of layer deepening during the timestep Δt by

$$\Delta h = \frac{E \Delta t}{b - b_l} . \qquad (2.2)$$

The difficulty with this formulation is that b_l must change during any timestep when a new isopycnic layer is reached by the deepening mixed layer. This makes it desirable to replace (2.2) by an expression that takes the isopycnic layer structure below the mixed layer explicitly into account. To do so in a simple and consistent manner, we now let b_1 and z_1 denote the buoyancy b and the depth h of the mixed layer at time t. The lower interfaces of the isopycnic layers $k = 2,3,4...N$ are denoted by a series of z_k's. The potential energy of the whole ensemble of layers prior to mixing to a depth z_N is then given by

$$PE = \frac{1}{2} \sum_{k=1}^{N} b_k \left(z_k^2 - z_{k-1}^2 \right) . \qquad (2.3)$$

This expression includes any massless layers for which $z_k = z_{k-1}$.

After a timestep Δt and vertical mixing down to the depth z_N the new mixed layer buoyancy will be

$$b_{mix} = \frac{1}{z_N} \left[\sum_{k=1}^{N} b_k(z_k - z_{k-1}) - B\Delta t \right] , \tag{2.4}$$

and the new potential energy is

$$PE_{mix} = b_{mix} \, z_N^2/2 = \left[\sum_{k=1}^{N} b_k(z_k - z_{k-1}) - B\Delta t \right] z_N/2 . \tag{2.5}$$

The depth z_N denotes now not the bottom interface of the Nth layer, but the depth to which water from layer N has been mixed with all the upper layers to give the new mixed layer depth $h(t+\Delta t) = z_N$.

The difference between (2.5) and (2.3) must be equal to the energy input $E\Delta t$. Solving the resulting equation for z_N yields

$$z_N = \frac{2E\Delta t + G(N-1) - b_N z_{N-1}^2}{F(N-1) - b_N z_{N-1} - B\Delta t} , \tag{2.6}$$

where

$$F(N) = \sum_{k=1}^{N} b_k(z_k - z_{k-1}) \tag{2.7}$$

and

$$G(N) = \sum_{k=1}^{N} b_k(z_k^2 - z_{k-1}^2) . \tag{2.8}$$

Before $z_N \equiv h(t+\Delta t)$ can be computed, N has to be known. In practice both tasks are solved simultaneously by computing trial values of z_N for $N=2,3,...$ This is continued until (2.6) yields a new layer depth that is smaller than the upper interface of the Nth isopycnic layer z_{N-1}. In other words the correct value of N and the new layer depth $h(t+\Delta t)$ become known as soon as the process of incrementing N has gone one step too far. The algorithm automatically disregards isopycnic layers with temporary zero thickness between the mixed layer base and the ocean interior.

3 ADJUSTMENT DURING MIXED LAYER SHALLOWING

The depth of mixed layer tends to increase most of the time from the height of summer to the beginning of the following spring. During this entrainment period, relatively cold and dense water is incorporated into the deepening mixed layer. The thickness of the underlying isopycnal layer decreases correspondingly. When that layer disappears altogther, it is carried on in the model with zero thickness and water from the next layer below is then entrained into the surface layer.

During spring, or during the morning hours in the daily cycle, the wind stirring is often not strong enough to balance the potential energy increase caused by surface warming. In other words, the working of the wind is generally insufficient to maintain mixing down to great depth. A new, warmer mixed layer with a shallower depth is then formed. The new layer depth h can be determined diagnostically by setting E in Eq. (2.1) equal to 0. As $m_2 = 1$ in periods of zero entrainment, one has then

$$h = -m_1 u_*^3 / B = m_1 L .$$ (3.1)

The proportionality factor m_1 is of order unity and the ratio L is commonly known as the Monin-Obukhov length.

With the salinity assumed constant, the surface buoyancy flux B becomes a unique function of the sum H of the fluxes of latent and sensible heat and radiation

$$B = \frac{g \alpha}{\rho c} H .$$ (3.2)

(g = gravitational acceleration and c = specific heat). The thermal expansion coefficient α is a function of the temperature

$$\alpha = -\rho^{-1} \partial \rho / \partial T .$$ (3.3)

In terms of H the depth of the retreating mixed layer is therefore specified by

$$h = -m_1 \frac{\rho c u_*^3}{g \alpha H} .$$ (3.4)

The corresponding local change of temperature is given by

$$\frac{\partial T}{\partial t} = \frac{H}{\rho c h} = \frac{g \alpha}{m_1 \rho^2 c^2} \frac{H^2}{u_*^3} .$$ (3.5)

If T, h, H and u_*^3 represent conditions at time t and if T_2 and T_1 denote the mixed layer temperatures at times $t_{2,1} = t \pm \Delta t$, application of the mean value theorem yields:

$$T_2 = T_1 + \frac{g \alpha}{m_1 \rho^2 c^2} \frac{H^2}{u_*^3} \Delta t .$$ (3.6)

An additional term has to be added to this expression to allow for advective changes associated with horizontal gradients of the mixed layer temperature.

The model does not allow us to deal with water masses that have T and ς values intermediate between those pertaining to the mixed layer and the assigned T_n and ς_n that characterize the immediately underlying n^{th} isopycnal layer. The volume of water left below the receding mixed layer has to be apportioned, therefore, in the model between the surface layer and the layer n. It is this detrainment and apportioning process that adds water of different composition to the various isopycnal layers in the interior ocean.

The need for such a partitioning of water volume and the associated adjustment in layer depth affects not only isopycnal models. Analogous adjustments would have to be made in models of mixed layers above interior oceans composed of horizontal or isobaric layers. However, in such models the adjustment would involve a change in the layer temperature and density, which is not desirable when interactive air-sea processes are to be simulated.

The apportioning or adjustment process can be carried out in a number of ways. We required conservation of the overall amount of internal energy in the water column and preservation of the true theoretical surface temperature computed from (3.6). These requirements can be met if the surface heat input H during the time interval Δt is identified with the heat gain of a layer with the adjusted depth h_2', which is warmed from T_1 to T_2, plus the heat loss from a stratum of depth $h_1' - h_2'$, which is cooled from T_1 to the temperature T_n (see Fig. 1a):

$$H = \rho c \left[h_2'(T_2 - T_1) - (h_1' - h_2')(T_1 - T_n) \right] . \tag{3.7}$$

Solving algebraically for h_2' yields easily

$$h_2' = \frac{h_1'(T_1 - T_n) + H\Delta t/(\rho c)}{T_2 - T_1} . \tag{3.8}$$

The evolution of the model mixed layer depth and temperature during warming periods when no new isopycnal layer is uncovered (that is, when none of these layers emerges from zero thickness) is determined completely by Eqs. (3.6), and (3.8).

A new isopycnal layer $n-1$ will be uncovered when $T_2 > T_{n-1}$ for the first time. At the end of the time step when this occurs, the model layer depth is identified with the true layer depth $h_2' = h_2$ and the adjustment procedure is used to establish the depth of the interface h_n between the isopycnal layers $n-1$ and n. The procedure keeps the subsequent departures of the model mixed layer energy from their true (theoretical) value within bounds. It is illustrated in Fig. 1b and described by the equation:

$$H/(\rho c) = h_2(T_2 - T_1) + (h_n - h_2)(T_{n-1} - T_1) - (h_1' - h_n)(T_1 - T_n) . \tag{3.9}$$

This yields

$$h_n = \frac{H\Delta t/(\rho c) + h_1'(T_1 - T_n) - h_2(T_2 - T_{n-1})}{T_{n-1} - T_n} \tag{3.10}$$

The indicated scheme preserves the overall thermal energy and the theoretical mixed layer temperature. It does not preserve the depth or the heat content of the mixed layer per se. It also was pointed out by Gaspar (verbal communication) that it does not conserve the potential energy. This is an error that occurs inevitably in any scheme that attempts to represent a non-linear vertical temperature or density profile in the framework of a discrete set of layers or strata.

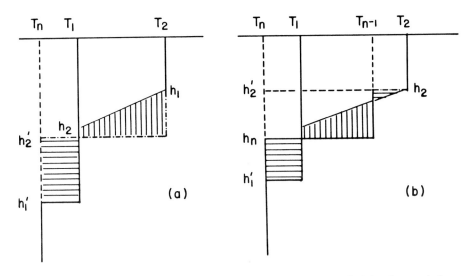

Fig. 1.   Adjustment schemes for mixed layer depth.   When the underlying isopycnic layer n does not change (a), the two shaded areas must be equal to conserve thermal energy. Here, h, h_2 and h_1 are the true theoretical layer depths at times t and $t \pm \Delta t$; h'_1 and h'_2 are adjusted depths.

In (b) the adjustment scheme illustrates the specification of the interface h_n between layer n and a newly emerging isopycnic layer n-1.

The depth of the adjusted layer can be very sensitive to the particular choice of values for the specification of the isopycnic layers. In practice this causes the difference between the true and adjusted layer depth to vary with time and latitude. This difference becomes, however, always very small at the time of the summer solstice when the temperature of the layer still tends to increase while its true depth becomes temporarily nearly constant.

After the summer solstice the mixed layer tends to deepen again.   The rate of entrainment and layer deepening depends on the density difference between the mixed layer and the underlying isopycnic stratum. On the other hand, the change in layer temperature depends both on the entrainment rate and on the temperature difference between the two layers. The temperature dependence of the expansion coefficient α prevents the ratio of the temperature and density differences from remaining constant during the development.

In nature, the mixed layer tends to deepen into a continuously stratified interior ocean below, but in the model it expands into a series of discrete, constant density and temperature strata. The model layer deepening occurs therefore at a somewhat irregular rate. This does not affect the ultimate layer characteristics, including the layer temperature, in late winter significantly, but it can cause additional departures from real conditions during fall.

4 PRELIMINARY RESULTS

In the first production runs with our model, we simulated the circulation in a hypothetical ocean bounded by two meridians and by zonal boundaries at 12° and 60°N. Fields within this domain were represented by 32x32 lateral grid points evenly spaced on a Mercator projection. Along the vertical, the model consisted of a mixed layer and 5 isopycnic layers below. The ς values of these layers were specified to be 10^{-4} times 247, 254, 260, 265, 270 and 273.

Initially all the layers were assumed to be horizontal with zero motion throughout the domain. In retrospect, we know that these initial conditions did not permit the model ocean to develop a realistic meridional slope and that is probably the most serious flaw in the results presented below.

Another weakness was caused by the adjustment procedure. In these early model runs, it was somewhat different from that described above and that caused the summer mixed layer to remain unduly deep near the northern boundary. However, the resulting error is relatively small over most of the domain and it does not seem to invalidate our preliminary deductions.

The model was forced by a kinetic energy input proportional to the third power of the wind velocity, a windstress vector proportional to the square of the wind velocity, radiation and a Newtonian-type turbulent heat flux. In the first instance, these forcing fields were assumed to be only time and latitude dependent. Longitudinal variations were not considered at this stage. Actual values of the forcing functions were chosen to be similar to the zonal averages observed over the North Atlantic, which are illustrated by Figs. 2-3. The model was integrated over a simulated period of 45 years and some of the output for the last year is shown in the following figures.

Fig. 4 shows the velocity and ς fields for the mixed layer in the months of March and September in the year 45. The mid-latitude ς contours have been displaced by about 10 degrees of latitude during these six months. Notice also the somewhat more pronounced eastern boundary currents in March.

The barotropic stream function for the model ocean in March and September is shown in Fig. 5. The northern cyclonic gyre is noticeably stronger in March than in September. We are not sure whether the same happens in the real ocean and whether it can be associated with the persistently larger cyclonic wind stress curl in that region during winter.

Fig. 6 represents North-South cross-sections across the middle of the domain. The isopycnal layers 2 and 3 extend much farther north in September than in March. This is due to the injection of water from the receding mixed layer into these isopycnal layers during the spring heating season.

To study the mixing of surface waters into the interior ocean, we seeded the surface with a tracer at the beginning of year 10. The concentration of this tracer in the mixed layer was maintained continuously at a constant value of 1.00. Fig. 7 illustrates the distribution of this tracer in layer 4 after 34 years of simulation. Layer 4 has a ς value of .0265 and an initial depth of 700-1500 m. The tracer can only reach it in the northern region of

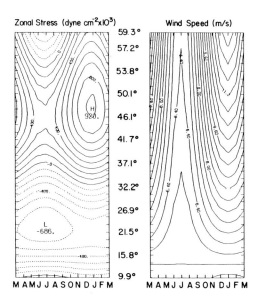

Fig. 2. Annual cycle of zonal wind stress (a) and 10-meter wind speed (b), from the Comprehensive Ocean-Atmosphere Data Set (COADS), zonally averaged.

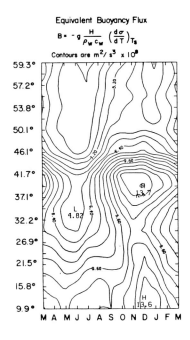

Fig. 3. Annual cycle of zonally averaged surface buoyancy flux.

60

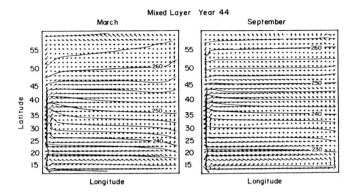

Fig. 4. Mixed layer ς and velocity fields for March (a) and September (b).

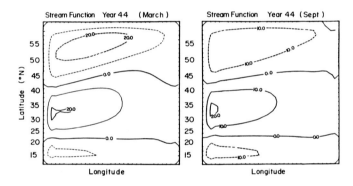

Fig. 5. Stream function for layer 4 in March (a) and September (b).

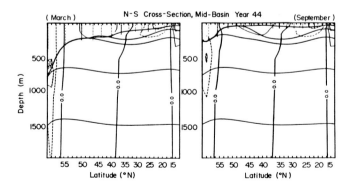

Fig. 6. Meridional cross-section in March (a) and September (b).

the domain during spring. It is then transported southward by the western boundary current and subsequently spreads across the ocean in the mid- oceanic eastward drift. It is from this drift current that it is injected into the eastern part of the anticyclonic gyre.

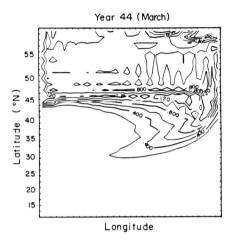

Fig. 7. Tracer distribution in layer 4.

The thermocline ventilation produced by this process does not depend directly on the local wind stress curl as does the subduction mechanism stipulated by Luyten, Pedlosky and Stommel (1983). It is likely that both mechanisms act in the ocean, with one being more effective in the lower layers and the other more important in the upper layers of the thermocline. We cannot simulate the subduction process with the present flat slope of the isopycnic surfaces, but we should be able to do so with future versions of the model that are now under development and that will start from more realistic initial conditions.

This brings us to the topic of the utility of this type of model. The problem of thermocline ventilation is obviously one topic for which the isopycnal model is particularly appropriate. The subduction mechanism depends crucially on the relative position of the outcropping of the various isopycnal surfaces and the line of zero wind stress curl. The latitudes of both processes vary seasonally and they may do so with different amplitudes and phase. We can investigate the results of these variations by an experiment involving different forcing configurations.

Our hybrid model might also serve as a useful tool for the investigation of the rate of changes and of the lag in atmospheric CO_2 content. This rate is probably affected by the seasonal variations of the mixed layer depth and temperature and by the injection or mixing of surface waters into the lower strata.

Isopycnic surfaces concentrate naturally in regions of water mass boundaries and oceanic fronts. This makes the isopycnic approach particularly useful for the study of these features. We will be interested in simulating the formation, maintenance and structure of oceanic fronts as a function of the surface heat flux, Ekman convergence, internal deformation fields etc.

Before we can make much progress with these tasks, it will probably be necessary to develop a new version of our model with a higher vertical resolution of the upper ocean in tropical and subtropical latitudes and possibly with a more realistic basin topography. These are routine technical modifications that are known to be feasible with sufficient effort. A more fundamental limitation - shared by all types of ocean models - is due to our limited understanding of oceanic turbulence.

All numerical considerations of isopycnal and diapycnal mixing in the interior ocean involve a not very well founded empiricism or outright arbitrariness. The present model is no exception to this rule. If it has any particular merit, that has to be found in the model's ability to describe a physical relationship between large-scale or meso-scale advection and mixing in parts of the upper ocean, where turbulent kinetic energy is supplied directly by the atmosphere. An improved assessment of the same relationship in the interior ocean probably depends less on modellers and their machines, than on the work of people who try to gain a quantitative understanding of turbulent processes in the oceanic environment.

Acknowledgments: The research leading to this paper has been supported by the U.S. Office of Naval Research under contract N00014-86-C-0363 and by the National Science Foundation under Grant OCE 85-00860 (EBK, HPH) and Grant OCE 86-00593 (RB).

5 REFERENCES

Bleck, R. and Boudra, D.B., 1981. Initial testing of a numerical ocean circulation model using a hybrid (quasi-isopycnic) vertical coordinate. *J. Phys. Oceanogr., 11:* 755-770.

Bleck, R. and Boudra, D.B., 1986. Wind-driven spin-up in eddy-resolving ocean models formulated in isopycnic and isobaric coordinates. *J. Geophys. Res. C, 91:* 7611-7622.

Cox M.D., 1987: Isopycnal diffusion in a z-coordinate ocean model. *Ocean Modelling* (in press).

Gargett, A.J., 1986. Small-scale parameterization in large-scale ocean models. In: J.J. O'Brien (Editor) *Advanced Physical Oceanographic Numerical Modelling*, D.Reidel Publishing Co., pp. 145-154.

Luyten, J.R., Pedlosky, J. and Stommel, H., 1983. The ventilated thermocline. *J. Phys. Oceanogr., 13:* 292-309.

Niiler, P.P. and Kraus, E.B., 1977. One-dimensional models of the upper ocean. In: E.B.Kraus (Editor) *Modelling and Prediction of the Upper Layers of the Ocean.* Pergamon Press pp.143-172.

Redi, M.H., 1982. Oceanic isopycnal mixing by coordinate rotation. *J. Phys. Oceanogr., 12:* 1154-1158.

Sarmiento, J.L., 1983. A simulation of bomb tritium entry into the Atlantic Ocean. *J. Phys. Oceanogr., 13:* 1924-1939.

Stull, R.B., 1984. Transilient turbulence theory. Part I: The concept of eddy-mixing across finite distances. *J. Atoms. Sci., 41:* 3351-3379.

MODELLING TURBULENCE IN SHALLOW SEA REGIONS

A.M. DAVIES and J.E. JONES
Proudman Oceanographic Laboratory, Bidston Observatory, Birkenhead, Merseyside
L43 7RA (England)

ABSTRACT
    A point turbulence energy model is used to examine profiles and time series
of current, viscosity and turbulence energy in a shallow homogeneous sea
region, under a range of physical conditions. A no slip bed condition is
applied and motion is induced by forcing at wind-wave or tidal periods.
    Calculations show that in the case of wind waves, a highly turbulent bottom
boundary layer having a thickness of the order of 10cm exists. Above this
layer, flow is in essence inviscid.
    For tidal flow the turbulent layer is much thicker and can extend to the
sea surface in water depths of the order of 20m, although decreasing rapidly
with height above the sea bed. In the case of wind induced flow, a region of
maximum turbulence exists at the sea surface, decreasing below the surface
layer.
    Although the model described here is for a single point, it can be readily
incorporated into a full three dimensional hydrodynamic model.

1 INTRODUCTION

    Although many oceanographic processes in the shelf seas, and shelf edge
region have traditionally been studied using two dimensional vertically
integrated models, there is now an increasing effort to develop and verify
three dimensional models.

    In the study of the transport of pollutants in shelf sea regions (e.g.
water quality models) it is necessary to accurately determine currents through
the vertical, and also the magnitude of bed stress under a range of
conditions. An accurate calculation of bed stress under the combined action
of tidal and wind induced currents together with wind waves is particularly
important in determining the fate of pollutants adhering to sediments.

    In view of the well known deficiencies of the bed stress formulation in
vertically integrated models (Jamart and Ozer 1987), and the importance of
wave-current interaction (Davies, A.G. 1986, Grant and Madsen 1986,
Christoffersen and Jonsson 1985), it has been necessary in recent years to
develop three dimensional shallow sea models (e.g. Backhaus and Hainbucher
1987, Davies, A.M. 1986a), which will ultimately include the generation of

turbulence, particularly in the near bed region, due to winds, waves, and currents.

The importance of various internal processes (Huthnance 1981) in particular internal tides and waves (Pingree et al 1986), in enhancing mixing processes at the shelf edge (New 1988) and moving shelf edge sediments (Heathershaw et al 1987) has recently been recognised. Modelling these processes will require three dimensional ocean-shelf models (Davies 1981), which can take into account vertical stratification (Davies A.M. 1986b, Heaps and Jones 1985). Ultimately three dimensional shelf edge models will be required which will predict internal mixing processes, through some form of turbulence model, and hence changes in the density field.

To date three dimensional shelf and shelf edge models have in general parameterized the vertical diffusion of momentum, via a coefficient of eddy viscosity which has been related to the flow field (Davies 1981, 1986a), or to the wind field (Davies 1985), and degree of stratification (Davies A.M. 1986b). However with increased computing power, and measuring techniques capable of examining the temporal and spatial variation of turbulence energy (see papers in this volume), there is now a strong incentive to develop three dimensional models, which can compute variations in turbulence energy. The coefficient of eddy viscosity used in such three dimensional models is related to the turbulence energy; a method involving a degree of physics which was not included in earlier models. Also by computing the turbulence energy there is the possibility of a direct comparison of predicted and measured turbulence.

It is not the intention in this paper to give a detailed review of the various turbulence models that exist in the literature; such a review would occupy a large volume and is not appropriate here. Nor is it the intention to present a detailed discussion of each of the major mechanisms, waves, tides, wind forcing which give rise to turbulence in a shallow sea region; such would be the topic of several research papers. Rather the objective of the present paper is to draw the attention of scientists concerned with measuring turbulence in the ocean, and modelling ocean circulation, to the application of turbulence energy models within shallow sea simulations.

The results presented here are preliminary calculations, using a one dimensional model through the vertical, aimed at determining the important mechanisms influencing turbulence intensity in shallow seas, prior to full three dimensional simulations. (Such a simulation is presently in progress and preliminary comparisons of computed and observed tidal currents at over one hundred current meters look particularly interesting, and results will be reported in due course).

In this paper we briefly describe the solution of the hydrodynamic and turbulence energy equations at a single point, using a finite difference grid through the vertical. Due to the importance of the bottom boundary layer in generating turbulence, a transformed coordinate system is used giving high resolution in the nearbed region.

In the calculations described here, the order of one hundred grid points were used in the vertical, in a water depth typically of order 100m or less. This resolution is significantly higher than that used in large scale ocean circulation models and is necessary to accurately resolve the high shear boundary layers which are a major source of turbulence production.

Examples are presented of bed generated turbulence due to short period wind waves, and long period tidal waves. Also shown are profiles of turbulence and associated eddy viscosity, together with currents, produced by a surface wind stress.

2 MODEL EQUATIONS AND SOLUTION METHOD

2.1 Equations

Using a Cartesian Coordinate system, with z, the vertical coordinate, having its origin at the sea bed, and increasing upwards, the linear hydrodynamic equations of motion at a point can be written as

$$\frac{\partial u}{\partial t} - fv = \frac{\partial P}{\partial x} + \frac{\partial}{\partial z} (\mu \frac{\partial u}{\partial z}) \tag{1}$$

$$\frac{\partial v}{\partial t} + fu = \frac{\partial P}{\partial y} + \frac{\partial}{\partial z} (\mu \frac{\partial v}{\partial z}) \tag{2}$$

In these equations u, v, are the x and y components of current, f the Coriolis parameter, with $\partial P/\partial x$ the externally specified pressure forcing which drives the flow, and μ the coefficient of vertical eddy viscosity.

At the sea bed, $z=z_o$, with z_o the roughness length, a no slip boundary condition is applied, thus

$$u=0, \quad v=0 \quad \text{at} \quad z=z_o \tag{3}$$

At the sea surface,

$$\mu \frac{\partial u}{\partial z} = \frac{\tau_x}{\rho}, \qquad \mu \frac{\partial v}{\partial z} = \frac{\tau_y}{\rho} \tag{4}$$

with τ_x, τ_y, the x and y components of the externally specified stress. These stresses are zero in the case of tidal motion or wave induced currents, and equal the imposed wind stress for wind driven flow.

The closure of equations (1) and (2) is achieved by relating the eddy viscosity μ, to a mixing length ℓ and turbulence energy b, computed from

$$\frac{\partial b}{\partial t} = \mu\left\{\left(\frac{\partial u}{\partial z}\right)^2 + \left(\frac{\partial v}{\partial z}\right)^2\right\} + \alpha_b \frac{\partial}{\partial z}\left(\mu\frac{\partial b}{\partial z}\right) - \varepsilon \tag{5}$$

In eqtn. (5), turbulence energy dissipation ε is determined by,

$$\varepsilon = \frac{c_1 b^{3/2}}{\ell} \tag{6}$$

with eddy viscosity μ computed from,

$$\mu = c_o \ell b^{\frac{1}{2}} \tag{7}$$

Various formulations exist in the literature for the mixing length ℓ.

Vager and Kagan (1969), give an expression for ℓ, which when integrated (Johns 1978, Davies A.G. 1986) gives

$$\ell = Kb^{\frac{1}{2}}\left\{\int_{z_o}^{z} b^{-\frac{1}{2}}dz + z_o b_o^{-\frac{1}{2}}\right\} \tag{8}$$

where b_o is the turbulence energy at the bed level ($z=z_o$).

An alternative expression (Mofjeld and Lavelle 1984) is,

$$\ell = \frac{Kz}{1+Kz/\ell_o} \tag{9}$$

with

$$\ell_o = \frac{\gamma \int_{z_o}^{h} zb^{\frac{1}{2}}dz}{\int_{z_o}^{h} b^{\frac{1}{2}}dz} \tag{10}$$

The sea bed boundary condition in eqtn. (5), is

$$\frac{db}{dz}\bigg|_{z=z_o} = 0, \text{ with a surface boundary condition}$$

$$\alpha_b \mu \frac{\partial b}{\partial z} = m_2 u_*^3, \tag{11}$$

where u_* is the frictional velocity at the sea surface, given by $u_* = (\tau/\rho)^{\frac{1}{2}}$, with τ wind stress magnitude and ρ density of sea water.

In these equations, K denotes Von Karman's constant, with a value K=0.4. Other coefficients have values, α_b=0.73, c_o=(0.046)$^{1/4}$, c_1=(0.046)$^{3/4}$, γ=0.2,

m_2=2.6 (values suggested by Vagner and Kagan (1969); Mofjeld and Lavelle (1984)).

2.2 Solution

In order to solve equations (1), (2) and (5) to (10), finite difference techniques were used in the vertical together with time stepping to advance the solution through time. In order to achieve a high resolution in the near bed region, a log-linear or log transformation was applied in the vertical (Davies, A.G. 1986, Johns 1978).

A detailed discussion of the numerical methods used is beyond the scope of this paper. However calculations revealed the importance of using a very high grid resolution through the vertical, with greatly enhanced resolution in the near bed region. Also the exact form of the time differencing, particularly of terms in the turbulence energy equation was important in attaining a numerically stable solution particularly at time steps as large as 1800 seconds.

3 EXAMPLES OF COMPUTED CURRENTS, VISCOSITIES AND TURBULENCE ENERGIES

In this section we will use the turbulence energy model described previously, to investigate the magnitude and vertical profile of turbulence energy in a shallow homogeneous sea. The influence of turbulence upon current profiles will also be considered.

Here we will examine turbulence arising from wind waves, tides, and wind forcing in water depths of 20m (a typical Southern Bight of the North Sea depth) and 100m (e.g. Northern North Sea or Malin Shelf region). Such preliminary calculations are necessary before performing full three dimensional calculations in shelf sea regions, to ensure the computational accuracy and stability of the model in a range of water depths.

In all the calculations presented here the roughness length was fixed at z_0=0.005m, a typical shallow sea value (Davies A.G. 1986). In the case of wind-wave and tidal flow, motion was started from rest and forced by an oscillatory pressure gradient $\partial P/\partial x$, having a magnitude that in an inviscid calculation would give a current of 1m/s. The pressure gradient term $\partial P/\partial y$ was zero. For wind waves, rotational effects are not particularly important, and Coriolis was set to zero (i.e. f=0). For consistency the effect of Coriolis was also neglected in the other calculations, i.e. the tidal flow was assumed rectilinear. In the case of wind induced flow, motion was started from rest and driven by a wind stress τ_x=1.0N/m², τ_y=0.0.

In the calculations presented here, one hundred grid points were used in the vertical together with either a logarithmic or a log-linear

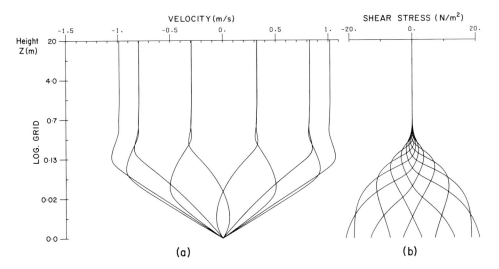

Fig. 1 (a) Current velocity profiles and (b) shear stress profiles over a wave period T=10s.

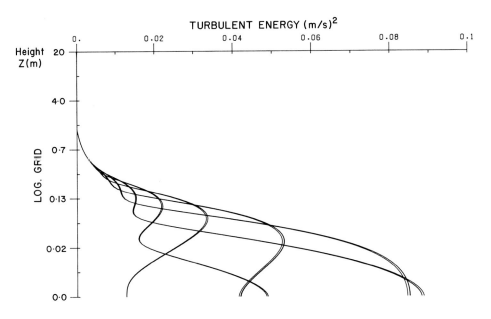

Fig. 2 Turbulence energy profiles at various times over a wave period T=10s.

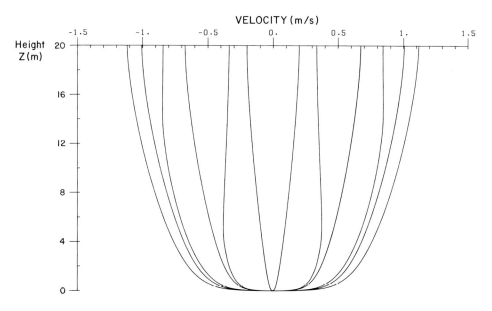

Fig. 3   Current velocity profiles at various times over an M_2 tidal
cycle.

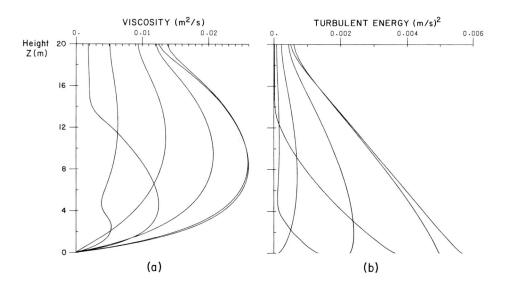

Fig. 4   Profiles of (a) eddy viscosity and (b) turbulence energy, over an
M_2 tidal cycle; water depth h=20m.

transformation. A time step of 900s. was employed in the calculations of tidal and wind induced flow, with a time step of 0.1s. for the wind wave problem, in which the wave had a period of 10s.

3.1 Wind wave of period 10s, water depth 20m

In this calculation, a log. grid was used in the vertical in order to give maximum resolution in the near bed region. The mixing length formulation of Vager and Kagan (1969), equation (8) was employed.

Fig. 1(a) shows computed wave currents over a wave period, plotted on a logarithmic grid, with the vertical profile of stress given in Fig. 1(b). It is apparent from Fig.1(a) that the wave induced current increases logarithmically with height above the sea bed. Also it is evident from these figures that the high shear bottom boundary layer only occupies the bottom 10 to 15cms of the water column, [in good agreement with observational results, Grant and Madsen (1986)], with free stream flow above this bottom layer. However the magnitude of the turbulence energy in this bottom boundary layer is high and can exceed 0.08 (m^2/s^2) (see Fig 2) although it is confined to the near bed region.

3.2 Tidal wave of M_2 period

For this longer period wave, in a water depth h=20m, the bottom boundary layer is comparable with the water depth. Consequently a log-linear grid was used in order to resolve the near bed region, though still maintaining sufficient resolution in the upper part of the water column.

Tidal current profiles, eddy viscosity profiles, and turbulence energy levels over a tidal cycle are shown in Figs 3 and 4a,b plotted on a z coordinate. It is clear from these figures that tidal currents (Fig. 3) exhibit a very high shear near bed region, where turbulence intensity is a maximum at flood tide. A detailed examination of the vertical variation of tidal current in the near bed layer, revealed a logarithmic profile close to the sea bed. Also the region of significant turbulence energy (Fig. 4b) extends much further above the sea bed than in the wind wave problem, although maximum bed turbulence is below 0.006(m^2/s^2), an order of magnitude smaller than that found in the wave problem.

It is apparent from Fig. 3 that the tidal current exhibits some shear throughout the water column (except at the sea surface where it is forced to zero by the boundary condition). This shear, together with turbulent diffusion is responsible for the non-zero levels of turbulence in the near surface layer, shown in Fig. 4(b), and the corresponding eddy viscosity profile Fig. 4(a).

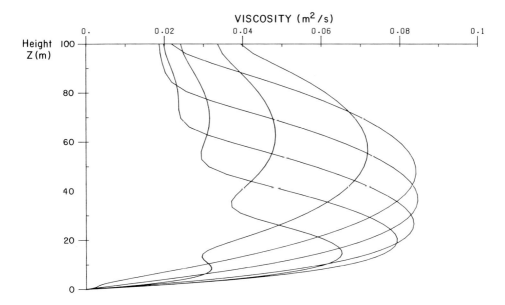

Fig. 5a Profiles of eddy viscosity over an M_2 tidal cycle; water depth h=100m.

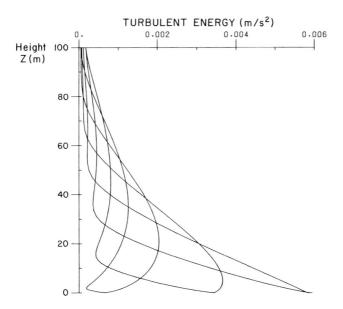

Fig. 5b Profiles of turbulence energy over an M_2 tidal cycle; water depth h=100m.

72

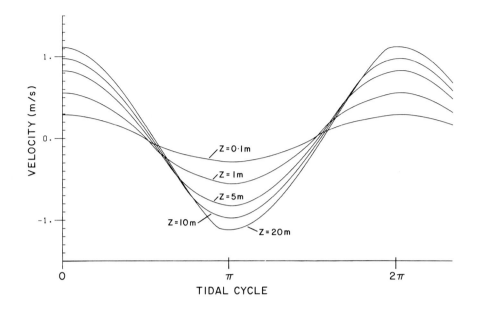

Fig. 6   Time variation of current over an M_2 tidal cycle, at five
heights above the sea bed.

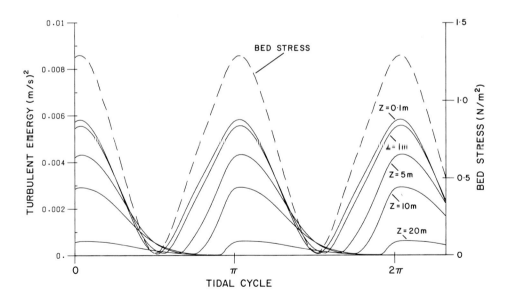

Fig. 7   Time variation of turbulence energy over an M_2 tidal cycle at
five heights, above the sea bed. Also shown is the bed stress.

In this calculation the mixing length formulation of Mofjeld and Lavelle (1984), eqtns (9),(10) was used. With this mixing length, the computed eddy viscosity profile (Fig. 4a) had a maximum below mid-water. Such a maximum at this depth has been deduced from observed shear measurements made in the Irish Sea (Wolf 1980), and appears to be physically realistic. The alternative mixing length formulation of Vager and Kagan (1969) did not give a maximum in the viscosity profile.

In deeper water, h=100m, a near mid-water maximum in the profile of eddy viscosity is still present, although the height above the sea bed at which this maximum occurs, varies over the tidal cycle (Fig. 5a). Although bed turbulence energy intensities are similar in water depths of 20m and 100m (compare Figs. 4b,5b), in the case of h=100m, surface turbulence is significantly less than in the shallower case. The reason for this is probably that in the shallower water, the source of turbulence production (namely the sea bed) is closer to the surface, and consequently turbulent diffusion leads to increased surface values.

Time series of tidal current, and turbulence energy in a water depth h=20m, at a number of levels in the vertical, (approx. heights above the sea bed are given in the figure), together with bed stress, over a tidal cycle, are shown in Figs. 6 and 7. It is evident from these figures that turbulence energy changes significantly over a tidal cycle, and that both it and the bed stress are highly correlated with the current magnitude.

3.3 Steady wind induced flow

Steady state velocity, viscosity and turbulence energy profiles in a water depth h=100m, induced by a wind stress τ_x=1.0N/m² are shown in Fig. 8. Again a high shear near bed region is evident in the velocity profile. However due to the large value (approx. 0.3m²/s) of surface eddy viscosity (Fig. 8), the current only decreases slowly with depth below the sea surface, and there is no evidence of a high shear surface region.

Comparing eddy viscosity and turbulence energy profiles in this calculation, with those computed for tidal flow in a water depth of 100m (Fig. 5a,b), it is evident that in the wind problem, both eddy viscosity and turbulence increase towards the sea surface (Fig. 8), rather than decreasing as in the tidal or wave problem. This increase in near surface turbulence, in the wind problem, arises directly from wind induced turbulence at the air-sea interface, the magnitude of which is influenced by the size of m_2 in Eqtn (11). A parameter which can also significantly change the intensity of the surface current.

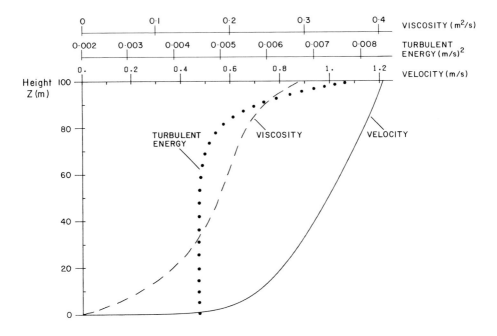

Fig. 8   Steady state vertical profiles of wind induced (τ_x=1.0N/m^2),
current, viscosity and turbulence energy in a water depth h=100m.

4 CONCLUDING REMARKS

It is clear that the turbulence energy model described in this paper is a
powerful tool in studying the various physical processes that produce
turbulence in a shallow sea region.   New developments in turbulence
measurement both in the sea (see papers in this volume) and in the laboratory,
will be particularly valuable in verifying computed turbulence levels.

The model developed here was found to be computationally stable even when
fine grids (over a hundred vertical points) and long time steps (of up to
1800s.) were used in the integration. Although various finite differencing
methods had little effect upon the accuracy of the solution, they did, as one
would expect significantly influence the stability of the solution and hence
the length of time step that could be used.

Although density effects have not been included in the present model (since
these calculations have been for homogeneous regions), their effects upon
turbulence levels in stratified regions is very significant and calculations
to examine this are presently in progress.

The model described here is for a single point in the vertical, but its
inclusion in a three dimensional sea model can be readily accomplished.

Naturally a full three dimensional sea model using a turbulence energy model
to compute eddy viscosity at each grid point and moment in time, with a
vertical grid of order one hundred points, and horizontal grid of several
thousand points will be a significant calculation. However with the new range
of computers such as the CRAY XMP/48 such calculations can be readily
performed. Consequently future computations will involve the use of three
dimensional hydrodynamic models with turbulence energy closure applied to
problems both within the North Sea, and shelf edge regions where mixing
processes are particularly important.

5 ACKNOWLEDGEMENTS

The authors are indebted to Drs. B. Jamart and J. Ozer for useful
discussions concerning the accuracy of turbulence energy models. The care
taken by Mrs. J. Huxley in typing the text and Mr. R. A. Smith in preparing
diagrams is appreciated.

The work was funded in part by the Procurement Executive, Ministry of
Defence.

6 REFERENCES

Backhaus, J.O. and Hainbucher, D., 1987. A finite difference general
    circulation model for shelf seas and its application to low frequency
    variability on the north European Shelf, pg 221-244, in Three-
    dimensional models of marine and estuarine dynamics, ed J.C.J. Nihoul
    and B.M. Jamart, No 45 in Elsevier Oceanography Series.
Christoffersen, J.B. and Jonsson, I.G., 1985. Bed friction and dissipation
    in a combined current and wave motion. Ocean Engineering, 12: 387-423.
Davies, A.G., 1986. A numerical model of the wave boundary layer.
    Continental Shelf Research, 6: 715-739.
Davies, A.M., 1981. Three dimensional hydrodynamic numerical models. Part 1.
    A homogeneous ocean-shelf model. Part 2. A stratified model of the
    northern North Sea. pp 306-426 in Vol 2. The Norwegian Coastal Current,
    (ed R. Saetre and M. Mork) Bergen University, 795pp.
Davies, A.M., 1985. A three dimensional modal model of wind induced flow in
    a sea region. Progress in Oceanography, 15: 71-128.
Davies, A.M., 1986a. A three-dimensional model of the northwest European
    continental shelf, with application to the M_4 tide. Journal of Physical
    Oceanography, 16: 797-813.
Davies, A.M., 1986b. Application of a spectral model to the calculation of
    wind drift currents in an idealized stratified sea. Continental Shelf
    Research, 5: 579-610.
Grant, W.D. and Madsen, O.S., 1986. The continental-shelf bottom boundary
    layer. Annual Review of Fluid Mechanics, 18: 265-305.
Heaps, N.S. and Jones, J.E., 1985. A three-layered spectral model with
    application to wind-induced motion in the presence of stratification and
    a bottom slope. Continental Shelf Research, 4: 279-319.
Heathershaw, A.D., New, A.L. and Edwards, P.D., 1987. Internal tides and
    sediment transport at the shelf-break in the Celtic Sea. Continental
    Shelf Research , 7, 485-517.

Huthnance, J.M., 1981. Waves and currents near the continental shelf edge. Progress in Oceanography, 10: 193-226.

Jamart, B.M. and Ozer, J., 1987. Comparison of 2-D and 3-D models of the steady wind-driven circulation in shallow waters. To appear in Coastal Engineering.

Johns, B., 1978. The modelling of tidal flow in a channel using a turbulence energy closure scheme. Journal of Physical Oceanography, 8: 1042-1049.

Mofjeld, H.O. and Lavelle, J.W., 1984. Setting the length scale in a second-order closure model of the unstratified bottom boundary layer. Journal of Physical Oceanography, 14: 833-839.

New, A.L., 1988. Internal tidal mixing in the Bay of Biscay. Submitted to Deep Sea Research.

Pingree, R.D., Mardell, G.T. and New, A.L., 1986. Propagation of internal tides from the upper slopes of the Bay of Biscay. Nature, 321: 154-158.

Vager, B.G. and Kagan, B.A., 1969. The dynamics of the turbulent boundary layer in a tidal current. Atmospheric and Ocean Physics, 5: 88-93.

Wolf, J., 1980. Estimation of shearing stresses in a tidal current with application to the Irish Sea, in, Marine Turbulence, ed, J.C.J. Nihoul, Elsevier Scientific Publishing Company, Amsterdam, pp 319-344.

TURBULENT FIELDS ASSOCIATED WITH THE GENERAL CIRCULATION
IN THE NORTHERN BERING SEA

E. DELEERSNIJDER[*] and Jacques C.J. NIHOUL
GHER, University of Liège, B5, Sart Tilman, 4000 Liège (Belgium)

The GHER 3D Mathematical Model has been described in previous reports and publications (e.g. Nihoul et al 1986, Nihoul and Djenidi 1987). The model has been applied to the calculation of the general circulation in the Northern Bering Sea under different winds in the scope of a preliminary study aimed at assessing the variability of the flow pattern.

In the present study, a special attention is paid to the case of zero (negligible) local wind which emphasizes the water circulation induced by the inflows and outflows of regional shelf currents through the straits (Anadyr, Shpanberg and Bering). In this case, if the inputs are maintained constant at values typical of the seasonal climatology, the simulated flow tends fairly rapidly towards a steady pattern approximating the general seasonal circulation.

One may argue indeed that, if the general circulation is defined as an average over a time of several days or weeks (i.e. much larger than the characteristic time of tides, passing storms and other mesoscale processes) the mean wind stress over that period will be comparatively small and negligible in a first approximation.

One must remember however that, if the mesoscale fluctuations are eliminated by the averaging, their mean products still affect the macroscale flow through the non-linear terms and in particular, the production of turbulence in the upper layer of the sea, - which behaves as the square of the wind stress -, is effectively much larger than what one computes from the mean shear in the absence of wind.

This however can easily be corrected by a modification of the boundary conditions imposed on the turbulent energy equation, taking into account the actual mean energy production calculated from the climatology of the mesoscale wind field variability. This modification has been introduced in the numerical model.

[*] Research Assistant , National Fund for Scientific Research (Belgium)

The model was calibrated and run with (initial and boundary) data typical of the summer season's climatology (Coachman et al, 1975). A total flow of 1.8 Sv was assumed through the Bering Strait. The flow was distributed along the southern boundary, ten grid points away from the limits of the Northern Bering Sea, according to observations and results from larger scale models. This resulted in a partition of inflows between the Anadyr Strait and the Shpanberg Strait in the ratio approximately two to one. The value of 1.8 Sv was derived from field measurements and actually interpolated between 1.9 Sv for a mean wind stress of 10^{-1} dyne cm^{-2} north in early July 1968 and 1.7 for a mean wind stress of the same order of magnitude south at the end of July 1972 (Coachman et al, 1975). The value of 1.8 Sv is not in itself a determinant factor of the general circulation pattern. At the general circulation scale, the non-linear terms are not dominant in the equations and the main effect of reducing or increasing the in-going transport is essentially a reduction or increase of the velocity scale, atlthough some localized discrepancies are not inconceivable (Nihoul at al, 1986).

The product of the simulation is the values of all state variables and auxiliary variables and parameters at all grid points as functions of depths.

These include characteristics of the turbulent field such as the turbulent kinetic energy and the turbulent dissipation rate.

These results however are often not exploited and regarded merely as intermediate calculations in the production of larger scale features of the velocity, temperature, buoyancy ... fields.

The purpose of this paper is, on the contrary, to display some illustrative examples of turbulent fields associated with specific representative features of the Bering Sea's general summer circulation.

The results show that the distributions of turbulent kinetic energy and turbulent dissipation rate agree well - both qualitatively and quantitatively - with essential features of the larger scale flow such as upwellings and downwellings, bottom friction, stratification, etc ... .

This would seem to be an argument for a broader use of three-dimensional models with turbulent energy closure - sofar restricted to large scale forecasts - in the investigation of marine turbulence.

79

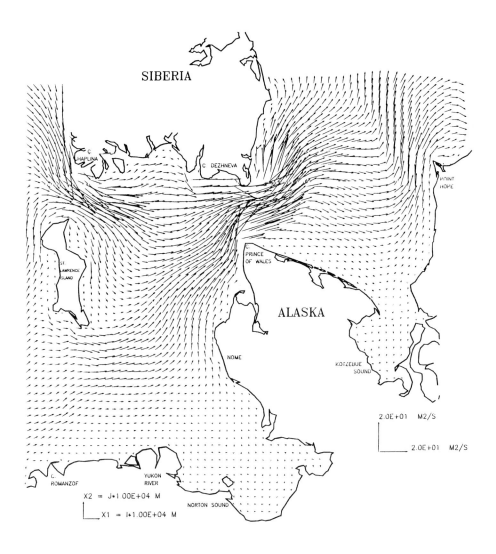

Fig. 1. General Summer Circulation in the Northern Bering Sea.
1.8 Sv through Bering Strait.
Total transport calculated by the 3D Model (m^2/sec).
One can see the essential contribution of the "Anadyr Stream" flowing in
through the Anadyr Strait (West of St. Lawrence Island) and deploying in
the Northern Bering Sea.

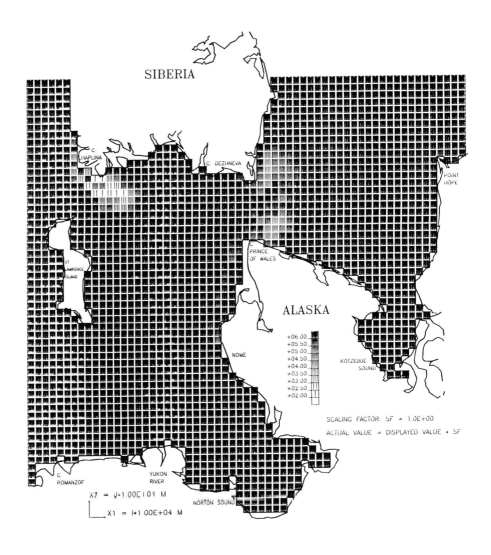

Fig. 2.   General Summer Circulation in the Northern Bering Sea.
1.8 Sv through Bering Strait.
Horizontal distribution of temperature at 5 m below the surface.

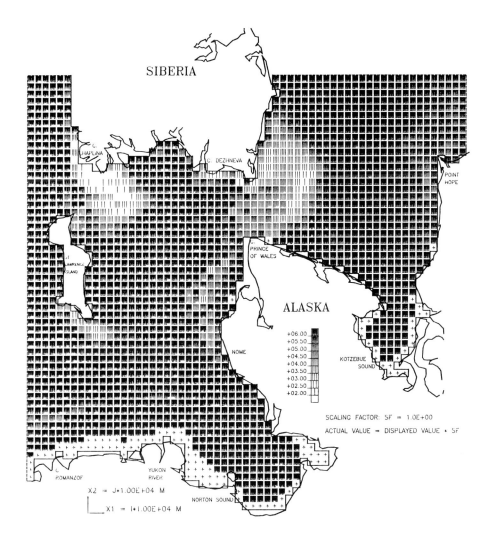

Fig. 3. General Summer Circulation in the Northern Bering Sea.
1.8 Sv through Bering Strait.
Horizontal distribution of temperature at 10 m below the surface.
Comparisons with figs 2 and 4 show distinct regions of upwelling along the
coast of Siberia and regions of downwelling along the coast of Alaska on
both sides of Cape Prince of Wales.

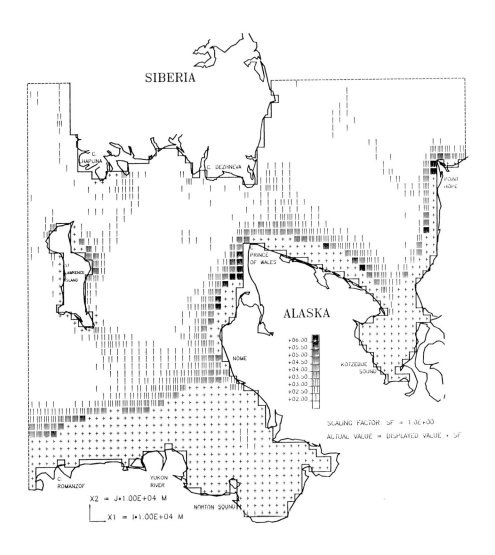

Fig. 4. General Summer Circulation in the Northern Bering Sea.
1.8 Sv through Bering Strait.
Horizontal distribution of temperature at 20 m below the surface.

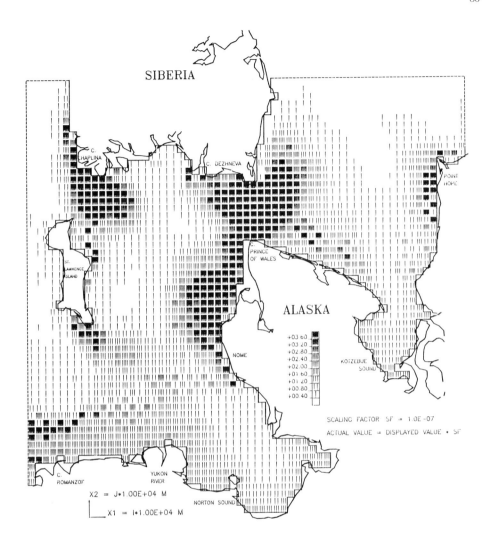

Fig. 5. General Summer Circulation in the Northern Bering Sea.
1.8 Sv through Bering Strait.
Horizontal distribution of the depth-averaged turbulent energy dissipation
rate (m^2/sec^3).

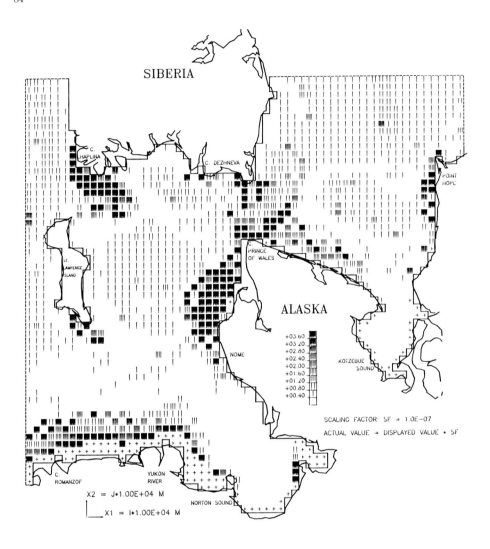

Fig. 6.   General Summer Circulation in the Northern Bering Sea.
1.8 Sv through Bering Strait.
Horizontal distribution of turbulent energy dissipation rate at 10 m below
the surface (m^2/sec^3).

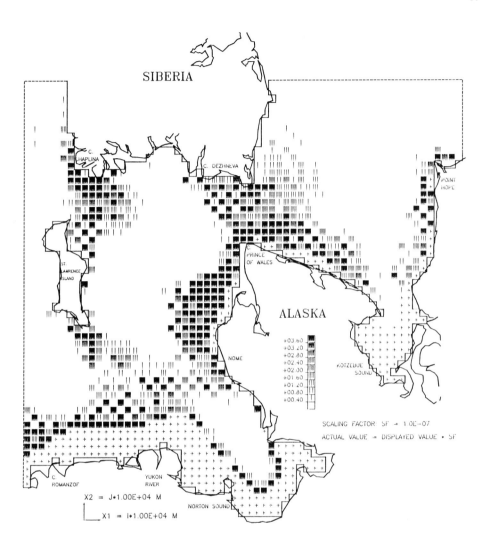

Fig. 7. General Summer Circulation in the Northern Bering Sea.
1.8 Sv through Bering Strait.
Horizontal distribution of turbulent energy dissipation rate at 15 m below
the surface (m^2/sec^3).

86

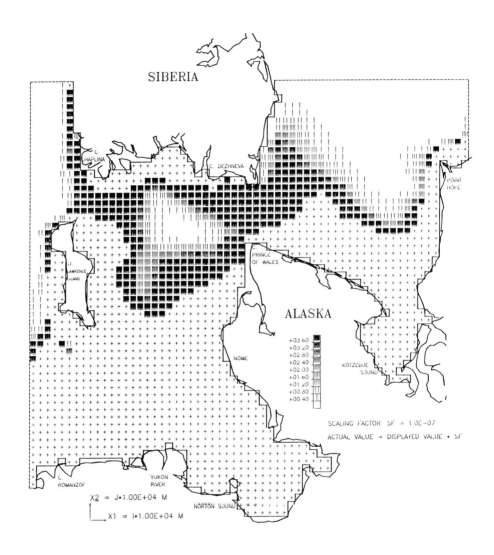

Fig. 8. General Summer Circulation in the Northern Bering Sea.
1.8 Sv through Bering Strait.
Horizontal distribution of turbulent energy dissipation rate at 35 m below
the surface (m^2/sec^3).
Figs 6, 7 and 8 are representative of the surface layer, the thermocline and
the bottom layer, respectively.

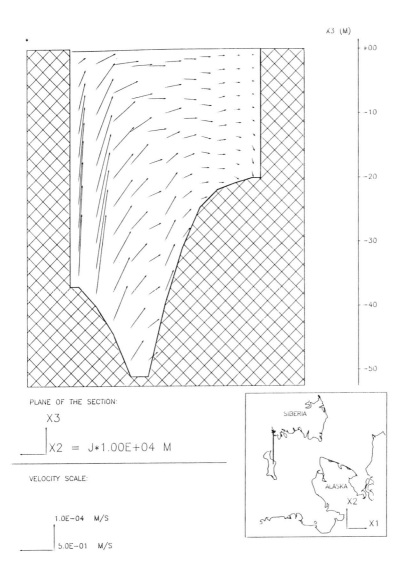

X3 (M)

+00

-10

-20

-30

-40

-50

PLANE OF THE SECTION:

X3

X2 = J*1.00E+04 M

VELOCITY SCALE:

1.0E-04 M/S

5.0E-01 M/S

SIBERIA

ALASKA

X2

X1

Fig. 9. General Summer Circulation in the Northern Bering Sea.
1.8 Sv through Bering Strait.
Anadyr Strait cross section : velocity vectors in the cross section plane.

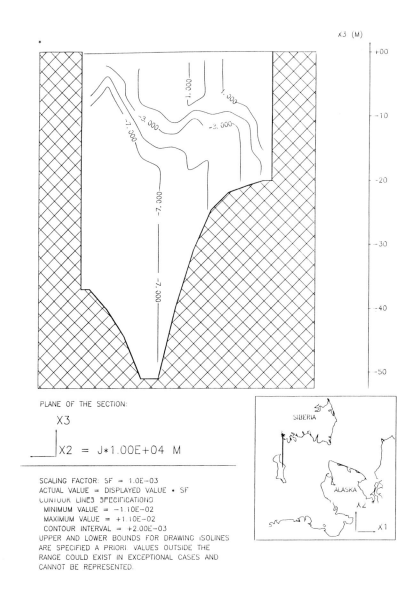

PLANE OF THE SECTION:

X3

X2 = J*1.00E+04 M

SCALING FACTOR: SF = 1.0E-03
ACTUAL VALUE = DISPLAYED VALUE * SF
CONTOUR LINES SPECIFICATIONS
    MINIMUM VALUE = -1.10E-02
    MAXIMUM VALUE = +1.10E-02
    CONTOUR INTERVAL = +2.00E-03
UPPER AND LOWER BOUNDS FOR DRAWING ISOLINES
ARE SPECIFIED A PRIORI. VALUES OUTSIDE THE
RANGE COULD EXIST IN EXCEPTIONAL CASES AND
CANNOT BE REPRESENTED.

Fig. 10.   General Summer Circulation in the Northern Bering Sea.
1.8 Sv through Bering Strait.
Anadyr Strait cross section : buoyancy (10^{-3} m/sec^2).

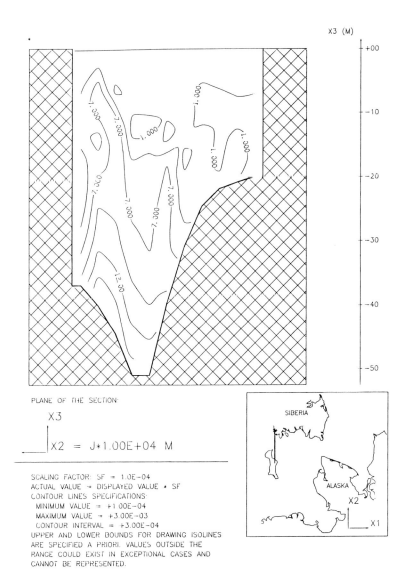

Fig. 11.   General Summer Circulation in the Northern Bering Sea.
1.8 Sv through Bering Strait.
Anadyr Strait cross section : turbulent kinetic energy (10^{-4} m^2/sec^2).

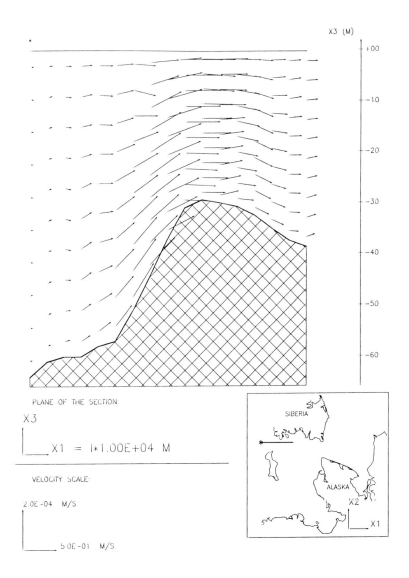

Fig. 12. General Summer Circulation in the Northern Bering Sea.
1.8 Sv through Bering Strait.
Anadyr Strait transect : velocity vectors in the plane of transect.

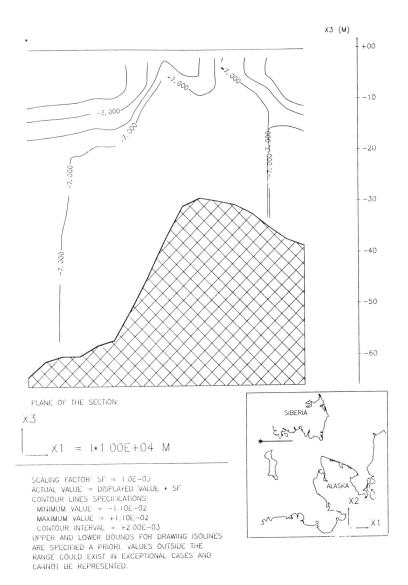

Fig. 13. General Summer Circulation in the Northern Bering Sea.
1.8 Sv through Bering Strait.
Anadyr Strait transect : buoyancy (10^{-3} m/sec^2).

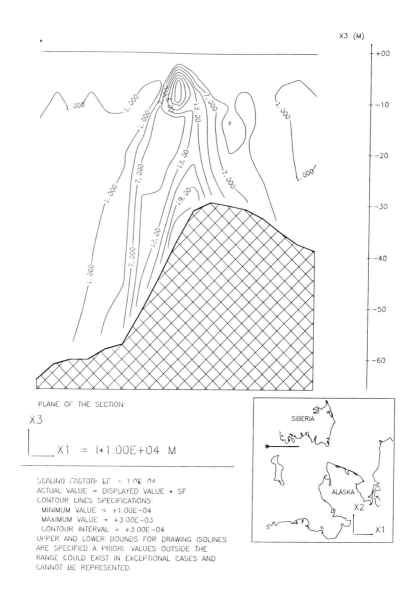

Fig. 14.   General Summer Circulation in the Northern Bering Sea.
1.8 Sv through Bering Strait.
Anadyr Strait transect : turbulent kinetic energy (10^{-4} m^2/sec^2).

REFERENCES

Coachman, L.K., Aagaard, K. and Tripp, R.B., 1975. Bering Strait. The Regio-
   nal Physical Oceanography, University of Washington Press, 172 pp.
Nihoul, J.C.J., Waleffe, F. and Djenidi, S., 1986. A 3d numerical model of
   the Northern Bering Sea. Environmental Software, 1 : 76-81.
Nihoul, J.C.J. and Djenidi, S., 1987. Perspective in three-dimensional
   modelling of the marine system. In : J.C.J. Nihoul and B.M. Jamart
   (Editors), Three-dimensional Models of Marine and Estuarine Dynamics,
   Elsevier Publ., Amsterdam : 1-33.

EMBEDDING STRATIFICATION MODELS IN OCEAN GENERAL CIRCULATION CLIMATE MODELS

Dr B. Henderson-Sellers, *Department of Mathematics and Computer Science, University of Salford, Salford M5 4WT, England*

ABSTRACT

For ocean general circulation models (OGCMs), vertical resolution of water temperature is required: either in terms of a two box approach — an embedded integral mixed layer (ML) model; or in terms of a higher resolution methodology in which a multi-layer eddy diffusion thermal stratification model is incorporated. For OGCM applications, appropriate values for the ML depth; either as an annual average or prespecified geographically and seasonally, remain dubious. Furthermore, even in ML models with a predictive capability, the long timescale of OGCM simulations becomes problematic since many ML models were designed for short period ($\lesssim 6$ months) simulations and, consequently, few existing ML models remain numerically stable over periods of decades and centuries. However, the main advantage of ML models, which has led to their consideration in OGCMs, is their relative computational simplicity. In comparison, eddy diffusion thermal stratification models appear to be much more stable over long time periods, whilst being more demanding of computer resources. One such (low order turbulent closure) model is utilized here to demonstrate some of the attributes of eddy diffusion models. The long term stability is readily demonstrable, suggesting its appropriateness for climatic studies. The results of a carbon dioxide warming perturbation experiment demonstrate the advantages for climate simulations. In addition preliminary results of forcing this eddy diffusion model directly by AGCM-produced surface parameters (radiation, wind etc.) demonstrate the necessity of avoiding oversimplistic ML models for OGCM applications. Nevertheless the large computational demands of eddy diffusion models have yet to be reconciled with current computer resources, suggesting the need for the development of a partially simplified eddy diffusion model for OGCM application in preference to ML models.

1. Introduction: mixed layer and multi-layer differential models

For ocean general circulation models (OGCMs), vertical resolution of water temperature is required: either in terms of a two box approach — an embedded integral mixed layer (ML) model; or a multi-layer resolution obtained by incorporating an eddy diffusion thermal stratification model. In mixed layer models, an assumption is made that a well-mixed, quasi-homogeneous, upper ocean layer exists so that a *single*, bulk equation can be written to describe changes in temperature and depth of this single slablike layer. In eddy diffusion models, the full heat transfer equation is solved at a relatively large number of depths using a finite difference scheme. Such models, which both *predict* the existence of a mixed layer and can resolve details *within* the layer, describe the vertical heat transfer in terms of an eddy diffusion coefficient of heat, K_H. Inter-model differences result from the level of complexity of the closure assumption (see e.g. Mellor and Yamada, 1974). Since the two model types are beginning to be able to be shown to be in good agreement (e.g. preliminary results of Martin, 1985), and furthermore since their future implementation is likely to be as part of a larger model (e.g. the U.S. Navy's ocean forecast model and coupled ocean-atmosphere general circulation models), it is perhaps apposite that the best features of both should be amalgamated into a new generation of definitive stratification models which would, ideally, be easily and cheaply implemented numerically (i.e. perhaps using the ML ideas for the upper layers) whilst retaining the spatial resolution of the differential models.

Although few three-dimensional coupled ocean–atmosphere general circulation climate simulations have been undertaken (partly because of the cost of extensive computer time needed — see discussion in Bryan, 1984), many authors have indeed oriented their developments of stratification models to take into account the inherent spatial and temporal constraints of GCMs. For example, Marchuk *et al.* (1977) suggested that the inherent mathematical simplicity of the integral approach is well-suited to GCM applications; whereas Garrett (1979) investigated the implementation of an eddy diffusion (differential) model.

Whilst many early ML models investigated mixed layer deepening over relatively short time scales (frequently less than 6 months), for GCM applications it has long been realised that an inherent year-to-year numerical stability is necessary. For example, Niiler (1977), in a comparative study of several stratification models, identified only one model (at that time), that of Warren (1972), which was numerically stable over a period of several years. At about the same time, Haney and Davies (1976) presented climatological simulations for the last year of a 100 year simulation by which time a stable and repeating thermal structure was evident. A similar concern for the applicability of ML models to longer period simulations was expressed by de Szoeke (1980) who extended the Niiler (1975) model to encompass horizontal variations in wind stress. This permitted a detailed study of upwelling/downwelling regions to be undertaken by means of the inclusion of the wind stress curl. However he noted deficiencies in the approach with regard to the role of surface heating restabilizing the upper mixed layer, suggesting it to be an area ripe for further developments.

Furthermore, concern over the role of the oceans in a warming world, likely to result from increasing atmospheric carbon dioxide, has prompted analysis using much simpler models.

Results from two recent, independent studies (Wigley and Schlesinger, 1985; Hansen *et al.*, 1985) using a simple box–diffusion model suggest that over the last 130 years, observations and models are not in disagreement. Although a diffusive deep ocean was indeed included in this analysis, it has been suggested (Henderson-Sellers, 1987) that the resulting ocean model offers an oversimplistic view of the oceanic contribution to the overall global atmosphere–ocean climate system and that it is necessary to add a convective capability to the water column model. Using this latter approach it has been suggested that, on a timescale of a century, the sea surface temperature increase is likely to be less by a factor of at least 3 than that previously forecast.

In addition preliminary results of forcing this eddy diffusion model directly by AGCM-produced surface parameters (radiation, wind etc.) demonstrate the necessity of avoiding oversimplistic ML models for OGCM applications. Nevertheless the large computational demands of eddy diffusion models have yet to be reconciled with current computer resources, suggesting the need for the development of a partially simplified eddy diffusion model for OGCM application in preference to ML models. It should also be noted that although the embedding of one-dimensional stratification models into dynamic codes is currently "popular", this approach cannot immediately be used in regions of upwelling/downwelling (unless a vertical advective term is added to the model) and is unlikely to be able to describe the details of thermocline ventilation (Luyten *et al.*, 1983; Cox and Bryan, 1984).

2. An hierarchy of stratification models for OGCMs

Parallel development of ocean circulation models and mixed layer models have led to the need to embed the latter within the former to produce an efficient three-dimensional ocean circulation model. Initial attempts (e.g. Haney, 1980) utilised crude representations of mixed layer depth; whilst recommending the integration of these two model types (see also Elsberry and Garwood, 1980). The embedding of the Garwood (1977) ML model within the Haney (1980) circulation model is described by Elsberry and Garwood (1980). Some simplification of the embedded mixed layer approach is proposed by Haney (1985), based on the "dynamic adjustment" of Adamec *et al.* (1981) in which ML deepening is controlled by a local gradient Richardson number — thus providing a link with the differential model approach.

Recently, Meehl (1984) has reviewed the coupling of ocean–atmosphere models in terms of an hierarchy of oceanic components. He suggests that even oversimplistic oceanic representations (e.g. (i) the "swamp" representation of e.g. Manabe and Wetherald, 1980, Washington and Meehl, 1983 — with no heat storage capacity and (ii) fixed depth ocean surface layer models where there is a heat capacity but no dynamics e.g. Manabe and Stouffer,

1980; Spelman and Manabe, 1984; Washington and Meehl, 1984) may be useful in helping to understand the processes acting in air-sea interaction. He identifies the advantage of utilising a fixed depth ML as permitting the inclusion of a full seasonal cycle in the atmospheric GCM (not possible with the oceanic "swamp") (see also Hansen *et al.*, 1983; Meehl and Washington, 1985). As a modification to the constant depth ML, Meehl (1984) then utilises a model in which ML depth is allowed to vary in a prespecified way; whilst acknowledging that more complete (and hence more complex) models of oceanic dynamics and thermodynamics exist — either in terms of a full computation of the oceanic surface layer or in terms of a three-dimensional oceanic circulation model (e.g. Washington *et al.*, 1980; Spelman and Manabe, 1984; Cox and Bryan, 1984). Meehl (1984) concludes that use of prespecified ML depth and calculated sea surface temperatures is inadequate for the simulation of the annual cycle of zonal heat storage, especially in the tropics, recognizing the need for a fully interactive coupling between atmosphere and ocean. Preliminary results of such a coupling are presented by Semtner (1984), using the Held and Suarez (1978) atmospheric model. He compares three oceanic representations: a simple diffusive convective two-layer model; a two layer model, but including a prognostic model of the ML; and a three layer primitive equation model. His conclusions are based upon only a 12 year integration for his third model type, which he acknowledges is inadequate for full equilibrium. Nevertheless he feels it appropriate to comment that, as might be expected, with this hierarchy accuracy/usefulness increases with increasing complexity of oceanic ML model — but so does the computational time required. However no sea-ice is included, an omission partially rectified by Pollard *et al.* (1983); but in this oceanic model, the atmospheric forcing is prescribed. Once again, by testing several models with differing degrees of complexity, it is concluded that fixed depth ML models are inadequate.

For ocean general circulation model (OGCM) applications, appropriate values for the ML depth; either as an annual average or prespecified geographically and seasonally, therefore remain dubious. Since one major concern of global modellers relates to heat storage terms, use of a two layer schematization, even if using well-simulated ML depths, will necessarily underestimate the total thermal capacity of the ocean by neglecting the heat storage in the water at depths below the thermocline. Such arguments are detailed by Meehl (1984) who notes that ML depths from various sources need to be modified (as a modelling approximation) and replaced by an "effective mixed layer depth". Such effective depths are used extensively (e.g. Manabe and Stouffer (1980) who quote a value of 68 m as an annual value; Wigley and Schlesinger (1985) who undertake simulations with mean ML depths of 70 m and 110 m; Van den Dool and Horel (1984) who use 25–50 m). This method of obviating the restrictions of the implicit step function profile of a slablike ML model can be contrasted with the addition of an intermediate layer by Harvey and Schneider (1985a,b); the exponential slab ML of Pollard *et al.* (1983); and the Richardson-number-induced "smoothing" of Price *et al.* (1986) — again illustrating one of the limitations of the (computationally more efficient) bulk modelling approach (cf. differential models).

However, there remains one obvious problem of integrating a variable depth mixed layer model like that of Price *et al.* (1986) into a fixed grid dynamic model: how should a mixed layer depth which is predicted to be at a depth intermediate to the dynamic model's grid be reconciled with this fixed grid, whilst ensuring that both realism and energy conservation are obtained (Adamec *et al.*, 1981)? Certainly the answer is non-trivial (Adamec, personal communication, 1987). On the other hand, eddy diffusion models are designed on the same concepts of finite difference grids and so can be more easily implemented. In this case the conflict between the demands of the stratification model for high resolution in the upper ocean and the inherently coarse vertical resolution of dynamic models (such as the Bryan-Cox-Semtner code) remains a problem. Such problems exist independently of the complexity of the chosen eddy diffusion model.

3. Climate Simulations

3.1 *The impact of increasing atmospheric CO_2 on SSTs*

Although the effects on climate of increasing CO_2 have been extensively commented on in the literature, full ocean–atmosphere simulations with a model which includes sea-ice thermodynamics (Manabe and Bryan, 1985), have been, to date, rarely undertaken whilst providing an attainable scientific goal (Washington and VerPlank, 1986); largely as a consequence of the enormous computational demands such a study would entail. However, simple attempts to include the oceanic effects in such a study have recently been made using a simplified analysis of ocean thermodynamics by Wigley and Schlesinger (1985) and by Hansen *et al.* (1985). They utilise a simple column energy balance approach, representing the ocean–atmosphere system by only four "compartments" or "boxes": two atmospheric (one over land, one over ocean), an oceanic mixed layer of fixed depth and a deeper diffusive ocean — a model based on that of Oeschger *et al.* (1975) and Siegenthaler and Oeschger (1984). (However it should be noted that application of such models in a more detailed analysis of carbon cycling (Bacastow and Björkström, 1981) suggests that the constant depth chosen for the mixed layer must be a function of the assumed fossil fuel production curve). The heating rate of the mixed layer is calculated by assuming a constant depth in which the temperature increment due to some perturbation, ΔT, increases as a result of (i) the change in the thermal (surface) forcing, ΔQ; (ii) the atmospheric feedback, expressed in terms of a climate feedback parameter, λ, and, additionally to many fixed depth mixed layer ocean models, leakage of energy is permitted into the underlying waters. This energy, ΔF acts as a surface boundary condition to a diffusive ocean, in which the turbulent diffusion coefficient, K_H, is assumed to be a constant. In the deeper ocean, the temperature increment ΔT_0 is a function of depth as well as time and satisfies the heat transfer equation. Hence, the differential equations describing the rates of heating in the two "layers" of this box–diffusion model are thus:

$$\text{Mixed layer}: \quad C_m\frac{d\Delta T}{dt} = \Delta Q - \lambda\Delta T - \Delta F \tag{1}$$

$$\text{Deeper waters}: \quad \frac{\partial\Delta T_0}{\partial t} = K_H\frac{\partial^2\Delta T_0}{\partial z^2} \tag{2}$$

This latter equation may be evaluated at depth, z, (measured vertically upwards from a zero at the interface) or calculated numerically using a finite difference grid, with grid spacing Δz.

However, the simplifications of the annual timestep, fixed mixed layer depth, box–diffusion model (Equations 1, 2) may invalidate its applications for climate simulations; as illustrated in a recent study (Henderson-Sellers, 1987) using an eddy diffusion model, EDD1 (**Eddy Diffusion Dimension 1**), based on numerical solution of the heat transfer equation

$$\frac{\partial T}{\partial t} = \frac{\partial}{\partial z}\left[(K_H + \alpha)\frac{\partial T}{\partial z}\right] + \frac{\frac{\partial\phi}{\partial z}}{\rho c_p} \tag{3}$$

where the temperature T is a function of both depth, z, and time, t, and determined by the values of the eddy diffusion coefficient, K_H, and the molecular diffusion coefficient, α. The shortwave radiation penetration, ϕ, is given by

$$\phi = (1 - \beta)\phi_0 e^{-\eta z} \tag{4}$$

where η is the extinction coefficient and β the (infrared) portion of the solar radiation, ϕ_0, absorbed within the uppermost layers.

The major characteristic of this model is that the eddy diffusion coefficient is given as a function of surface friction velocity (related to the wind speed) and water stability (characterized by the Richardson number, R_i):

$$K_H = (1/P_o)ku_* z\, exp(-k^* z)\left(1 + 37R_i^2\right)^{-1} \tag{5}$$

(Henderson-Sellers, 1985) where P_o is the neutral value of the turbulent Prandtl number ($\simeq 1$), k is von Kármán's constant ($=0.4$), u_* is the surface friction velocity and k^* is given as a function of latitude θ and wind speed U by the empirical equation

$$k^* = 6.6(\sin\theta)^{0.5}U^{-1.84} \tag{6}$$

(in which the coefficient is dimensional). The parameter k^* can be shown essentially to be inversely proportional to the Ekman depth. The derivation of Equation 5 is based on boundary layer theory and hence is a conceptually-based prognostic submodel for the turbulent mixing process. The observational data and theoretical arguments supporting this approach are discussed in Henderson-Sellers (1985). This eddy diffusion parametrisation represents turbulent mixing at all times of year when $R_i \geq 0$. However when the water column becomes thermodynamically unstable $(R_i < 0)$, convection is permitted using a simple heat balance algorithm in which heat is redistributed throughout a newly formed isothermal upper mixed layer interatively until the instability is eliminated (Ryan and Harleman, 1071).

The bottom boundary condition is that of no flux across the interface and the surface boundary condition is evaluated in terms of the energy fluxes for solar radiation, longwave radiation exchanges, evaporative loss and sensible heat. These are detailed in Henderson-Sellers (1986). In summary, the longwave terms are based on blackbody calculations, including emissivity terms; the calculation of evaporation includes both convective and mechanical removal of water vapour (Sill, 1983) such that the evaporative flux depends on the wind speed, humidity gradient and also on the air-sea temperature difference; the sensible heat transfer is calculated from the evaporation using a Bowen ratio approach (see e.g. Brutsaert, 1982). The prescribed meteorological data for these surface flux calculations are therefore air temperature, wind speed, relative humidity (or, equivalently, vapour pressure) and cloud cover, all of which can be prescribed as functions of time of year and latitude. In the experiments described here, climatological values, derived from various sources, are used i.e. the variables are cyclic over a period of one year for a given latitude; although in the experiments described here the wind speed is taken to be 5 m s^{-1} independently of latitude and season. This is a good first approximation (see e.g. Oort, 1983). No tuning to specific observational data is required once these latitude dependent parameters have been specified for the location being studied.

Since the differential model includes specification and calculation of a full surface energy budget, use of the model permits a detailed analysis of the feedbacks associated with perturbations to the sea surface temperature and the atmospheric temperature; and, consequently, an evaluation of the appropriateness of undertaking increasing CO_2 experiments with simple box–diffusion models.

Since the schematisation is that of a multi-level model, the mixed layer may occupy several layers, as can the pycnocline between the mixed layer and the deeper (yet still diffusive) ocean. Consequently, there is a higher vertical spatial (as well as temporal) resolution.

Both the mixed layer depths and the heat storage term are modelled reasonably success-fully (Henderson-Sellers, 1987). Furthermore when heat storage is plotted against surface temperature, there is an hysteresis effect, in agreement with observations (Gill and Turner, 1976), which is impossible to reproduce with simple box models. Such simulations lend credence to the increasing CO_2 experiment using the model EDD1. The results (detailed in Henderson-Sellers, 1987) for experiments designed to mimic those of Wigley and Schlesinger (1985) and Hansen *et al.* (1985), suggest that the values of ΔT, for two contrasting latitudes, are significantly smaller than for the box–diffusion model (Table 1). These lower surface temperatures can probably be best understood in terms of the convective capacity of the latter model. During winter, the stratification breaks down and the mixed layer deepens considerably. This deepening, coupled with surface heat loss, results in a cooler surface tem-perature. The second important difference lies in the higher spatial and temporal resolution of the convective–diffusive model. For example, the surface fluxes are non-linear in temperature and hence changes in the surface flux components induced (by the ΔQ perturbation) on a daily timescale cannot be equal to changes in a value of a "mean annual" value for the surface fluxes — as essentially included in the box–diffusion and box models. It should be noted that although the comparison of different values of K_H and ML depth given in figure 1 of Wigley and Schlesinger (1985) suggests that the ocean simulation is not over-sensitive to ML depth and eddy diffusion coefficient, these conclusions would only be globally valid if the water body was stratified throughout the year. Consequently this cannot be used to substantiate the box–diffusion model in preference to the convective–diffusion model since ML models lack the ability to display the non-linear hysteresis of stored heat content vs. surface temperatures described by Gill and Turner (1976).

TABLE 1

Values of ΔT for Experiments 1 and 2 for the global box–diffusion model and the convective–diffusion model at two selected latitudes (after Henderson-Sellers, 1987).

	Global box–diffusion model	Convective–diffusion model	
		(15°N)	(54°N)
Experiment 1 (no feedbacks)	0.31 K	0.005 K	0.148 K
Experiment 2 (with feedbacks*)	0.60 K	0.012 K	0.212 K
$= \frac{\Delta T \text{(feedbacks)}}{\Delta T \text{(no feedbacks)}}$	1.9	2.4	1.4

* In the global box–diffusion model experiment with feedbacks, a value of 3.58 W m^{-2} K^{-1} is taken for λ. This corresponds to a value of about 2.4 for the feedback factor of Hansen *et al.* (1985) — a value calculated in the convective–diffusive model experiments.

These results suggest that the convective dynamics of an ocean with a seasonal cycle contribute an ameliorative effect on increasing oceanic surface temperatures. This may be related to energy released from the ocean to the atmosphere immediately following convective overturning, as well as to the greater capability of the ocean to absorb heat in winter when the thermocline depth is considerably deeper than the assumed annual mean of ~100m (see earlier discussion) used in earlier models (Oeschger et al., 1975; Siegenthaler and Oeschger, 1984; Wigley and Schlesinger, 1985; Hansen et al., 1984, 1985). However a fully coupled atmosphere–ocean model would be needed to assess the impact on the atmospheric component of the climate system of this convectively released energy. Indeed, preliminary experiments in such coupling have been recently completed, although in general using specified mixed layer depth (see e.g. Bryan et al., 1984; Washington and Meehl, 1984; Nihoul, 1985).

3.2 Thermodynamic simulations of SST using AGCM data

A second evaluation of the potential for the use of an eddy diffusion (differential) in OGCM simulations has been undertaken by using the one-dimensional model EDD1 to represent the water column of the ocean at specific, yet different, locations. Three locations were chosen as representative of difference forcing conditions: a location in mid-Pacific at $20°$ N; in the North Atlantic at $20°$ N; and in the South Atlantic at $30°$ S. Wind fields and net radiation data for these three locations were provided from the Canadian Climate Centre (CCC) GCM (Boer et al., 1984a,b). In the initial experiment (reported here) only the wind forcing from the GCM results was utilised. This permits the simulations to be more site-specific, since the interactive surface energy budget is also appropriate for a range of latitudes. The results, shown in Table 2, suggest that this type of model appears to be well-suited to describe summer (maximum) sea surface temperatures (SSTs). Minimum winter (February) SSTs are, however, overestimated at the two lower latitudes, whilst being accurate at $40°$ N. Furthermore the annual cycle is well represented for all cases, as is the mixed layer depth (with the caveats discussed above). However, further experiments are required, not only to ensure a closer coupling between an AGCM and a thermal stratification model, but also to assess the sensitivity of the parametrisation of the chosen model (for example, the formulation for K_H and $f(R_i)$ in low order closure models such as EDD1 and in higher order closure schemes — see Mellor and Yamada, 1974).

4. Conclusions

Further developments of thermal stratification models for incorporation into large and expensive coupled ocean–atmosphere GCMs, are likely to favour computational speed at the

104

TABLE 2

Observed and simulated maximum and minimum SSTs

Location	Minimum (February)		Maximum (August)	
	Observed	*Simulated*	*Observed*	*Simulated*
20° N	293–298K	291K	298–300K	299K
30° S	290–292K	292K	295–298K	299K
40° N	283K	283K	293K	294K

expense of a high degree of accuracy. Two examples for illustration are: (i) the development at Oregon State University (OSU) of a coarse resolution synchronously coupled model which has only two atmospheric layers and six layers in the ocean, but an hourly temporal resolution (Gates *et al.*, 1985; Han *et al.*, 1985). In addition, it is intended that this model will utilise an embedded Kraus–Turner mixed layer model. Preliminary results (Gates *et al.*, 1985) highlight the need to simulate the SEB realistically. (ii) Foreman (1986) describes a 5–day synchronous coupling scheme and identifies several areas of error in his ten month simulation using a modified Cox–Bryan model, but without sea ice. (He does not, however, describe the type of stratification model used in this project). Such aims (to build coupled ocean–atmosphere GCMs) will encourage the development of one-dimensional models which themselves then become spatially dependent upon their grid location within a larger model. Discontinuities of thermocline depth between neighbouring grid cells is likely to be a source of difficulty in the numerical schematization if oversimple mixed layer models are utilised.

5. Acknowledgements

I am grateful to Drs Boer and McFarlane and to Mr Lazare of the Canadian Climate Centre (AES, Downsview, Ontario) for providing me with the AGCM-generated wind and radiation fields.

6. References

Adamec, D., Elsberry, R.L., Garwood, R.W. jr. and Haney, R.L., 1981, An embedded mixed layer–ocean circulation model, *Dyn. Atmos. Oceans,* **5**, 69–96.
Bacastow, R.B. and Björkström, A., 1981, Comparison of ocean models for the carbon cycle, Chapter 2 in *SCOPE 16: Carbon Cycle Modelling* (ed. B. Bolin), J. Wiley and Sons, Chichester, 29–79.

Boer, G.J., McFarlane, N.A., Laprise, R., Henderson, J.D. and Blanchet, J-P., 1984a, The Canadian Climate Centre spectral atmospheric general circulation model, *Atmos.-Ocean,* **22**, 397–429.

Boer, G.J., McFarlane, N.A. and Laprise, R., 1984b, The climatology of the Canadian Climate Centre general circulation model as obtained from a five-year simulation, *Atmos.-Ocean,* **22**, 430–475.

Brutsaert, W.H., 1982, *Evaporation into the Atmosphere: Theory, History, and Applications,* D. Reidel, Hingham, Mass, 299pp.

Bryan, K., 1984, Accelerating the convergence to equilibrium of ocean–climate models, *J. Phys. Oceanogr.,* **14**, 666–673.

Bryan, K., Komro, F.G. and Rooth, C., 1984, The ocean's transient response to global surface temperature anomalies, in *Climate Processes and Climate Sensitivity,* (eds. J.E. Hansen and T. Takahashi), American Geophysical Union, Washington DC, 29–38.

Cox, M.D. and Bryan, K., 1984, A numerical model of the ventilated thermocline, *J. Phys. Oceanogr.,* **14**, 674–687.

Davidson, K.L. and Garwood, R.W., 1984, Coupled oceanic and atmospheric mixed layer model, *Dyn. Atmos. Oceans,* **8**, 283–296.

De Szoeke, R.A., 1980, On the effects of horizontal variability of wind stress on the dynamics of the ocean mixed layer, *J. Phys. Oceanogr.,* **10**, 1439–1454.

Elsberry, R.L. and Garwood, R.W., jr., 1980, Numerical ocean prediction models — goal for the 1980s, *Bull. Amer. Meteor. Soc.,* **61**, 1556–1566.

Foreman, S.J., 1986, Ocean and atmosphere interact!, *Meteorol. Mag.,* **115**, 358–361.

Garrett, C., 1979, Mixing in the ocean interior, *Dyn. Atmos. Oceans,* **3**, 239–265.

Garwood, R.W., jr., 1977, An oceanic mixed layer model capable of simulating cyclic states, *J. Phys. Oceanogr.,* **7**, 455–468

Gates, W.L., Han, Y-J. and Schlesinger, M.E., 1985, The global climate simulated by a coupled atmosphere–ocean general circulation model: preliminary results, in *Coupled Ocean–Atmosphere Models* (ed. J.C.J. Nihoul), Elsevier, Amsterdam, 131–151.

Gill, A.E. and Turner, J.S., 1976, A comparison of seasonal thermocline models with observations, *Deep-Sea Res.,* **23**, 391–401.

Han, Y-J., Schlesinger, M.E. and Gates, W.L., 1985, An analysis of the air-sea-ice interactions simulated by the OSU-coupled atmosphere–ocean general circulation model, in *Coupled Ocean-Atmosphere Models* (ed. J.C.J. Nihoul), Elsevier, Amsterdam, 167–182.

Haney, R.L., 1980, A numerical case study of the development of large-scale thermal anomalies in the central North Pacific Ocean, *J. Phys. Oceanogr.,* **10**, 541–556.

Haney, R.L., 1985, Midlatitude sea surface temperature anomalies: a numerical hindcast, *J. Phys. Oceanogr.,* **15**, 787–799.

Haney, R.L. and Davies, R.W., 1976, The role of surface mixing in the seasonal variation of the ocean thermal structure, *J. Phys. Oceanogr.,* **6**, 504–510

Hansen, J., Russell, G., Rind, D., Stone, P., Lacis, A., Lebedeff, S., Ruedy, R. and Travis, L., 1983, Efficient three-dimensional global models for climate studies: models I and II, *Mon. Wea. Rev.,* **111**, 609–662.

Hansen, J., Lacis, A., Rind, D., Russell, G., Stone, P., Fung, I., Ruedy, R. and Lerner, J., 1984, Climate sensitivity: analysis of feedback mechanisms, in *Climate Processes and Climate Sensitivity,* (eds. J.E. Hansen and T. Takahashi), American Geophysical Union, Washington DC, 130–163.

Hansen, J., Russell, G., Lacis, A., Fung, I., Rind, D. and Stone, P., 1985, Climatic response times: dependence on climate sensitivity and ocean mixing, *Science*, **229**, 857–859.

Harvey, L.D.D. and Schneider, S.H., 1985a, Transient climate response to external forcing on 10^0–10^4 year time scales Part 1: experiments with globally averaged, coupled, atmosphere and ocean energy balance models, *J. Geophys. Res.*, **90**, 2191–2205.

Harvey, L.D.D. and Schneider, S.H., 1985b, Transient climate response to external forcing on 10^0–10^4 year time scales 2. Sensitivity experiments with a seasonal, hemispherically averaged coupled atmosphere, land, and ocean energy balance model, *J. Geophys. Res.*, **90**, 2207–2222.

Held, I.M. and Suarez, M.J., 1978, A two-level primitive equation atmospheric model designed for climatic sensitivity experiments, *J. Atmos. Sci.*, **35**, 106–229.

Henderson-Sellers, B., 1985, New formulation of eddy diffusion thermocline models, *Appl. Math. Model.*, **9**, 441–446.

Henderson-Sellers, B., 1986, Calculating the surface energy balance for lake and reservoir modeling: a review, *Rev. Geophys.*, **24**, 625–649.

Henderson-Sellers, B., 1987, Modelling sea surface temperature rise resulting from increasing atmospheric carbon dioxide concentrations, *Climatic Change*, **11**, 349–359.

Luyten, J.R., Pedlosky, J. and Stommel, H., 1983, The ventilated thermocline, *J. Phys. Oceanogr.*, **13**, 292–309.

Manabe, S. and Bryan, K., 1985, CO_2-induced change in a coupled ocean–atmosphere model and its paleoclimatic implications, *J. Geophys. Res.*, **90**, 11689–11707.

Manabe, S. and Stouffer, R.J., 1980, Sensitivity of a global climate model to an increase of CO_2 concentration in the atmosphere, *J. Geophys. Res.*, **85**, 5529–5554.

Manabe, S. and Wetherald, R.T., 1980, On the distribution of climate change resulting from an increase in CO_2 content of the atmosphere, *J. Atmos. Sci.*, **37**, 99–118.

Marchuk, G.I., Kochergin, V.P., Klimok, V.I. and Sukhorukov, V.A., 1977, On the dynamics of the ocean surface mixed layer, *J. Phys. Oceanogr.*, **7**, 865–875

Martin, P.J., 1985, Simulation of the mixed layer at OWS November and Papa with several models, *J. Geophys. Res.*, **90**, 903–916.

Meehl, G.A., 1984, A calculation of ocean heat storage and effective ocean surface layer depths for the Northern Hemisphere, *J. Phys. Oceanogr.*, **14**, 1747–1761.

Meehl, G.A. and Washington, W.M., 1985, Sea surface temperatures computed by a simple ocean mixed layer coupled to an atmospheric GCM, *J. Phys. Oceanogr.*, **15**, 92–104.

Mellor, G.L. and Yamada, T., 1974, A hierarchy of turbulent closure models for planetary boundary layers, *J. Atmos. Sci.*, **31**, 1791–1806.

Nihoul, J.C.J. (ed.), 1985, *Coupled Ocean–Atmosphere Models*, Elsevier, Amsterdam.

Niiler, P.P., 1975, Deepening of the wind-mixed layer, *J. Mar. Res.*, **33**, 405–422.

Niiler, P.P., 1977, One-dimensional models of the seasonal thermocline, in *The Sea*, Vol. 6: Marine Modeling), ed. I.N. McCave, J.J. O'Brien and J.H. Steele, J. Wiley, New York, 97–115.

Oeschger, H., Siegenthaler, U., Schotterer, U. and Gugelmann, A., 1975, A box diffusion model to study the carbon dioxide exchange in nature, *Tellus*, **27**, 168–192.

Oort, A.H., 1983, Global Atmospheric Circulation Statistics 1958–1973, NOAA Prof. Paper No. 14, U.S. Govt Printing Office

Pollard, D., Batteen, M.L. and Han, Y.J., 1983, Development of a simple oceanic mixed-layer and sea ice model, *J. Phys. Oceanogr.*, **13**, 754–768.

Price, J.P., Weller, R.A. and Pinkel, R., 1986, Diurnal cycling: observations and models of the upper ocean response to diurnal heating, cooling, and wind mixing, *J. Geophys. Res.*, **91**, 8411–8427.

Ryan, P.J. and Harleman, D.R.F., 1971, *Prediction of the annual cycle of temperature change in a stratified lake or reservoir: mathematical model and user's manual*, MIT Tech Rept 137, MIT, Cambridge, Mass.

Semtner, A.J. jr., 1984, Development of efficient, dynamical ocean-atmosphere models for climatic studies, *J. Clim. Appl. Meteor.*, **23**, 353–374.

Siegenthaler, U. and Oeschger, H., 1984, Transient temperature changes due to increasing CO_2 using simple models, *Ann. Glaciol.*, **5**, 153–159.

Sill, B.L., 1983, Free and forced convection effects on evaporation, *J. Hyd. Eng.*, **109**, 1216–1231.

Spelman, M.J. and Manabe, S., 1984, Influence of oceanic heat transport upon the sensitivity of a model climate, *J. Geophys. Res.*, **89**, 571–586.

Van den Dool, H.M. and Horel, J.D., 1984, An attempt to estimate the thermal resistance of the upper ocean to climatic change, *J. Atmos. Sci.*, **41**, 1601–1612.

Warren, B.A., 1972, Insensitivity of subtropical model water characteristics to meteorological flucatuations, *Deep Sea Res.*, **19**, 1–20.

Washington, W.M. and Meehl, G.A., 1983, General circulation model experiments on the climatic effects due to a doubling and quadrupling of carbon dioxide concentrations, *J. Geophys. Res.*, **88**, 6600–6610.

Washington, W.M. and Meehl, G.A., 1984, Seasonal cycle experiment on the climate sensitivity due to a doubling of CO_2 with an atmospheric general circulation model coupled to a simple mixed layer ocean model, *J. Geophys. Res.*, **89**, 9475–9503.

Washington, W.M. and VerPlank, L., 1986, *A description of coupled general circulation models of the atmosphere and oceans used for carbon dioxide studies*, NCAR Technical Note NCAR/TN–271+EDD, 29pp.

Washington, W.M., Semtner, A.J., Meehl, G.A., Knight, D.J. and Mayer, T.A., 1980, A general circulation model experiment with a coupled atmosphere, ocean and sea ice model, *J. Phys. Oceanogr.*, **10**, 1887–1908.

Wigley, T.M.L. and Schlesinger, M.E., 1985, Analytical solution for the effect of increasing CO_2 on global mean temperature, *Nature*, **315**, 649–652.

SIMULATION OF THE MIXED LAYER IN A GLOBAL OCEAN GENERAL CIRCULATION MODEL

S.J. FOREMAN and K. MASKELL
Meteorological Office, Bracknell, Berks. UK

ABSTRACT

   Mixed layer depths are diagnosed from an integration of a global ocean
general circulation model. Two methods of calculating the mixed layer depth
are used: one based on temperature differences between the base of the mixed
layer and the sea surface, and the other on the vertical density gradient.
Both sets of calculations show that the model produces mixed layers which are
too shallow at mid-latitudes in summer, and too deep in winter.

1 INTRODUCTION

   Modelling of the sea surface temperature (SST) is an essential task for an

ocean model which is to be used as part of a coupled ocean-atmosphere general

circulation model. In ocean only integrations the sea surface temperatures

arise as a result of the general oceanic flow and play only a minor part in

determining its evolution (depending upon the technique used to specify the

upper boundary condition), but in the coupled system the SST influences the

heat and fresh water fluxes at the ocean surface, and thus the circulation in

the atmosphere model. These same fluxes, when applied to the ocean model,

influence the calculated SSTs and feed back on the flow in a local fashion.

However, the influence of SST on the circulation patterns of the atmosphere

model may result in larger anomalies than would be derived from purely local

balances, such as the global impact of El Nino events (e.g. Quiroz, 1983)

(warm sea surface temperatures in the tropical Pacific) on the large scale

atmospheric flow.

   Although the evolution of SST involves large scale advective processes,

vertical mixing within the mixed layer also governs the rates of heat storage

and release by the ocean. It is therefore important that ocean general

circulation models to be used for investigating the climate system should

represent the mixed layer realistically. This paper describes an investigation of the distribution of mixed layer depth in an ocean general circulation model driven by surface fluxes calculated during an integration of a general circulation model of the atmosphere. In addition to a conventional definition of mixed layer depth based on temperature differences, a second derivation in terms of quantities used directly in the vertical mixing calculations is developed. The latter definition is under consideration for use as a discriminant between different forms of the parametrization of lateral mixing to be used by the ocean model.

2 THE MODEL

The model is based on that of Bryan (1969); the version due to Cox (1984) was used. In the experiments discussed here, lateral mixing was implemented using the horizontal scheme of the standard model, although integrations in which mixing of tracers was confined to isopycnal surfaces, following the scheme of Redi (1982), have also been performed.

TABLE

Distribution of model layers in the ocean model

Level	Depth of centre (m)	Thickness (m)
1	5.0	10.0
2	15.0	10.0
3	25.0	10.0
4	35.1	10.2
5	47.4	15.3
6	66.5	23.0
7	95.8	34.5
8	138.9	51.8
9	203.7	77.8
10	301.0	116.8
11	447.1	175.3
12	665.3	263.2
13	995.6	395.3
14	1490.0	593.5
15	2232.2	891.0
16	3295.0	1236.2
17	4707.0	1586.0

A uniform 2.5° latitude by 3.75° longitude grid was used in the horizontal, with 17 unequally spaced layers in the vertical. The layer depths and thicknesses are shown in the table, and were chosen to give greatest resolution near the ocean surface, in a manner consistent with the suggestion of Gill (1984). Calculations were performed with a timestep of half an hour, and Fourier filtering was used to reduce the effective resolution of the model at latitudes poleward of 72°.

Pacanowski and Philander (1981) proposed a form for the vertical diffusion scheme in ocean general circulation models in which the diffusivity depended upon the local Richardson number of the flow. This model was designed for use in the tropics, and in particular for the modelling of the equatorial undercurrent. In the integrations described in this paper the parametrization was only slightly modified. Use of a 10 m thick top layer made it necessary to use a shorter timestep for calculating the effects of vertical mixing than for the other terms in the equations of motion, in order to prevent the numerical stability criteria being violated. The alternative approach of arbitrarily limiting the diffusion coefficients to the maximum permitted for stability was not followed as it was considered that this would reduce the effectiveness of the scheme. Code is now available to solve the vertical diffusion equation implicitly, and this will be used in future integrations.

The Richardson number is defined by

$$
Ri = \frac{-g \, \rho_z}{\rho (U_z^2 + V_z^2)}
\tag{1}
$$

where the velocity components are U and V, g is the acceleration due to gravity, ρ the density, and the subscript z indicates a vertical derivative.

Following Pacanowski and Philander, the coefficients of viscosity (χ), (for currents) and diffusivity (Υ), (for temperature and salinity) are

$$
\chi = \frac{\chi_0}{(1 + \alpha \, Ri)^n} + \chi_b
\tag{2}
$$

and

$$
\Upsilon = \frac{\chi}{(1 + \alpha \, Ri)} + \Upsilon_b
\tag{3}
$$

respectively. The values chosen for the disposable parameters were:

χ_0 = 55.0 cm^2 s^{-1}
χ_b = 0.50 cm^2 s^{-1}
γ_b = 0.05 cm^2 s^{-1}
and

α = 5.0.

The value of n was 2, and a minimum value of 10 cm^2 s^{-1} was imposed for each of γ and χ between the first and second model levels.

In developing a lateral diffusion scheme which mixes tracers along isopycnal surfaces, care must be taken with the treatment of the mixed layer. Within the mixed layer, isopycnal surfaces are nearly vertical, and therefore the primarily lateral diffusion scheme with its large diffusion coefficients would create a large amount of vertical mixing, dominating that produced by the Richardson number dependent scheme which had previously been introduced to model the mixed layer. It was therefore necessary to derive a definition of mixed layer depth which could be readily calculated from within the model, and which was relevant to the vertical mixing scheme, in order to enable the isopycnal mixing scheme to be imposed only outside the mixed layer. As may be seen from (1), the vertical density gradient is a major factor in determining the amount of mixing in the vertical diffusion scheme, and is therefore a physically meaningful quantity to use to define the mixed layer depth in this context. This study was performed in order to assess the relevance of this diagnostic for mixed layer depth in a model integration as a discriminant between the mixed layer and the underlying ocean.

3 THE MODEL INTEGRATION

The integration described in this paper was initialized with the Levitus (1982) values of salinity and temperature appropriate for the month of September, with the ocean at rest. The surface forcing was derived from an individual year of a multi-year control integration of the Meteorological Office 11 layer general circulation model (an earlier version of the atmosphere model is described in Slingo (1985) and this version includes the parametrization of gravity wave drag described by Palmer et al (1986)). The upper boundary condition for tracers consisted of the net surface fluxes calculated by the atmospheric model (averaged over a month) together with a relaxation towards climatological values of the tracer at the surface, at the rate of 35 W m^{-2} K^{-1} for temperature (in the manner of Haney, 1971) and with a coefficient chosen to give the same e^{-}folding time for salinity. Figure 9 shows the net surface heat flux applied to the model ocean during the first September of the integration. The monthly mean windstress calculated by the atmospheric model was used directly.

4 MIXED LAYER DEPTH

As there is no universal definition of mixed layer depth, two definitions are illustrated in this paper. The first, used by Woods (1984), defines the mixed layer depth to be the depth at which the temperature becomes 0.5°C less than the surface value; the other is based on the vertical density gradient, and defines the mixed layer depth to correspond to the depth of the first maximum of vertical density gradient working downwards from the ocean surface; another criterion of choosing the absolute maximum of density gradient was found to give similar results, except where a new mixed layer was forming over an existing deep one. All calculations were performed on the grid of the numerical model, and monthly averaged values of temperature and salinity were used.

Observational studies frequently use a temperature difference rather than a density difference to define the mixed layer depth. This enables a comparison of many individual profiles, as temperature measurements are made more frequently than those of salinity, which are required for the determination of density. In a numerical model (and in climatological datasets) both temperature and salinity are available, and the derivation of density is central to the solution of the equations of motion.

A single year of forcing data from an integration of an atmosphere model was used to provide the surface boundary conditions for an integration of the ocean model, and the results for March (after seven months integration) and September (after 13 months) were compared with the corresponding data from the Levitus (1982) datasets. The resulting ocean simulation is strongly influenced by the interannual variability of the atmospheric model, and by differences between the climatologies of the atmosphere in the model and in the real world. Also, the coarse resolution of the ocean model results in systematic errors in the ocean circulation, which might be expected to introduce differences between the observed and modelled mixed layer depths, in addition to those due to inadequacies of the mixed layer formulation. The large-scale

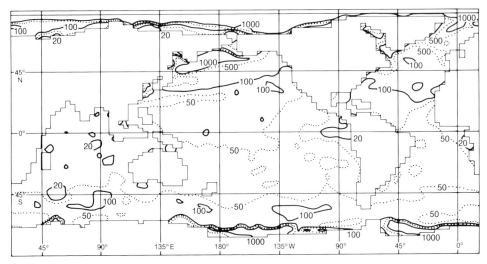

Fig.1 Mixed-layer depths calculated from Levitus (1982) data for March based on a temperature difference of 0.5°C. Contours at 50 m, 100 m, 500 m and 1000 m; 500 m contour omitted near Antarctic continent for clarity.

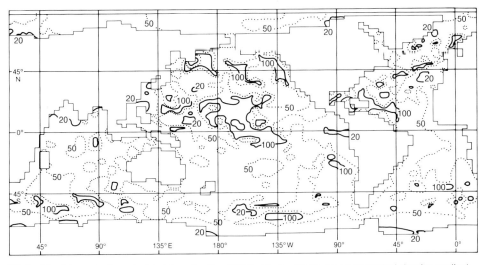

Fig. 2 Mixed-layer depths calculated from Levitus (1982) data for March based on vertical density gradient. Contours as Fig. 1.

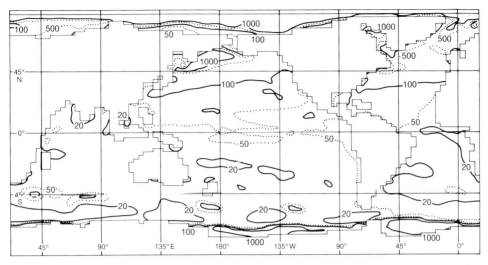

Fig. 3   Mixed-layer depths calculated from ocean-model data for March based on a temperature difference of 0.5° C.
Contours as Fig. 1.

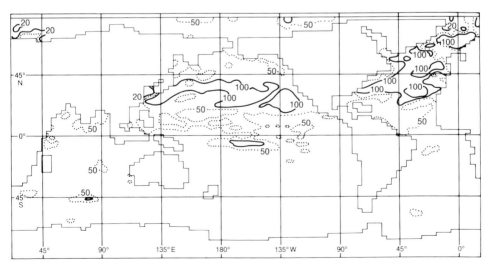

Fig. 4   Mixed-layer depths calculated from ocean-model data for March based on vertical density gradient.
Contours as Fig. 1.

ocean circulation takes centuries to reach equilibrium with the atmosphere forcing, whereas the mixed layer forms and erodes on an annual cycle within the broader flow. Experiments with coupled ocean-atmosphere models (eg Gates et al, 1985) have shown that short integrations give a useful indication of the systematic errors of the model system. Because the behaviour of the mixed layer, not the dynamics of the general circulation of the model, was being investigated it was desirable that the large scale temperature and salinity fields closely resemble those of the climatology being used for comparison; use of a longer term integration would have resulted in a strong influence of the circulation of the ocean model on the results, perhaps masking the characteristics of the mixed layer parametrization.

Figures 1 and 2 show the mixed layer depths calculated using the temperature difference and density gradients respectively for the March Levitus data. Before the fields were calculated the data were interpolated onto the grid of the ocean model and adjusted by the convective adjustment scheme to ensure that the profiles were statically stable. These are compared below with Figures 3 and 4, which show the same fields calculated from the ocean model results.

Comparing the global fields for March (after seven months of the integration) based on temperature differences (Figure 1 for observations and Figure 3 for the model), the large scale structure of the mixed layer depth may be seen to be similar, with a shallow layer in the southern (summer) hemisphere, and a much deeper layer in the northern (winter) hemisphere. Over much of the southern oceans, the mixed layer was between 20 m and 50 m thick in both the model and climatology. Within the Antarctic Circumpolar Current the model produces a shallow layer, the observations suggesting a deeper mixed layer of between 50 m and 100 m. However, at some points the model produces far deeper mixed layers seemingly of convective origin, corresponding to similar, although less deep, features in the time and space smoothed observational climatology in different locations. The deep mixed layer of the northwest Pacific is well represented in the model.

In September (not shown) the seasons are reversed, and the shallow mixed layer was found in the relatively well-obsrved northern hemisphere. Once again, the model (after 13 months of integration) produced a mixed layer which was too shallow, although not to the same extent as in the southern hemisphere during March. In the western tropical Pacific the mixing did not extend sufficiently deep. Deepening of the mixed layer in the southern winter was well represented. Together, these two months suggest that the vertical mixing scheme did not provide sufficient mixing of heat during the integration.

Using the alternative definition of mixed layer depth based on the vertical density gradient produced fields which were less smooth when the climatological data were used (Figure 2). The patterns, when compared with those of the numerical model (Figure 4), lead to conclusions similar to those based on the temperature calculations. In the summer hemisphere and western tropical Pacific, the mixed layer was too shallow, but the model depths were similar to those of the observations in the winter hemisphere. Major differences between the mixed layer depths produced by the two definitions may be seen in both the model and climatological fields within the Arctic and around Antarctica (eg Figures 1 and 2). The definition based on temperature differences fails to detect the strong stable layer near the surface produced by the strong halocline, resulting in differences of over 1000 m between the two definitions. This illustrates the importance of salinity in determining density at high latitudes.

Figures 5 to 8 illustrate the mixed layer depths in the north Atlantic in more detail for the month of March. Comparing the two methods of calculation for the observed data (Figures 5 and 6) shows that the two diagnostics select different aspects of the ocean structure. Relying on temperature to specify mixed layer depths leads to the conclusion that the deepest mixed layers are found south of Newfoundland, while the use of density gradient shows the maximum depth to lie along a line from 30°N 60°W to 50°N 15°W. These features are reflected in the corresponding fields for the model (Figures 7 and 8). Both diagnostics show that the model mixed layer in the area bounded by (45°N, 30°W), (45°N, 45°W), (60°N, 45°W) and (60°N, 30°W) was markedly deeper than in the observations. The cause of this has yet to be determined. Care should be taken in interpreting the very shallow mixed layer produced by the model between Greenland and Scandinavia in Figure 8: it is possible that in the area enclosed by the 10 m contour there was no maximum of density gradient within the water column, in which case the diagnostic would arbitrarily define the mixed layer to be shallower than 10 m. Comparison with Figure 7 in that area suggests that the mixed layer is indeed shallow in that region.

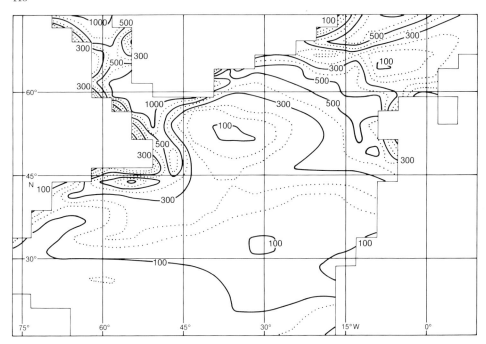

Fig. 5   Mixed-layer depths calculated from Levitus (1982) data for March based on a temperature difference of 0.5°C. Contours at 10 m, 20 m, 30 m, 50 m, 100 m, 150 m, 200 m, 300 m, 400 m, 500 m, 750 m and 1000 m.

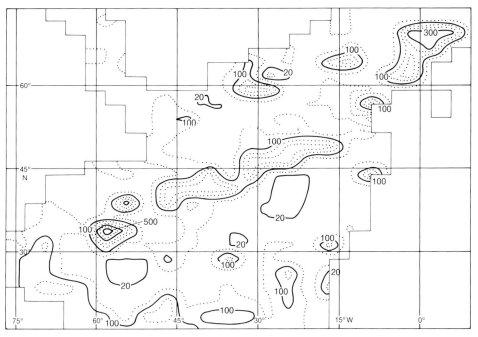

Fig. 6   Mixed-layer depths calculated from Levitus (1982) data for March based on vertical density gradient. Contours as Fig. 5.

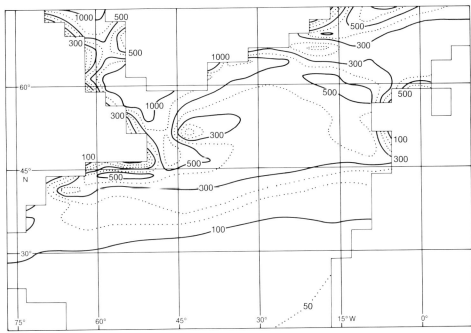

Fig. 7   Mixed-layer depths calculated from ocean-model data for March based on a temperature difference of 0.5° C. Contours as Fig. 5.

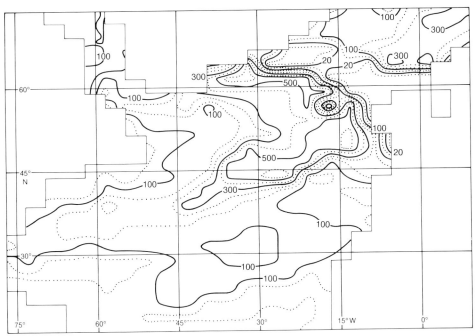

Fig. 8   Mixed-layer depths calculated from ocean-model data for March based on vertical density gradient. Contours as Fig. 5.

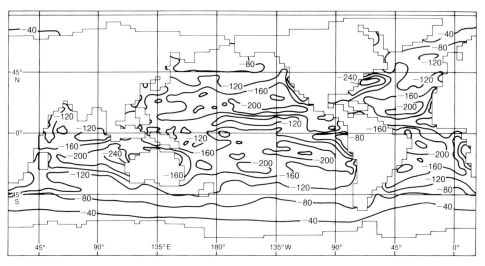

Fig. 9   Net surface heat flux applied to the model during the first September. Contour interval 20 W m^{-2}.

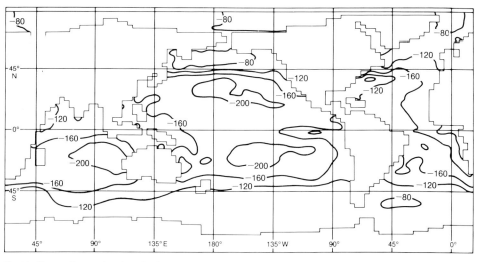

Fig. 10   Observed net surface heat flux for September (after Esbensen and Kushnir, 1981). Contour interval 20 W m^{-2}.

5 CONCLUSIONS

Use of a stability dependent vertical mixing coefficient in a global ocean model has produced a realistic representation of the gross features of the annual cycle of mixed layer depth, although aspects such as the timing of the spring shallowing of the mixed layer have not yet been considered. Systematic errors which have been identified are: shallow mixed layers in summer months; excessive mixed layer depth in the north Atlantic in winter; and shallow mixed layers in the western tropical Pacific. Possible contributions to these errors arise from several factors: the use of forcing derived from an atmosphere model (as a result of using an individual year rather than a mean climatology and of systematic errors in the climatology of the atmosphere model, effects which may be seen by comparing the model-derived fields for September (Figure 9) with a climatological estimate (Figure 10)); the form of the dependence of mixing coefficients on stability (Pacanowski and Philander (1981) confined their attention to the tropical oceans); and the use of monthly meaned forcing datasets, as well as deficiencies in the large scale flow in the coarse resolution ocean model.

It has been demonstrated that use of the vertical density gradient in diagnosing mixed layer depth results in fields which are coherent on large space scales and which show the seasonal variations present in more conventional calculations. This diagnostic is now being included in the ocean model in preparation for improvements to the lateral mixing scheme based on the isopycnal diffusion scheme of Redi (1982).

6 REFERENCES

Bryan, K., 1969. A Numerical Method for the Study of the Circulation of the World Ocean. J. Comp. Phys., 4, 347-376.

Cox, M.D., 1984. A Primitive Equation, Three Dimensional Model of the Ocean. GFDL Ocean Modelling Group Technical Report Number 1.

Esbensen, S.K. and Kushnir, Y., 1981. The Heat Budget of the Global Ocean: An Atlas Based on Estimates From Surface Marine Observations. Climatic Research Institute Report No. 29, Oregon State University, Corvallis, Oregon.

Gates, W.L., Han, Y.-J. and Schlesinger, M.E., 1985. The Global Climate Simulated by a Coupled Ocean-atmosphere General Circulation Model: Preliminary Results. In: J.C.J. Nihoul (Editor) 'Coupled Ocean-atmosphere Models' (Elsevier Oceanographic series, 40) Elsevier, Amsterdam, pp131-149.

Gill, A.E., 1984. On the Behaviour of Internal Waves in the Wakes of Storms. J. Phys. Oceanogr., 14, 1129-1151.

Haney, R.L., 1971. Surface Thermal Boundary Condition for Ocean Circulation Models. J. Phys. Oceanogr., 1, 241-248.

Levitus, S., 1982. Climatological Atlas of the World Ocean. NOAA Prof.
Paper 13, US Government Printing Office, Washington DC, 173pp.

Palmer, T.N., Shutts, G.J. and Swinbank, R., 1986. Alleviation of a Systematic
Westerly Bias in General Circulation and Numerical Weather Prediction
Models Through an Orographic Gravity Wave Drag parametrization. Q. J. R.
Meteorol. Soc., 112, 1001-1040.

Pacanowski, R.C. and Philander, S.G.H., 1981. Parametrization of Vertical
Mixing in Numerical Models of Tropical Oceans. J. Phys. Oceanogr., 11,
1443-1451.

Quiroz, R.S., 1983. The Climate of the "El Nino" of 1982-83 — a Season of
Extraordinary Climate Anomalies. Mon. Wea. Rev., 111, 1685-1706.

Redi, M.H., 1982. Oceanic Mixing by Coordinate Rotation. J. Phys. Oceanogr.,
12, 1154-1158.

Slingo, A., 1985. Simulation of the Earth's Radiation Budget With the 11-layer
General Circulation Model. Met. Mag., 114, 121-141.

Woods, J.D., 1984. The Warmwatersphere of the Northeast Atlantic. A
Miscellany. Institut für Meerskunde, Universität Kiel Report No. 128.

A THERMODYNAMIC MODEL OF THE GLOBAL SEA-SURFACE TEMPERATURE AND
MIXED-LAYER DEPTH

Jean-Yves Simonot[**][(*)], Hervé Le Treut[*] and Michel Crépon[**]
[*] Laboratoire de Météorologie Dynamique, Ecole Normale Supérieure
   24, Rue Lhomond 75231 Paris Cedex 05, France
[**] Laboratoire d'Océanographie Dynamique et de Climatologie
   Université P. et M. Curie Paris VI, Tour 14 - 2ème étage
   4 Place Jussieu, 75252 Paris Cedex 05, France

ABSTRACT

   We have built a model of the oceanic upper layers. It consists
of a Mixed-Layer model embedded in an horizontal advective and
diffusive process, and uses observed quarterly climatological currents
and a parameterization of the diffusion due to mesoscale processes.
We have also introduced a simple parameterization of the equatorial
upwelling as a function of the local wind-stress. Outside the
equatorial belt, the same formulation allows us to take into account
the Ekman pumping. The aim is to obtain a simple but realistic model
of the Sea-Surface Temperature (SST) in order to make air-sea
interaction experiments. In the tuning phase, the model is forced
with one year of daily surface heat fluxes and stresses. In the
present paper we show the results of various simulations we have
achieved.
The deepening of the mixed-layer is too shallow during the summer :
this is very probably due to an underestimation of the wind-stress in
the data used to force our model. Also the simulated meridional
oceanic transport depends critically on the assumed vertical profile
of the currents in the oceanic surface layers.

1. INTRODUCTION

   The ocean plays a key-role in the evolution of the climatic system

at all time-scales over some weeks or months. In many experiments,

however, the oceanic boundary conditions used for Atmospheric General

Circulation Models (AGCM) (for example, sensitivity experiments to a

doubling of the atmospheric CO_2 content) have been over-simplified.

In some experiments (Manabe and Stouffer 1979, 1980, Manabe et al.

1981), the ocean was considered as an ensemble of independant boxes,

exchanging heat with the atmosphere at the air-sea interface, and

where the Mixed-Layer Depth (MLD) was annually or seasonally

prescribed. In other experiments (Spelman and Manabe 1984), the

horizontal and vertical advection of heat is taken care by means of a

coupling with a low resolution Oceanic General Circulation Model

(OGCM), but these results are very sensitive to the formulation being used to relate the upper ocean vertical diffusion of momentum to the model generated Richardson number.

In the last few years, there has been a renewal of interest in establishing models of the oceanic boundary layer, able to accurately represent oceanic diapycnal mixing. They have been successfully tested at different Ocean Weather Stations (P, R, N) where long series of meteorological and oceanic measurements are available. These models are not able to predict realistic SST values in regions where the large-scale advection is important. For the equatorial oceans a resolution of the dynamical equations by a numerical model is possible and allows a successfull prediction of the evolution of the SST field averaged over larger time-scales. At high latitudes, numerical models need a very high spatial resolution to simulate realistic currents and cannot be used for climatological purposes. As a first step in understanding the main mechanisms which control the SST at mid- and high latitudes, it might be attempted to model the SST at these latitudes by embedding a Mixed-Layer model in an advective and diffusive process determined from observed climatological currents. This type of model seems suitable to reproduce SST at mid-and high latitudes, since changes in the ocean circulation are very slow with respect to changes due to air-sea interaction processes and thus they will be considered as a second order effect in our studies. The feasability of such an approach has been investigated in a preliminary experiment on the North-Atlantic Ocean (Le Treut et al. 1985). As will be shown below this approach is not valid for the equatorial oceans. However, as low latitudes SST changes do not play a dominant role in long-term climatic changes, we have used a simple method to take into account the mean contribution of the equatorial upwelling. Although we do not intend to simulate a realistic equatorial ocean at a time scale of some months, the model has been implemented on the global ocean in order to allow heat to be exported from the equator towards higher latitudes, where surface heat losses occur in the annual mean.

Hereafter, we describe the model and present the results of some seasonal simulations. We have investigated the sensitivity of the model to some of the prescribed parameters which enter its formulation.

2 THE MODEL

2.1 The Mixed-Layer model

We use Gaspar's (1985, 1987) model. It is a thermodynamical model and predicts the evolution of the temperature vertical profile. It may also be used for salinity. No attempt is made to compute the surface currents. The closure equations are the conservation equations for the total eddy kinetic energy in the Mixed-Layer E_m, and for the "vertical eddy kinetic energy" in the Mixed-Layer W_m, defined as.

$$E_m = \frac{1}{2}(\overline{u'}^2 + \overline{v'}^2 + \overline{w'}^2) \quad \text{and} \quad W_m = \frac{1}{2}\overline{w'}^2$$

The coefficients involved in the various terms of production/destruction of these energies have been obtained by Gaspar from a review of oceanic measurements, laboratory experiments and higher order closure turbulence numerical experiments. Gaspar was able to simulate successfully SST and MLD seasonal cycles at Ocean Weather Station (OWS)P ("Papa"), using these arbitrary coefficients. We were also able to simualte successfully SST at (OWS) R ("Romeo") with Gaspar's model (Figure 1).

In the entrainment case, the MLD change $\partial h / \partial t$ is explicitely computed, while in the detrainment case, h is obtained from an implicit equation which is iteratively solved. The model is formulated so as to conserve potential energy. The vertical profile has been discretized over 20 layers, the thicknesses of which are displayed in Table 1.

Layer thickness	Layers
15m	upper layer
5m	9 following layers
20m	9 following layers
240m	bottom layer

Table 1: Discrete profile used

126

The diurnal cycle has not been taken into account ; the upper layer is therefore rather thick to prevent excessive heating. The lowest one is very thick and allows the damping of large temperature changes at the bottom of the model. The maximum depth is 800m. The Mixed-Layer model is run with a Δt = 1day time-step.

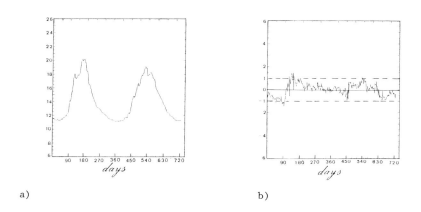

a)   b)

Figure 1: a) Simulated Sea-Surface-Temperature at Point R
b) Difference between simulated and measured SSTs at Point R
Day 1 is March 1, 1979

2.2 Horizontal transport

We have implemented the Mixed-Layer model on the 64x50 grid of the LMD-AGCM, which is regular in longitude and sine of latitude. The transported quantity is the heat content H intergrated over a depth h_1, roughly corresponding to an Ekman layer:

$$H = \int_{-h_1}^{o} \rho_o . c_p . T(z) . dz \qquad (1)$$

where ρ_o is the standard specific mass of the water, c_p is its heat capacity, T(z) is the temperature profile and z is the vertical coordinate (positive upwards).

The heat content H is horizontally advected by non-divergent quarterly climatological surface currents, which can be down-scaled by a coefficient 1/A, to account for the decrease of the currents with respect to depth, and is diffused following different schemes which account for climatological mesoscale processes. For each time-step, the heat content change ∂H/∂t resulting from advection and diffusion

is redistributed on the vertical, according to an homothetic transform
of the temperature profile, where the bottom temperature and the MLD
are kept unchanged (Figure 2).

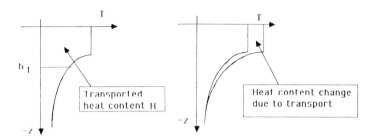

Figure 2: Vertical redistribution scheme for the
heat content change due to horizontal transport

Therefore, we neglect the seasonal temperature change for the
bottom layer, as inferred from Levitus' (1982) climatology at 800m
depth. For numerical stability imposed by the currents and diffusion
orders of magnitude, we use a 6hrs time-step.   The numerical
integration scheme is a simple Euler scheme and we use a staggered
grid.

2.3 Currents

We use Meehl's (1980) quarterly climatological currents
supplemented with Patterson's (1985) annual mean currents in the
Southern Ocean. The divergence of these large scale climatological
currents (not presented here) appears to be noisy and does not seem to
represent known features of large-scale upwellings and downwellings.
We have therefore chosen not to represent any vertical advection and
to use the non-divergent part of these currents only.

We compute first the quarterly stream-function Φ defined by:

$$\Delta\Phi = \mathrm{curl}_z(\vec{U}) \qquad\qquad (2)$$

where $\vec{U}$ is the original current, and Δ is the horizontal Laplacian
operator.

As a boundary condition we have to specify the value of Φ over
the continents. It is set to zero on the American, European, Asian
and African continents.   For Madagascar, Borneo, New-Guinea and
Australia (resolved by the grid we use), Φ is computed via the spatial

integration of :

$$
\left[
\begin{array}{l}
u = -\dfrac{\partial \Phi}{\partial y} \\[2em]
v = \dfrac{\partial \Phi}{\partial x}
\end{array}
\right.
\tag{3}
$$

where (u,v) are the current components in the climatology, and where integration occurs from a coast with a known value for Φ to a coast with an unknown one. On the Antarctica Φ has been estimated from Whitworth and Peterson's (1985) study in the Drake Passage. Once the values of Φ known for each continent resolved by our grid, Equation (2) is solved by an iterative over- relaxation numerical method.

The obtained quarterly stream-functions enable us to compute non-divergent currents (that are parallel to the coasts). This procedure tends to slightly decrease the amplitude of the currents in the western parts of the oceanic basins, but conserves all known features of the oceanic general circulation, such as equatorial counter-currents, Guinea Current, Monsoon reverse gyres (Figure 3).

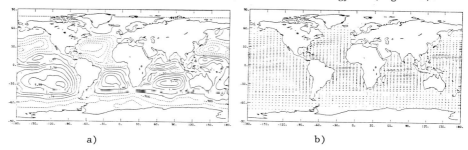

a)   b)

Figure 3: a) Obtained winter stream function
b) Obtained winter non-divergent currents

2.4 Diffusion

In the simulations which we have performed, we have used two different diffusive schemes. One is a Laplacian scheme, and the other is an upstream scheme.

a) The Laplacian scheme is defined by:

$$\frac{\delta H}{\delta t} = \vec{\nabla} \cdot (K \cdot \vec{\nabla} H) \qquad (4)$$

where H is the transported heat content (see section 2.2), and where K is a diffusion coefficient.

We have determined the value of K from the following equation, which is a result of bi-dimensional turbulence numerical experiments with Lagrangian tracers (Babiano et al. 1987), and is valid both in the barotropic and baroclinic cases, for spatial scales above the internal deformation radius:

$$K = 2.26 \cdot E \cdot Z^{-1/2} \qquad (5)$$

where $Z = 1/2(curl_z(\vec{U}))^2$ is the enstrophy of the mean flow in a grid-mesh and E is the horizontal eddy kinetic energy (EKE) in the same mesh.

An EKE climatology is then required. We use Wyrtki's chart of the EKE distribution on the World Ocean (Wyrtki et al. 1976). This chart based on historical ship-drift data, may overestimate low EKE values and underestimate high ones, as discussed by Richardson (1983), but spatial structures agree with EKE fields from drifting buoys, at our resolution. To correct the values given by Wyrtki, we have used estimates from drifting buoys over the North Atlantic Ocean, obtained and analysed by Ollitrault (personal communication) at Institut Français pour la Recherche et l'Exploitation de la Mer (IFREMER). The processed buoy trajectories include all the buoys used by Richardson (1983) plus 18 others, and the EKE field is very similar to that given by Richardson.

The corrected EKE field was then supplemented over the Southern Ocean with Patterson's (1985) EKE field. From this global field we have substracted the EKE part which is due to quarterly changes of the mean currents (see section 2.3). We have then assumed that the remaining EKE can be equally shared among the four quarters.

When using Equation (5), we have to impose a minimum value for $curl_z(\vec{U})$ (we chose $5.10^{-5} s^{-1}$), as the equation diverges for zero enstrophy. The resulting diffusion coefficients vary from 0.1 to 15. times $10^5 m^2 . s^{-1}$, which is the order of magnitude for mesoscale diffusion in the ocean, and have been interpolated to our grid with a condition of zero diffusion across the shores. We have used these diffusion coefficients as thermal diffusion coefficients (Figure 4).

130

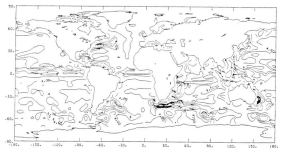

Figure 4 : Winter diffusion coefficient obtained
(unit: $10^5 m^2.s^{-1}$, increment: $2.10^5 m^2 . s^1$)

b) <u>The upstream diffusion scheme</u> (Sadourny 1986) is defined by:

$$\frac{\delta H}{\delta t} = \vec{U} . \vec{\nabla}(\theta . \vec{U} . \vec{\nabla}H) \tag{6}$$

where $\vec{U}$ is the advective current (see section 2.3), H is the transported heat content (see section 2.2), and θ is a time-scale given by:

$$\theta = 2.26 . Z^{-1/2} \tag{7}$$

This scheme uses the same input data than the advection, but does not represent lateral diffusion (perpendicular to the mean current).

In our simulations we have combined the two schemes.

2.5 <u>Turbidity</u>

The sea-water optical properties have a significant influence on MLD and SST long-term simulations, because vertical stratification - due to depth-dependance of the solar heating - interacts with vertical turbulent mixing. In Gaspar's Mixed-Layer model, Paulson and Simpson's (1977) approximation is used to compute solar irradiance versus depth. A set of 3 parameters, which depend on the Jerlov's water-type (Jerlov 1976) needs to be specified. We have used the guess-field of these quantities presented by Simonot and Le Treut (1986).

2.6 Surface heat flux and stress forcing

We have used one year (April 84 - May 85) of daily surface heat fluxes and stresses from the European Center for Medium Range Weather Forecasts (ECMWF). A description of these fluxes may be found in Simonot (1986) and Le Treut et al. (1987). The corresponding oceanic meridional heat transports have been diagnosed in Simonot and Le Treut (1987).

2.7 Initial state

The initial state is set on the first of April. It is a linear time-interpolation of the winter and spring 1° x 1° climatological fields of oceanic temperatures from Levitus (1982). After horizontal interpolation onto our grid, the vertical profiles are interpolated on our model vertical levels by the use of a procedure due to Gaspar (personal communication). It ensures vertical stability by convective mixing, while conserving potential energy. The initial MLD has been computed as the depth of 0.5°C difference with the surface temperature. Figures 5a and 5b show the initial SST and MLD field.

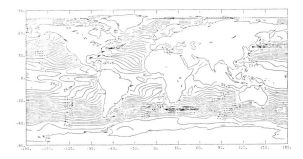

Figure 5.a : Initial SST field (unit : °C, increment : 1°C)

The SST initial field agrees to less than 1°C over 80% of the ocean, with a linear time-interpolation of the National Meteorological Center (NMC) climatologies for March and April.

132

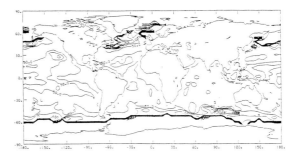

Figure 5.b : Initial MLD field (unit: m, increment: 25m)

2.8 Parameterization of the tropical upwelling

We have run a series of seasonal integrations with our model as
described herebefore. In the equatorial oceans, the mean horizontal
heat transport H.$\vec{U}$ is not able to compensate the surface heat gain
and, in order to obtain reasonable SSTs, it is necessary to include
some parameterization of the tropical upwelling.

We have added a cooling term for the Mixed-Layer. At the MLD h,
the cooling due to an upwelling would write:

$$\rho_o \cdot c_p \cdot h \cdot \frac{\delta T}{\delta t} = - W_E \cdot \frac{\delta T}{\delta z} \tag{8}$$

where W_E is the upwelling velocity and where $\delta T/\delta z$ is a vertical
gradient.

From integration of the continuity equation from a chosen depth
h' to the surface, W_E is the divergence of the (h'U, h'V) vector,
where (U,V) is the averaged drift over h' and is given by :

$$
\left[
\begin{array}{l}
rU - fV = \dfrac{\tau_X}{\rho_o h'} \\[3ex]
rV + fU = \dfrac{\tau_Y}{\rho_o h'}
\end{array}
\right.
\tag{9}
$$

with ρ_o the water standard specific mass, (τ_X, τ_Y) the surface stress
vector, f the Coriolis parameter and r a lapse-rate.

The resulting Ekmann velocity is thus dominated by the β effect in the divergence of the $\vec{\tau}/f$ vector near the equator, an by its curl at middle latitude (with a lapse rate of the order of 1 over a few days).

Outside the equatorial belt, this term is computed at the bottom of the Mixed-Layer, the corresponding $\partial T/\partial z$ is computed as the vertical temperature gradient between the simulated Mixed-Layer and the underlaying layer.

In the equatorial belt, we have defined:

$$\frac{\partial T}{\partial z} = \frac{T - 20}{h_{20}} \tag{10}$$

where T is the simulated SST (in °C) and h_{20} is the imposed climatological depth of the 20°C isotherm. This term provides a relaxation of the system towards observations, in the form of a cooling.

3. RESULTS

We first give the results of a control integration of the model over one year, starting from the first of April. The following values are chosen for the model main parameters : for the upwelling coefficient r = 1/3days, for the scaling of the currents A = 1 and for the advected layer depth h_1=50m. Figure 6.a,b shows the mean SST for the months of September (after 6 months of model integration) and March (after 12 months of model integration). The differences between these fields and the corresponding NMC climatologies are shown in Figure 7.a,b. Finally the MLD for the same months is given in Figure 8.a,b.

There appear locally some large differences between our results and the observations. Some of them are due at least partly to the differences between the climatology used for the initialization (chosen for its good vertical resolution) and that used for the validation (which had to be monthly). This is for example the case off the coasts of Labrador. In most areas, however, the simulated errors consist in too warm SSTs corresponding to too shallow MLDs, this occuring in the summer hemisphere. In the Southern Hemisphere in March, the MLD never reaches 50m, although in reality it is deeper than 75m in the latitudinal belt around 50°S. The corresponding SSTs may be 10°C larger than the observations. It must be recalled that

134

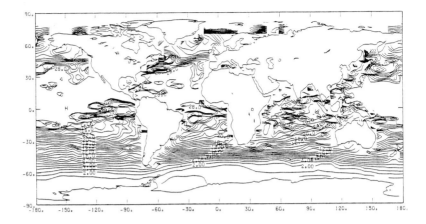

a) Month of September

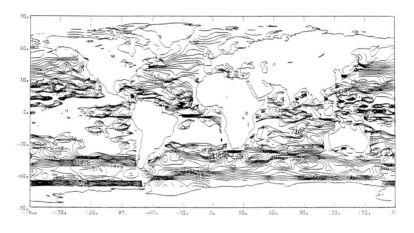

b) Month of March

Figure 6: Mean monthly SST simulated by the model in the control case

Gaspar's model was validated at OWS Papa and Romeo, where the local measured wind was available. The surface stress from the ECMWF analyses correspond to the motion of the atmosphere at synoptic scales and may not have the same statistical properties than the local stress. This is supported by the fact that model warmer temperatures appear in regions of cyclonic activity. One may also think that the conditions at the surface of the ocean may vary from one region to the

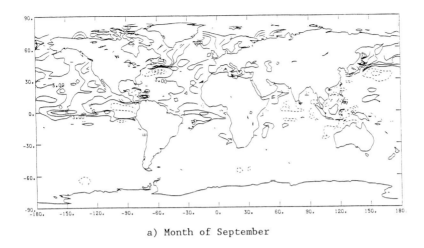

a) Month of September

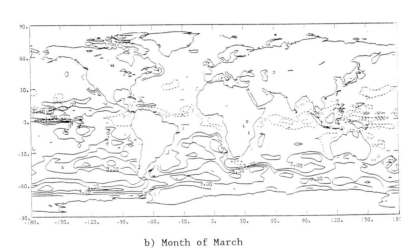

b) Month of March

Figure 7: Difference between simulated and observed SST mean monthly
distribution

other (larger amplitude of the waves), which is not taken into account
in our model.

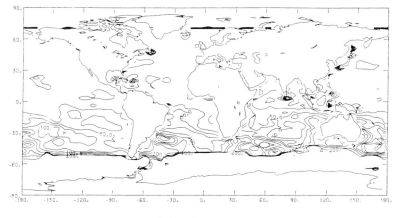

a) Month of September

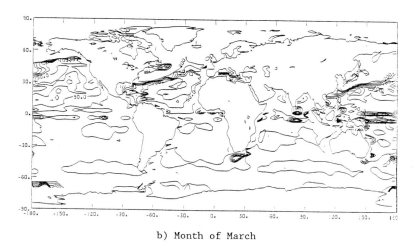

b) Month of March

Figure 8: Mean monthly MLD simulated by the model in the control case

---

4. SENSITIVITY TO MODEL PARAMETERS

We have made various sensitivity experiments to test the sensitivity of our model to the main model parameters. The most important results are summarized below.

Results in the equatorial belt are very sensitive to the parameterization of the tropical upwelling. The omission of this parameterization in the model formulation causes a continuous increase of the temperature near the Equator, which reaches maxima of 45° after 6 months of integration. In a coupled ocean atmosphere model this would not happen due to the cooling of the ocean by the evaporation, but the surface energy fluxes would then probably be seriously wrong.

The values of A (scaling of the currents) and h_1 (depth of the advected layer) have been changed in various sensitivity experiments. With the values used in the control case, the North-South simulated mean annual energy transport is larger than the transport necessary to equilibrate the radiative forcing. This indicates that A is probably underestimated and h_1 probably overestimated. This may partly explain the high temperatures found during the autumn in the Northern Atlantic or the Northern Pacific : reducing the energy transport would reduce the temperature in those regions. In an experiment where A was changed from 1 to 1.5 this temperature decrease reached 2°C in the atlantic and 1°C in the Pacific, after 8 months of integration.

The sensitivity of the model results to the prescription of the oceanic optical properties has also been investigated. We have run two short (2 months) sensitivity experiments where we have changed the turbidity : in a first-one, Jerlov's water-type was set to type IA everywhere and in second-one, it was set to type II everywhere.

For the clear case (water-type IA) the model simulates cooler SSTs and for the turbid case (water-type II) the model simulates warmer SSTs. This sensitivity is of the same order of magnitude as the error of the reference simulation when compared to climatologies, for the two months considered in these experiments. Interesting is that this sensitivity is not higher than expected (less than 0.5° at the locations of OWS P (Papa), R (Romeo) and N (November), where Mixed-Layer models have been usually tested on the local mode, but can reach much higher values in other regions of the ocean. This indicates that turbidity is an important actor for SST modelling. In particular, it may be important to take into account its seasonal cycle, which in the open ocean mainly results from phytoplanctonic blooms and life cycle.

5. CONCLUSIONS

We have presented a thermodynamic model of the global ocean. Although optimal results have not yet been reached, the model is able to simulate realistic MLD and SST patterns over large oceanic regions, at middle latitudes and in the eastern parts of the tropical oceans.

Before coupling the model to the LMD Atmospheric GCM we have used climatological fields of wind stress, surface energy fluxes, surface currents, eddy kinetic energy and oceanic turbidity to force it. It turns out to be an interesting working tool to provide information about the consistency between these data, the currently available SST distributions, and the physical mechanisms which are thought to be important for the thermodynamics of the oceanic upper layers. In particular we have found that coupling a mixed layer model with an atmospheric GCM would cause serious problems in the equatorial belt : a parameterization of the mean upwelling effect is necessary. Other key issues are the determination of the vertical profile of the surface currents and of the local wind stress. In this latter case forcing directly the model through satellite retrievals of the surface stress might be an interesting approach to increase our understanding of the upper ocean thermodynamics.

ACKNOWLEDGMENTS

This work was supported by funds of the Centre National de la Recherche Scientifique, the Centre National d'Etudes Spatiales and the Plan National d'Etude de la Dynamique du Climat. We are grateful to two unknown referees for their comments on the original text. We thank M.C. Cally for preparing the camera-ready version of the manuscript.

REFERENCES

Babiano A., Basdevant C., Le Roy P. and R. Sadourny, 1987 : Single-particle dispersion, Lagrangian structure function and Lagrangian energy spectrum in two-dimensional incompressible turbulence, J. Mar. Res., 45, pp. 107-131, 1987.

Gaspar P., 1985 : Modèles de la couche active de l'océan pour des simulations climatiques, Dissertation Doctorale, Université Catholique de Louvain, Belgium.

Gaspar P., 1987 : Modelling the seasonal cycle of the upper ocean, J. Phys. Oceanogr. in press.

Jerlov N.G., 1976 : Marine Optics, Elsevier Oceanogr. Ser., 14, 231 pp., Elsevier, New-York.

Le Treut H., Simonot J.Y. and Crépon M., 1985 : A model for the sea-surface temperature and heat content of the North Atlantic Ocean, in Coupled Ocean-Atmosphere Models, J.C. Nihoul ed., Elsevier Science Publishers B.V., Amsterdam, pp. 439-445

Le Treut H., Simonot J.Y., Piniot H. and Portes J., 1987 : Colour processing of sea surface fluxes from an operational system, Ocean-air Interactions, Vol 1, pp. 191-194.

Levitus S., 1982 : Climatological atlas of the World Ocean, NOAA Prof. Paper, 13.

Manabe S. and Stouffer R.J., 1979 : A CO2 - climate sensitivity study with a mathematical model of the global climate, Nature, 282, pp. 491-493.

Manabe S. and Stouffer R.J., 1980 : Sensitivity of a global climate model to an increase of CO_2 concentration in the atmosphere, J. Geophys. Res., 85, pp. 5529-5554.

Manabe S., Wetherald R.T. and Stouffer R.J., 1981 : Summer dryness due to an increase of atmospheric CO_2 concentration, Clim. Change, 3, pp. 347-386.

Meehl G.A., 1980 : Observed World Ocean seasonal surface currents on a 5° grid, NCAR Technical Note.

Patterson S.L., 1985 : Surface circulation and kinetic energy distribution in the southern hemisphere oceans from FGGE drifting buoys, J. Geophys. Res., 90, pp. 865-884, 1985.

Paulson C.A. and J.J. Simpson, 1977 : Irradiance measurements in the upper ocean, J. Phys. Oceanogr., 7, pp. 952-956.

Richardson P.L., 1983 : Eddy kinetic energy in the North Atlantic Ocean from surface drifters, J. Geophys. Res., 88, pp. 4355-4367.

Sadourny R., 1986 : Turbulent diffusion in large-scale flows, in Large-scale transport processes in oceans and atmosphere, J. Willebrand and D.L.T Anderson ed., NATO-ASI Series, p. 359-372.

Simonot J.Y., 1986 : Atlas of ECMWF surface stresses and heat fluxes (July 83-April 85) Laboratoire de Météorologie Dynamique, Centre National de la Recherche Scientifique, Paris, France, 258 pp., 1986.

Simonot J.Y. and Le Treut H., 1986 : A climatological field of mean optical properties of the World Ocean, J. Geophs. Res., 91, pp. 6642-6646.

Simonot J.Y. and Le Treut H., 1987 : Sea surface fluxes from a numerical weather prediction system, Climate Dynamics, 2, pp. 11-28.

Spelman M.J. and Manabe S., 1984 : Influence of oceanic heat transport upon the sensitivity of a model climate, J. Geophys. Res., 89, pp. 571-586, 1984.

Whitworth T. III and Peterson R.G., 1985 : Volume transport of the Antarctic Circumpolar Current from bottom pressure measurements, J. Phys. Oceanogr., 15, pp. 810-816.

Wyrtki K., Magaard L. and Hager J., 1976 : Eddy energy in the Oceans, J. Geophys. Res., 81, pp. 2641-2646.

COASTAL OCEAN RESPONSE TO ATMOSPHERIC FORCING

B. RAJKOVIC[1] and G.L. MELLOR
GFD Program, Princeton University, Princeton, 08540 N.J. (USA)

ABSTRACT

In order to examine the response of a coastal ocean to atmospheric forcing, successive integrations of 2-D atmospheric and oceanic models are performed. The atmospheric model has a prescribed sea surface temperature that is independent of time and a prescribed, time-dependent land surface temperature. The oceanic model is forced by the wind obtained from the atmospheric model.

In the case of constant sea surface temperature, the wind stress distribution is fairly constant in the cross-shore direction except in the vicinity of the coastline. With a sea surface temperature distribution corresponding to a well-developed upwelling situation, the atmosphere model develops a wind stress distribution with a pronounced decrease in a 40 km band next to the coast.

The model ocean, forced with the wind stress obtained from the atmospheric run with homogeneous sea surface temperature, develops a strong upwelling zone and a strong equatorward current with an embedded jet near the coast. Forced with the wind stress from the atmospheric run with inhomogeneous sea surface temperature, the ocean run has a much weaker upwelling and a double structured alongshore current with poleward flow in the vicinity of the coastline and equatorward flow in the region away from the coast.

1. INTRODUCTION

This paper is concerned with the response of a coastal ocean to the atmospheric forcing. Whether or not an atmospheric flow will influence the ocean circulation depends strongly on the horizontal, cross-shore derivative of the wind stress, the wind stress curl. There is some observational though not conclusive evidence that the wind stress does have structure within 50 km or less from the coast. If that is the case, then the recent work by Mellor (1986) shows that the wind stress curl will have a significant influence on the upwelling and on the alongshore flow structure.

Our approach is to examine the response of the ocean to the wind stress obtained from an atmospheric model. Both models are two-dimensional and have high resolution in the horizontal direction and very high resolution in the vertical direction, such that the bottom boundary layer in the atmospheric model and both the top and bottom boundary layers in the oceanic model are resolved.

The oceanic model is forced with two different distributions of the wind stress that are obtained in atmospheric runs with different sea surface temperature distributions. The first

---

[1] Present affiliation : Belgrade University, Department of Physics and Meteorology, Belgrade, P.O. Box 505, 11000 Belgrade (Yugoslavia).

atmospheric run has a homogeneous sea surface temperature. Upwelling is then quite pronounced and the alongshore flow is equatorward with an embedded jet near the coast. Both surface and bottom boundary layers are well developed. The second atmospheric run has a inhomogeneous sea surface temperature as the lower boundary condition. In that case, upwelling is much weaker and the alongshore flow develops a double structure with poleward flow near the coast and equatorward flow away from it.

2. DESCRIPTION OF THE MODELS

2.1. The governing dynamic and thermodynamic equations

The oceanic model is a version of the 3-D model developed by Blumberg and Mellor (1987) and adapted by Mellor (1986) for the solution of 2-D problems. For this work, the oceanic model was also converted to an atmospheric model. The differences between the oceanic and the atmospheric models include the formulation of the thermodynamic equation and new lateral boundary conditions.

The governing equations for both the ocean and the atmosphere are for a hydrostatic, Boussinesq fluid. The coordinate system is a sigma system such that the geometrical height z is transformed to σ according to the formula :

$$\sigma = \frac{z - \eta(x)}{\eta(x) + H(x)} \tag{1}$$

where H(x) is the height/depth of the unperturbed fluid and $\eta(x)$ is the free surface elevation.

Both the atmospheric and the oceanic models have a free surface at the upper boundary. The main reason for this choice was to include tidal effects in the oceanic model. However, since it appeared that the computational efficiency was increased (the solution of the Poisson equation required by a rigid lid model is bypassed in this way), it was also kept in the atmospheric model.

With $D = H + \eta$, we can write the x-component of the equation for the conservation of momentum as :

$$\partial_t (u D) + \partial_x (u u D) + \partial_\sigma (u w^T) - f v D = - g D \partial_x \eta - \partial_\sigma <u' w'>$$
$$- D^2 \partial_x \int_\sigma^0 b \, d\sigma + D \, \partial_x D \int_\sigma^0 \sigma \, \partial_\sigma b \, d\sigma + \partial_x (A_M D \, \partial_x u) \tag{2}$$

and the y-component as :

$$\partial_t (v D) + \partial_x (u v D) + \partial_\sigma (v w^T) + f u D = - g D \partial_y \overline{\eta} - \partial_\sigma <v' w'>$$

$$- D^2 \partial_y \int_\sigma^0 b \, d\sigma + D \partial_y D \int_\sigma^0 \sigma \partial_\sigma b \, d\sigma + \partial_x (A_M D \partial_x v) \qquad (3)$$

where A_M is the momentum horizontal diffusion coefficient, $\overline{\eta}$ is a prescribed mean elevation, b is buoyancy defined as $b = g \, \rho/\rho_0$, $<w'u'>$ and $<w'v'>$ are the Reynolds stress terms (brackets denote ensemble average), and w^T is a new vertical velocity in the sigma system defined as :

$$w^T = w - u \cdot \left[\sigma \frac{\partial D}{\partial x} + \frac{\partial \eta}{\partial x} \right] - \left[\sigma \frac{\partial D}{\partial t} + \frac{\partial \eta}{\partial t} \right] \qquad (4)$$

The continuity equation takes the form :

$$- \partial_\sigma w^T = \partial_t \eta + \partial_x (uD) \qquad (5)$$

while the thermodynamic equation is :

$$\partial_t (\Theta D) + \partial_x (u \Theta D) + \partial_\sigma (w^T \Theta) = - \partial_\sigma <w' \Theta'> + \partial_x (A_H D \partial_x \Theta) \qquad (6)$$

where A_H is the temperature horizontal diffusion coefficient.

The oceanic model equations have a β-term. The main reason for the inclusion of the β-term is the specification of the y variation of the pressure gradient force. For more details about the modeling of the ocean part, see Mellor (1986).

We integrate the equations of motion from bottom to top to get the external model, for which we then carry out separate computations. If we define a vertical average according to :

$$A = \overline{a^V} = \int_{-1}^0 a \, d\sigma \qquad (7)$$

the governing equations for the external mode are as follows:

$$\partial_t \, \eta = - \, \partial_x \, (U \, D) \tag{8}$$

$$\partial_t \, (U \, D) - f \, V \, D = - \, g \, D \, \partial_x \, \eta - D^2 \int_{-1}^{0} d\sigma \int_{\sigma}^{0} \partial_x b \, d\sigma' + D \, \partial_x \, D \int_{-1}^{0} d\sigma \int_{\sigma}^{0} \sigma' \, \partial_{\sigma'} \, b \, d\sigma'$$
$$- <u' \, w'>_0 + <u' \, w'>_{-1} - \partial_x \, \overline{(u \, u \, D)}^v + \partial_x \, (A_M \, D \, \partial_x U) \tag{9}$$

$$\partial_t \, (V \, D) + f \, U \, D = - \, g \, D \, \partial_y \, \overline{\eta} - D^2 \int_{-1}^{0} d\sigma \int_{\sigma}^{0} \partial_y b \, d\sigma' + D \, \partial_y \, D \int_{-1}^{0} d\sigma \int_{\sigma}^{0} \sigma' \, \partial_{\sigma'} \, b \, d\sigma'$$
$$- <v' \, w'>_0 + <v' \, w'>_{-1} - \partial_x \, \overline{(u \, v \, D)}^v + \partial_x \, (A_M \, D \, \partial_x V) \tag{10}$$

In order to close this system of equations, one has to make some assumptions about the Reynolds stress terms. The models for both the ocean and the atmosphere follow Mellor and Yamada (1974) and include the so-called level 2.5 closure. For a complete description, see the review paper by Mellor and Yamada (1982).

For purely numerical reasons, we have also included a horizontal diffusion term in both the external and the internal mode calculations. Diffusion is harmonic with a constant coefficient of diffusion for the external mode while the internal mode has a variable coefficient of the Smagorinsky type.

2.2. Boundary Conditions and Model Forcing

(i) Lateral Boundary Conditions.

Since the ocean currents are much weaker and the internal-gravity phase speed is much smaller in the ocean than in the atmosphere, the lateral boundary conditions pose less of a problem for the oceanic model. We use a radiation type boundary condition for the elevation:

$$\partial_t \, \eta + c \, \partial_x \, \eta = \frac{\eta}{\tau} \tag{11}$$

where c denotes the external mode gravity wave phase speed $\pm\sqrt{gH}$, and τ is a time constant which ensures that on long time scales ($\gg L_x / (\sqrt{gH})$) the value of the elevation at the boundary is close to zero.

For the internal mode, the oceanic model has a slightly different boundary condition than the atmospheric model. For the u component the boundary condition is :

$$\partial_t \, u + C_{Or} \, \partial_x u = 0 \tag{12}$$

where C_{Or} is the phase speed calculated according to the method proposed by Orlanski (1976). Boundary conditions for v, temperature and salinity for the outflow points are of the form :

$$\partial_t \psi + u \, \partial_x \psi = 0 \tag{13}$$

For the inflow points, temperature and salinity are fixed at their initial values.

(ii) Lower and Upper Boundary Conditions.

The ocean model's surface and bottom boundary conditions for the x and y components of momentum are :

$$K_M \, (\partial_z u , \partial_z v) \approx (\tau_x^0 , \tau_y^0) \;\; , \;\; z \to \eta \tag{14}$$

$$K_M \, (\partial_z u , \partial_z v) \approx C_D \, |u| \, (u,v) \;\; , \;\; z \to -H \tag{15}$$

where τ_x^0 and τ_y^0 are the momentum fluxes obtained from the atmospheric model. The bottom stress is determined by matching the velocities to the logarithmic law of the wall such that the value of the drag coefficient C_D given by :

$$C_D = \frac{1}{\kappa} \ln \left[\frac{H + z_b}{z_o} \right]^{-2} \tag{16}$$

where z_b is the height of the grid point nearest the bottom. The value for z_0 is 0.01 meter if the depth is more than 100 meters. For the shallower parts of the domain, z_0 increases inversely proportional with depth. This variation is intended to simulate enhanced bottom drag due to nearshore wave action (see Mellor, 1986).

The boundary conditions for heat and salinity at the surface have the same form as for momentum. The heat and salinity fluxes have values required to provide thermodynamic steady state. They were evaluated from the integrated heat and salinity transports across the seaward boundary. Exported heat flux in terms of equivalent surface heat flux was 95 Wm^{-2}. At the bottom, we prescribe zero flux for both heat and salinity.

3. OCEAN RESPONSE TO THE ATMOSPHERIC FORCING AND COASTAL UPWELLING

3.1. Background

We are concerned only with the generation of upwelling or downwelling by the local forcing and thus we assume that the wind stress is uniform along the coast. In the model, the coastline is in the south-north direction.

Following the work of Mellor (1986), a balance relevant for the long time averages (we neglect nonlinear terms) can be expressed as :

$$-\int_{-H}^{0} v \, dz = -\frac{g\,H}{f} \partial_x \eta - \frac{1}{f} \int_{-H}^{0} \int_{z}^{0} \partial_x b \, dz' \, dz + 0 - \frac{\tau_x^b}{f} \tag{17}$$

$$\int_{-H}^{0} u \, dz = 0 = -\frac{g\,H}{f} \partial_y \eta - \frac{1}{f} \int_{-H}^{0} \int_{z}^{0} \partial_y b \, dz' \, dz + \frac{\tau_y^0}{f} - \frac{\tau_y^b}{f} \tag{18}$$

$$(A) \qquad\qquad (B) \qquad\qquad (C) \quad (D)$$

Thus, at steady state, a balance must exist between the onshore-offshore pressure gradient force (sum of A and B), the Ekman surface transport (term C), and the bottom Ekman transport (term D). This balance explains the existence of poleward or equatorward flows. A change in the sign of the A+B+C results in the transition from poleward flow and strong upwelling to equatorward flow and weak upwelling.

3.2. Analysis of the model runs

The y-component of the pressure gradient and mean elevation are prescribed in the same way as in Mellor (1986) and, as a continuation of his analysis on the influence of the offshore wind stress distribution on the flow structure in the coastal regions two cases will be examined. Fig. 1 (top panels) presents two characteristic sea surface temperature distributions that were imposed in the two atmospheric runs. Case I has a homogeneous sea surface temperature, while case II has a temperature drop in the region 20 km next to the coast resembling an upwelling situation. The next row of panels shows the wind stress distributions calculated in those two cases (24 hours means, solid lines). Presented in the same panels are the vertically-integrated y-component of the pressure gradient used in the ocean model (dashed lines). The pressure gradient is prescribed and determined from observations by Mellor (1986). The third row panels are temperature fields after 40 days of integration. The bottom panels present the alongshore flow structure also after 40 days of integration (solid lines for poleward direction and dashed lines for the equatorward direction).

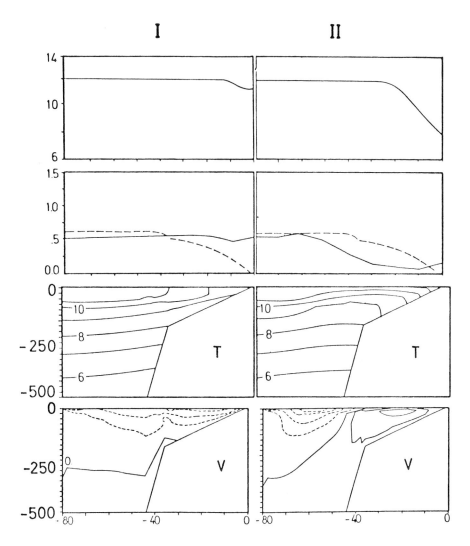

Fig.1. Response of the model ocean to the wind stress distribution obtained with the atmospheric model. In case I, the wind stress is calculated with a homogeneous sea surface temperature while in case II, a nonhomogeneous sea surface temperature resembling an upwelling situation is prescribed.

- The sea surface temperature distributions (in deg C) are presented in the top panels; the abscissa shows the off-shore distance in km.

- The next two panels present the calculated wind stress distributions (24 hours means, solid lines) and the prescribed integrated y-component of the pressure gradient (dashed line), both in dyn/cm^2.

- The third row of panels shows the potential temperature fields, as functions of depth versus off-shore distance, after 40 days of integration; the contour interval is 1 deg C.

- The bottom panels show the alongshore component of the velocity for the same position and time. Dashed lines designate equatorward flow, solid lines designate poleward flow. The contour interval is 1m/s.

In case I, the forcing is calculated in the atmospheric model with a homogeneous sea surface temperature. The main feature of the results obtained with the oceanic model is that the flow above 250 meters is equatorward, while at lower depths, a weak poleward current exists. The separation point occurs where the wind stress becomes larger than the integrated pressure gradient. The sloping of the zero line is presumably induced by the sloping of the shelf. Because the wind stress is stronger than the imposed pressure gradient in the region near the coast, an intense equatorward current forms and quite intense upwelling occurs. Both the surface and the bottom boundary layers are well developed. This flow structure is similar to the flow structure for the case I of Mellor (1986).

Since upwelling is well developed, the resultant sea surface temperature differs from the imposed, homogeneous sea surface temperature used in the atmospheric run. Thus the atmosphere and the ocean states are inconsistent in that sense. Therefore, the next case illustrates the difference in the ocean response when the atmospheric forcing is calculated with a inhomogeneous sea surface temperature.

The ocean response in case II shows a different flow structure from the previous case because the relation between the calculated wind stress and the imposed integrated pressure gradient is different. In this case, the balance of alongshore terms requires a change of sign of bottom stress and therefore a poleward flow. Since the integrated pressure gradient in the nearshore region is much larger than the wind stress, the region with poleward flow extends all the way to the surface with an embedded coastal jet. Beyond the point where the wind stress is equal to the pressure gradient, the surface flow becomes again equatorward. The upwelling is much weaker than that observed in case I, with surface waters next to the coastline 4° warmer. The sea surface temperature is almost homogeneous with a weak minimum occurring about 40 km offshore. As in case I, the imposed and the resultant sea surface temperatures differ. In case II, the calculated sea surface temperature exhibits weaker horizontal gradients than the imposed ones.

4. CONCLUSIONS

For the atmospheric model, the inclusion of a variable temperature at the lower boundary, over the ocean part of the model domain, changes qualitatively the shape of the wind stress distribution. Instead of a constant value with a weak increase approaching the coast, a strong decrease starts 60 km away from the coastline. The minimum value of the wind stress is only 20% of the value at the seaward boundary of the model domain.

The oceanic model, when forced with the wind stress calculated with a horizontally homogeneous reference state, produces a well developed upwelling with a reduction from the initial temperature of 4.5° C near the coast. The alongshore flow is equatorward with an embedded nearshore coastal jet.

When forced with the wind stress calculated with a inhomogeneous sea surface temperature distribution, the oceanic model produces very different flow and upwelling structures. The alongshore component of the flow develops a double structure with a poleward flow near the coast in the band where the y-component of the integrated pressure gradient is stronger than the y-component of the wind stress. In the region where the relation between integrated pressure gradient and wind stress is the opposite, equatorward flow develops. The resultant upwelling is very weak compared to the first case.

5. ACKNOWLEDGMENTS

We thank S. Nickovic and S. Misic for drafting the figure. This work forms a part of the doctoral thesis of B. Rajkovic who wishes to acknowledge the support of the National Science Foundation through Grant ATM 8218761, of the Republic Science Fund through Grant 6623/1 of S.R Serbia, and of the Academy of Sciences and Arts of S.R. Serbia.

6. REFERENCES

Blumberg, A.F., and G.L. Mellor, 1987. A description of a three-dimensional coastal ocean circulation model. In: Three-Dimensional Coastal Ocean Models (Ed. N.S. Heaps), *Coastal and Estuarine Sciences*, 4, 1-16.

Mellor, G.L., 1986. The numerical simulation and analysis of coastal circulation off California. *Continental Shelf Research*, 6, 689-713.

Mellor, G.L., and T. Yamada, 1974. A Hierarchy of turbulence closure models for planetary boundary layers. *J. Atmos. Sci.*, 31, 1791-1806. (Corrigenda, *J. atmos. Sci.*, 34, 1482).

Mellor, G.L., and T. Yamada, 1982. Development of a turbulence closure for geophysical problems. *Rev. Geophys. Space. Phys.*, 20, 851-875.

Orlanski, I., 1976. A simple boundary condition for unbounded hyperbolic flows. *J. Comput. Phys.*, 21, 251-269.

PARAMETERIZATION OF VERTICAL DIFFUSION IN A NUMERICAL MODEL OF THE BLACK SEA

H. J. FRIEDRICH[1] and E. V. STANEV[2]
[1] Institute of Oceanography, University of Hamburg, Troplowitzstr. 7, 2000 Hamburg 54 (F.R. Germany)
[2] Faculty of Physics, University of Sofia, 5 Anton Ivanov Blvd., 1126 Sofia, (Bulgaria)

ABSTRACT

The thermohaline stratification of the Black Sea depends strongly on vertical diffusion and on convection. By numerical experiments with a general circulation model of the Princeton type it is shown that this stratification can be simulated at least qualitatively in such a model. This is not possible however with a constant coefficient of vertical diffusivity. Here, convection is parameterized by convective adjustment as usual, and the vertical diffusivity coefficient is taken as a function of a model Richardson number. By carefully choosing the parameters of this function and by prescribing realistic boundary conditions the present model reproduces such sensitive details of the stratification as the shape of the permanent halocline and the basinwide cold intermediate layer.

1 INTRODUCTION

The upper 200-300 meters of the water column in the Black Sea are formed by several sublayers: a surface mixed layer, a seasonal thermocline (in summer only), a cold intermediate layer (CIL) and a permanent halocline (see fig. 1). The temperature minimum of the CIL can be observed throughout the Black Sea at depths 50-150 meters, whenever the sea surface temperature (SST) is above 8 - - 9°C. Since in the most parts of the sea the SST and the temperature of the deep water are higher than 8°C one can conclude that the water of the CIL originates from a small cooling area in the north-west, and spreads on an isopycnal surface.

The strong permanent halocline separates the surface water from the deep water. As can be seen from figure 1, the mean salinity of the surface water changes between 17.5ppt in winter and 18.5ppt in summer. Only in the northwestern part of the Black Sea anomalously low salinity values are observed as a consequence of the large amount of river inflow into that area. The salinity of the deep water has values around 23ppt. The strongest vertical gradients are

found in a relatively thin layer in the upper 150-300m. The shape of the halo-

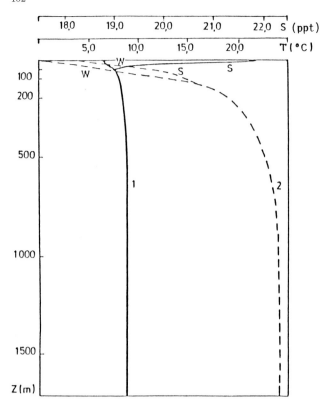

Fig. 1. Vertical profiles of temperature (1) and salinity (2) in summer (S) and in winter (W), according to Sorokin's (1982) data.

cline depends strongly on vertical exchange processes. Its depth decreases towards the central Black Sea due to large scale upwelling.

The deep water below the halocline - that is below a depth of 200-300m - is very homogenous in temperature and in salinity. It is not affected by seasonal change.

Vertical gradients of temperature are restricted to an even thinner surface layer, and density gradients beneath the seasonal thermocline are primarily governed by salinity gradients. Numerical experiments with appropriate boundary conditions confirmed, that the salinity distribution is the main source of baroclinicity in the Black Sea (Stanev et al., 1987).

The exceptionally high stability, connected with the salinity gradient of the halocline, acts as a barrier for convective penetration of cold water, which is formed in the northwestern Black Sea during winter cooling. Spreading horizontally in the CIL this water can still be detected in the eastern-most part of the sea.

The maintenance of the salinity field in the Black Sea is a result of the total water and salt budgets but depends strongly on the local distribution of sources and sinks, and also on the geometry of the basin, on its openings and on the windfield. The water and salt budgets of the Black Sea can be expressed by simple relations:

$$Q_B^S + Q_B^C + Q_R + P + E = 0, \tag{1}$$

$$Q_B^S S_B + Q_B^C S_M = 0, \tag{2}$$

where Q_B^S and Q_B^C are the water fluxes of the Bosphorus surface current and of the Bosphorus countercurrent, respectively, Q_R is the river flux, P and E are precipitation and evaporation, S_B and S_M are the mean salinities of the Black Sea and Mediteranean Sea. The components of the water and salt budget have been considered in a number of previous studies. Here we refer to Bulgakov and Korotaev (1984).

TABLE 1

Inflow	km³/year	Outflow	km³/year
Bosphorus subsurface current	180	Bosphorus surface current	360
From Azov Sea	50	To Azov Sea	30
Precipitation	120	Evaporation	320
From rivers	360		
Total	710	Total	710

The flow through the Kerch Strait is small in comparison to the other terms of the water budget. Also the net salt flux through the Kerch Strait is zero. Thus without making a big error we simplify the water and salt budget of Bulgakov and Korotaev by ommiting the exchange through the Kerch Strait and correcting the surface water budget correpondingly: P + E = -180km³/year. Keeping in mind that the Mediterranean water has twice the salinity value of the Black Sea water (S_M=36ppt, S_B =18ppt) we obtain that $Q_B^S = - 2Q_B^C$. Then (1) can be written as:

$$Q_B^C = Q_R + PLE, \tag{3}$$

where PLE = P + E. This relationship will be used further as a water budget equation in the numerical model.

There are two main ideas concerning the penetration of Mediterranean water into the Black Sea. The first one was presented by Spindler and Wrangel. As quoted by Filippov (1968) they assume that the relatively heavy Mediterranean water sinks without substantial entrainment of surrounding water and reaches the very bottom, where it spreads horizontally and mixes with the bottom water. A general upward motion compensates the sinking.

According to the second idea, given by Skalovski, quoted by Filippov (1968), the Mediterranean water undergoes substantial entrainment and does not sink down to the very bottom. Instead after reaching some level it spreads horizontally along isopycnic surfaces. On the basis of existing data it is not possible to confirm or to reject definitively one of the two hypotheses. One can expect however that a coarse resolution GCM supports Skalowski's hypothesis; but this would primarily be due to the insufficient resolution of the salinity source at the Bosphorus. In order to simulate realistically the entrainment processes and the distribution of the Bosphorus plume a separate local model would have to be developed. However, this is not the scope of the present study.

As it has been already pointed out, the establishment of the CIL results from the winter convection in the northwestern part of the Black Sea and the subsequent transport of cold water on isopycnic surfaces. This layer has a thickness of only 20-50 m, and a numerical model to study its establishment and seasonal variability would have to resolve this vertical scale.

The CIL is sandwiched in summer between the permanent halocline and the seasonal thermocline. The temperature of the upper mixing layer then is more than 10° C higher than in the CIL. If the vertical mixing coefficient would be constant with depth at the value of the upper mixing layer the CIL would be destroyed by the strong heat flux from the surface. This however does not happen, which demonstrates, that the vertical exchange is highly reduced by the stability of the seasonal thermocline. Thus, both the seasonal thermocline and the halocline shield CIL from severe mixing with the layers above and below, and the properties of the CIL are highly conservative. To realistically simulate this feature in a numerical model vertical exchange coefficients would have to depend on stability.

Pacanowski and Philander (1981) have discussed such a parameterization of vertical mixing for numerical models of the tropical ocean.

In the following sections of this work we discuss some numerical results obtained with a Black Sea circulation model with different assumptions for the parameterization of vertical mixing.

2 NUMERICAL MODEL

2.1 Description of the model

Bryan's (1969) oceanic general circulation model and Semter's (1974) FORTRAN code have been used in this study. A detailed description of the physics of the model and of the finite difference scheme may be found in the references giver above. The equation of motion are simplified using the Boussinesq and hydrostatic approximations. The governing system of equations in spherical coordinates (the vertical axis is directed upwards) can be written as:

$$u_t + L(u) - 2\Omega nv = -ma^{-1}(P/\rho_o)_\lambda + F^\lambda, \tag{4}$$

$$v_t + L(v) + 2\Omega nu = - a^{-1}(P/\rho_o)_\phi + F^\phi, \tag{5}$$

$$L(1) = 0, \tag{6}$$

$$g\rho = -P_z, \tag{7}$$

$$(T, S)_t + L(T, S) = F^{T,S}, \tag{8}$$

$$\rho = \rho(P, T, S). \tag{9}$$

The first two equations are the momentum equations, (6) is the continuity equation, (7) is the equation of hydrostatics, (8) is the equation of heat and salt conservation, (9) is the equation of state. The following symbols are used in (4)-(9):

$$L(\mu) = ma^{-1}\{(u\mu)_\lambda + (v\mu m^{-1})_\phi\} + (w\mu)_z, \tag{10}$$

$m=\sec\phi$, $n=\sin\phi$, ϕ and λ are latitude and longitude, a is the radius of the earth. The turbulent terms are denoted by F:

$$F^\lambda = A_{MV}u_{zz} + (A_{MH}a^{-2})\Delta u, \tag{11}$$

$$F^\phi = A_{MV}v_{zz} + (A_{MH}a^{-2})\Delta v, \tag{12}$$

$$F^{T,S} = (A_{HV}\delta^{-1}(T, S)_z)_z + A_{HH}a^{-2}\Delta(T, S). \tag{13}$$

In (11)-(13)

$$\Delta\mu = m^2 \mu_{\lambda\lambda} + m(\mu_\phi m^{-1})_\phi, \tag{14}$$

and $A_{a,b}$ is mixing coefficient corresponding to the scheme:

$$a = \begin{cases} M - \text{momentum} \\ H - \text{temperature, salinity} \end{cases} \qquad b = \begin{cases} V - \text{vertical} \\ H - \text{horizontal.} \end{cases}$$

Vertical convection is incorporated in the model through the simple parameterization using in (13) the delta function, where

$$\delta = \begin{cases} 1 \text{ if } \rho_z < 0 \\ 0 \text{ if } \rho_z > 0 \end{cases} \tag{15}$$

The following boundary conditions are used at the sea surface, z=0, at the bottom z=-H(λ,ϕ), and at the closed lateral boundaries:

$$w = 0$$
$$\rho_\circ A_{MV}(u_z, v_z) = (\tau^\lambda, \tau^\phi),$$
$$\rho_\circ A_{HV}(T_z, S_z) = (F^T, F^S), \qquad \text{at } z=0, \tag{16}$$

$$u_z = v_z = 0,$$
$$w = -ua^{-1}mH_\lambda - va^{-1}H_\phi, \tag{17}$$
$$(T_z, S_z) = 0 \qquad \text{at } z=-H,$$

$$u = v = 0,$$
$$(T_n, S_n) = 0 \qquad \text{at the lateral boundaries,} \tag{18}$$

where n is normal to the boundary. In (16) $(\tau^\lambda, \tau^\phi)$ are the wind stress components, and (F^T, F^S) are heat and salinity fluxes.

The horizontal grid intervals are $\Delta\lambda = 1°$ and $\Delta\phi = 0.5°$. In the vertical direction irregular grid steps are used: Δz_i (i=1,...,12) = (20, 20, 20, 20, 40, 60, 120, 280, 320, 480, 520) m.

The time steps used for the integration of the momentum equation $\Delta t_{u,v}$ and of the heat (salinity) conservation equation $\Delta t_{T,S}$ are 0.5 and 12 hours respectively. The horizontal mixing coefficients are chosen as: $A_{MH}=A_{HH}=4.0\times10^6 \text{cm}^2/s$.

Hellermann's (1983) 2° wind data are interpolated to the grid points. Heat fluxes are determined from Makerov's (1961) climatic data and precipitation and evaporation are obtained from the climatic handbook of the Black Sea (1974). All data was determined for the four months: January, April, July, October. The actual values for the boundary conditions at each time step were calculated by linear interpolation. The real bottom relief is resolved approximately by using different numbers of levels at different locations.

2.2 Parameterization of salt fluxes

In Semtner's (1974) model salinity input at the sea surface is calculated as the product of net difference between precipitation and evaporation with a constant salinity.

We use the same approach for parameterization of salt transport through the Bosphorus and of the diluting effect of river flow. We introduce these flows as diffusive flows in order not to involve additional problems connected with mass conservation. The Bosphorus inflow taken across the area of one grid element would correspond to a current velocity of only 3.10^{-1}cm/s, which is negligible. At each time step the salinity at one land point is computed such, that the net diffusion flow A_{HH} $\vec{n}.\nabla S$ is equal to the salinity flux at this point SQ_B^C , where $\vec{n}$ is normal to the boundary. Thus we control at every time step the salinity inflow and the total salt content. The inflow of fresh water from rivers corresponds to a negative salinity inflow and the inflow with the Bosphorus subsurface current to a positive salinity inflow.

The observations show, that the salinity difference between the Black Sea and the Marmara Sea is substantially smaller at the surface than at the level of inflowing Mediterranean water. For this reason we also neglect the diffusive salt flux at the outflow level and prescribe only a diffusive salinity source at the level of the Bosphorus deep counter current.

The equation of the salinity balance can be written in the form:

$$Q_B^C S = Q_R S + S a^2 \int ple.\cos(\phi) \, d\phi d\lambda , \tag{19}$$

where ple (λ, ϕ) is the horizontal distribution of the local difference between precipitation and evaporation.

The diffusion flow corresponding to the Bosphorus subsurface current we introduce at the 70 m level and the river flux at the sea surface. As shown by Altmann and Kumish (1986), 79% of the river inflow enters the Black Sea in the northwest. We introduce this flow in a single "Danube delta" and express its time dependence by a simple trigonometric function:

$$q_R(t) = Q_R \left(1 + 0.5 \sin \frac{2\pi (t-\theta)}{T}\right) , \tag{20}$$

where θ = 1 month and T = 1 year. Also we approximate the Bosphorus subsurface current as

$$q_B^C (t) = Q_B^C (1 - 0.5 \sin \frac{2\pi (t-\theta)}{T}) \ , \tag{21}$$

The comparison of (20) and (21) with the experimental data of Altman and Kumish (1986) and Bogdanova (1972) show that the errors of these approximations are small.

2.3 Parameterization of vertical diffusion

As it is shown by Stanev et al. (1987) the integration of the Black Sea model with prescribed surface data for temperature and salinity and without taking into account the Bosphorus inflow leads to rapid change of the vertical stratification. In the present investigation we introduce the salinity balance and use Neumann boundary conditions in order to maintain the salt supply for the deep water. In the next section, however, it will be shown that this improvemnt of boundary conditions is not enough to obtain realistic model results. The parameterization of vertical mixing has to be realistical as well. It is known that Bryan's (1969) general circulation model is very sensitive to variations in vertical mixing. Some empirical studies in the tropical Atlantic indicate a dependence of the mixing coefficient on the Richardson number R_i= $g/\rho \ (\partial \rho / \partial z)/(\partial v / \partial z)^2$. As it is shown by Shaffer (1986) the calibration of A_{HV} leads to a substantial improvement of results concerning the diapycnal exchange in the shelf upwelling zone off Peru. There are some indications (see Kolesnikov and Boguslavski, 1978), that in the Black Sea vertical mixing is strongly dependent on the vertical salinity gradient.

A series of numerical experiments have been carried out by Pacanowski and Philander (1981) to tune a numerical model of the tropical Atlantic with respect to vertical mixing. They propose the empirical formula:

$$A_{HV} = \frac{A_{MV}}{1 + \alpha R i} + K_b \ , \tag{22}$$

where α and K_b are constants.

The numerically determined constants for the tropical Atlantic according to Pacanowski and Philander are α =5, K_b=0.1cm^2/s. One can expect these constants to depend on the model and on the problem under investigation. In the present study we examine the applicability of (22) for parameterizing the vertical diffusion in the Black Sea. Unlike Pacanowski and Philander (1981) we take the coefficient of vertical mixing of momentum constant in all our experiments: A_{MV}=10cm^2/s.

Numerical experiments

Three experiments with time dependent boundary conditions were carried out. The model was initialized with a state of rest and integrated in time over a period of 60 years. The first experiment (A) is done with a constant vertical diffusion coefficient $A_{HV} = 1$ cm^2/s and with constant temperature. The other two experiments involved mixing coefficients according to (22) and were carried out in two stages: The temperature was kept constant during the first 40 years of integration. Then, for the next 20 years, time dependent heat fluxes were imposed at the sea surface. In the experiment B the parameters in (22) corres- pond to those of Pacanowski and Philander (1981) and in experiment C $\alpha = 0.008$, $K_b = 0.01$cm^2/s. The results of the three experiments are discussed in the next section.

3 NUMERICAL RESULTS

The horizontally averaged vertical profile of salinity and its evolution to a stationary state in experiment A is shown in fig. 2 and fig. 3. As can be seen from fig. 2 salinity increases with increasing depth, which corresponds qualitatively to the observed data. However, the sharpness of the halocline is underestimated, and the surface salinity is too large. Obviously this is a consequence of overestimating vertical mixing and shows that substantial improvements in tuning the model are necessary.

Horizontally averaged winter profiles of salinity after 40 years of integra- tion are shown in figures 4 and 5 for experiments B and C respectively. As can be seen, a good agreement exists between both salinity profiles as well as with the observed profile of figure 1. The profiles of the coefficient of vertical diffusivity, however, differ considerably for the two experiments. In par- ticular, there is a secondary subsurface maximum in the profile of figure 5. In the depth of this maximum there are some minor differences between the two salinity profiles. The difference in the background diffusivity between the two experiments, appearing as the profile values in the deep water, does not seem to affect critically the shape of the main halocline. It may be argued then, that this basic feature of the Black Sea stratification is not too sensitive to changes of the vertical mixing coefficient for values less than 0.1 cm^2/s.

The results of the three experiments give some insight into the sensitivity of the model to changes in the parameterization of vertical diffusion. The coefficient of vertical diffusivity resulting from experiment B (figure 4) is almost constant with depth, whereas it changes substantially in the upper layers for experiment C (figure 5). This difference is easily explained by the difference in the values of α and K_b in formula (22) for the two experiments.

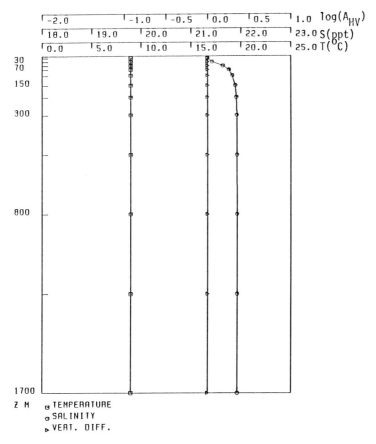

Fig. 2. Horizontally averaged vertical profile of salinity for experiment A after 60 years of integration. ($A_{HV}=1cm^2/s$).

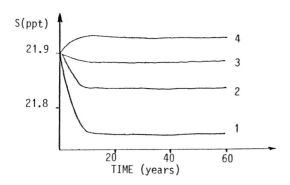

Fig. 3. Time changes of salinity during the integration of experiment A: 1 - 10m, 2 - 50m, 3 - 100m, 4 - 1700m.

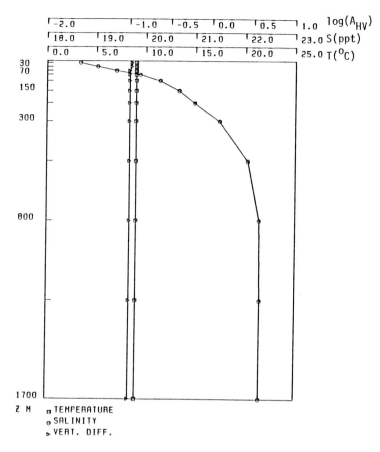

Fig. 4. Horizontally averaged winter vertical profiles of salinity and of the coefficient of vertical diffusion after 40 years of integration - experiment B.

In the Black Sea the stability of the stratification is larger than in the tropical Atlantic. This leads to an increase of the Richardson number, and the first term on the right hand side of (22) tends to decrease accordingly. The value of α given by Pacanowski and Philander (1981) is too high in this case to obtain the necessary dependence of this term on the Richardson number. The vertical diffusion in experiment B remains unsensitive to changes of the Richardson number in the Black Sea model. In order to make the model more sensitive to these changes, we had to choose a smaller value for α. Our value of $\alpha = 0.008$ leads to a large decrease of A_{HV} from the surface down to a depth, where in summer the base of the seasonal thermocline is observed. A_{HV} reaches an intermediate minimum in a layer corresponding to the CIL. Below this layer a subsurface maximum of A_{HV} occurs due to an increase in the vertical shear of

162

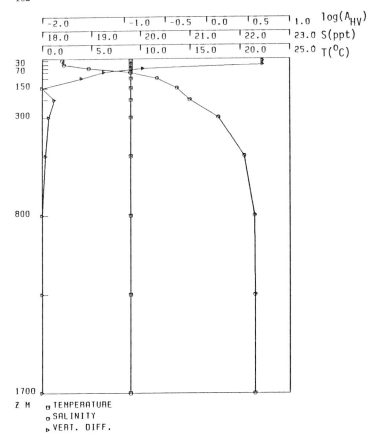

Fig. 5 As in Fig. 3 a but for experiment C.

the currents. The deep layers of the Black Sea are almost barotropic, and as a consequence A_{HV} tends to the value of the background diffusivity coefficient K_b.

The larger sensitivity of A_{HV} to changes of the Richardson number in experiment C together with the smaller background diffusivity gives a better agreement between the computed and the observed salinity profiles than the one in experiment B. In particular, due to the reduced mixing below a depth of 100m the exchange between deep and surface layers is severly restricted in experiment C. This leads to a difference between salinity values at the surface and at the bottom, which is larger than in experiment B and much closer to the observed value (compare figure 1). An examination of the horizontal distribution of salinity simulated in experiment B (not shown here) reveals unrealistically large values for the salinity at the sea surface, especially in the central part of the Black Sea.

A comparison between figures 2 (A_{HV}=1cm^2/s) and Figures 4 and 5 [A_{HV}=f(Ri)] shows, that a substantial improvement of the salinity profile can be obtained, when A_{HV} below a depth of 100m is less than 0.1cm^2/s. This, however, does not mean that a constant coefficient of this order of magnitude should be used at all depths. As will be shown below, the stability dependent parameterization of diffusion is particularly important for a realistic modelling of the CIL.

Horizontally averaged summer profiles of temperature, salinity and the coefficient of vertical diffusivity after the second stage of experiments B and C are shown in figures 6 and 7. Obviously here again there is a good qualitative agreement between the profiles of temperature and salinity for the two experiments. But still it is true that the surface salinity in experiment B is higher than in C and it is too high in comparison to experimental data. This

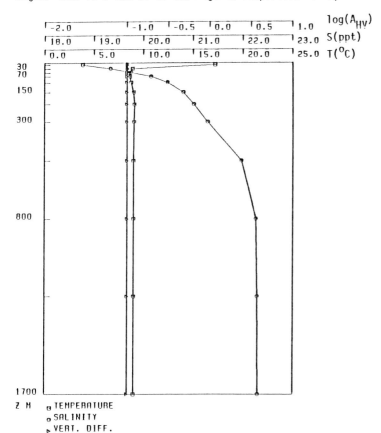

Fig. 6 Horizontally averaged summer vertical profiles of salinity, temperature and of coefficient of vertical diffusion, - second stage of experiment B.

unrealistically high surface salinity occurs again predominantly in the central

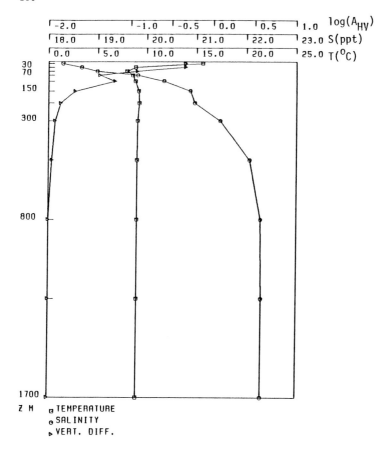

Fig. 7 As in Fig. 6 a but for the second stage of experiment C.

Black Sea. Some differences between the profiles exist in the CIL. This layer is more pronounced and in better agreement with observations (compare figure 1) for experiment C. With the diffusivity constants of this experiment the vertical diffusivity profile shows an intermediate minimum with a value of only 0.02cm²/s just in the core of the CIL.

The small subsurface temperature maximum at a depth of about 200m modelled by both experiments is unrealistic. It is the result of some deep penetration of warm water from the surface layer in a very small area close to the Bosphorus Strait, and perhaps this is due to small errors in the boundary conditions. As can be seen from figures 6 and 7 in comparison to figures 4 and 5 this feature also affects the salinity profiles. With the second stage of integration of experiments B and C the shape of the permanent halocline has changed considerably. The changes, however, are larger and develop more rapidly in experiment B. Notice that the profiles for experiment B refer to an integration period of only 5 years with heat flux boundary conditions.

The penetration of the seasonal temperature signal into the main halocline produces some erosion of the salinity profile especially in experiment B. Convection is parameterized in the model by convective adjustment which mixes completely and instantaneously the heat and the salt between statically unstable layers.

The different role of vertical diffusivity in establishing the thermohaline stratification in experiments B and C can also be demonstrated with the zonal temperature sections shown in figure 8a and 8b. With values for the coeffi-

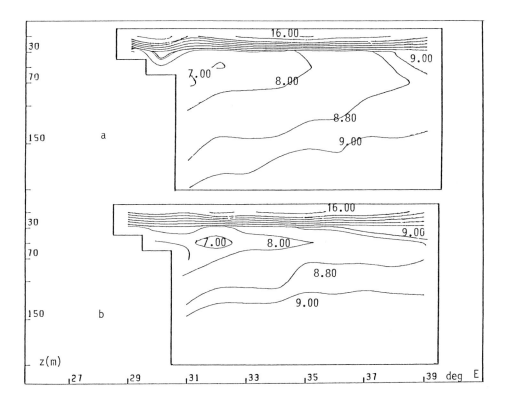

Fig. 8 Vertical cross-sections of temperature along 44° N. (a) experiment B, (b) experiment C.

cients α and K_b as given in Pacanowski and Philander (1981) (figure 8a) the seasonal temperature signal penetrates deeper, and the lower boundary of the CIL settles about 50m deeper than in experiment C with the adjusted values for these coefficients. This is also due to less vertical stability in experiment B leading to deeper convective penetration of cold water from the cooling sources.

The simulated temperature in the core of the CIL is in general agreement with experimental data. However, the data indicate (see Blatov et al. 1984), that the 8°C-isotherm could be used to identify the upper and lower boundaries of the CIL throughout the section. In the model results it is rather the 8.8°-isotherm which could serve for such a purpose. This shows that some additional decoupling of this layer from the warmer layers above and below is necessary to improve the model. Also, perhaps the rate of spreading of cold water within the CIL depends critically on advection, which may not be simulated realistically by our coarse resolution model. Here we do not discuss the contribution of advection to the maintenance of the CIL in our experiments, but we expect that for instance higher horizontal resolution will intensify the circulation in the model of the Black Sea, and will thus provide a better supply of the easternmost part of the sea with cold water.

4 REFERENCES

Altman, E.N. and Kumish, H.I., 1986. Interannual and seasonal variability of the fresh water balance in the Black Sea. Trudy GOIN, Moscow, 176, 3-8. (Russian)

Blatov, A.S., Bulgakov, N.P., Ivanov, Y.A. and Kosarev, A.N., 1984. Variability of the hydrophysical fields in the Black Sea, Hydrometeoizdat, Leningrad, 240 pp. (Russian).

Bogdanova, A.K., 1972. Seasonal and interannual variations of the water-exchange through the Strait of Bosphorus. In Marine Biology, Izd, Naukova dumka, Kiev, 41-54. (Russian).

Bryan, K., 1969. A numerical method for the study of the circulation of the World Ocean. Journal of Computational Physics, v. 3, 3, 347-378.

Bulgakov, S.N. and Korotaev, G.K., 1984, Possible mechanism of water stationary circulation in the Black Sea. Complex research in the Black Sea, Sevastopol, 32-40. (Russian).

Climatic Handbook of the Black Sea, 1974, Hydrometeoizdat, Moscow, 406 p. (Russian).

Filippov, D.M., 1968. Circulation and structure of waters in the Black Sea. Nauka, Moscow, 136 pp. (Russian).

Hellerman, S. and Rosenstein, M., 1983. Normal monthly wind stress over the World Ocean with error estimates. Journal of Physical Oceanography, 13, 1093-1104.

Kolesnikov, A.G. and Boguslavski, S.G., (1978). Vertical transport in the Black Sea. Marine Hydrophysical Investigations, N2. Sevastopol. (Russian).

Makerov, Yu.V., (1961). Heat balance of the Black Sea. Trudy GOIN, Moscow, 61, 169-183. (Russian).

Pacanovski, R.C. and Philander, S.G.H., 1981. Parameterization of vertical mixing in numerical models of tropical oceans. Journal of Physical Oceanography, 11, 1443-1451.

Semtner, A.J., 1974. An oceanic general circulation model with bottom topography. Numerical simulation of weather and climate. Technical Report, N9. UCLA, Los Angeles, 99p.

Shaffer, G., 1986. On the upwelling circulation over the wide shelf off Peru: 2. Vertical velocities, internal mixing and heat balance. Journal of Marine Research, 44, 2, 227-266.

Sorokin, Yu. I., 1982. The Black Sea. Nauka, Moskow, 217pp. (Russian).

Stanev, E.V., Troukchev, D.I. and Roussenov, V.M., 1987. The Black Sea circulation and numerical modelling of the Black Sea currents. Sofia University Press, Sofia, (submitted). (Russian)

Zaz, V.I. and Rosmann, L.D., 1979. The structure of turbulent mixing in the central part of the Black Sea. Complex studies of the Black Sea, Marine Hydrophysical Institut, Sevastopol, 71-76. (Russian).

LONG-TERM SIMULATIONS OF UPPER OCEAN VERTICAL MIXING USING MODELS OF DIFFERENT
TYPES

Ph. GASPAR and Y. GREGORIS
Centre National de Recherches Météorologiques (DMN/EERM)
31057 Toulouse Cedex (France)
R. STULL
Department of Meteorology, University of Wisconsin
Madison WI 53706 (USA)
Ch. BOISSIER
Service Hydrographique et Océanographique de la Marine
75200 Paris (France)

ABSTRACT

Different models for vertical mixing of the upper ocean are now available.
The integral, local turbulence closure and transilient models are briefly
reviewed. Models of these three types are then tested for multiannual
simulations of the upper ocean at weathership Papa. The simulated sea surface
temperatures and vertical temperature profiles are analyzed and compared. The
computational efficiency of the different models is also investigated.

1 INTRODUCTION

Vertical mixing in the upper ocean plays a major role in modulating all
exchanges between the atmosphere and the underlying ocean. Models simulating
this mixing are essentially one-dimensional boundary layer models. They are
thus clearly identified as process models that can be embedded into more
comprehensive oceanic models so as to yield improved simulations of the upper
ocean mixing (e.g. Schopf and Cane, 1983 ; Miyakoda and Rosati, 1984).

Among the mixing models available, three types will be studied here :
integral models, local turbulence closure models and transilient models.
Because of their different original conceptions, these models clearly have
specific merits and faults. Martin (1985) first compared several integral and
turbulence closure models by simulating one annual cycle of the upper ocean at
weathership stations November and Papa. Martin (1986) added to this work a
comparison on shorter term simulations for which more detailed observations
were available. Stull and Kraus (1987) first applied the transilient
turbulence theory to the ocean. They compared the results of the transilient
model with those of an integral and a turbulence closure model. The comparison
was restricted to short term simulations of a few hours to a few days.

In the present paper, results from the three different model types are compared for multiannual simulations of the upper ocean evolution at station P. Similar simulations were performed by Gaspar (1987) to test several integral models. Such long-term integrations provide a severe test under a wide range of surface conditions. In particular, they reveal if the simulation errors tend to grow or disappear from year to year or if some errors tend to repeat at the same period of the year.

In the next section, the basic principles of the three types of models are briefly presented. For more details, in addition to the papers cited in the text, the reader is referred to Niiler and Kraus (1977) for the integral models, Mellor and Yamada (1982) for the local turbulence closure models and Stull (1984) for the transilient models. In the presentation and in the subsequent comparison of the models, the emphasis lays on the temperature simulation because :

(i) it is the main result to be analyzed in long-term simulations that are essentially of climatic interest.

(ii) long time series are more readily available for the upper ocean temperature profiles than for any other variable. Consequently, the analysis of the simulated temperatures should be the most enriching.

2 DESCRIPTION OF THE MODELS

In the one-dimensional case, the heat equation assumes the form :

$$\frac{\partial \bar{T}}{\partial t} = \frac{F_{sol}}{\rho_o c_p} \frac{\partial I}{\partial z} + \frac{\partial}{\partial z} \left(K_a \frac{\partial \bar{T}}{\partial z} \right) - \frac{\partial}{\partial z} \overline{T'w'} \tag{1}$$

where T is the temperature, w the vertical velocity, t the time, z the vertical coordinate, ρ_o and c_p the reference volumic mass and specific heat of sea water, F_{sol} the absorbed solar irradiance, I(z) the fraction of F_{sol} that penetrates to the depth z and K_a the coefficient for ambient diffusion. In the upper mixed layer (ML), the effect of diffusion is overwhelmed by an intense turbulent mixing. Diffusion regains importance in the underlying more quiescent thermocline. In (1), the distinction can be made between the "non turbulent change" of temperature, i.e. the change caused by the absorption of the solar irradiance and the ambient diffusion, and the "turbulent change" caused by the divergence of the vertical turbulent heat flux. The determination of this turbulent change is a critical part of any vertical mixing model.

2.1 Integral models

Basic to the integral models is the assumption that the upper oceanic boundary layer is mixed to homogeneity by turbulence. Therefore, the evolution of all properties in this ML can be predicted using equations integrated over the ML depth (h), and the turbulent fluxes have to be determined only at the base of the ML. The integration of (1) yields :

$$h \frac{\partial T_m}{\partial t} = \frac{F_{sol}}{\rho_o c_p} [1-I(-h)] - \overline{T'w'}(0) - K_a \frac{\partial \overline{T}}{\partial z}(-h) + \overline{T'w'}(-h) \qquad (2)$$

where T_m is the ML temperature. The turbulent surface heat flux is a forcing term given by the sum of the latent, sensible and net infra-red heat fluxes. The flux at z = -h is determined from the so-called jump equation :

$$-\overline{T'w'}(-h) = w_e \, \Delta T \qquad (3)$$

where ΔT denotes the temperature jump at the base of the ML and w_e is the entrainment velocity ($w_e = \partial h/\partial t$ if $\partial h/\partial t > 0$, otherwise $w_e=0$). This entrainment velocity is classically obtained from a parameterized form of the turbulent kinetic energy (TKE) budget of the ML. Since the early works of Ball (1960) and Kraus and Turner (1967) several expressions of this budget have been proposed (for a review see Zilitinkevich et al., 1979). Different numerical simulations presented in section 3 use either the formulation of Niiler and Kraus (1977) or that of Gaspar (1987).

Because of the temperature discontinuity at z=-h, the diffusive heat flux at the base of the ML is estimated as follows :

$$-K_a \frac{\partial \overline{T}}{\partial z}(-h) = -K_a \frac{\Delta T}{\delta} \qquad (4)$$

where δ represents a "pseudo-thickness" of the transition zone between the ML and the underlying thermocline.

Below the ML, the turbulence is assumed to vanish so that the evolution of the temperature is predicted using equation (1), neglecting the divergence of the turbulent heat flux.

2.2 Local turbulence closure models

a. K-theory

The most usual closure is based on the eddy-diffusion concept, or K-theory, that relates the turbulent fluxes to the local gradient of the mean

172

variables. For example,

$$-\overline{T'w'} = K_t \ \partial \overline{T}/\partial z \tag{5}$$

where K_t is the eddy-diffusivity of heat. Ekman (1905) first used constant values of the eddy-diffusivities. Later on, more elaborate expressions for these coefficients were proposed (e.g. Munk and Anderson, 1948). An appealing, and widely used, possibility is to relate the eddy-diffusivities to the TKE, e.g.

$$K_t = a \ l_k \ \overline{e}^{\ 1/2} \tag{6}$$

where a is a coefficient to be determined, l_k a mixing length and $\overline{e}$ the TKE. The models using this extension of the original K-theory have to solve an equation for the TKE. They are therefore called KE-models or 1.5-order closure models.

Numerical experiments are performed in this paper using the KE-model of Therry and Lacarrère (1983). This model, first devised for atmospheric simulations, was adapted to the ocean by André and Lacarrère (1985). The mixing length used here is slightly different from that originally proposed by the authors. It assumes the general form :

$$l_k = \text{Min} \ (l_B, l_s) \tag{7}$$

where l_B is the Blackadar's length (see e.g. André and Lacarrère, 1985) and $l_s = \overline{e}^{\ 1/2}/N$, N being the Brunt-Vaisala frequency. This last length is defined to be the vertical distance travelled by a fluid particle in converting all of its TKE into potential energy in a stably stratified environment. It is thus an upper limit for the vertical size of the largest turbulent eddies in the presence of stratification. In the stratified region below the boundary layer, l_s is always smaller than l_B so that (7) reduces to $l_k = l_s$.

Accordingly, (6) becomes

$$K_t = a \ \overline{e} \ N^{-1} \tag{8}$$

A similar N^{-1} dependency of the diffusivity in the ocean interior was deduced by Gargett and Holloway (1984) from arguments on the diffusion by internal wave breaking. Using (8) and imposing that $\overline{e}$ does not fall below a minimal prescribed value, an adequate N^{-1} dependent vertical diffusivity can be maintained below the turbulent ML.

b.High-order turbulence closures

Rather than using the K-theory, some models compute the second-order moments from their rate eqations. For the vertical turbulent heat flux the rate equations is :

$$\frac{\partial \overline{T'w'}}{\partial t} = -\frac{\partial}{\partial z} \overline{T'w'^2} - \overline{w'^2}\frac{\partial \overline{T}}{\partial z} - \alpha g \overline{T'^2} - \frac{1}{\rho_0}\overline{T'\frac{\partial p'}{\partial z}} \qquad (9)$$

where α is the thermal expansion coefficient of sea water, g the gravity and p the pressure. Depending on the closure used, all the second- and third-order moments on the right-hand side of (9) can be either parameterized as a function of lower-order moments or determined from their own rate equations. For example, the third-order turbulence closure model of André and Lacarrère (1985) has some thirty rate equations for the second- and third-order moments, the basic variables being the temperature and the two horizontal components of the velocity vector. A limited numerical experiment has been performed with this model (see section 3.4).

2.3 Transilient models

The concept of transilient turbulence has recently been developed by Stull (1984) and applied to oceanic simulations by Stull and Kraus (1987). This non-local turbulence closure model allows eddies of different sizes to transport fluid over a range of different distances. More formally the water column is splitted into n equally-spaced grid boxes and $T_i(t)$ is used to denote the average value of T within grid box i at time t. If C_{ij} represents the fraction of fluid going from box j to box i in a time step Δt, then the turbulent change of T_i in one time step is given by

$$T_i(t+\Delta t) = \sum_{j=1}^{n} C_{ij} T_j(t) \qquad (10)$$

The conservation laws of mass and heat require respectively

$$\sum_{j=1}^{n} C_{ij} = 1 \qquad \text{for each i} \qquad (11)$$

$$\sum_{i=1}^{n} C_{ij} = 1 \qquad \text{for each j} \qquad (12)$$

Different parameterizations of the transilient coefficients C_{ij} have been proposed. All use the exchange hypothesis : $C_{ij}=C_{ji}$. The first solution proposed by Stull (1984) was based on critical Richardson number arguments. However Stull and Driedonks (1987) pointed out that the so-defined coefficients were ill-defined or undefined when the Richardson number was approaching infinity in near-zero shear. They thus proposed a new formulation which we used for oceanic simulations. Stull and Driedonks (1987) first define Y_{ij}, a "potential" for mixing between i and j :

$$Y_{ij} = \frac{T_o \, \Delta t}{\Delta z_{ij}^2} \, [\, \Delta u_{ij}^{-2} + \Delta v_{ij}^{-2} - \frac{\Delta b_{ij}}{R_c} \Delta z_{ij}] - \frac{D \, \Delta t}{T_o} \qquad \text{for } i \neq j \qquad (13)$$

$$Y_{ii} = \text{Max} \, (Y_{i,i-1} ; \, Y_{i,i+1}) + Y_{ref} \qquad (14)$$

where $a_{ij} = a_j - a_i$ for any available a, (u,v) is the horizontal velocity vector, b the buoyancy, T_o a characteristic time scale of turbulence, D a dimensionless dissipation factor, R_c is analoguous to a critical Richardson number and Y_{ref} is a reference potential that accounts for internal mixing within each grid box. Stull and Driedonks (1987) choose $R_c = 0.21$ and argue that D should be on the order of one. Based on sensitivity experiments they recommend : $100s \leqslant T_o < 1000s$ and $Y_{ref} = O(10^2 \ldots 10^3)$. We used $R_c = 0.21$, $D = 1$, $T_o = 1000s$ and $Y_{ref} = 100$.

The relation (13) can be viewed as a nonlocal simplified form of the TKE equation. The potential for mixing between i and j is proportional to the shear production of TKE between i and j and to the buoyancy difference between these two boxes. There is also a dissipation factor that always reduces the potential for mixing. The transilient coefficients are finally obtained as normalized values of the potentials :

$$c_{ij} = Y_{ij} \, / \, ||Y|| \qquad i \neq j \qquad (15)$$

$$c_{ii} = 1 - \sum_{\substack{j=1 \\ j \neq i}}^{n} c_{ij} \qquad (16)$$

where $||Y|| = \text{Max}_i \, (\sum_{j=1}^{n} Y_{ij})$

3 COMPARISON OF THE MODELS

The integral models of Niiler and Kraus (1977) and Gaspar (1987), the KE-model of Therry and Lacarrère (1983) and the transilient model of Stull and Driedonks (1987) were used to simulate the evolution of the upper ocean at station P (50°N, 145°W) during the years 1969 and 1970. These models will

hereafter be referred to as NK, GA, TL and SD respectively. The station P was chosen for testing the models because :
- the advective effects are known to be weak at this location ; in particular, Gaspar (1987) showed that the changes of the upper ocean heat content at station P are well accounted for by the surface heat fluxes during the two years studied ;
- extended data sets are available, including surface radiation measurements and frequent bathythermographic soundings.

3.1 Numerical characteristics of the simulations

Starting from the vertical temperature profile given by the bathythermographic data on the first of January 1969, all models were integrated over two years using the three-hourly values of the surface forcing (wind stress, solar irradiance, latent, sensible and infra-red heat fluxes) obtained by Tricot (1985) from the standard meteo-oceanographic data of station P (including pyranometer data for F_{sol}). Gaspar (1987) also uses this data set and gives more details about it. As Martin (1985) showed, the surface salinity fluxes at station P can only have a small effect on the vertical mixing. These fluxes were therefore neglected and a time-independent climatological salinity profile (Tabata, 1964) was used. The absorption law I(z) was parameterized according to Paulson and Simpson (1977) for a Jerlov's (1976) optical water type II.

In the TL and SD models, the ambient diffusion coefficient was held at zero. In the TL model, the prescribed minimal value of $\overline{e}$ (see Eq.8) was so small that K_t was virtually equal to zero below the ML. As a consequence, the TL and SD model had no diffusion outside the turbulent region. Accordingly, the results of the integral models discussed in the following section 3.2 were obtained with $K_a=0$.

The integral models were integrated using a time step of 3 hours. The KE and transilient models had a time step of 15 minutes. All models had the same uniform vertical resolution of 5 m from the surface down to 150 m.

3.2 Results for the sea surface temperature (SST)

The difference between the simulated and observed monthly mean SST (ΔSST) is plotted on Fig. 1 for the GA, TL and SD models. The error on the SST predicted by the GA model is always smaller than 0.5 K but has a well-marked annual cycle : the SST is systematically overestimated in summer and underestimated in autumn. The error of the NK model (not shown here) has the same cycle but with an amplitude that is approximately doubled. This sort of

176

error is typical of most integral models (see e.g. Gaspar, 1987). It is due to an underestimation of the summer ML depth and a too rapid autumnal erosion of the thermocline.

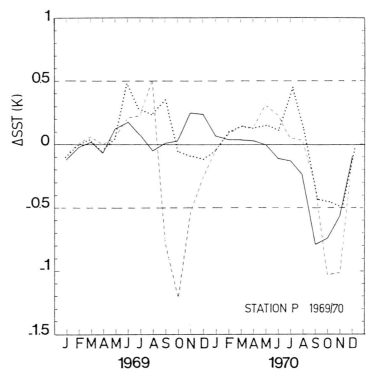

Fig.1. Difference between the simulated and observed monthly mean SST for the GA (dotted line), TL (solid line) and SD (dashed line) models. Station P, years 1969 and 1970.

The TL model performs very well during most of the integration. However, Δ SST reaches an extremum of -0.8K in August 1970. The error does not seem to have a cyclical behaviour.

The error of the SD model has an annual cycle similar to that of the integral model but with a larger amplitude. The autumnal underestimation of the SST is particularly marked, with SST reaching an extremum of -1.2K in October 1969. The reason for this error is discussed in the following section 3.3. It should also be pointed out that only a few sensitivity experiments were performed to determine the "best" values of the parameters T_0 and Y_{ref} while R_c and D were fixed at 0.21 and 1 respectively. We selected the calibration for which the maximal value of SST was the smallest.

Additional tuning of the parameters would probably lead to a reduced error though we noticed that for all tested combinations of (T_0, Y_{ref}), the error on the simulated SST always had the same type of annual cycle.

3.3 Results for the thermal structure

The observed evolution of the upper ocean thermal structure at station P during the year 1969 is presented in Fig. 2.

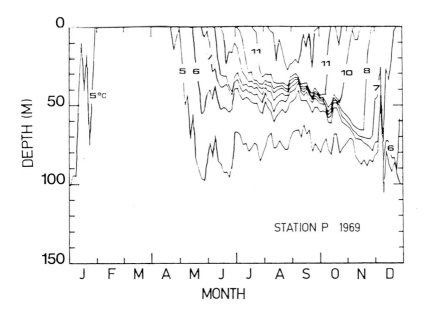

Fig.2 Isotherm depths at station P for the year 1969 determined from 3-day mean BT observations.

From January to April, there is almost no stratification. Temperature gradients appear in May. They remain quite weak until the end of spring. Then, the summer thermocline becomes well-marked but it spreads over a depth of 15 to 30 m. The isotherms really squeeze-up only during autumn when the upper ones are pushed down by the deepening ML. This typical behaviour repeats from year to year.

The evolution of the thermal structure (station P, year 1969) simulated by the integral NK model, with no ambient diffusion below the ML, is represented in Fig. 3.

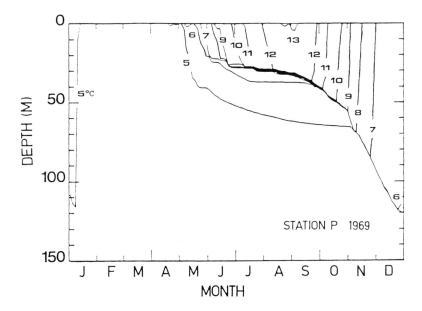

Fig. 3 Isotherm depths at station P for the year 1969 simulated with the NK model. No ambient diffusion below the ML $(K_a=0)$.

The stratification effectively shows up in May but then the isotherms stay packed at the base of the ML for the rest of the year. This is simply a picture of the basic assumptions used to build any integral model : one has a perfectly mixed upper layer with a discrete temperature jump at its base. In the absence of diffusion below the ML, this temperature gradient cannot spread.

To investigate the effects of heat diffusion within the thermocline, two additional integrations were performed with $K_a=10^{-5}m^2s^{-1}$ and $K_a=510^{-5}m^2s^{-1}$ respectively. A fixed value of $\delta=10m$ was used. For $K_a=10^{-5}m^2s^{-1}$ the spacing between the isotherms just below the ML seems to be quite correct, though slightly insufficient (Fig. 4). For $K_a=510^{-5}m^2s^{-1}$, the downward diffusive heat flux appears to be overestimated and causes an exaggerated deepening of the lower isotherms (Fig. 5).

Obviously, more elaborate diffusion schemes could have been used instead of a constant diffusivity. However, it does not seem advisable that the simulation of the thermocline be refined while a simple bulk model of the ML is used. Furthermore, the introduction of ambient heat diffusion does not improve the simulation of the SST. Indeed, the SST is almost unaffected by heat diffusion between January and May, as the thermocline stratification is very weak throughout this period. The downward diffusive heat flux becomes important in June but the maximum resulting decrease of the SST is obtained only in September or October. For $K_a=10^{-5}m^2s^{-1}$, the September mean SST is decreased by 0.6 K relative to the case with $K_a=0$. As a consequence, the simulation of the ambient heat diffusion below the ML essentially reinforces the systematic underestimation of the SST in autumn while it only slightly reduces the overestimation of the summer SST (Gaspar, 1985).

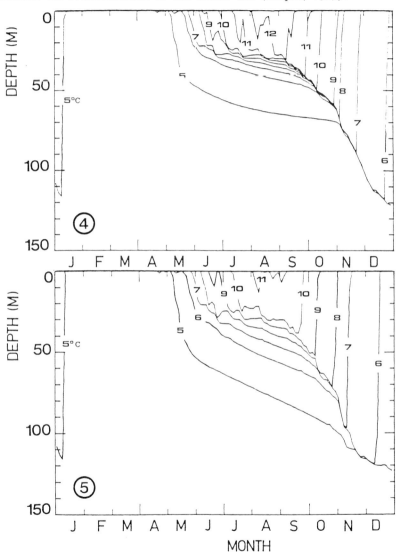

Figs. 4 and 5. As in Fig.3 but for $K_a=10^{-5}m^2s^{-1}$ and $K_a=5 \ 10^{-5}m^2s^{-1}$ respectively

The TL model produces too large temperature gradients below the summer ML but gives a realistic simulation of the temperature profiles during autumn (Fig. 6). Works are still in progress on the formulation of the mixing length in order to improve the simulation of the transition zone between the turbulent boundary layer and the underlying thermocline.

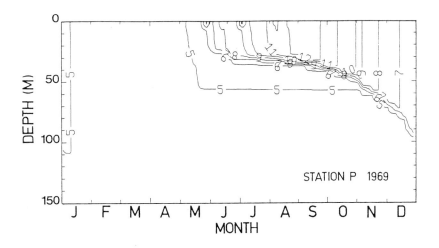

Fig.6 Isotherm depths at station P for the year 1969 simulated with the TL model.

The SD model also has a too strong summer thermocline (Fig. 7). In autumn, it simulates a spreading of the isotherms contrary to what is observed. This spreading is due to some turbulent mixing reaching the thermocline itself from September forward. The 5°C-isotherm is already affected at the end of September. This mixing entrains important masses of cool water from below into the upper layers. Consequently, the SST decreases too rapidly as noticed in the previous section. This problem is probably related to errors in the velocity field forecast by the model. Indeed, in the SD approach, the amount of turbulent mixing is strongly coupled to both the mean velocity and temperature fields. Errors or insufficient data regarding the momentum forcing result in prognostic errors of the current which in turn affect the turbulence which alters the temperature forecast. To reduce this problem, more work is certainly needed on the parameterization of the transilient coefficients. More data concerning the momentum forcing and momentum advection should also be helpful.

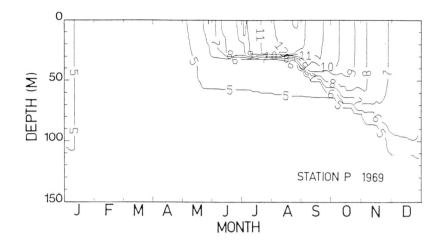

Fig. 7 Isotherm depths at station P for the year 1969 simulated with the SD model.

3.4 Computational efficiency of the models

The central processor (CP) time used by the different models for one time step (CPS) and for a one-day simulation (CPD) are listed in Table 1.

TABLE 1

CP-time used by the models for one time step (CPS) and a one-day simulation (CPD) on a CYBER 175 computer. All models have 30 discrete levels along the vertical.

Model	Time step	CPS (10^{-3}s)	CPD (10^{-3}s)
Integral (GA)	3 hr	1.6	13.
Kinetic Energy (TL)	15 min	4.7	451.
Transilient (SD)	15 min	22.8	2189.
3rd order closure (AL)	4 min	17.6	6336.

The tested version of the GA model includes an explicit finite-difference solution of the heat diffusion equation below the ML. Results from the third-order turbulence closure model of André and Lacarrère (1985) (AL) have been added to the comparison. These results were obtained from a short integration of a thousand time steps of 4 minutes each, using the same vertical resolution as that used with the other models. All simulations were performed on a CYBER 175 (nonvectorizing) computer.

With the shortest CPS and the largest time step, the integral model is clearly the fastest. The CPS of the KE model is only 3 times larger than that of the integral model. However, its time step is 12 times shorter so that the KE-model is about 40 times slower than the integral for a one-day simulation. The transilient and third-order closure models have an almost equally large CPS. For a one-day simulation, the SD and AL models are respectively 170 and 490 times slower than the integral. Finally, notice that for a vertical resolution of n discrete levels, the SD model has to compute $n^2/2$ transilient coefficients. Accordingly, for a fixed value of the time step, the CPS of SD grows like n^2. Since the CPS of all other tested models grows like n, the transilient model is relatively less efficient for higher resolutions.

4. SUMMARY AND CONCLUSIONS

Among all tested models, none is definitely superior to the other ones with regard to the simulation of the SST. With a minimal demand on CP-time, the GA integral model maintains the monthly mean simulated SST within 0.5 K. of the observations. However, the upper ocean thermal structure simulated by any integral model is only schematic. Such models are thus particularly useful for climatic studies in which an efficient simulation of the SST is of major interest while the shape of the underlying temperature profile does not matter very much.

The KE-model of TL generally produces a good simulation of the SST. It also predicts quite well the evolution of the vertical temperature profile though this could still be improved, specially in the summer thermocline. Because of its reasonable computational efficiency, this model can be integrated over long periods at a relatively low cost.

The simulation with the SD model reported here is the first oceanic long term integration performed using the transilient turbulence theory. These first results, though fairly realistic, indicate that more work is needed to improve the parameterization of the transilient coefficients. At present, the SD model has relatively high computational requirements. However, the code could significantly be speeded up by vectorization since ample use is made of matrix calculus. On the other hand, the transilient model is certainly handicapped at high resolutions as the number of transilient coefficients to compute is proportional to the number of discrete levels squared. To obtain reasonable results with a relatively modest vertical resolution, the transilient scheme should be adapted to allow for an unequal grid spacing. This has recently been done by Kraus (1987).

ACKNOWLEDMENTS

The authors wish to thank J.C. André and P. Lacarrère for helpful comments and suggestions on this study. For part of this work, Ph. Gaspar was supported by the Belgian National Fund for Scientific Research. This support is gratefully acknowledged.

5 REFERENCES

André, J.C., and P. Lacarrère, 1985 : Mean and turbulent structures of the oceanic surface layer as determined from one-dimensional, third-order simulations. J. Phys. Oceanogr., 15, 121-132.
Ball, F.K., 1960 : Control of inversion height by surface heating. Quart J. Roy. Meteorol. Soc., 86, 483-494.
Ekman, V.W., 1905 : On the influence of earth's rotation on ocean currents. Ark. Mat. Astr. Fys., 2, 1-52.
Gargett, A.E., and G. Holloway, 1984 : Dissipation and diffusion by internal wave breaking. J. Mar. Res., 42, 15-27.
Gaspar, Ph., 1985 : Modèles de la couche active de l'océan pour des simulations climatiques. Ph. D. Thesis, Catholic University of Louvain-La-Neuve, Belgium. 187 pp.
Gaspar, Ph., 1987 : Modelling the seasonal cycle of the upper ocean. Submitted to J. Phys. Oceanogr.
Jerlov, N.G., 1976 : Marine Optics. Elsevier Oceanography series, 14, 231 pp.
Kraus, E.B., and J.S. Turner, 1967 : A one-dimensional model of the seasonal thermocline. II. The general theory and its consequences. Tellus, 19, 98-105.
Kraus, E.B., 1987 : The structure of the ocean surface layers near the equator (this volume).
Martin, P.J., 1985 : Simulation of the mixed layer at OWS November and Papa with several models. J. Geophys. Res., 90, 903-916.
Martin, P.J., 1986 : Testing and comparison of several mixed layer models. NORDA Report 143, 34 pp.
Mellor, G.L., and T. Yamada, 1982 : Development of a turbulence closure model for geophysical fluid problems. Rev. Geophys. Space. Phys., 20, 851-875.
Miyakoda, K., and A. Rosati, 1984 : The variation of sea surface temperature in 1976-1977. 2. The simulation with mixed layer models. J. Geophys. Res., 89, 6533-6542.
Munk, W.H., and E.R. Anderson, 1948 : Notes on a theory of the thermocline. J. Mar. Res., 7, 276-295.
Niiler, P.P., and E.B. Kraus, 1977 : One-dimensional models of the upper ocean. Modelling and prediction of the upper layers of the ocean. E.B. Kraus (Ed.), Pergamon Press, 143-172.
Paulson, C.A., and J.J. Simpson, 1977 : Irradiance measurements in the upper ocean. J. Phys. Oceanogr., 7, 952-956.
Schopf, P.S., and M.A. Cane, 1983 : On equatorial dynamics, mixed layer physics and sea surface temperature. J. Phys. Oceanogr., 13, 917-935.
Stull, R.B., 1984 : Transilient turbulence theory. Part I : The concept of eddy-mixing across finite distance. J. Atmos. Sci., 41, 3351-3367.
Stull, R.B., and A.G.M. Driedonks, 1987 : Applications of the transilient turbulence parameterization to atmospheric boundary layer simulations. Bound. Layer Meteorol. In press.
Stull, R.B., and E.B. Kraus, 1987 : The transilient model of the upper ocean. J. Geophys. Res. (Oceans). In press.

Tabata, S., 1984 : A study of the main physical factors governing the oceanographic conditions at ocean station P in the Northeast Pacific Ocean. Ph. D. Thesis, University of Tokyo, 264 pp.

Therry, G., and P. Lacarrère, 1983 : Improving the eddy kinetic energy model for planetary boundary layer description. Bound. Layer Meteorol., 25, 63-88.

Tricot, Ch., 1985 : Estimation des flux de chaleur en surface à la station météo-océanographique Papa. Scientific Report 1985/9. Institute of Astronomy and Geophysics G. Lemaître, Catholic University of Louvain-La-Neuve, Belgium.

Zilitinkevich, S.S., D.V. Chalikov and Yu. D. Resnyanskiy, 1979 : Modelling the oceanic upper layer. Oceanol. Acta, 2, 219-240.

SOME DYNAMICAL AND STATISTICAL PROPERTIES OF EQUATORIAL TURBULENCE

H. PETERS and M. C. GREGG

Applied Physics Laboratory and School of Oceanography, College of Ocean and Fishery Sciences, University of Washington, Seattle, WA 98105 (USA)

ABSTRACT

In November 1984 the dissipation rate, ε, on the equator in the central Pacific varied by a factor of 100 in a diurnal cycle extending much deeper than the surface mixed layer. Vertically coherent bursts of strong dissipation, occurring most often during nighttime convective deepening of the mixed layer, were the major mechanism causing the diurnal cycle. Approximate statistical confidence limits for $\overline{\varepsilon(p)}$ in this region are a factor of 3. Dissipation rates were much lower in the core, owing to the low shear, and the confidence limit decreased to a factor of 1.6. In the shear zone below the core, the mixing was dominated by too few events to establish confidence limits. Because ε departed significantly from lognormality over most of the profile, lognormal statistics cannot be used with confidence to predict the likelihood that the sampling failed to sample adequately the rare intense events that dominate the average. Rather, the physics of the diurnal cycle and of the vertically coherent bursts must first be determined and related to forcing mechanisms.

Equatorial turbulence exhibited a vast range of dynamic states not paralleled in mid-latitude oceans, nor in tank or numerical experiments: the "activity" parameter $\varepsilon/\nu N^2$ varied over 5½ decades. The mixing efficiency, γ_{mix}, the ratio of the buoyancy flux to ε, could be estimated only with low accuracy. As a function of $\varepsilon/\nu N^2$, γ_{mix} nevertheless showed no systematic trend. This result also constrains the flux Richardson number; its mean was close to the critical value.

1 INTRODUCTION

Year around, the equatorial undercurrent is a fascinating, ocean-scale laboratory for stratified turbulence in a shear flow. Unlike most of the open ocean, the upper equatorial waters have steady, large vertical shears and low Richardson numbers. Although intense turbulence was invariably found on previous microstructure cruises, the complexity of processes associated with the turbulence was not revealed until November 1984, when intense surveys increased the total number of equatorial microstructure samples by nearly two orders of magnitude. The 1984 cruises of RV *Thompson*, carrying the authors, and Oregon State's *Wecoma* were part of the Tropic Heat experiment, which was designed to study the upper ocean momentum, heat and mass budgets in the equatorial central Pacific (Eriksen, 1985).

Given this background, analysis has been focused on the patterns of variability, on the diapycnal turbulent fluxes of momentum and heat, and on the parameterization of turbulent mixing (Gregg et al., 1985; Moum and Caldwell, 1985; Chereskin et al., 1986; Moum et al., 1986; Toole et al., 1987; Peters et al., 1988). The data, procedures, and uncertainties are described in the latter two references.

In this paper, we investigate two turbulent characteristics: statistics of the rate of kinetic energy dissipation, ε, (Section 3), and the mixing efficiency (Section 4). The equator showed strong, quasi-deterministic variations of ε in time and in space. We show that these can be determined with adequate precision. Peculiarities of the distributions of ε can be related to its systematic changes, demonstrating uniquely the connection between the physics of mixing and the statistics of turbulence parameters. As the ratio of the buoyancy flux to the dissipation rate, the mixing efficiency is closely related to the flux Richardson number and is a key parameter of turbulence in stratified fluids. The equator is a unique regime for testing the dependence of γ_{mix} for a dependence on $\varepsilon/\nu N^2$. The latter, where ν is the kinematic viscosity, is used as an indicator for the "activity" of mixing (Stillinger et al., 1983; Rohr et al., 1987). According to laboratory experiments, the dissipation rate has to exceed the transition value $\varepsilon_{tr} = \hat{c} \, \nu N^2$ in order to produce the negative buoyancy flux necessary to change potential energy. (Values of $\hat{c}$ are 25 and 16, respectively, in the two papers.)

Before we proceed to the statistics of dissipation and the mixing efficiency, however, we need to describe the patterns found in stratification, shear, and turbulence.

2 SUMMARY OF DISSIPATION VARIABILITY

The most important aspect of the mean stratification and current profiles (Figs. 1a-c) is the very low mean gradient Richardson number, $Ri \equiv \overline{N^2} [(\partial \overline{u}/\partial z)^2 + (\partial \overline{v}/\partial z)^2]^{-1}$, which makes the undercurrent dynamically unique among open ocean currents; e.g., $Ri < 0.5$ above 0.85 MPa. ($\overline{N^2}$ is the square of the Brunt-Vaisala frequency, and $\overline{u}$ and $\overline{v}$ are the components of the mean horizontal current.) The dissipation rate of kinetic energy, ε, and the dissipation rate of temperature variance, χ, varied inversely with Ri (Fig. 1d,e), being high in the upper shear zone, low in the core, and moderate in the lower shear zone.

In the upper shear zone, the turbulent dissipation rate exhibited a diurnal cycle with hundredfold changes between nighttime highs and afternoon lows (Fig. 2c).

The nighttime heat loss of the ocean, causing convection in the surface mixed layer, corresponded to the high-ε phase. Surface forcing, however, does not explain the diurnal variability in the stratified zone, which was probably associated with high-frequency internal waves (Peters et al., 1988). The strong mixing reached to between 0.7 and 0.9 MPa, far into the stratified zone below the surface mixed layer (compare Figs. 2b and 2c). The daytime values of $\varepsilon/\nu N^2$ ranged from below 20 to above 200, indicating a marginally "active" regime, whereas the nighttime values exceeded 2×10^3 in the stratified region and 2×10^4 in the mixed layer (Fig. 3). Each night, two or three vertically coherent bursts, each lasting several hours, dominated the mixing. During the 4½ days of observations, about 15 bursts were observed. Although a few occurred in the day, their greater frequency at night was a major component of the diurnal cycle.

In the core, defined as the region where $Ri > 1$ (1.05–1.2 MPa), $\varepsilon/\nu N^2 < 20$ throughout our observations (Fig. 3). A few events of intense mixing occurred in the lower shear zone (1.2–1.5 MPa), but there were no periodic changes in ε and χ (Fig. 2c).

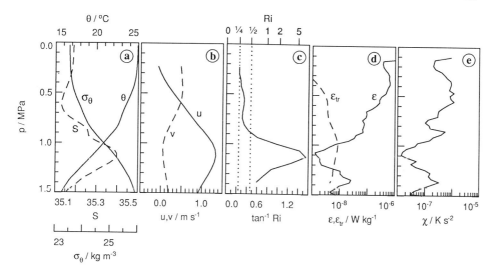

Fig. 1. Average profiles from the 4½-day time series on the equator. Pressure is in megapascals; 1 MPa = 100 decibar.

(a) Potential temperature θ, salinity S, and potential density σ_θ. Salinity showed a maximum near the undercurrent core, characteristic of the location.

(b) Mean velocity profiles from the acoustic Doppler data (u=east, v=north), shifted by a constant to match moored current meter data taken simultaneously 20 km distant (D. Halpern, personal communication, 1986). The undercurrent core was at 1.1 MPa. Above 0.3 MPa, the South Equatorial Current prevailed.

(c) Gradient Richardson number, Ri, computed from the mean N^2 and shear. The Richardson number was high in the core and low above and below. Note that $Ri < 0.5$ above 0.85 MPa. Such low values make the undercurrent unique among open-ocean currents.

(d) Mean kinetic energy dissipation rate ε (solid) and transitional dissipation rate $\varepsilon_{tr} = 25 \, v \, N^2$ (dashed; Stillinger et al., 1983). The dissipation rate was low in the core, similar in magnitude to average values found in the mid-latitude thermocline. Toward the surface, ε rose by 2½ decades and increased by 1 decade in the lower shear zone. Where $Ri > 1$, ε dropped below ε_{tr}.

(e) Mean temperature dissipation rate χ. The vertical structure was similar to that of ε.

3 STATISTICS OF DISSIPATION

Finding confidence limits for the systematic variations of ε with depth and with the diurnal cycle is complicated because all the statistical properties, not just the average values described in the previous section, showed complex variations. This was especially true above 0.7 MPa, where the distributions of ε deviated strongly from lognormality. As described above, the turbulence was nonstationary and inhomogeneous, and varied in well-defined patterns characteristic of the equatorial environment and different from conditions at mid-latitude (e.g., Gregg et al., 1986).

In this section, we first examine the ε patterns and distributions in an intuitive manner. After introducing maximum likelihood estimates, we discuss confidence limits based on lognormal statistics.

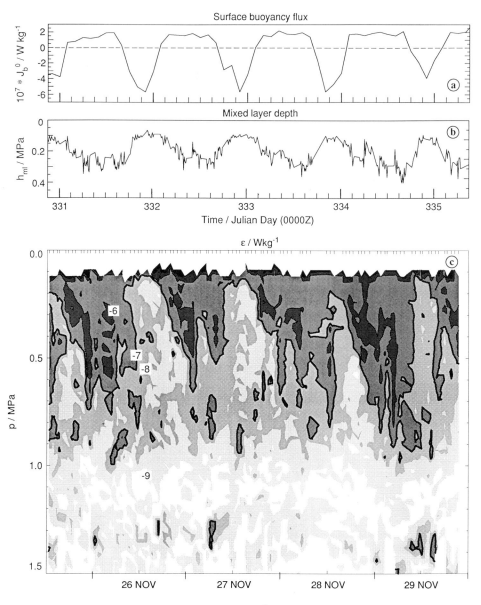

Fig. 2. (a) Time series of the surface buoyancy flux J_b^0, based on the 2-hourly routine meteorological observations using bulk formulas. The ocean gained heat during daytime ($J_b^0 < 0$) and lost heat at night ($J_b^0 > 0$). (b) Time series of the mixed layer depth h_{ml}, defined as the point where the surface density was exceeded by 0.01 kg m^{-3}. Nighttime deepening and daytime restratification corresponded to the changes in J_b^0. (c) Contours of log ε. Shades range from white, $\varepsilon < 10^{-9}$ W kg^{-1}, to black, $\varepsilon > 10^{-5}$ W kg^{-1}. The heavy 10^{-7} W kg^{-1} contour outlines the intense mixing in the "night" phase of the diurnal cycle. Vertically coherent, narrow fingers of high ε (bursts) appear to be an integral part of the diurnal cycle. In the core zone, between 1.05 and 1.2 MPa, dissipation rates mostly stayed below 10^{-9} W kg^{-1}, whereas a few intense mixing events occurred in the shear zone, below. Hourly averages of ε in 3-m depth bins were used for contouring.

$$\varepsilon / \nu N^2$$

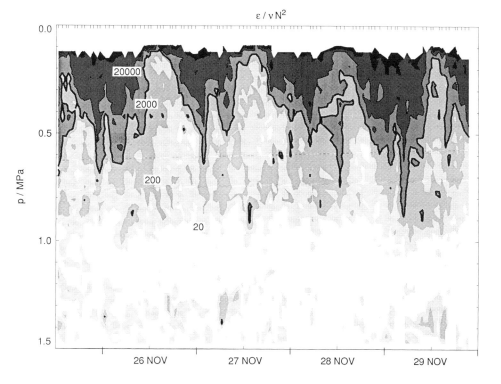

Fig. 3. Contours of log $\varepsilon/\nu N^2$ characterizing the activity of the turbulence. Shades range from white, $\varepsilon/\nu N^2 < 20$, to black, $\varepsilon/\nu N^2 > 2 \times 10^5$. The white area approximately coincides with ε dropping below ε_{tr}, and is thus considered "inactive." The range $20 < \varepsilon/\nu N^2 < 200$ is considered "marginally active." The upper shear zone exhibited a large variation of $\varepsilon/\nu N^2$ in a diurnal cycle; the highly active zone in the "night" sector is bounded by a heavy line for $\varepsilon/\nu N^2 = 2000$. The core was inactive nearly all the time, but a few events of more than marginal activity occurred below.

3.1 Time series of ε between 0.3 and 0.6 MPa

The largest variability occurred between 0.3 and 0.6 MPa, where the diurnal cycle extended into the continuously stratified profile. Because the vertically coherent bursts dominated the variability, dissipation rates were averaged across the 30-m-thick section, resulting in a series of 379 values taken, on average, every 17 minutes. To illustrate the sampling problem, one value, 2×10^{-5} W kg^{-1}, accounted for 11% of the ensemble average, $\bar{\varepsilon} = 4.9 \times 10^{-7}$ W kg^{-1} (Fig. 4). Ten profiles that large would have doubled $\bar{\varepsilon}$, but 100 would have been needed to increase $\bar{\varepsilon}$ by a decade. In view of a sample every 17 minutes, compared with $2\pi/N = 14$ minutes, missing nine events with $\bar{\varepsilon} = 2 \times 10^{-5}$ seems possible but unlikely—missing 99 would have been impossible without a major change in mixing regime.

The cumulative contributions to $\bar{\varepsilon}$ in Fig. 4a illustrate the role of the bursts and the dominance of nighttime mixing. The bursts, defined approximately by the 20% of the data having ε greater than $\bar{\varepsilon}$, contributed 70% of $\bar{\varepsilon}$. Consequently, variability in the frequency and intensity of

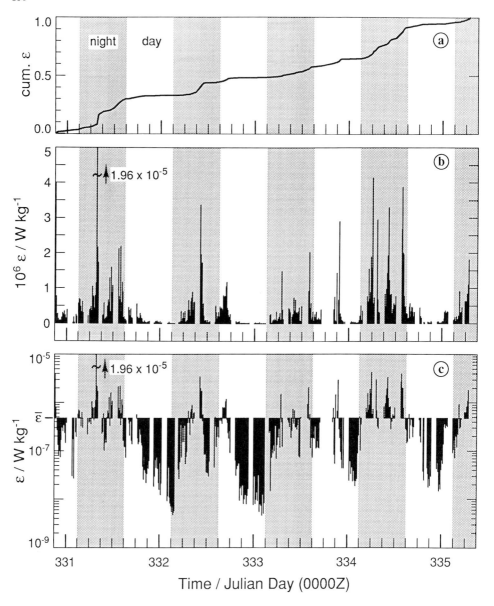

Fig. 4. Time series of ε averaged from 0.3–0.6 MPa. The shading indicates the local night 1800-0600.

(a) Cumulative contribution to the mean demonstrates the dominant role of nighttime mixing. Several jumps result from individual bursts. Note day-to-day variations in the dominant night segments.

(b) Time series of ε on a linear scale, emphasizing the event-like bursts which contributed most of the variance.

(c) Time series of ε on a logarithmic scale, showing the full variability. The bars are folded around the mean. The 20% of the samples above the mean contributed 70% of the variance; they roughly constitute the bursts. The "base" of the bursts varied by a factor of 10 between different nights.

the bursts was the major factor controlling $\bar{\epsilon}$. For example, the dissipation was relatively weak during the night of Julian day 333, owing to a decreased intensity of the bursts. Until the mechanism producing the bursts is understood, the relative importance of randomness and deterministic factors cannot be ascertained.

3.2 ε distributions

Traditional statistical analysis uses the standard deviation for the spread of random variables having stationary distributions. Because ε distributions are more lognormal than normal, the standard deviation of $\ln \epsilon$ is used and denoted by $\sigma_{\ln}$. Before estimating the uncertainty of our average dissipation profile, we must first assess the lognormality of the ε samples and compute $\sigma_{\ln}$.

As evident in Fig. 5, the shape of the $\ln \epsilon$ distributions changed markedly with pressure. Those at 0.18 and 0.36 MPa were skewed toward large values, that at 0.72 MPa was nearly sym-

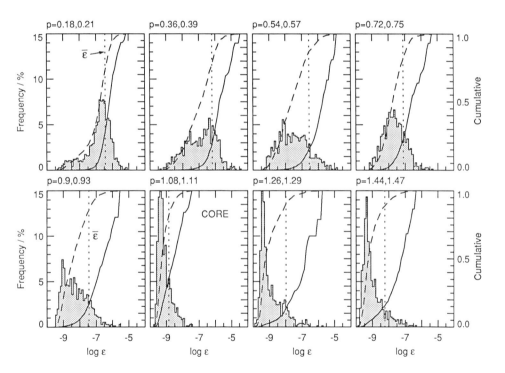

Fig. 5. Distributions of ε at 0.18-MPa intervals. Each distribution contains two samples, both averaged over 0.03 MPa, from each profile. Each panel contains the distribution (shaded), the cumulative distribution (dashed), and the cumulative contribution to the average (solid). The arithmetic mean of ε is marked as a vertical dotted line. The large spread of distributions above 0.9 MPa was produced by the diurnal variability. Owing to this nonstationarity, the distributions at 0.18 and 0.36 MPa deviate strongly from lognormality. The low ends of distributions below 0.9 MPa were limited by instrumental noise. The spread of the distributions had a minimum at the level of the undercurrent core (1.08 MPa), where little mixing occurred. In contrast, the spread in the lower shear zone (1.26 and 1.44 MPa) was large due to a few high-ε events producing long tails in the distributions.

metric, and those deeper were skewed to low values. The strong bias to large values in the upper two distributions was produced by the longer duration of the nighttime phase of the mixing cycle (Fig. 2c). By 0.72 MPa the day/night disparity was much weaker, and the distribution was nearly symmetric. The further decrease in intensity at greater pressures resulted in the peak of the distribution being established by instrumental noise. Owing to the scarcity of mixing events in the core, the 1.08 MPa distribution had a small high ε tail. The longer tails at greater pressure reflect the few strong events.

Only the data from 0.72 MPa and above were used in testing the full distributions for lognormality, owing to the large effect of noise on the deeper distributions. All failed tests for lognormality except the one from 0.72 MPa (Fig. 6). Sorting by time of day did not remove the deviations from lognormality. At 0.3–0.5 MPa, midway in the pressure domain of the diurnal cycle, the nighttime data were skewed toward high ε (Fig. 7a), whereas the daytime data had a slight excess at lower ε (Fig. 7b). As a result, the total distribution was flat in the center (Fig. 7c).

At the core and below, the scarcity of mixing events resulted in the modes of the distributions being set by instrumental noise. In these cases, plots of cumulative distributions using probability coordinates reveal portions of the distributions having lognormal behavior; they plot as straight lines with a slope of $\sigma_{\ln}^{-1}$. This is found between 60% and 95% at the core (Fig. 8). Deviations from lognormal statistics were much more dramatic in the lower shear zone, where the cumulative distributions showed two linear regions having distinctly different slopes, a behavior related to the tails noted above. The deviation from this pattern and the change in the slope shown in the data from the lower shear zone were so strong that lognormal statistics cannot be expected to yield reasonable results.

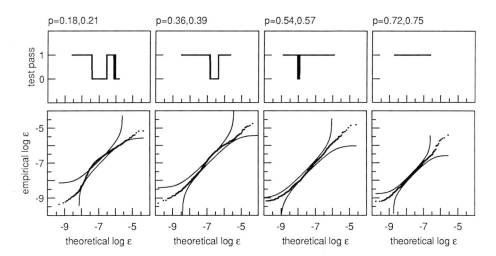

Fig. 6. Quantile-quantile tests of log ε for lognormality are shown in the lower panels. The mirror image solid curves are the Kolmogorov-Smirnov 95% confidence limits, constructed from sample means and standard deviations. Where the data lie within the confidence limits, the test is passed and the indicators in the upper panel are set to 1; otherwise they are set to 0. Only distributions unaffected by instrumental noise are shown, and of these only that at 0.72 MPa passed the test.

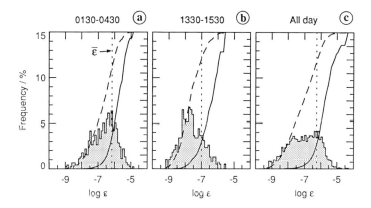

Fig. 7. Diurnal variation of ε distributions between 0.3 and 0.51 MPa: (a) "night" = 0130–0430 local time, (b) "day" = 1330–1530 local time, (c) all day. The skewness of the distributions changed with the time of day, with an excess of high ε at nighttime and an excess of low values at daytime. As a consequence, the resulting overall distribution had a flat center. The distributions were formed from 0.03-MPa estimates of ε.

Along the profile, σ_{ln} varied from 1 to 2.2 (Fig. 9b). These values were only moderately affected by the vertical averaging, owing to large vertical "patch" sizes (Table 1 and Fig. 2). The minimum of σ_{ln} occurred in the core as a result of weak mixing.[1] This value is similar to observations found in relatively inactive thermoclines at mid-latitudes (Gregg et al., 1986). The maximum of σ_{ln} was found well within the zone strongly affected by the diurnal cycle, reflecting the nonstationary nature of the mixing in addition to the intermittence of steady turbulence. Sorting data by time of day yields smaller spreads where the diurnal cycle was most intense (Fig. 10), again demonstrating the nonstationarity of the distributions.

3.3 Lognormal statistics

Estimating the mean of a lognormal distribution requires more samples than needed for a normal distribution, owing to the much larger contribution of the high-magnitude tail. Small data sets tend to cluster around the mode, at low magnitudes, and to have few samples from the tail. To provide a means of estimating the statistical uncertainty of lognormal data, Baker and Gibson (1987) developed a maximum likelihood estimator (mle),

---

[1]Gregg et al. (1985) report $\sigma_{ln} = 2.3$ for 0.5-m ε data from a rather thick isopycnal layer around the core. The high spread is due to the inclusion of some high-ε events from above and below the core. Using 3-m ε estimates from the potential density range $\sigma_\theta = 25.1$ to $\sigma_\theta = 25.4$ yields $\sigma_{ln} = 1.1$, consistent with the isobaric analysis of this paper.

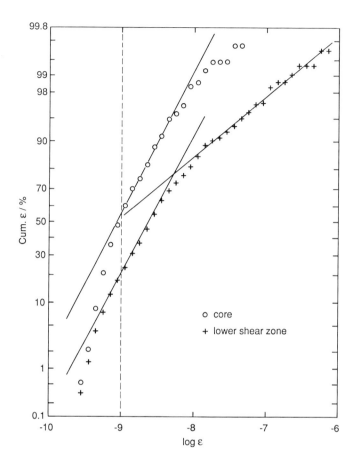

Fig. 8. Cumulative distributions of log ε in probability coordinates in the core (1.05–1.2 MPa, "o") and in the lower shear zone (1.23–1.45 MPa, "+"). In this type of plot, portions of distributions having lognormal behavior plot as straight lines. Because data with log ε < −9 are affected by instrumental noise, lognormal behavior was expected for larger values. The core showed a slight departure from lognormality, and the lower shear zone a large deviation.

$$\varepsilon_{mle} = \exp(\overline{\ln\varepsilon} + \tfrac{1}{2}\sigma_{ln}^2) \tag{1}$$

and a multiplicative factor giving the 95% confidence interval of ε_{mle}

$$\delta = \exp\left[1.96^2\left[\frac{1}{n}\sigma_{ln}^2 + \frac{1}{2(n-1)}\sigma_{ln}^4\right]^{1/2}\right], \tag{2}$$

where n is the number of independent samples.

The arithmetic average, $\overline{\varepsilon}$, was within a factor of 3 of ε_{mle} throughout the profile (Fig. 9a). Because of the pronounced deviations from lognormality, it is not obvious which is the better estimate. Above 0.5 MPa, $\varepsilon_{mle} > \overline{\varepsilon}$ because of the skewness toward small values; $\varepsilon_{mle} \approx \varepsilon$ between 0.5

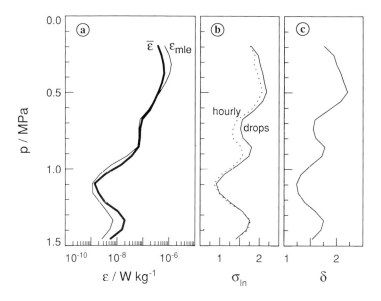

Fig. 9. Statistical parameters of ε:
(a) Arithmetic mean $\bar{\varepsilon}$ (solid, heavy), and maximum likelihood estimate ε_{mle} (solid, thin). Though the mle method is questionable because distributions of ε were not lognormal in the top 0.7 MPa, $\bar{\varepsilon}$ and ε_{mle} agreed to within a factor of 2 or better. Departures from lognormality also occurred in the lower shear zone, where ε_{mle} was smaller than $\bar{\varepsilon}$ by a factor of 3.
(b) Spread, σ_{ln}, of the ε distributions computed from all drops (solid) and from hourly averages (dotted). The spread was large in the upper shear zone because of the diurnal cycle, whereas it was small in the core. Though a local maximum occurs in the lower shear zone, it underestimates the variability. Hourly averages had nearly the same σ_{ln} as the values from the drops, indicating coherence on an hourly time scale.
(c) δ, the 95% confidence limits of ε_{mle}, from (2). Owing to the temporal coherence of ε in the upper and lower shear zones, δ is an underestimate. The graph shows a smaller uncertainty in the lower shear zone than above the core, which is inconsistent with the mixing patterns (Fig. 2c) and is a result of strong departures from lognormality. Note the minimum of δ at the core, resulting from small intermittence.

TABLE 1

Effect of vertical bin size on the spread of distributions of ε, σ_{ln}. The data are from 0.09 MPa (9 m) depth segments in the upper shear zone (centered at $\bar{p} = 0.495$ MPa), from the core (1.095 MPa), and from the lower shear zone (1.345 MPa). The vertical resolutions are $\Delta p = 0.005$ MPa, 0.03 MPa, and 0.09 MPa. The bin size had little influence on the spread of the ε distributions; it decreased only slowly with decreasing resolution.

$\bar{p}$ (MPa)	Δp (MPa)	σ_{ln} -	$\bar{p}$ (MPa)	Δp (MPa)	σ_{ln} -	$\bar{p}$ (MPa)	Δp (MPa)	σ_{ln} -
0.495	0.005	2.3	1.095	0.005	1.0	1.345	0.005	1.8
0.495	0.03	2.0	1.095	0.03	0.9	1.345	0.03	1.7
0.495	0.09	1.8	1.095	0.09	0.8	1.345	0.09	1.6

196

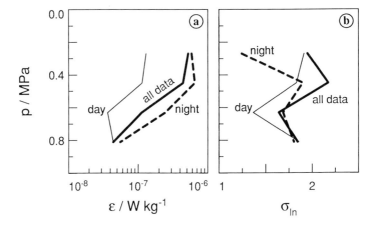

Fig. 10. Diurnal variability of statistical parameters of ε:
(a) $\overline{\varepsilon}$, overall mean (solid, thick), nighttime (dashed, thick), daytime (solid, thin). Nighttime includes the local times 0130–0430; daytime includes 1330–1530. The overall average stems from all data, not only from the combined daytime and nighttime values. The diurnal signal is clearly seen.
(b) Spread of the distributions of ε and σ_{ln} for daytime (solid, thin), nighttime (dashed, thick), and all data (solid, thick), based on 3-m averages contained in 18-m depth bins. Stratifying the data by time of day led to a reduction in σ_{ln}, though to different degrees at different depths.

and 0.85 MPa owing to more lognormal distributions. Below 1.2 MPa, the mle method underestimated the mean considerably owing to the extended high-end tails of the ε distributions (Figs. 5 and 8). Over most of the profile, the differences between ε_{mle} and $\overline{\varepsilon}$ and the range of δ (Fig. 9c) are no more than the uncertainty in $\overline{\varepsilon}$ resulting from instrumental and data processing uncertainties (Peters et al., 1988). Consequently, we find no advantage to using ε_{mle} in lieu of the arithmetic average.

The fairly good coincidence of $\overline{\varepsilon}$ and ε_{mle} seems inconsistent with Baker and Gibson's (1987) claim that the mle method should yield much smaller confidence limits than the arithmetic average. We note that Washburn and Gibson's (1984) "graphical" method of computing σ_{ln} and ε_{mle} cannot be applied to most of our data because of the strong deviations from the lognormal distribution.

Confidence limits of ε_{mle} based on (2) are expected to be reasonably accurate only where the deviations from lognormality are minor or moderate. We compute rough estimates of δ from 3-hour averages of ε, which are statistically independent according to the "Run" test (Bendat and Piersol, 1971, pp. 122–125). For the 0.3–0.6-MPa mean, we find $\delta = 3.3$, consistent with our heuristic analysis in Section 3.1. Averages from 1.05–1.2 MPa, the core zone, yield $\delta = 1.6$. Combining the uncertainty estimates from the upper shear zone and from the core shows that the hundredfold decay of $\overline{\varepsilon}$ (or ε_{mle}) from the surface toward the core was well resolved. The difference between daytime and nighttime averages of ε was less well determined. Taking windows of 8 hours length (1000–1800; 2200–0600) and 2-hour, 0.3–0.6 MPa averages leads to $\delta_{day} \approx 5$ and

$\delta_{night} \approx 3$, which is large compared with $\overline{\varepsilon_{day}} = 1.5 \times 10^{-7} \, \mathrm{W \, kg^{-1}}$ and $\overline{\varepsilon_{night}} = 9.1 \times 10^{-7} \, \mathrm{W \, kg^{-1}}$. The poor resolution of the average day–night differences in ε is surprising, considering how outstanding the diurnal cycle is in the drop data with more than hundredfold changes between minima and maxima (Fig. 4c). The large confidence limits of $\overline{\varepsilon_{day}}$ and $\overline{\varepsilon_{night}}$ do not indicate that there was no diurnal signal; they reflect the strong day-to-day changes in the cycle (Fig. 4a).

4 MIXING EFFICIENCY

The mixing efficiency, γ_{mix}, is central to the dynamics of stratified turbulence. In stationary, homogeneous turbulence, the turbulent kinetic energy equation reduces to a local balance of shear production, buoyancy flux, and dissipation. The mixing efficiency is the ratio of ε to the buoyancy flux. The flux Richardson number, R_f, is the ratio of the buoyancy flux to the shear production. Therefore,

$$\gamma_{mix} = R_f / (1 - R_f) . \tag{3}$$

Ellison (1957) found $R_f \leq 0.15$. The eddy diffusivity for mass can be expressed as

$$K_\rho = \gamma_{mix} \varepsilon / N^2 . \tag{4}$$

If all density variance is due to vertical overturning, and if the stratification is dominated by temperature (following Oakey, 1982), then γ_{mix} can be determined from ε and χ as

$$\gamma_{mix} = \frac{\frac{1}{2} N^2}{(\partial \overline{T}/\partial z)^2} \frac{\chi}{\varepsilon} . \tag{5}$$

Oakey obtained $\gamma_{mix} = 0.35$ in fairly energetic turbulence from the base of an active mixed layer, whereas Gregg et al. (1986) found $\gamma_{mix} = 0.2$ from the less active thermocline of the California Current. Gargett (1987) proposed that γ_{mix} may vary systematically with the turbulent activity, $\varepsilon/\nu N^2$.

The activity parameter spanned the enormous range of 5½ decades (Fig. 11), encompassing highly energetic turbulence with 10-m overturns and motions too weak to be turbulent. Thus, the equatorial data provide a unique test for a possible variation of γ_{mix} with $\varepsilon/\nu N^2$. However, because of possibly large systematic uncertainties in ε and χ, we were only able to resolve major trends in γ_{mix}.

As explained in detail by Peters et al. (1988), the accuracy of ε was most strongly affected by insufficiently resolved spectra at high ε and shallow depth, and by noise contamination at and below the undercurrent core. Estimates of the unresolved variance were based on the "universal" spectrum (Nasmyth, 1970; Oakey, 1982). In 98% of the original 3-m averages, at least 50% of the total variance was measured; in 86% of the samples, 75% or more of the variance was recovered. Insufficient wavenumber resolution also affected the χ data, where at least 50% of the variance was measured in 91% of the samples (75% variance recovery in 70% of the data). A correction of the unrecovered variance was based on the Batchelor (1959) spectrum. The largest contribution to the uncertainty in χ was the insufficiently known spectral response of the sensors.

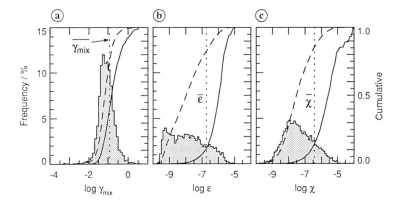

Fig. 11. Mixing versus $\varepsilon/\nu N^2$ using the data in Fig. 12. Over the 5½-decade range of $\varepsilon/\nu N^2$, γ_{mix} showed no statistically significant departure from the overall mean, as indicated by the continuous lines giving the mle value in various bands of $\varepsilon/\nu N^2$. The mle was used rather than the mean because the latter is more sensitive to outliers.

Peters et al. (1988) estimate ε and χ to have overall accuracies of factors of 2 and 3, respectively. However, the majority of the γ_{mix} values is accurate to better than a factor of 6, because they are based on well-resolved shear and temperature spectra, eliminating some of the error sources. We estimate that we can determine averages and trends of γ_{mix} to within a factor of 3.

Hourly values of γ_{mix}, computed using our χ and ε data in Eq. (5), had an approximately lognormal distribution, with $\sigma_{ln} = 0.8$ and an arithmetic mean of 0.12 (Fig. 12). Relative to the parent

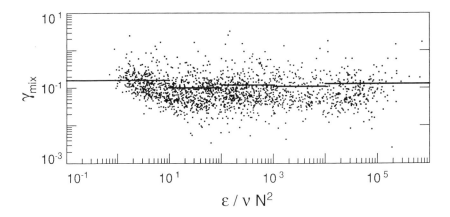

Fig. 12. Frequency distribution of γ_{mix} computed using (5) from hourly means of ε and χ, shown to the right. The arithmetic mean is 0.12, and $\sigma_{ln} = 0.8$. The maximum likelihood estimate is 0.11, with $\delta = 1.1$. Although γ_{mix} varied more than 2 decades, the range is much less than the 4 to 5 decades of the parent distributions. The dissipation rates were determined over intervals of 0.064 MPa between 0.23 and 1.44 MPa. The dashed line is the cumulative distribution; the solid line is the cumulative contribution to the average.

distributions, the scatter of γ_{mix} was small, demonstrating a strong tendency toward limited variability. The mixing efficiency showed no statistically significant variations over the full range of $\varepsilon/\nu N^2$ (Fig. 11). Especially, γ_{mix} did not change in the diurnal cycle of mixing.

Because of the systematic uncertainty discussed above, our estimate of the mean γ_{mix} is not significantly different from the values reported earlier. Though we can exclude only major systematic variations of γ_{mix}, our findings demonstrate that on average γ_{mix} was about 10^{-1} and did not drop as low as 10^{-2}. We expect that the variability in the flux Richardson number was constrained to the same degree as the mixing efficiency. The value $\gamma_{mix} = 0.12$ corresponds to $R_f = 0.11$, which is close to $R_f \leq 0.15$ obtained by Ellison (1957).

5 SUMMARY

We conclude that the $\bar{\varepsilon}(p)$ we obtained was most likely within a factor of 2 or 3 of the true value during the time of our observations, and certainly was within a decade. The major uncertainty of the representativeness for that season and site comes from ignorance of the mechanisms responsible for the energetic bursts, not from inadequate sampling. Once the frequency of occurrence and intensity of the bursts can be predicted, statistical analysis of turbulent variability within the bursts can be used to provide better estimates of $\bar{\varepsilon}(p)$.

The mixing efficiency had a value of about 0.1 and showed no trend as a function of $\varepsilon/\nu N^2$.

Acknowledgments

This work was supported by the National Science Foundation as part of the Tropic Heat program. The acoustic Doppler current meter data were taken by John Toole of the Woods Hole Oceanographic Institution. David Halpern gave us generous access to moored current meter data. We profited from discussion with Jim Moum, Tom Dillon, Doug Caldwell, Ann Gargett, and Nordeen Larson. Tom Shay contributed greatly to the software. Contribution No. 1732 of the School of Oceanography, University of Washington, Seattle.

6 REFERENCES

Baker, M. A. and Gibson, C. H., 1987. Sampling turbulence in the stratified ocean: Statistical consequences of strong intermittency. J. Phys. Oceanogr. (in press)

Batchelor, G. K., 1959. Small-scale variation of convected quantities like temperature in turbulent fluid. Part 1. General discussion and the case of small conductivity. J. Fluid Mech., 5: 113-133.

Bendat, J. S. and Piersol, A. G., 1971. Random Data: Analysis and Measurement Procedures. Wiley, New York.

Chereskin, T. K., Moum, J. N., Stabeno, P. J., Caldwell, D. R., Paulson, C. A., Regier, L. A. and Halpern, D., 1986. Fine-scale variability at 140°W in the equatorial Pacific. J. Geophys. Res., 91: 12,887-12,897.

Ellison, T. H., 1957. Turbulent transport of heat and momentum from an infinite rough plane. J. Fluid Mech., 2: 456-466.

Eriksen, C. C., 1985. The Tropic Heat Program: An overview. EOS, 66: 50-52.

Gargett, A.E., 1987. The scaling of turbulence in the presence of stable stratification. J. Geophys. Res. (submitted).

Gregg, M. C., Peters, H., Wesson, J. C., Oakey, N. S. and Shay, T. J., 1985. Intensive measurements of turbulence and shear in the equatorial undercurrent. Nature, 318: 140-144.

Gregg, M. C., D'Asaro, E. A., Shay, T. J. and Larson, N., 1986. Observations of persistent mixing and near-inertial internal waves. J. Phys. Oceanogr., 16: 856-885.

Moum, J. N. and Caldwell, D. R., 1985. Local influences on shear-flow turbulence in the equatorial ocean. Science, 230: 315-316.

Moum, J. N., Caldwell, D. R., Paulson, C. A., Chereskin, T. K. and Regier, L. A., 1986. Does ocean turbulence peak at the equator? J. Phys. Oceanogr., 16: 1991-1994.

Nasmyth, P. W., 1970. Oceanic turbulence. Ph.D. dissertation, University of British Colombia, Vancouver, 69 pp.

Oakey, N. S., 1982. Determination of the rate of dissipation of turbulent energy from simultaneous temperature and velocity shear microstructure measurements. J. Phys. Oceanogr., 12: 256-271.

Peters, H., Gregg, M. C. and Toole, J. M., 1988. On the parameterization of equatorial turbulence. J. Geophys. Res. (in press).

Rohr, J.J., Helland, K.N., Itsweire, E.C. and Van Atta, C.W., 1987. Turbulence in a stably stratified shear flow: A progress report. In Turbulent Shear Flows, Vol. 5, Springer-Verlag, Berlin.

Stillinger, D. C., Helland, K. N. and Van Atta, C. W., 1983. Experiments on the transition of homogeneous turbulence to internal waves in a stratified fluid. J. Fluid Mech., 131: 91-122.

Toole, J. M., Peters, H. and Gregg, M. C., 1987. Upper ocean shear and density variability at the equator during Tropic Heat. J. Phys. Oceanogr., 17: 1397-1406.

Washburn, L. and Gibson, C. H., 1984. Horizontal variability of temperature microstructure at the base of a mixed layer during MILE. J. Geophys. Res., 89: 3507-3522.

TURBULENT DISSIPATION IN THE STRAIT OF GIBRALTAR AND ASSOCIATED MIXING

J.C. WESSON and M.C. GREGG
Applied Physics Laboratory and School of Oceanography, College of Ocean and Fishery Sciences, University of Washington, Seattle, WA 98105 USA

ABSTRACT

The Strait of Gibraltar is a channel with a permanent two-layer flow strongly modulated by energetic tidal currents. Compared with the open ocean thermocline, turbulent dissipation rates, ε, are high throughout the strait. Two flow regimes, however, have particularly intense dissipation: the lee west of Camarinal Sill during strong outflow of Mediterranean water, and the internal bore or solitary wave propagating eastward from the sill after reversal of the tide. Overturns exceeding 50 m and ε greater than 10^{-3} W kg^{-1} were observed in the lee of the sill. Dissipation rates of 10^{-5} to 10^{-4} W kg^{-1} were found behind the leading edge of the bore as it moved eastward through Tarifa Narrows, dropping the interface more than 100 m in less than 15 minutes.

1 INTRODUCTION

The Strait of Gibraltar is an energetic tidal channel with two-layer flow. At the surface, Atlantic water enters the Mediterranean, and at depth Mediterranean water flows out. In the eastern end the strait is a relatively uniform "U" that narrows at Tarifa Narrows and then widens and shallows at Camarinal Sill in the west end (5° 45' W; Fig. 1). Another sill is located even farther west (5° 57' W), opposite Cape Spartel. Both the sills and the constriction at the narrows cause internal flow transitions between subcritical and supercritical states (Armi and Farmer, 1986; Farmer and Armi, 1986). The dynamics of exchange at the strait have been studied by many investigators, both experimentally and theoretically (e.g., Frassetto, 1964; Lacombe and Richez, 1982; Bryden and Stommel, 1984; Armi and Farmer, 1986; Farmer and Armi, 1986). The effects of mixing and turbulent dissipation of kinetic energy, however, have not previously been addressed. Although Bryden and Stommel (1984) discuss "overmixing," they assume it occurs in the Mediterranean basin, not in the strait itself. The theory of Armi and Farmer is explicitly non-dissipative, based on a Bernoulli formulation.

The mean inflow of Atlantic water is driven by net evaporation over the Mediterranean of 1 m per year, while the deep outflow is supplied by wintertime formation of dense water south of France and in the Levant (Bryden and Stommel, 1984). Tides, bathymetry, and meteorological

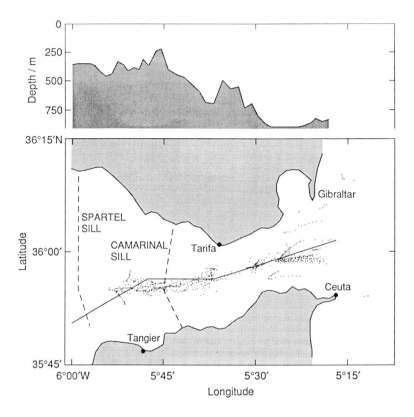

Fig. 1. Observations in May 1986; each dot corresponds to a profile. Dashed lines locate the two sills. Camarinal Sill is the shallower of the two; the hydraulic jump and bores discussed are generated there. East of Camarinal Sill the strait is deep and "U" shaped: the 400 m contour is 8 km or wider. The bathymetry shown in the upper panel corresponds to depths along the axis of the strait.

effects modulate the mean flow (Lacombe and Richez, 1982). For example, long term (6-12 months) current meter records from the Gibraltar Experiment (Kinder and Bryden, 1987) show complete stopping of the outflow at Camarinal Sill during falling tides.

As part of the Gibraltar Experiment, we measured turbulence and mixing 13–20 October 1985 from USNS *Lynch* and 3–16 May 1986 from RV *Oceanus*. The instruments and procedures used in that difficult environment are described in Section 2, followed in Section 3 by an along-strait view of the average layers and turbulence. Section 4 shows some of the observations near Camarinal Sill during strong outflow, and the internal bores at Tarifa Narrows are illustrated in Section 5.

2 INSTRUMENTS AND OPERATIONS

Heavy, and sometimes aggressive, ship traffic, strong currents, levanters up to 25 m s^{-1}, and intense fishing often interrupted our attempts to obtain time series, particularly at Camarinal and

Spartel sills. Nevertheless, two 12-hour series of slow tow-yo's over Camarinal Sill were made, and records several days long were obtained in the Tarifa Narrows and between Ceuta and Gibraltar. In all, 275 profiles were taken during the first cruise, and nearly 1000 during the second. Figure 1 shows the locations of drops during May 1986.

Turbulent dissipation, temperature, and salinity were measured with the Advanced Micro-structure Profiler (AMP) (Gregg et al., 1982). It falls freely, attached to the ship by a 2 mm diameter Kevlar tether containing a fiber-optic data link. The airfoil probes used to measure small scale horizontal velocity variations cannot resolve scales less than 1 cm (Osborn and Crawford, 1980; Ninnis, 1984). Therefore dissipation rates exceeding 10^{-4} W kg^{-1} were estimated using a universal dissipation spectrum (Nasmyth, 1970; Oakey, 1982).

During the second cruise, continuous velocity profiles were made with a ship-mounted RD Instruments Acoustic Doppler Current Profiler (ADCP), operated with a pulse rate of 1 Hz and an acoustic frequency of 150 kHz. Averaged velocity profiles with 0.08 MPa bin size were recorded every minute. Typically, useful profiles were obtained to 4 MPa. The only velocity data for the first cruise is from 19 XCPs taken at the end of the cruise.

A Biosonics 120 kHz acoustic backscatter echosounder (provided by David Farmer) revealed a rich variety of structures and flow regimes. These images were essential for interpreting the AMP profiles, which could not be made often enough to avoid aliasing some rapidly evolving structures.

3 ALONG-STRAIT AVERAGE

The Atlantic layer becomes thinner to the east, particularly after passing Camarinal Sill (Fig. 2), while the Mediterranean layer thins from east to west. Lacombe and Richez (1982) give the mean transport in either direction as $1-1.5 \times 10^6$ m^3 s^{-1}. Because the time averaged transport is uniform along the strait, the velocity structure is essentially that thinning layers accelerate. The averaged velocity profiles show this in Fig. 2b (the Doppler profiler did not function on the October cruise). The isopycnal contours in both October and May have major similarities: the largest variations in isopycnal depths occur near Camarinal Sill (5° 45′W), and the upper layer thickness is almost constant east of Camarinal Sill in both panels. However, the average obviously cannot reflect the large variations we observed in layer thicknesses along the strait.

The asymmetry of the along-strait isopycnals implies that the reservoir conditions for the two-layer flow are different on each side of the strait and that the flow is hydraulically controlled at the strait (Armi, 1986). The control consists of one or more points of transition between subcritical and supercritical flow. Critical internal flow occurs when

$$G^2 \equiv Fr_1^2 + Fr_2^2 = 1 , \tag{1}$$

where G is the overall Froude number and $Fr_i \equiv U_i(g'h_i)^{-1/2}$ are layer Froude numbers. The transition from supercritical to subcritical flow can only take place with a loss of energy. Large energy losses are dissipative; smaller energy losses may be accomplished by wave radiation (Chow, 1959).

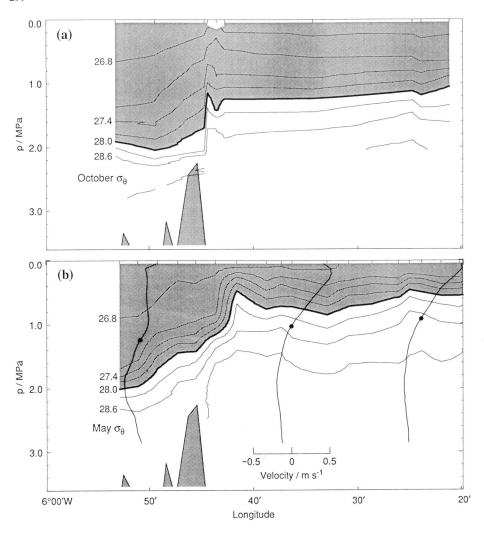

Fig. 2. Contours of σ_θ for October 1985 and May 1986. East of Camarinal Sill the upper layer was 40–50 m thicker in October 1985 than in May 1986. The heavy contour, $\sigma_\theta = 28.0$ kg m^{-3}, was in the middle of the interface. The contours were formed from average profiles of drops grouped in longitude bins. Centers of the bins are at top; each bin contained 20 or more drops in October and 40 or more drops in May. The effects of internal tides are reduced but may not be eliminated. The velocity (u) profiles for May are averages relative to the ship and are demeaned. They show the velocity structure but are not absolute. Solid circles mark zero velocity for each profile.

Based on a two-layer inviscid flow model, Armi and Farmer (1985, 1986) predict that maximal flow will produce a 105 m thick upper layer at Tarifa Narrows and a 37 m thick upper layer at Gibraltar; slightly submaximal flow will have a 105 m upper layer at Tarifa and a 150 m upper layer at Gibraltar. The maximal flow is critical at both Camarinal Sill and Tarifa Narrows, but the submaximal flow is not critical at Tarifa. Bormans et al. (1986) state that high interfacial friction (quadratic drag coefficient $C_d = 3 \times 10^{-3}$) could raise the interface in the submaximal case by 50 m into the observed range near 100 m.

The Froude numbers of the flow east of Camarinal Sill indicate there may be a seasonal cycle in flow conditions, as predicted by Bormans et al. (1986). They predict a maximum interface depth in fall and a minimum in spring due to wintertime formation of Mediterranean deep water. To estimate upper layer Froude numbers, $\sigma_\theta = 28.0$ kg m^{-3} defines the interface, $\Delta\rho = 1.6$ kg m^{-3}, and the mean velocity is

$$U_1 = \overline{q}_1/h_1 W_1, \tag{2}$$

where $\overline{q}_1 = 1\text{--}1.5 \times 10^6$ m^3 s^{-1} is the mean transport, h_1 is the layer thickness, and the width $W_1 \approx 12\text{--}15$ km. In October, the upper layer thickness of 125 m gives $U_1 = 0.5\text{--}0.8$ m s^{-1}, so $\mathrm{Fr}_1 = 0.5\text{--}0.85$, which is subcritical. In May, $h = 75$ m and $U_1 = 0.9\text{--}2.0$ m s^{-1}, so the upper layer is supercritical, $\mathrm{Fr}_1 = 0.8\text{--}1.85$. Individual ADCP and XCP profiles support these estimates of upper layer velocities and thickness.

The average dissipation rate for all the AMP drops is $\approx 1 \times 10^{-5}$ W kg^{-1}, which is huge compared with 3×10^{-10} to 10^{-8} W kg^{-1} in the open ocean thermocline (Moum and Lueck, 1985). If the average of two cruises is representative, total energy dissipation in the strait (200 m × 15 km × 40 km) is the same as in a region of thermocline (1 km × 1100 km × 1100 km) where $\overline{\varepsilon} = 1 \times 10^{-9}$ W kg^{-1}. The average ε is high, but there is large variability; some drops have very low dissipation throughout ($\varepsilon < 10^{-8}$ W kg^{-1}). Although we have taken more than 1200 drops in the strait, the number of drops in the regions of highest dissipation is far fewer, and the sampling is inadequate for good statistics. Nevertheless, some patterns are evident. The Atlantic layer is least turbulent west of Camarinal, and the lower layer is least turbulent at the east end of the strait (Fig. 3). The particular patterns of interest are the locations of maximum dissipation. Maxima occurred at Camarinal Sill and in Tarifa Narrows. Hydraulic jumps, or lee waves, during strong outflow of the lower layer produced the former; propagating internal bores, or solitary waves, produced the latter. The high dissipation in the strait is dominated by hydraulic jumps at Camarinal Sill and bores observed at Tarifa Narrows; examples of both features are examined in the following sections.

4 STRONG OUTFLOW OVER CAMARINAL SILL

Strong outflow during a rising tide is dramatically shown by the echosounder image in Fig. 4. The section was begun west of Camarinal Sill 4 hours before high water at Gibraltar and took over an hour to complete. To the east (right) several strong scattering surfaces mark the interface, approaching the sill near 1 MPa. Starting at the east edge of the sill, the interface sharpens and plunges downward with a slope of 7°. Interfacial oscillations, beginning over the crest, quickly grow into a broad zone of turbulence that spreads the acoustic scatterers below 0.8 MPa. The intense mixing thickens the interface to more than 100 m (Fig. 5b), demonstrating water mass conversion in action. Overturns up to tens of meters thick are common west of the sill crest.

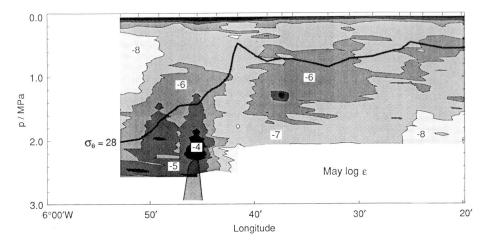

Fig. 3. Log($\overline{\epsilon}$) for the average of nearly 1000 profiles in May 1986. Contours were generated as in Fig. 2. Peak dissipations occurred at Camarinal Sill in hydraulic jumps formed during strong outflow of Mediterranean water. Observations at the sill were two 12-hour sequences of AMP tow-yo's from west to east. The other dissipation maximum was at Tarifa Narrows (5° 37′ W). It was produced by propagating bores that evolved into trains of solitary waves. Although many more profiles were taken in 6 days of profiling east of Camarinal, only three bores were seen. Therefore, the average in that region was determined by a few events. The heavy solid line shows the 28.0 σ_θ contour.

The velocity profiles in Fig. 5a show diffuse, and relatively small, velocities at 1600 m along the section (and farther west), while velocities are highest from 2800 m to 3300 m just east of the jump. Dissipation rates are extremely high, rising to 1 W kg^{-1} near the bottom of the mixing zone (Fig. 5b); in the upper layer the low values are near 10^{-9} W kg^{-1}, similar to mid-ocean dissipation rates. The region of high ϵ, integrated over the width of the sill, dissipates 0.4–4.0 × 10^8 W.

In general, a hydraulic jump occurs when a supercritical flow changes to subcritical with a large loss of energy. Here it appears (Fig. 4) that the lower layer is supercritical at the sill crest; as it goes farther west the rapid deepening of the topography forces it to thicken and become subcritical. Starting in the east, the 28.0 σ_θ contour (Fig. 5a) drops as the lower layer thins over the sill and rebounds west of the sill. The velocity profiles are demeaned and indicate that the flow is subcritical throughout. With these demeaned profiles, the largest overall Froude number, $G^2 = 0.8$, is at the sill ($h_1 = 128$ m, $u_1 = 0.6$ m s^{-1}, $h_2 = 50$ m, $u_2 = -0.7$ m s^{-1}, $\Delta\rho/\rho = 0.0016$). However, because of the stationary topography, absolute velocities must be used to determine the Froude numbers, so the demeaned profiles may give substantial underestimates. The high dissipation, large overturns, variation of isopycnals along the section, and spreading of the interface west of the sill all imply that a hydraulic jump, or very high amplitude lee wave, exists.

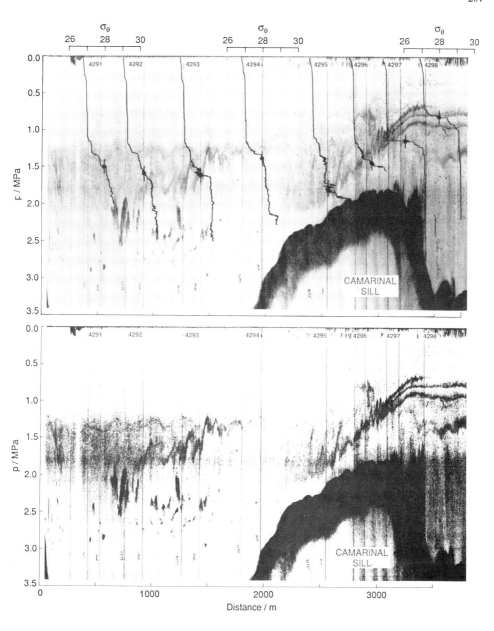

Fig. 4. Lower panel: Echosounder image from a section at Camarinal Sill. West is to the left, and east to the right. Vertical lines indicate AMP drops. The thin interface in the east plunges down the west side of Camarinal Sill and then becomes diffuse after an internal hydraulic jump. Upper panel: Overlay of σ_θ profiles on the echosounder image reveals several steps in the sharp interface approaching the sill from the east. High shear over the sill thinned the principal step to less than 1 m. Farther west the interface broadens and contains density instabilities. Solid circles with vertical lines mark $\sigma_\theta = 28.0$

208

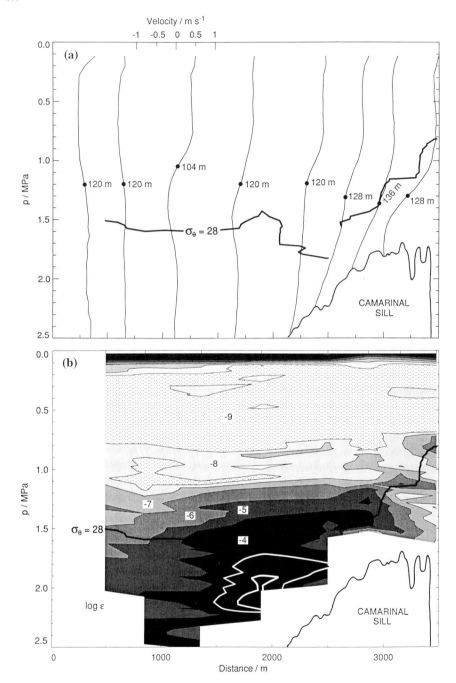

Fig. 5. (a) Velocity profiles (u) for the AMP drops of Fig. 4 obtained with a ship-mounted Doppler profiler. Profiles are demeaned and relative to the ship. (b) Log(ε) for the drops of Fig. 4. Extremely high dissipation is at the base of the hydraulic jump, 1 km west of the sill crest, in the region where the interface is thickened by overturns and the scatterers are dispersed vertically.

5 AN INTERNAL BORE PASSING THROUGH TARIFA NARROWS

High-frequency phenomena in the strait consist of bores and trains of internal oscillations. Observations of internal bores have been reported by Frassetto (1964), Zeigenbein (1969, 1970), and Kinder (1984). They describe trains of large internal waves accompanied by surface temperature fronts, and the visual evidence at the surface consisting of bands of ripples along the wavefronts. The internal wave trains decay as they propagate away from the sill, but they are observed at moorings east of Gibraltar. Satellite photographs show the surface manifestations of the waves 50 km east of Gibraltar. The bores are tidal, as indicated by the congregation of "most dramatic" observations (including those here) at or near spring tides. At neap tides, bores are much smaller or nonexistent.

On 9 May from 0413 to 0420 GMT in Tarifa Narrows (near 5° 37′W), the interface between the Atlantic and Mediterranean layers plunged from 0.2 MPa to 1.2 MPa between successive AMP drops (Fig. 6a). The leading edge of this bore had an apparent overshoot followed by smaller oscillations of the isopycnals; by 0700 GMT the interface had risen to 0.6 MPa and the sharp density gradient had become more diffuse. The velocity structure of the bore in Fig. 6b showed essentially quiescent water before the arrival of the bore, followed by strong baroclinic flow when the bore arrived. Peak velocity differences across the interface were >0.9 m s^{-1}, with strong eastward flow confined to the upper layer defined by the 28.0 σ_θ contour. Dissipation rates jumped when the bore passed the ship (Fig. 7).

We can estimate some of the characteristics of the bore from these observations. The speed of a bore is amplitude dependent and greater than the long internal wave speed. Taking the formula for the solitary wave speed, the 100 m bore in Fig. 6a has a calculated speed

$$c = c_o \left(1 + \frac{\eta}{2\,h_1}\,(1 - h_1/h_2)\right) = 1.6 \text{ m s}^{-1}, \tag{3}$$

where c_o is the long internal wave speed and η is the amplitude (Osborne and Burch, 1980). This speed implies a release from Camarinal Sill 40 minutes after high water. The wavelength of the first crest is about 600 m. Maximum accelerations on the leading edge are as large as $(\Delta\rho/\rho)g$ for the two-layer system. The energy contained in the 2 km long wave packet is

$$E \approx \Delta\rho\, g\, \eta^2\, L\, W \approx 2 \times 10^{12} \text{ J}, \tag{4}$$

where $\eta = 100$ m, length $L = 2$ km, and width $W = 12$ km (Osborne and Burch, 1980).

The pattern of dissipation in Fig. 7 shows that ε was very low before the arrival of the bore (as low as 10^{-9} W kg^{-1}) and very high only while peak speeds were greater than 0.6 m s^{-1}. Peak dissipation rates of 1×10^{-2} W kg^{-1} were recorded at the "crest" of the inverted bore. The mean dissipation before the arrival of the bore was 4×10^{-7} W kg^{-1}; after its arrival, $\bar{\varepsilon}$ jumped to 2×10^{-5} W kg^{-1}. Because the peak dissipation strongly influences the overall mean and the peak is not well sampled (only two drops in 15 minutes), the uncertainty in the estimate of mean dissipation is large. Nonetheless, the lifetime of the wave packet, estimated using $\bar{\varepsilon} = 2 \times 10^{-5}$ W kg^{-1}, is about 1 day.

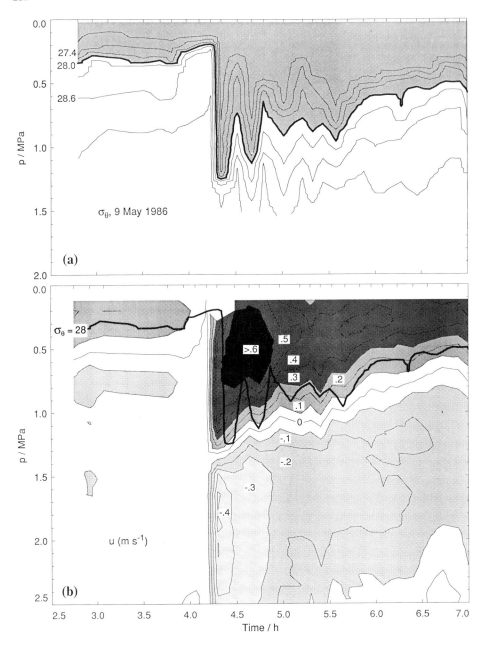

Fig. 6. (a) Density, σ_θ, in Tarifa Narrows during the passage of an internal bore. Ticks on the upper axis correspond to drops. The ship drifted 2.6 km in the 3 hours between the arrival of the bore and the end of the section, so this is a mixed time and space series. (b) Velocity (east-west) contours from the same set of drops. Peak shears are 0.024 s^{-1}. Peak accelerations of almost g' are found on the leading edge. The arrival of the bore is marked by the jump in velocity magnitudes, and the strong baroclinic flow is simultaneous with the other observations.

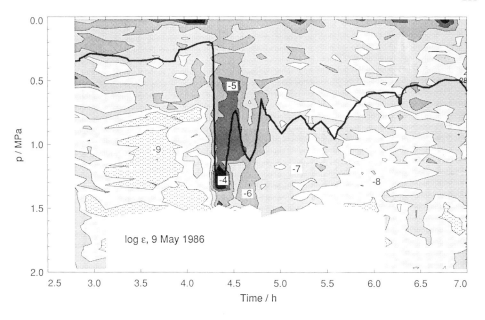

Fig. 7. Dissipation, ε, contours for drops of Fig. 6. High dissipation accompanied the arrival of the bore. Dissipation rates have returned approximately to background values after 2 hours. The heavy solid line marks $\sigma_\theta = 28$.

Bores were not observed in the May cruise farther east in the strait. This may be due to the tidal variation (observations at Gibraltar-Ceuta were not at spring tides), or because the time series at Gibraltar-Ceuta was not dense enough. There may also be seasonal modulation of the generation of bores that affected the May observations.

6 SUMMARY

There are three main points to these observations. First, $\bar{\varepsilon}$ in the Strait of Gibraltar is many orders of magnitude larger than in most open ocean environments. Second, the very high mean dissipation is dominated by two particular regimes: turbulence that is localized at the sills and the intermittent passage of internal bores. The Camarinal Sill region of high ε is caused by two-layer flow over the sill and the subsequent instability and mixing of the interfacial waters. The high dissipations observed at Tarifa Narrows are due to the passage of turbulent internal bores of 100 m or greater amplitude with short time scales. Third, both regimes, as well as the overall dissipation in the Strait of Gibraltar, are strongly modulated by the semidiurnal tides. Reversing tides vary the flow at the sill, so hydraulic jumps exist intermittently; similarly, the semidiurnal tides generate bores that were observed at Tarifa Narrows.

Acknowledgments

The Gibraltar Experiment was made possible with the help of the Governments of Spain and Morocco. We thank Major General Joseph Franklin (USA) for his assistance. We are grateful to Captains Grey and Howland of the *Lynch* and *Oceanus* for their expertise and cooperation in the hectic Strait of Gibraltar and to Wayne Nodland and Dale Hirt for technical operations, especially Dale for retrieving the AMP profiler when its tether was caught in the propeller of the *Oceanus*. This research is supported by the Coastal Sciences Division of the Office of Naval Research as part of the Strategic Straits Program. Contribution No. 1735 of the School of Oceanography, University of Washington, Seattle.

7 REFERENCES

Armi, L., 1986. The hydraulics of two flowing layers with different densities. J. Fluid Mech., 163: 27–58.

Armi, L. and Farmer, D., 1985. The internal hydraulics of the Strait of Gibraltar and associated sills and narrows. Ocean. Acta, 8: 37–46.

Armi, L. and Farmer, D.M., 1986. Maximal two-layer exchange through a contraction with barotropic net flow. J. Fluid Mech., 164: 27–51.

Bormans, M., Garrett, C. and Thompson, K.R., 1986. Seasonal variability of the surface inflow through the Strait of Gibraltar. Ocean. Acta, 9: 403–414.

Bryden, H.L. and Stommel, H.M., 1984. Limiting processes that determine basic features of the circulation in the Mediterranean Sea. Ocean. Acta, 7: 289–296.

Chow, V.T., 1959. Open-Channel Hydraulics. New York, McGraw-Hill, pp. 39–59.

Farmer, D. and Armi, L., 1986. Maximal two-layer exchange over a sill and through the combination of a sill and contraction with barotropic flow. J. Fluid Mech., 164: 53–76.

Frassetto, R., 1964. Short-period vertical displacements of the upper layers in the Strait of Gibraltar. Saclant ASW Res. Center, La Spezia, Italy, Tech. Rep. No. 30, 49 pp.

Gregg, M.C., Nodland, W.E., Aagaard, E.E. and Hirt, D.H., 1982. Use of a fiber-optic cable with a free-fall microstructure profiler, Oceans '82 Conference Rec., IEEE Cat. No. 82CH1827-5, Marine Tech. Soc., Washington, D.C., pp. 260–265.

Kinder, T.H., 1984. Net mass transport by internal waves near the Strait of Gibraltar. Geophys. Res. Lett., 11: 987–990.

Kinder, T.H. and Bryden, H.L., 1987. The 1985-1986 Gibraltar Experiment: Data collection and preliminary results. EOS, in press.

Lacombe, H. and Richez, C., 1982. The regime of the Strait of Gibraltar. In: J.C.J. Nihoul (Editor), Hydrodynamics of Semi-Enclosed Seas. Elsevier Oceanogr. Ser., 34: 13–73.

Moum, J.N. and Lueck, R.G., 1985. Causes and implications of noise in oceanic dissipation measurements. Deep-Sea Res., 32: 379–390.

Nasmyth, P., 1970. Oceanic Turbulence. Ph.D. Thesis, Univ. Brit. Columbia, Vancouver, 69 pp.

Ninnis, R., 1984. The effects of spatial averaging on airfoil probe measurements of oceanic velocity microstructure. Ph.D. Thesis, Univ. Brit. Columbia, Vancouver, 109 pp.

Oakey, N.S., 1982. Determination of the rate of dissipation of turbulent energy from simultaneous temperature and velocity shear microstructure measurements. J. Phys. Oceanogr., 12: 256–271.

Osborn, T.R. and Crawford, W.R., 1980. An airfoil probe for measuring turbulent fluctuations in water. In: F. Dobson, L. Hasse, and R. Davis (Editors), Air-Sea Interaction: Instruments and Methods. Plenum, New York, pp. 369–386.

Osborne, A.R. and Burch, T.L., 1980. Internal solitons in the Andaman Sea. Science, 208: 451–460.

Zeigenbein, J., 1969. Short internal waves in the Strait of Gibraltar. Deep-Sea Res., 16: 479–487.

Zeigenbein, J., 1970. Spatial observations of short internal waves in the Strait of Gibraltar. Deep-Sea Res., 17: 867–875.

MICROSTRUCTURE AND VERTICAL VELOCITY SHEAR DISTRIBUTION IN MONTEREY BAY

E. C. ITSWEIRE and T. R. OSBORN†
Chesapeake Bay Institute, The Johns Hopkins University,
771 West 40th Street, Suite 340, Baltimore, MD 21211, USA.

ABSTRACT

High-resolution velocity shear and microstructure measurements were made simultaneously from the research submarine *Dolphin* in Monterey Canyon during October 1984. Inside the thermocline, the velocity shear appeared to be concentrated at set depths in layers 10 to 15 meters thick extending several kilometers horizontally.

During the October 19 dive, the *Dolphin* depth-cycled between 20 and 60 m and followed a predetermined track 10 miles off Monterey. The track started with three parallel, seven mile east-west legs one mile apart and ended with a five mile southwest leg. During the last leg, five very distinct salinity and temperature ramps were observed between two strong shear layers ($\partial U/\partial z \approx 0.02$ sec^{-1}). The ramps were estimated to be 5 m thick and 50 m horizontally. The corresponding T-S diagram did not show any anomaly indicating that the ramps are caused by a local mechanism, maybe some shear instability. Thereafter, the submarine encountered a different water mass and no more ramps were observed near the shear layers.

1. Introduction

Past studies of oceanic mixing have shown the need for simultaneous observations of the mean vertical shear and the microstructure. In order to parameterize vertical eddy diffusivities it is important to relate mixing events (regions with high dissipation rates) to the local mean shear and stratification. Strong shear layers are a source of kinetic energy that can be extracted from the mean flow through instability mechanisms and entrainment. The resulting turbulent kinetic energy is converted into heat and potential energy.

Thorpe (1987) reviewed the different types of shear instabilities that might occur in the ocean. For example, Kelvin-Helmoltz billows can be created by small disturbances when the local Richardson number is below a critical value of 0.25. Divers (Woods, 1968) observed billows in the Mediterranean thermocline using dye to follow the motion of water parcels. Thorpe & Hall (1974) and Thorpe et *al.* (1977) measured the temperature field associated with Kelvin-Helmoltz instability in the thermally stratified Loch Ness. There, breaking billows exihibited a ramp-like temperature signal. Temperature ramps are a common feature of the atmospheric boundary layer and have been observed both over land (e.g., Kaimal et *al.*, 1972, Champagne et *al.*, 1977) and water (e.g., Pond et *al.*, 1972, Gibson et *al.*, 1977, Schmitt et *al.*, 1978) under stable and unstable conditions. Under stable conditions, ramps

---

† also Department of Earth and Planetary Sciences, The Johns Hopkins University.

have been associated with Kelvin-Helmholtz instability, while under unstable conditions, the underlying mechanism would be, either convective plumes sustained by buoyant acceleration, or horizontal roll vortices or both. Gibson et *al.* (1977) showed that temperature ramps in the unstable marine surface layer and laboratory shear flows such as jets, wakes and boundary layers were characteristic of turbulent mixing of a scalar in a mean velocity shear (with large scale organized structures).

Temperature fronts or ramps have also been associated with wind rows and Langmuir circulation (Thorpe & Hall, 1980, 1982). These observations were made in the mixing layer of the fresh water, thermally stratified Loch Ness. The narrow temperature fronts or ramps are usually normal to the wind direction. They are thought to be analogous to the ramps observed in the atmospheric surface layer.

2. Instrumentation

The measurements reported in this paper are from an October 1984 cruise of the U.S.S. *Dolphin*, a diesel-electric powered research submarine owned and operated by the U.S. Navy. A general description of the submarine and the turbulence instrumentation can be found in Osborn and Lueck (1985). Figure 1 shows a side view of the instrumentation lay-out. All the probes were mounted along the centerline of the submarine.

Three separate data acquisition systems made up the instrumentation. The ship status system recorded the ship heading, pitch and roll and shaft rpm on floppy discs on a portable Compaq computer every 30 seconds. Data from the Acoustic Doppler Current Profiler were recorded every 15 seconds (36 pings average) on an HP 9212 microcomputer and data from the turbulence and CTD packages were written on magnetic tapes with an LSI 11/23 microcomputer.

The Acoustic Doppler Current Profiler (ADCP) (①+②) was mounted on the hull 4.7 m below and 4.74 m forward of the turbulence package. The ADCP was an upward looking, four-beam, 1.2 MHz system made by RD Instruments. It had a 1 meter vertical resolution and a nominal range of 30 m. A problem with one of the beams reduced the actual range to about 25 m. Thirty bins were telemetered and 36 pings for each bin were ensemble averaged every 15 seconds. The rms velocity noise for one-minute averaged ADCP data was about 1.5 cm/s (above the manufacturer specified noise level of 1 cm/s). The noise level of the first four bins was twice the noise level of the upper twenty bins due to side lobes and flow distortion off and around the submarine. A potential flow calculation (Loeser & Chapman, 1979) showed that the flow distortion around the hull of the submarine increased the axial velocity (u) along the centerline by up to 7.4%. Off the centerline the axial flow acceleration is quickly reduced and is accompanied with an outward transversal flow. The axial velocities measured in bins 1 to 8 of the ADCP were estimated to be 6.9, 3.5, 2.0, 1.0, 0.7, 0.5, 0.4 and 0.2% higher than the true velocities. All four beams were at 30° to the vertical (in submarine coordinates) and the ADCP was rotated so that the two pairs of diagonal beams were at 45° with the mean speed of the submarine. This configuration canceled out most of the transerval velocity errors and some the axial velocity errors due to the flow distorsion around the hull of the submarine. An additional sidelobe reflection off the tripod contamined the returned signal of the third bin of the backward looking beams. Therefore, the first and third bins of the velocity profiles were eliminated from the data processing of the velocity and shear profiles. Bins 2 to 8 were corrected for flow distortion.

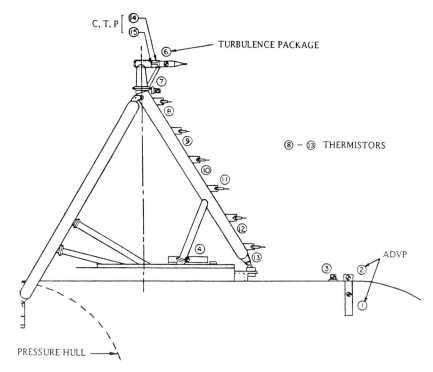

Fig. 1. Intrumentation lay-out on the *Dolphin*.

The turbulence package ((6)) comprised two airfoil probes (Osborn & Crawford, 1980), a fast response thermistor (Thermometrics FP07), and three orthogonal accelerometers. It was mounted atop a tripod, 4.88 m from the hull of the submarine. One airfoil probe was aligned to measure the vertical velocity shear ($\partial w/\partial x$) and the other to measure the transverse velocity shear ($\partial v/\partial x$). The accelerometers measured pitch, roll and heave of the submarine and high frequency accelerations caused by tripod vibrations. The water column properties were measured with a SEABIRD CTD package, i.e., a SBE-4 conductivity cell ((14)) (Pederson & Gregg, 1979) and a glass coated thermistor SBE-3 ((15)) which were located 0.75 m aft of the turbulence package. A Viatran pressure transducer was mounted on the pressure case of the turbulence package.

Finally, a six-thermistor (Thermometrics FP07) chain ((8)–(13)) was mounted on the front leg of the tripod. From top to bottom each thermistor is located 0.64 m below and 0.41 m forward of the previous thermistor. The top thermistor of the chain ((8)) is 0.91 m below and 0.26 m aft of the turbulence package.

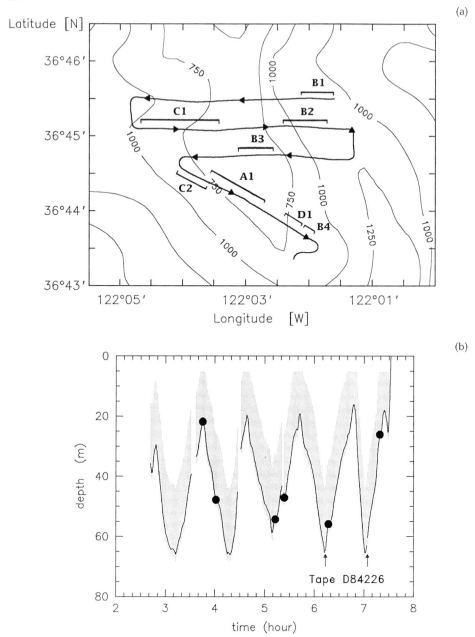

Fig. 2. Track and depth patterns of Dive 22 on October 19, 1984. (a) Track pattern with bottom topography (in meters). The ship speed was obtained by averaging the velocities from bins 4 and 5 of the ADCP and the heading was from the ship gyroscope. The shear layers encountered during the dive are labeled from **A1** to **D1**, as in Table 1. (b) Depth of the turbulence package as a function of local time. The shaded areas show the range of the ADCP data. The black circles indicate the heading changes in the track pattern.

3. Experimental conditions

On October 19, 1984, the operating area for dive 22 was centered around 122°03'W Long., 36°45'N Lat., 10 miles northwest of Monterey and 13 miles south of Santa Cruz, California (see Fig. 2a). The *Dolphin* track crossed a north-south ridge (depth 600 m) penetrating Monterey canyon. The deepest part of the submarine canyon (≈ 2000 m) lays to the southwest of the ridge, with its end curving northeast around the ridge. The predetermined track pattern consisted of three parallel east-west legs (10, 11 and 9.2 km long, respectively) followed by a 7.5 km southeast leg. The east-west legs were a kilometer apart. The average speed of the submarine, measured by the lower bins (bins 4 and 5) of the ADCP, was 1.5 m/s. During dive 22, the *Dolphin* operated in a depth-cycling mode between 15 and 65 m as shown in Fig. 2b. The average cycle period was 50 minutes. The turbulence and CTD data analysed in this paper come the later part of the dive (tape d84226) when the submarine changed course from east-west headings to a southeast heading (Fig. 2a).

4. Observations of shear layers

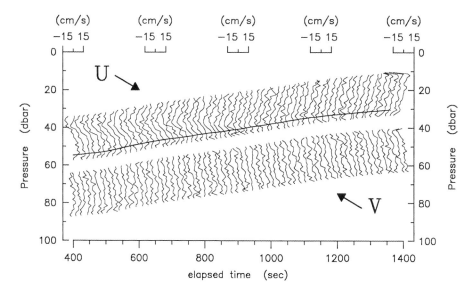

Fig. 3. Times series of the 15 second average velocity profiles for the upward-going leg of the southeast leg (Tape D84226). The horizontal axis is elapsed time from the beginning of the tape. *U* is along the axis of the submarine and *V* in the transversal direction (*V* profiles have been down-shifted by 30 m). The solid line across the plot marks the depth of the microstructure and CTD instrumentation.

An example of the longitudinal (*U*) and transversal (*V*) mean velocity profiles is shown in Fig. 3. The velocity profiles are 15 seconds apart, corresponding to a 36 ping ensemble average. The upper set of curves corresponds to the longitudinal velocity profiles at their true depth while the lower set of lines corresponds to the transversal velocity. Each bin of the velocity profiles has a rms noise level of 2.5 cm/s. In order to reduce this short-time error 4 profiles were averaged together to produce velocity profiles every minutes with a rms noise level less than 1.5 cm/s (slightly higher than the manufacturer value of 1 cm/s.).

A strong, 10 m thick shear layer (C2), the strongest in this dive, was observed between 35 and 45 m at the beginning of the southeast leg of the dive (219 minutes into dive 22 or 8 minutes after the start of tape D84226). The average mean shear in that layer was 0.02 sec^{-1}. As shown in Fig. 4, the shear was predominantly in the longitudinal velocity with a weaker shear in the transversal velocity. A close examination of Fig. 4 shows that the horizontal shear vector rotated 180° between 30 and 45 m.

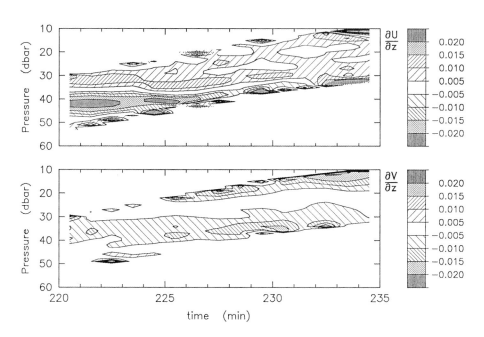

Fig. 4. Times series of the two components of the horizontal shear for the ADCP data of Fig. 3. Time is elapsed time from the start of the dive. Tape D84226 started 215 minutes into Dive 22.

The shear data of Fig. 4 are combined to get the total horizontal shear squared $S^2 \equiv (\partial U/\partial z)^2 + (\partial V/\partial z)^2$. S^2 is compared for the up and down-going leg of tape D84226 in Fig. 5. The core of the shear layer (located between 40 and 45 dbar) has a very large value ($> 5 \; 10^{-4}$), of the same order as values found in the equatorial undercurrents (Katz *et al.*, 1980, Gregg *et al.*, 1985). Best estimates of the stratification from the CTD data yield values of the Brunt-Väisälä frequency between 1.1 and 1.6 10^{-2} rad/sec (or 6.3 to 9.1 cycles/hour). The corresponding local Richardson numbers are between 0.2 and 2.5. These low values of

Richardson numbers imply that shear instabilities are viable. The shear layer observed in the up-going leg is not present in the following down-going leg. Shear is only present in the mixed layer (patch A1 between 10 and 20 m). As shown in the following section, the submarine entered a water mass with a very different T-S relationship after the up-going leg. No large scale circulation data is available for this dive to establish how different the flow patterns in Monterey Canyon and on the Continental Shelf are. A local front may exist near the southern of the ridge described in section 3 and create a complex flow pattern around the ridge.

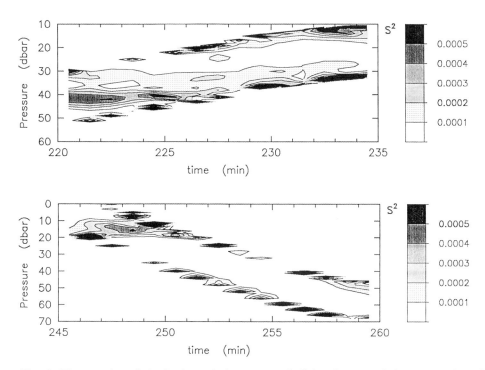

Fig. 5. Times series of the horizontal shear squared S^2 for the up and down going leg of tape D84226 (Dive 22). Time is elapsed time from the start of the dive. Tape D84226 started 211 minutes into Dive 22.

5. CTD data and T-S diagram

One-second averaged time series of temperature, salinity, density and pressure were computed from the Seabird CTD data. In the up-going leg, the *Dolphin* went from 65 to 15 m in 35 minutes. The average horizontal speed was 1.5 m/s, while the average rising speed, where the ramps were observed, was 2.5 cm/s. Five very well defined ramps were observed starting at 6:32 am (230 minutes into the dive) between 32 and 35 db (see Fig. 6). The time series of temperature, salinity and density shows that the ramps had a salinity and temperature structure which are not compensated in density. As a result, each ramp is made of water heavier than its surrounding and has an increased potential energy. This could be interpreted as an internal wave displacement or an overturning event, possibly a Kelvin-Helmholtz billow ready to collapse. The average horizontal scale of the ramps is 50 m (assuming they are stationary or moving slowly compared to the submarine speed), while their thickness can be estimated as 5 m (consistent with the thermistor chain observations discussed in Sec. 7). Some salinity spikes are due to the time constant missmatch between the thermistor and the conductivity probe. In the following down-going leg, the *Dolphin* descended from 20 to 65 m in 11 minutes, at an average descending rate of 6.8 cm/s.

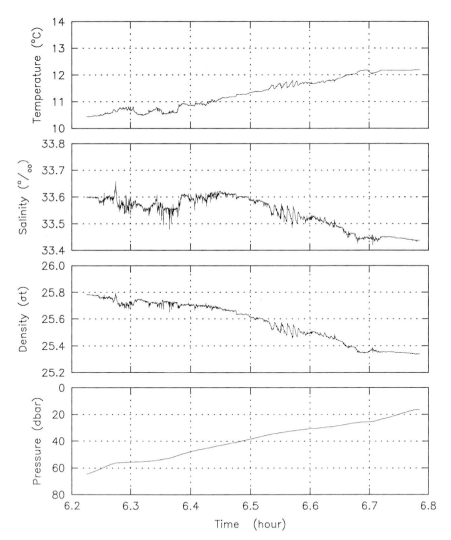

Fig. 6. One-second average time series of temperature, salinity, density and pressure from the Seabird CTD; dive 22, tape D84226. Time is local time on October 19, 1984.

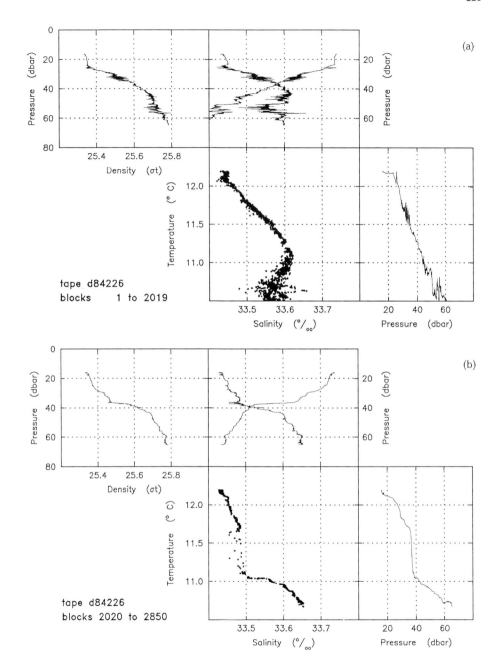

Fig. 7. T-S diagram and temperature, salinity and density vs. depth. (a) Upward going leg of tape D84226. The ramps were observed near 35 db. (b) Downward going leg of tape D84226. Note the different T-S relationship between the two profiles, indicating two distinct water masses.

The CTD data are plotted as a T-S diagram and temperature, salinity and density versus pressure plots in Fig. 7 (up-going leg in Fig. 7a and down-going leg in Fig. 7b). The central plot shows temperature versus salinity. To the right of the T-S diagram, temperature is plotted versus pressure, while above it, salinity is plotted versus pressure. In this last plot, the temperature trace has been added using the same scale as the left bottom plot to enable easy comparison of salinity and temperature traces at a given pressure level. Finally the upper right plot show density versus pressure which can be compared with the adjacent temperature and salinity traces. The temperature and salinity location on the pressure trace of any point of the T-S diagram can be obtained by going right and up respectively. A salinity maximum of 33.63 ppt occurs at 43 dbar below the depth where the ramps were observed. This T-S relationship is identical to the one measured in the earlier part of the dive.

6. Turbulence data

The temperature, temperature gradient and velocity shear time series from the turbulence package are shown in Fig. 8. The horizontal axis is the elapsed time in seconds from the start of tape D84226, i.e., an elapsed time of 0 correspond to an absolute time of 6:15 am or 215 minutes into dive 22. The data of Fig. 8 are a 32 point average of the original time series, corresponding to a Nyquist frequency of 8 Hz.

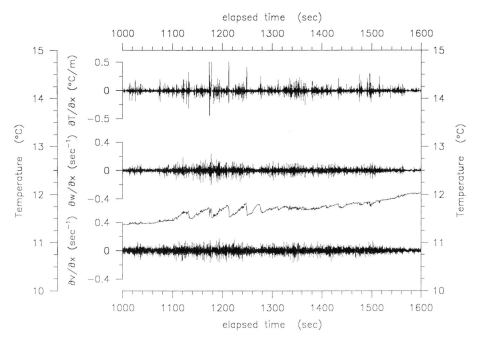

Fig. 8. Time series of microstructure temperature, temperature gradient $\partial T/\partial x$ and velocity shears $\partial w/\partial x$ and $\partial v/\partial x$ for dive 22, tape D84226.

The temperature trace from the fast thermistor (upward moving solid line scaled with the outmost axes) shows the details of the ramps observed between 1100 and 1300 seconds into tape d84226 and suggests that several more ramps between 1280 and 1500 seconds may have been partially sampled. This last observation in conjunction with the assumption that ramps are formed of surrounding water was used to estimate the thickness of the ramp as approximately 5 m. The typical shape of a ramp, as shown in the fourth ramp (between 1210 and 1240 seconds) consists of a slow increase in temperature of 0.25 °C followed by a very sharp temperature drop. The sharp trailing edge of the ramp is associated with a large mostly one sided spike in the temperature gradient (upper curve) and a decrease in the velocity shear variance (bottom two traces). In contrast the slow rising leading edge of the ramp is associated with increased velocity shear variance and probably dissipation and substantial temperature gradient variance. At this point it should be noted that if the submarine had been sampling in the opposite direction, i.e., going northwest the shape of the ramp would have been reversed (with a sharp increasing leading edge and a slow decreasing trailing edge). Some such observations have been made in another dive. No ramps were observed in the previous or following downward going legs.

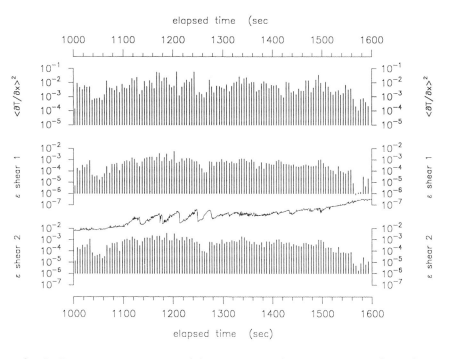

Fig. 9. Temperature variance and kinetic energy dissipation rates ϵ from the two shear probes for dive 22, tape D84226. (units are °C^{2}m^{-2} and cm^{2}sec^{-3} respectively).

Fig. 9 shows five second averages of the temperature variance and dissipation rates from the two shear probes for the data of Fig. 8. The temperature variance has maxima at the sharp edge of the ramps (trailing edge), followed by smaller values (by an order of magnitude) at the leading edge of the following ramp. A similar behavior is present, but less pronounced, in the kinetic energy dissipation rates ϵ. The average dissipation rates of kinetic energy are about 10^{-3}cm^2/sec^3 inside the first four ramps but drop by an order of magnitude in the last ramp. From Fig. 8 it can be observed that the last ramp has a much smoother temperature profile that is consistent with a low turbulence level. The noise level for the dissipation rates measured from the *Dolphin* are about $2 \cdot 10^{-7}$ cm^2/sec^3 (Osborn & Lueck, 1985).

7. Thermistor chain data

Finally the thermistor chain measurements of Fig. 10 provide an indication of the vertical temperature structure of the ramps. The top temperature trace of Fig. 10 is the fast temperature sensor of the turbulence package as shown in Fig. 8. The traces below it have zero mean, are from thermistors ⑧, ⑩, ⑪ respectively and have been offset by increments of 0.5 °C. No absolute temperature calibration was possible for the thermistors. The middle and lower thermistors of the chain were broken during dive 22.

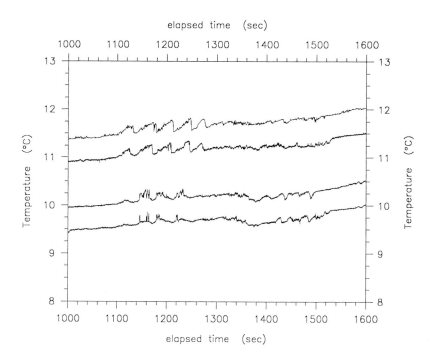

Fig. 10. Time series of the thermistor chain data for dive 22, tape D84226. Time is elapsed time from the start of the tape. The upper is the temperature signal from the thermistor of the turbulence package ⑥. The lower traces are from thermistors ⑧, ⑩ and ⑪ from the chain located on the front leg of the tripod. Thermistors ⑨, ⑫ and ⑬ of the chain were broken.

The most interesting aspect of Fig. 10 is the contrast between the shape of the top traces and the bottom traces. The shape of the ramps in the bottom traces is reversed, with the ramps starting with a sharp interface. They also contained blobs of warmer water. It would have been very interesting to have the temperature signal of the middle thermistor ⑨. Some weaker or decaying ramps are also observed between 1400 and 1500 sec. At that time the bottom thermistors reached the depth at which ramps were first observed by the turbulence package thermistor ⑥. The region between 1250 and 1400 sec could be a mixture of forming and dying ramps as suggested by the temperature gradient and shear time series of Fig. 7.

8. Conclusions

The shear measurements for dive 22 are summarized in Table 1. Table 1 shows that shear layers extend at least a thousand meters horizontally at the bottom of the mixed layer (class B) and in the seasonal thermocline (class C). The scales of the shear layers from Table 1 can be combined with the locations of the layers relative to the topography (Fig. 1) to infer an estimate of their overall horizontal scale. The observation of shear between 20 and 35 m (layers B1 through B3) indicate that there might be a continous (with some unknown intermittency) shear layer over several square kilometers. The elapsed time between the first and last observation was 3 hours. Therefore it is possible that the shear layers could persist an inertial period. The same conclusion could apply to the shear layers between 35 and 45 m (C1 and C2).

TABLE 1

Horizontal and vertical scales of velocity shear in Monterey Bay. The measurements shown in the table were taken on October 19, 1984 from the research submarine *USS Dolphin*. The depth range sampled during that dive was 20 to 65 m. The vertical range of the 1.2 MHz Acoustic Doppler Current Profiler (ADCP) was 25 m.

Location on the track	Relationship to the stratification	Depth range (m)	Vertical scale (m)	Horizontal scale (m)	Vertical shear S^2 (10^{-4} s^{-2})
A1	Mixed layer	10-20	10	>1500	2-5
B1	Bottom of mixed layer	20-30	10	>1000	2-4
B2	Bottom of mixed layer	25-35	10	>1600	2-4
B3	Bottom of mixed layer	20-30	10	>1100	3-6
B4	Bottom of mixed layer	20-35	15	> 340	3-6
C1	Seasonal thermocline	35-45	10	>2350	2-5
C2	Seasonal thermocline	35-45	10	>1100	2-6
D1	Main thermocline	40-48	8	> 500	1-4

Measurements of the vertical structure of the mean shear from a submarine showed that, in Monterey Canyon, the shear tends to be concentrated in layers 10 to 15 m thick. More measurements with similar vertical resolution are needed to establish whether this 10 m thickness is a typical oceanic value. Instruments with a vertical resolution better than 2 m are required to adequately sample the thickness and the maximum shear value of the layers.

Estimates of the horizontal scales of the shear layers was more difficult because of heading and depth changes of the submarine. Only a lower bound of the horizontal scales could be obtained. These shear layers extended more than 1000 m horizontally, possibly covering several square kilometers for several hours. The observations did not last long enough to clearly establish whether the shear layers were near-inertial features of the flow. The largest horizontal shear value measured with the Acoustic Doppler Current Profiler was comparable to the shear found in the equatorial undercurrents.

Some well-defined temperature and salinity ramps were observed in an upgoing leg of dive 22 during the October 1984 *Dolphin* cruise. The size of the temperature ramps is 5 to 10 times larger than similar ramps observed in the thermocline of Loch Ness (Thorpe and Hall, 1974) but their generating mechanism could be similar. The shear and density profiles measurements show that the local Richardson number *Ri* was below the critical value of ~0.25 in the shear layer, a condition for Kelvin-Helmholtz stability. The ramps were observed at the top of the shear layer where the Richardson number is greater than one. Therefore it is not clear whether the ramps are a manifestation of Kelvin-Helmholtz instability, a train of solitary or internal waves or entrainment at the edge of the shear layer. They appear to be an intermittent feature of the flow, thus unlikely to be observed very often. In order to determine whether these ramps are breaking Kelvin-Helmholtz billows, one can compare their vertical scale ($\sim$ 5 m from the vertical displacements in the temperature and density profiles) to the buoyancy length scale or Ozmidov scale L_O (Ozmidov, 1965). Near 35 m, the depth at which the ramps were observed, the background stratification was $N \approx 2 \cdot 10^{-2}$ sec^{-1} and the average dissipation rate inside the ramps was $\epsilon \approx 10^{-3}$ cm^2/sec^3, yielding a buoyancy scale $L_O \approx 11$ cm. Therefore it seems more likely that the ramps are steep internal waves trapped near the front, rather than breaking billows. The high dissipation rates inside the ramps could be the result of turbulence produced by secondary instabilities occuring where the waves are the steepest and the instantaneous shear (the sum of the internal wave shear and the mean current shear) is the largest.

If the ramps are breaking Kelvin-Helmoltz billows, then the thermistor chain measurements of Fig. 10 suggest that ramps 2, 3 and 4 are breaking billows. The fifth ramp could be a billow in the early phase of growth and the first ramp could be either a partially sampled billow or a dying billow. The sharp interfaces of the ramps are associated with high dissipation rates and temperature variances.

9. Acknowledgement

This work was supported by the Office of Naval Research under contracts N00014-86-K-0060 and N00014-87-K-0087.

10. References

Champagne, F.H., Friehe, C.A., LaRue, J.C., and Wyngaard, J.C., 1977. Flux measurements, flux estimation techniques, and fine scale turbulence in the unstable surface layer over land. *J. Atmos. Sci.*, **34:** 515-530.

Gibson, C.H., Friehe, C.A., and McConnell, S.O., 1977. Structure of sheared turbulent fields. *Phys. Fluids,* **20 (10):** s156-s167.

Gregg, M.C., Peters, H., Wesson, J.C., Oakey, N.S. and Shay, T.S., 1985. Intensive measurements of turbulence and shear in the equatorial undercurrent. *Nature,* **318, (6042):** 140-144.

Kaimal, J.C., Wyngaard, J.C., Izumi, Y., and Cote, O.R., 1972. Spectral characteristics of surface-layer turbulence. *Quart. J. Roy. Meteor. Soc.,* **98:** 563-589.

Katz, E.J., Bruce, J.G., and Petrie, B.D., 1980. Salt and mass flux in the Atlantic equatorial undercurrent. In *Equatorial and A-Scale oceanography, suppl. II Deep Sea Res.,* **26:** 137-160.

Loeser, D.J., and Chapman, R.B., 1979. Flow perturbations induced by the SS Dolphin. SAI Report SAI-052-79-848-LJ.

Osborn, T.R. and Crawford, W.R., 1983. An airfoil probe for measuring turbulent velocity fluctuations in water. In: F. Dobson, L. Hasse and R. Davis, (Editors), Plenum Press. *Air-Sea Interaction Instruments and Methods,* 369-386.

Osborn, T.R. and Lueck, R.G., 1985. Turbulence measurements with a submarine. *J. Phys. Ocean.,* **15:** 1502-1520.

Ozmidov, R.V., 1965. On the turbulent exchange in a stably stratified ocean. *Atmos. Ocean. Phys.,* **8,** 853-860.

Pederson, A.M. and Gregg, M.C., 1979. Development of a small *in situ* conductivity instrument. *J. Oceanogr. Eng.,* **4:** 69-75.

Pond, S., Phelps, G.T., Paquin, J.E., McBean, G. and Stewart, R.W., 1971. Measurements of the turbulent fluxes of momentum, moisture and sensible heat over the ocean. *J. Atmos. Sci.,* **28:** 901-917.

Schmitt, K.F., Friehe, C.A. and Gibson, C.H., 1979. Structure of marine surface layer turbulence. *J. Atmos. Sci.,* **36:** 602-618.

Thorpe, S.A., 1977. Turbulence and mixing in a Scottish loch. *Phil. Trans Roy. Soc. London,* **A286:** 125-181.

Thorpe, S.A., 1987. Transitional phenomena and the development of turbulence in stratified fluids: A review. *J. Geophys. Res.,* **92:** 5231-5248.

Thorpe, S.A. and Hall, A.J., 1974. Evidence of Kelvin-Helmholtz billows in Loch Ness. *Linmology & Oceanog.,* **19 (6):** 973-976.

Thorpe, S.A. and Hall, A.J., 1980. The mixing layer in Loch Ness. *J. Fluid Mech.,* **101:** 687-703.

Thorpe, S.A. and Hall, A.J., 1982. Observations of thermal structure of Langmuir circulation. *J. Fluid Mech.,* **114:** 237-250.

Thorpe, S.A., Hall, A.J., Taylor, C. and Allen, J., 1977. Billows in Loch Ness. *Deep-Sea Res.,* **24:** 371-379.

Woods, J.D., 1968. Wave induced shear instability in the summer thermocline. *J. Fluid Mech.,* **32:** 791-800.

MEDITERRANEAN SALT LENSES

DAVE HEBERT

Department of Oceanography, Dalhousie University, Halifax, Nova Scotia (Canada)

ABSTRACT

Small coherent eddies of Mediterranean water (commonly called Meddies) have been found in several places in the Atlantic Ocean. The importance of these eddies and other similar types of eddies to ocean mixing depends on several factors, such as production and decay rates. In this paper, I will attempt to estimate the effect of Meddies on salt transport for different geographical regions based on the few observations available. In the last section, a brief description of a recently completed cooperative experiment to study the evolution of a Meddy will be given.

1 INTRODUCTION

Small coherent eddies with anomalous water properties have been found far from their believed source waters. McWilliams (1985) has reviewed the basic knowledge of these features and speculated whether these eddies are an important mechanism in the transport of heat, salt and other chemical tracers in the ocean. To determine their importance, it is necessary to answer several questions:

1) Where and how are the eddies formed? We need to know the source water and initial composition of the eddies, to understand their transport properties.

2) What is the production rate of the eddies? If only a few eddies of a particular water mass are formed per year, those eddies are probably not important in the overall transport of tracers within the ocean.

3) Where and how do the eddies decay? If the eddies carry the source water to a particular region and then destroyed [e.g. by a large external shear (Ruddick, 1987)], parameterizing the transport will be different than the case where the eddy slowly loses its source water as it is advected.

At the present time, these questions cannot be answered for most types of eddies. One group of eddies with water properties of the Mediterranean (Meddies) have been observed in several locations of the Atlantic Ocean. McWilliams (1985) believes that this type of eddy may play an important role, if not the dominant one, in the transport of salt.

In this paper, the formation region of Meddies will be discussed first, followed by speculation on the formation mechanism and production rate of Meddies (based on weak evidence). This will be followed by a discussion of the salt transport in the Atlantic Ocean. Finally, a brief description of a recently completed experiment proposed by Larry Armi to track and observe the decay of a Meddy will be presented.

2 FORMATION OF MEDDIES

During Leg 3 of the Transient Tracers in the Ocean (TTO) North Atlantic Study (May - June 1981), three Meddies were found in the Canary Basin (Armi & Zenk, 1984). Each had a central core of relatively uniform water, presumably little changed since the formation of the Meddy. Along with salinity and temperature, nutrients were measured in the core of these Meddies (Physical and Chemical Oceanographic Data Facility, 1981; Armi and Zenk, 1984). Similar measurements were made during the second visit to another Meddy observed two years later (called Sharon*; see Section 4). Nutrients (PO_4, SiO_3 and NO_3) and salinity on the σ_1**$= 32$ density surface for each Meddy are given in Table 1. These properties of the Meddies are an indication that their source water is the salty and nutrient depleted Mediterranean. The distribution of salt and nutrients on this density surface for the eastern North Atlantic region from the archived NODC data set are shown in Figure 1. As we can see, the source of the Meddies must be Mediterranean outflow water after it has mixed with some North Atlantic Central Water along the southern coast of Spain and Portugal, reducing the salinity and increasing the nutrient content of the Mediterranean water.

	Position	P/db	S/PSU	$\theta_1/°C$	$PO_4/$ $(\mu M/kg)$	$SiO_3/$ $(\mu M/kg)$	$NO_3/$ $(\mu M/kg)$
TTO #1	$31°\ 0'N, 26°45'N$ (May 1981)	889	36.180	11.718	0.86	7.7	14.1
#2	$31°10'N, 22°30'N$ (June 1981)	973	36.156	11.629	0.88	7.8	14.1
#3	$33°\ 4'N, 21°35'W$ (June 1981)	1000	36.127	11.521	0.89	8.0	14.6
Sharon	$28°30'N, 23°\ 0'W$ (June 1985)	993	36.224	11.880	1.03	7.31	13.36

Table 1. List of salinity, potential temperature θ_1 and nutrients in the core of the Meddy on the $\sigma_1 = 32$ density surface.

Several mechanisms are possible candidates for the formation of Meddies. Käse and Zenk (1987) have suggested that the Azores current can cause the Mediterranean salt tongue at $33°N$, $22°W$ to become baroclinically unstable and produce Meddies. This seems unlikely since the Mediterranean salt tongue in this region has a smaller signal of Mediterranean water than the Meddies. From Figure 1, we see that this region has a salinity less than $35.8 PSU$. After the Mediterranean water flows through the Strait of Gibraltar, it follows the

---

* During the first cruise (October 1984) by Larry Armi to find and track a Meddy, it was decided that the Meddy would be named after the person who had picked the hour of day when the Meddy was found. The lucky winner was Sharon Yamasaki.

** σ_1 is the density of a parcel of water moved adiabatically to a pressure of 1000 db. That is $\sigma_1 = \rho(S, \theta(S, T, p, 1000\ db), 100\ db) - 1000$. Also, $\theta_1 = \theta(S, T, p, 1000\ db)$.

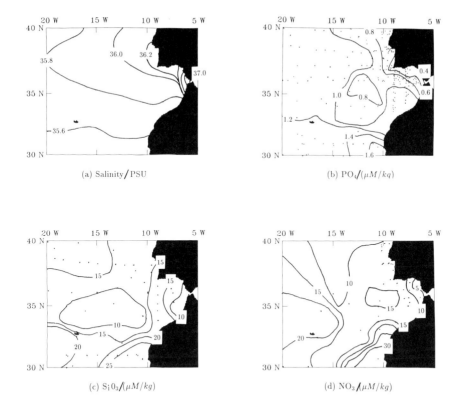

Figure 1. Distribution of (a) salinity, (b) phosphate, (c) silicate and (d) nitrate on $\sigma_1 = 32$ density surface. Dots are at the location of data points used in the contouring. (Salinity has enough data points to render useless the displaying their locations.) The distribution of temperature (θ_1) is related to salinity (*e.g.* $\theta_1 = 9.5°C$ when $S = 35.6\ PSU$ and $11.8°C$ when $S = 36.2\ PSU$).

bottom topography along the northern side of the Gulf of Cadiz, mixing with North Atlantic Central Water (NACW) and descending to its stable density level. There are several canyons along the shelf edge and the Mediterranean outflow has been observed to follow some of these canyons (Grünlingh, 1981). At Cape St. Vincent, the southwestern tip of Portugal, Ambar & Howe (1979) observed that the Mediterranean outflow meanders up a canyon after turning northward. It is possible that the current becomes unstable at this point. From the salt and nutrient distribution on the $\sigma_1 = 32$ density surface, we see that this region is a likely formation region for Meddies. There is more salinity and PO_4 data than SiO_3 and NO_3 data in the Gulf of Cadiz region; therefore the contour plots of salinity and phosphate show the Mediterranean outflow better than the plots of silicate and nitrate. There is still another possible formation mechanism. Gründingh (1981) observed a pulse of Mediterranean water traveling down some of the canyons and passing several current meter moorings located between $7°30'W$ and $8°40'W$. The duration of the pulse at any location was approximately two days. From the increased Mediterranean outflow flux and the size of the pulse, there is enough excess Mediterranean water to form a Meddy approximately the size of Meddies observed in the Canary Basin (Armi & Zenk, 1984). This increased flow due to the pulse was such that the outflow water in two adjacent canyons were connected. Therefore it is easy to see that part of the increased outflow can be pinched off and form an isolated eddy after undergoing geostrophic adjustment. Or this excess flow could cause the Mediterranean outflow, which is a density current, to become baroclinically unstable. Also, it is unknown how much extra, if any, entrainment of NACW occurs as this pulse travels through the Gulf of Cadiz. This pulse was observed with current meters; thus the water properties of the pulse is unknown. Without actual observations of a Meddy forming, which is unlikely in this complex region, any theory on the formation of Meddies is highly speculative.

3 TRANSPORT OF TRACERS BY MEDDIES

In the previous section, it was shown that Meddies contain warm, salty, nutrient-depleted water compared to the Atlantic water on the same density surface. As an example of the possible importance of Meddies in the transport of tracers, the transport of salt by Meddies will be examined in this section. The importance of Meddies in the transport of salt will depend on the geographic region. How may Meddies enter the region per year and how much salt and other tracers do they lose in the region. In this paper, I will discuss their importance for three regions: the Gulf of Cadiz, the Canary Basin and the β–triangle region. Estimates of the salt transport by Meddies are based on the few observations available; thus, any estimates presented here should be believable but with reservations.

Gulf of Cadiz

Gründlingh (1981) observed the transport of the Mediterranean outflow to increase by a factor of 4 during a solitary event. At the current meter mooring located at $36°30'N$, $8°40'W$, three events were observed over a 28 day period. Based on this occurrence rate, the salt flux due to these preformed Meddies is one-half of the Mediterranean outflow flux in the Gulf of Cadiz. Thorpe (1976) observed one event in 21 days from the data of a mooring located at approximately $36°N$, $8°W$. This occurrence rate would give a Meddy flux that is 1/4 of the Mediterranean outflow in this region. If this flux estimate is correct, Meddies (or preformed Meddies) appear to be an important transport mechanism of salt but not the dominant one in the Gulf of Cadiz.

Canary Basin

Most of the Meddies observed have been found in the Canary Basin (Armi & Zenk, 1984; Käse & Zenk, 1987). From the TTO cruise data, Armi & Zenk (1984) estimated that 8% of the region contains Meddies. Examining NODC historical data they found that 4% of the Canary Basin data contained anomalous high salinity samples believed to be Meddies. Käse & Zenk (1987) found evidence for three Meddies in their survey region at approximately

$34°N$, $22°W$. From their data, it is estimated that approximately 5% of the Canary Basin region contains Meddies. It is difficult to estimate the advective flux of salt and its divergence in this region because of the very variability in the velocity field (Käse and Siedler, 1982). From the observation of "Sharon" (see Section 4), Meddies lose a significant portion of their excess salt and heat in this region. Thus, they are probably a significant transport mechanism of salt and heat into the Canary Basin.

β–triangle (centered at $27°N$, $33°W$)

Armi & Stommel (1983) found one CTD cast in 143 taken in this region went through a Meddy. From this, they estimated that 0.7% of the region possibly contains Meddies. They calculated that if three Meddies entered the β–triangle region and died every year, then this flux divergence would equal the calculated advective salt flux divergence. The advective flux across the β–triangle region is $u \, \delta S \, h\ell$ where u is the mean advection speed ($2 \times 10^{-2} ms^{-1}$); δS, the salt difference across the β–triangle in the downstream direction ($0.1 kg \ m^{-3}$); h, the thickness of the water column that the Meddy's salt transport would affect ($10^3 m$); and ℓ, the cross-stream width of the β–triangle ($10^6 m$). Thus the advective salt flux through the β–triangle is $2 \times 10^6 kg \ s^{-1}$. The salt flux due to Meddies is $V \Delta S \ (n/t)$ where V is the volume of a Meddy ($2 \times 10^{12} m^3$); ΔS, the salt anomaly ($1 kg \ m^{-3}$); and (n/t), the number of Meddies per unit time passing through the β–triangle region. Thus one Meddy every 10 days must pass through the β–triangle to equal the advective salt flux. Or that 25% of the area covered by the β–triangle would contain a Meddy at any time (based on a Meddy radius of $35 km$). Since only two Meddies have been found in this region (Armi & Stommel, 1983; Berestov, et al., 1986), it seems unlikely that Meddies are an important salt transport mechanism . It is interesting to note that the Meddy described by Berestov et al. (1986) has a smaller Mediterranean Water signal than that of Armi & Stommel (1983) but its structure is similar to Meddy "Sharon" during the fourth survey. Also, this Meddy was found south of the β–triangle region($20°N$, $37°W$) and moving in a northwestern direction. It might have traveled completely through the Canary Basin and into the Cape Verde Abyssal Plain before being advected westward whereas the Armi & Stommel (1983) Meddy might have not traveled as far south before moving westward.

4 THE MEDITERRANEAN SALT LENS EXPERIMENT

After discovering three Meddies during Leg 3 of the TTO North Atlantic Study, Armi proposed to find a Meddy, track it using SOFAR floats, and observe the changes between two surveys a year apart. A third visit to the Meddy was proposed by Oakey and Ruddick to examine the small scale mixing processes occurring and compare these results to the large scale changes in the Meddy. Richardson and Price surveyed the same Meddy during the cruise to replace the SOFAR listening stations of the Mediterranean Outflow Experiment. Table 2 lists the participants in the experiment along with a list of measurements made during each survey. The trajectory of the Meddy (Figure 2) shows that it moved in an erratic pattern with variable speed and direction but in a generally southward direction (Price et al., 1986).

234

Date	Principal Investigators	Objectives
Oct. 84	Armi (Scripps) Rossby (URI)	- find a Meddy - send it with SOFAR floats and pop-up drifters - detailed CTD survey - PEGASUS velocity profiles
June 85	Oakey (BIO) Ruddick (Dal) Hebert (Dal)	- locate same Meddy - detailed CTD survey - nutrients & $^3H - He$ samples - velocity measurements (9 expendable current profilers (XCPs) and a current meter mooring) - EPSONDE microstructure measurements
Oct. 85	Armi (Scripps) Rossby (URI)	- locate Meddy - detailed CTD survey - PEGASUS velocity profiles
Oct. 86	Richardson (WHOI) Price (WHOI)	- locate Meddy - XBT survey - CTD survey - XCP profiles

Table 2. List of participants and measurements made during the Mediterranean Salt Lens Experiment.

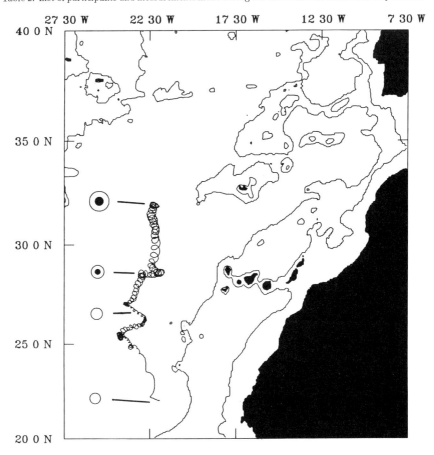

Figure 2. Trajectory of SOFAR float EB128 in Meddy "Sharon" for a two year period. The lcoation and size (to scale) of the core of the Meddy (solid circles) and total size (open circles) for the four surveys listed in Table 2.

The approximate size of the core (solid circle) and total size (open circle), along with the location of the survey, are shown in Figure 2. The size of the Meddy for each survey is given in Table 3. From these results, it appears that the core decreases at a constant rate, in terms

| Date | Radius/km | | Thickness/m |
	Core	Total	
Oct. 84	24	58	750
June 85	15	38	550
Oct. 85	—	34	425
Oct. 86	—	31	75

Table 3. Overall size of Meddy "Sharon" at the time of the four surveys. The radius of the core was defined as the radius where the peak-to-peak amplitude of the salinity interleaving becomes greater than 0.01 PSU. The radius of the total Meddy was defined as the radius where the peak-to-peak amplitude of the salinity interleaving becomes less than 0.05 PSU.

of area, which the total size decreases at a much slower rate as the Meddy decays. The thickness of the Meddy decreases at nearly a uniform rate. The decay mechanisms for the Meddy for the vertical and horizontal directions are likely to be different and the horizontal decay rate seems to depend on whether the Meddy core is present or not.

There are several mechanisms that could be responsible for the decay of a Meddy such as intrusions (eg. Joyce, 1977), double diffusion (Turner, 1973), shear-driven turbulent mixing (Oakey, 1988) and numerous larger scale instabilities (McWilliams, 1985). For "Sharon", we have evidence of intrusions eating away from the edge of the Meddy (see paper by Ruddick & Hebert, 1988) and double diffusive processes especially evident at the bottom of the Meddy (Figure 3). The actual change in the structure of the Meddy and comparison with mixing rates from different small scale processes will be made at a later date. Another Meddy found by Armi in October 1985 could not be found in 1986 by Richardson (pers. comm., 1986), but the SOFAR floats were found near the Cruiser and Great Meteor Seamounts. It is possible that this Meddy interacted with these seamounts and was torn apart by some unknown process.

5 CONCLUSIONS

From the very few observations available, the transport of salt and heat by Meddies appear to be important only in Gulf of Cadiz and Canary Basin region. Further west, they carry only a small portion of the total salt transport; therefore they are not a significant mechanism of mixing in the ocean basin, although the first published account of a Meddy (McDowell & Rossby, 1978) was of a Meddy found near the Bahamas. Also, the decay of Meddies appears to be a slow gradual death. There is not a deposit of salt and heat in a particular region far from the source waters. Therefore, the salt transport by Meddies can be included in the parameterization of the transport by the mesoscale eddies.

Tracking a particular Meddy over a period of time will allow us to examine models that

Salinity / PSU

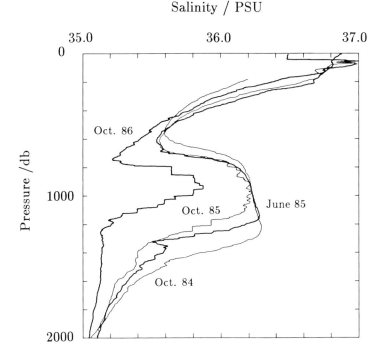

Figure 3. Typical salinity profiles through the centre of the Meddy for each survey.

parameterize small scale mixing processes and compare the results to the bulk changes in the Meddy. This area of research will likely be the most fruitful part of the experiment. A taste of this research can be found in the accompanying papers concerning data collected during the second survey of Meddy "Sharon" (Ruddick & Hebert, 1988; Oakey, 1988).

ACKNOWLEDGEMENTS
    I thank Larry Armi and Phil Richardson for allowing me the use of their CTD data. The useful discussions with them and the other collaborators, Neil Oakey, Barry Ruddick and Tom Rossby, is greatly appreciated. I also thank Chris Garrett and Barry Ruddick for the financial support which allowed me to attend the Liège Colloquium.

REFERENCES
Ambar, I. and M.R. Howe, 1979. Observations of the Mediterranean outflow – II The deep circulation in the vicinity of the Gulf of Cadiz. *Deep Sea Res.*, **26**: 555-568.
Armi, L. and H. Stommel, 1983. Four Views of a Portion of the North Atlantic Subtropical Gyre. *J. Phys. Oceanogr.*, **13**: 828-857.
Armi, L. and W. Zenk, 1984. Large Lenses of Highly Saline Mediterranean Water. *J. Phys. Oceanogr.*, **14**: 1560-1576.
Berestov, A.L., V.D. Egorikhin, Yu.A. Ivanov, V.G. Kort, M.N. Koshlyakov, Yu.F. Lukashev, A.S. Monin, E.G. Morozov, K.V. Moroshkin, I.M. Ovchinnikov, V.T. Paka, T.B. Tsybaneva, I.F. Shadrin, S.M. Shapovalov, AlD. Shcherbinin and M.I. Yaremchuk, 1986. Thermohaline, hydrochemical and dynamical characteristics of an intrusive lens

of Mediterranean water as based on the data collected during the "Mesopolygon – 85" expedition in the Tropical Atlantic. (in Russian) IN: K.N. Fedorov, Editor, Intrathermocline eddies in the ocean. (in Russian). Moscow: P.P. Shirshov Institute of Oceanology of the USSR Academy of Sciences, pp. 35-49.

Gründlingh, M. L., 1981. On the observation of a solitary event in the Mediterranean outflow west of Gibraltar. *"Meteor" Forsch. - Ergebn. A/B*, **23**: 15-46.

Joyce, T.M., 1977. A note on the lateral mixing of water masses. *J. Phys. Oceanogr.* **7**: 626-629.

Käse, R.H. and G. Siedler, 1982. Meandering of the subtropical front southeast of the Azores. *Nature* **300**(5889): 245-246.

Käse, R.H. and W. Zenk, 1987. Reconstructed Mediterranean Salt Lens Trajectories. *J. Phys. Oceanogr.*, **17**: 158-163.

McDowell, S.E. and H.T. Rossby, 1978. Mediterranean Water. An intense mesoscale eddy off the Bahamas. *Science*, **202**: 1085-1087.

McWilliams, J.C., 1985. Submesoscale coherent vortices in the ocean. *Rev. Geophys.* **23**: 165-182.

Oakey, N.S., 1988. Estimates of mixing inferred from velocity and temperature microstructure. (this book)

Physical and Chemical Oceanography Data Facility, 1981. Transient Tracers in the Ocean. Preliminary Hydrographic Report Leg 3, PACODF Publ. No. 215, Scripps Institution of Oceanography, La Jolla, CA, 268 pp.

Price, J.F., T. McKee, J.F. Vandez, P.L. Richardson and L. Armi, 1986. SOFAR Float Mediterranean Outflow Data from the first year, 1984-1985. Woods Hole Oceanographic Institution Technical Report WHOI-86-31, 199pp.

Ruddick, B.R., 1987. Anticyclonic Lenses in Large-Scale Strain and Shear. *J. Phys. Oceanogr.* **17**: 741-749.

Ruddick, B.R. and D. Hebert, 1988. The Mixing of Meddy "Sharon". (this book)

Thorpe, S., 1976. Variability of the Mediterranean undercurrent in the Gulf of Cadiz. *Deep Sea Res.* **23**: 711-727.

Turner, J.S., 1973. *Buoyancy Effects in Fluids*. Cambridge University Press, 373 pp.

ESTIMATES OF MIXING INFERRED FROM TEMPERATURE AND VELOCITY MICROSTRUCTURE

NEIL S. OAKEY
Physical and Chemical Sciences, Department of Fisheries and Oceans
Bedford Institute of Oceanography, Dartmouth, Nova Scotia, Canada, B2Y 4A2

ABSTRACT
Oakey, N.S.,1988. Estimates of mixing inferred from temperature and velocity
    microstructure, Proceedings of the 19th International Liège colloquium on
    Small Scale Mixing in the Ocean.

    Using the profiler EPSONDE, temperature and velocity microstructure
measurements have been made to depths of 1500 m in the Canary Basin during the
study of a Mediterranean Salt Lens (a MEDDY). In this region where conditions
are favorable for double diffusive processes, particularly in the periphery of
the MEDDY, one finds large differences in estimates of vertical diffusivities
depending on whether they are calculated using temperature or velocity
microstructure. Some of the implications of this are discussed.

1 INTRODUCTION

    The rate of vertical mixing in the ocean may be inferred from the
intensity of observed temperature microstructure or alternately velocity
microstructure. The vertical diffusivity of heat, K_θ, may be derived from the
turbulent heat equation assuming stationarity, homogeneity and isotropy,
resulting in

$$K_\theta = 3\ D\ \frac{\overline{(\partial T'/\partial z)^2}}{(\partial \bar{T}/\partial z)^2} \qquad m^2 s^{-1} \tag{1}$$

where D is the molecular diffusivity of heat $(1.4 \times 10^{-7} m^2 s^{-1})$, and fluctuating
(T′) and mean (T) temperatures can be inferred from a vertical profile.
Alternatively, from the turbulent kinetic energy equation with assumptions as
before, the vertical diffusivity of mass, K_ρ, may be obtained from

$$K_\rho = \Gamma\ \varepsilon/N^2 \qquad m^2 s^{-1} \tag{2}$$

where ε is the turbulent kinetic energy dissipation and N is the buoyancy
frequency. Γ is a constant of proportionality related to the flux Richardson
Number as follows

$$\Gamma = R_f/(1-R_f)\ . \tag{3}$$

From critical flux Richardson number arguments as presented by Osborn(1980),
one may expect Γ to be of the order of 0.25.

For shear-driven mixing one assumes that K_ρ is equal to K_θ to obtain an equation for Γ in terms of measured quantities

$$\Gamma = N^2 \chi_\theta / (2 \varepsilon (\partial T/\partial z)^2) \tag{4}$$

where $\chi_\theta = 6 D \overline{(\partial T'/\partial z)^2}$ $^\circ C s^{-1}$ $\tag{5}$

is the rate of dissipation of temperature variance. Using this relationship, Oakey(1985) determined a mean value of $\Gamma=0.26$ for turbulence in the mixed layer. The relationship, (4), should hold true for shear driven turbulence in regions where the rate of strain of the temperature field is a consequence of the turbulent velocity field. One finds that, indeed, the cut-off scale of the temperature spectrum in such regions is determined from the Batchelor scale obtained using the dissipation, ε, from the velocity turbulence. In other regions in the ocean such as those where salt fingering or double-diffusive convection are possible, the diffusion of heat and mass may be quite different, yet may still provide strong temperature and velocity micro-structure signals that (with appropriate assumptions) may be used to estimate vertical diffusivities.

It is the intention in this paper to examine some vertical microstructure profiles between the surface to deeper than 1400 m using the profiler EPSONDE in an experiment in the Canary Basin. The study focussed on a Mediterranean salt lens (a MEDDY) at 1000 m depth. At the top and at the underside of this lens there were pronounced density steps where double diffusion should occur. Temperature and velocity microstructure measurements in these steppy regions produced extremely high values of Γ which may be used to infer flux rates in a situation where the driving force is not shear but rather the potential energy released by double diffusion.

2 THE EXPERIMENT

During June 1985 a MEDDY at $(28^\circ 30N, 23^\circ W)$ in the Canary Basin was studied extensively using a CTD to study the large scale aspects of this feature (Hebert,1988) and EPSONDE for microscale phenomena. During the experiment, over 150 CTD stations and 100 EPSONDE profiles were obtained to study the energetics of this lens of water. The present paper, however, will look in detail at only two of the profiles, Station 212 near the center and Station 220 near the periphery.

EPSONDE is a 2 meter long free-fall profiler (Oakey,1987) deployed on a kevlar multiconductor tether and data link. During the MEDDY study, this profiler measured temperature and temperature microstructure with a platinum thin-film thermometer and velocity microstructure with two shear probes of the type described by Osborn and Crawford(1980). EPSONDE was allowed to free-fall to about 1500 m at 0.8 to 1.0 ms^{-1} with the data telemetered to the surface via the tether cable. At the end of each profile EPSONDE was recovered by its tether line. For the current discussion, the EPSONDE data of interest are temperature and velocity gradients (microstructure) sampled at 256 Hz and temperature and depth (at 32 Hz). Using standard spectral analysis techniques and correcting for sensor and electronic frequency response, temperature gradient variance, $\overline{(\partial T'/\partial z)^2}$, and velocity shear variance, $\overline{(\partial u_i/\partial z)^2}$, for each of the two shear probes were calculated. Dissipation of temperature variance, χ_θ, is calculated from equation (5) and the dissipation of turbulent kinetic energy, ε, is given by

$$\varepsilon = 7.5 \, \nu \, \overline{(\partial u/\partial z)^2} \qquad m^2 s^{-3} \tag{6}$$

where $\nu \cong 1.3 \times 10^{-6} \, m^2 s^{-1}$ is the kinematic viscosity.

For the current analysis, the data were analyzed in 2 second data segments corresponding to 1.5 to 2 m in the vertical. For each segment the values of the above quantities were calculated, and those above the noise level of the instrument or data system were accepted for the calculation of Γ from equation (4). The value of the buoyancy frequency, N, was obtained from a CTD station taken either just before or just after the EPSONDE profile. N was calculated from a 40 meter running average and was in fact nearly constant $(2 \text{ to } 4 \times 10^{-3} \, s^{-1})$ over all of the profiles.

The MEDDY is shown in Figure 1 as a temperature section from near the center along a radius to its edge . It is centered near 1000 m depth and is about 500 m thick with a core of warmer saltier water than the surroundings. The isotherm at 11 °C superimposed on the CTD profiles identifies the edge of the lens of anomalous water. Temperature microstructure profiles obtained using the tethered-free-fall profiler EPSONDE are plotted as well.

To examine the microstructure, profiles of χ_θ and ε (averaged over 50 m in the vertical) are shown in the composites of Figures 2 and 3. Station 212, Figure 2, near the MEDDY center has higher levels of χ_θ and ε near the surface, decreasing to a minimum at a depth of about 600 m. The microstructure levels increase sharply just above the MEDDY until the relatively homogeneous water is reached at 800 m. Both decrease considerably (often below the instrument noise level) in the core of the MEDDY until one reaches a depth of about 1100 m. In this region of the feature where the large scale structure is quite steppy, levels are again quite high. By comparison, station 220 (Figure 3) at a radius of the lens where the core becomes rather diffuse, shows uniformly higher levels between 600 m and 1300 m. These are similar to levels above and below the core near the center when compared to Figure 2; although above 600 m the levels in both stations are very similar.

242

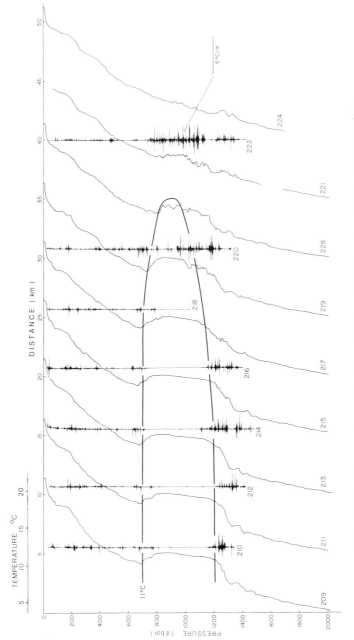

Fig. 1. The temperature structure (thin solid lines) and the corresponding temperature microstructure ('wiggly lines') is shown for a Mediterranean salt lens in the Canary Basin. The heavy 11 °C temperature contour marks the core of the MEDDY. Microstructure is most intense at the periphery.

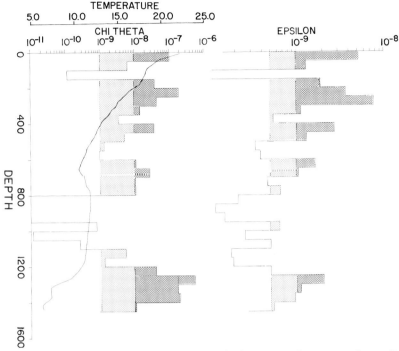

Fig. 2. For station 212 near the center of the MEDDY the vertical profile of χ_θ is shown superimposed on the mean temperature profile at the left. Values greater than 10^{-8} $^\circ C^2$ s^{-1} are shown darkly shaded, above 10^{-9} are lightly shaded. On the right, the corresponding dissipation, ε, is shown where values above $10^{-9} m^2 s^{-3}$ are shown heavily shaded and above 5×10^{-1} lightly shaded.

To show the detailed relationship of the microstructure intensities to the step structure below the MEDDY, the section of data from 1100 m to 1450 m for Station 212 is shown in Figure 4. There are several large steps between 1250 m and 1325 m. Very large values of χ_θ (the histogram) and ε (the circles) are associated with the high temperature gradient regions of these steps and yield values of Γ smaller than 1. Within the layers where the temperature gradients are small, there are by contrast, much smaller values of χ_θ and ε and many large values of Γ. The right hand panel of Figure 4 presents those Γ values calculated only when χ_θ and ε are greater than minimum values based on sensor noise levels. Gaps in the plot correspond not to small Γ values but to a segment where one of the parameters is less than its noise minimum. The Γ calculation assumes that the buoyancy frequency in the steppy region may be satisfactorily calculated from the CTD data averaged over 40 m. It is unfortunate that, at the time of this study, no CTD was installed on EPSONDE.

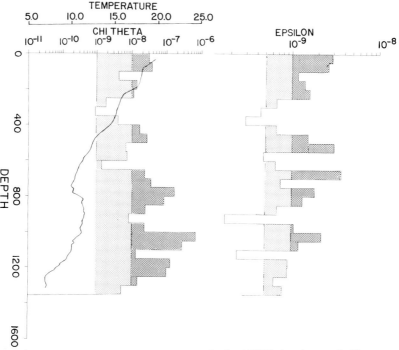

Fig. 3. For station 220 near the edge of the MEDDY is shown similar information to Fig 2. In this case, near the edge, there is much more microstructure than in the center.

2 THE DATA

We have discussed the microstructure quantities as if they were correctly modeled by ideas based on turbulence theories. Obviously, these models cannot be used in regions where double diffusion plays a major role. As pointed out by McDougall(1987) there are large differences in the diffusivities of heat, salt and mass in these regions.

In order to display the present data in a framework related to double diffusion, the Turner Angle (Ruddick,1983) is calculated. The angle is defined as follows (for z negative down).

$$Tu = Arctan((N_T^2 + N_S^2)/N^2) \qquad (7)$$

where $N_T^2 = \alpha \, g \, \partial T/\partial z$ and $N_S^2 = \beta g \, \partial S/\partial z$. The coefficients α and β are respectively the expansion coefficients for heat and salt, respectively. Tu is an angle from -90^0 to $+90^0$ for stable stratification. Angles from -90^0 to -45^0 are unstable to diffusive convection, -45^0 to $+45^0$ are stable and from 45^0 to 90^0 are unstable to salt fingering. Since $N^2 = N_T^2 - N_S^2$ this can be reformulated as follows

$$Tu = Arctan((2N_T^2 - N^2)/N^2) \qquad (8)$$

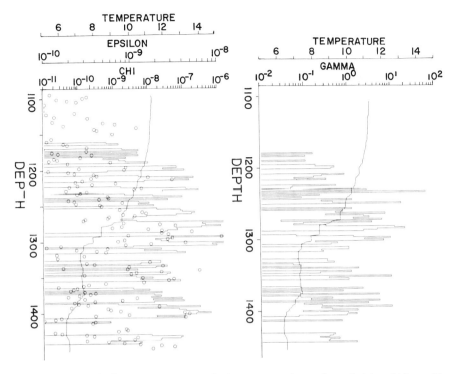

Fig. 4. The detailed microstructure features are shown for station 212 on the underside of the MEDDY. On the left, the temperature profile shows the steps and layers. There is increased temperature microstructure, χ_θ , in the higher gradient regions (the histogram plot) and corresponding higher levels of velocity microstructure, ε (circles). On the right, overlaid on a copy of the temperature profile is plotted Γ. The largest values are in general in the layers where the temperature gradient is lowest.

This form of the equation was used with N_T^2 obtained from the EPSONDE temperature gradient and with N^2 obtained from the CTD data. Figure 5 is a summary of all of the values of Γ for the two Stations 212 and 220 with $\varepsilon \geq 2 \times 10^{-10} \, m^2 s^{-3}$, $\chi_\theta \geq 5 \times 10^{-10} \, {}^0C^2 s^{-1}$ and $\partial T/\partial z \geq 3 \times 10^{-5} \, ({}^0C m^{-1})$ and N^2 (from 40 m averaged CTD data) plotted as a polar plot where the angles are the Turner angle and the radius is proportional to $\log_{10}(\Gamma)$.

There are several interesting features about this plot. Firstly, the only region where there are values of Γ significantly greater than 1 are in the regions where salt fingering or double diffusion are possible. A value of Γ greater than 1 is unlikely for turbulent mixing because it implies that more than half the energy that is being extracted from the turbulent field is going into potential energy. There are only a few (3) points which are slightly greater than 1 in the stable region and this is within the error in determining Γ . The other point of interest is that there are only a very few values of Γ in the region $-45^\circ \leq \Gamma \leq 45^\circ$. Almost all of the profile is characterized by warmer saltier water overlying colder fresher water and is therefore unstable in the salt fingering sense. Above the MEDDY there are regions of colder fresher water over warmer saltier water where diffusive convective instability is possible. Below the MEDDY it is predominantly salt fingering with small patches where diffusive convective processes can occur.

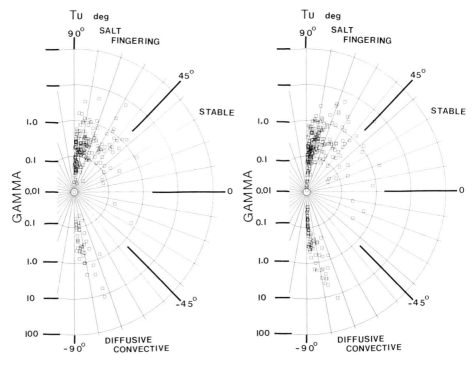

Fig 5. A polar plot of $\log_{10}\Gamma$ versus Turner Angle (Tu) is shown on the left for station 212. On the right is shown a similar plot for station 220.

4 DISCUSSION

The quantity Γ defined by (4) was obtained by assuming a turbulent mixing regime with $K_\theta = K_\rho$, in which case critical flux Richardson Number arguments (Osborn,1980) or measurements in the mixed layer (Oakey, 1985) suggest the value $\Gamma = 0.25$. If the assumption that $K_\theta = K_\rho$ is now relaxed and the value of Γ under this condition is denoted Γ^*, then this more general parameter can be evaluatedby calculating K_θ and K_ρ from (1) and (2) respectively using simultaneous temperature and microstructure measurements. The new parameter satisfies

$$\Gamma^*/\Gamma = K_\theta/K_\rho \qquad (9)$$

A large value of Γ^* implies that there is much more temperature variance than can be accounted for by the strain of the temperature field induced by the turbulent velocity field which we estimate by measuring dissipation. χ_θ is too large for the existing temperature gradient and the turbulent energy, ε, suggesting that another energy source is present. Note that, although a large Γ^* suggests that $K_\theta \gg K_\rho$ both quantities are calculated on the assumption of a turbulent field. The fact that there is a large temperature gradient variance does not imply that there is a large vertical heat flux $\overline{u'T'}$. This is a region where flux laws appropriate to double diffusion must be used.

The results presented here indicate that one must be very careful in calculating vertical eddy diffusivities from temperature and velocity microstructure alone. In large survey programs such as WOCE one must not simply measure ε, for example, and rely on automatic data processing to yield correct values for diffusivity.

5 ACKNOWLEDGEMENTS

I would like to thank Dave Hebert and Barry Ruddick for their many helpful discussions during the experiment and during the preparation of this paper.

6 REFERENCES

Hebert, D., 1988. Mediterranean Salt Lenses. Proceedings of the 19th International Liège Colloquium on Small Scale Mixing in the Ocean, Elsevier (in press, this issue).

McDougall, T.J., 1987. Thermobaricity, cabbeling and water-mass conversion. J. Geophys. Res., 92,5448-5464.

Oakey, N.S.,1985. Statistics of mixing parameters in the upper ocean during JASIN Phase 2. J. Phys. Oceanogr., 15, 1662-1675.

Oakey, N.S., 1987. EPSONDE: A deep ocean microstructure profiler. Oceans '87 Proceedings of the Marine Technology Society ,316-321.

Osborn, T.R., 1980. Estimates of the local rate of verticaldiffusion determined from dissipation measurements. J. Phys. Oceanogr., 10, 83-89.

Osborn, T.R., and Crawford, W.R., 1980. An airfoil probe for measuring velocity fluctuations in the water. Air-Sea Interaction: Instruments and Methods, F.W. Dobson, L. Hasse and R. Davis, Eds., Plenum, 369-386.

Ruddick, B., 1983. A practical indicator of the stability of the water column to double diffusive activity. Deep Sea Res., 30, 1105-1107.

THE MIXING OF MEDDY "SHARON"

BARRY RUDDICK and DAVE HEBERT
Dept. of Oceanography, Dalhousie University, Halifax, NS, CANADA

ABSTRACT

It is suggested that thermohaline intrusions are the dominant mechanism for mixing of a Mediterranean salt lens, and vertical and horizontal diffusivities for salt are estimated. Models of thermohaline intrusions are reviewed, and their predictions are compared to the lens observations. It was found that the "Meddy" front can be considered a wide front in the sense that the dynamics depend on the local value of the horizontal salinity gradient.

1 INTRODUCTION

In October, 1984, Larry Armi and Tom Rossby began a unique experiment: they found a lens of Mediterranean water drifting in the Eastern Atlantic, surveyed it with CTD and velocity profilers, and seeded it with SOFAR floats so that they and collaborators could return and re-observe the lens during its lifetime. This lens has been surveyed four times during two years. This is, to our knowledge, the longest experiment in which the movements and evolution of an identified parcel of seawater have been successfully traced.

One reason for tracking Mediterranean salt lenses ("Meddies") is to assess their role in lateral ocean stirring by the larger scale field of eddies. This aspect is examined in a companion paper (Hebert, 1988a). An alternate view is to consider the Meddy to be a very large laboratory tank in which the mixing behaviour of the deep ocean may be observed for an extended period. The swirling currents of the Meddy serve to contain the anomalous water in a coherent lens. The changes in lens properties can then be observed over long times, so that rates and mechanisms of exchange can be inferred.

In section 2 we suggest that intrusive interleaving was the dominant mixing mechanism of the lens, and in section 3 models of thermohaline intrusions are reviewed and compared to the Meddy observations.

2 OBSERVATIONS

In figure 1 a section of salinity profiles from the second survey, by Oakey, Ruddick and Hebert, is displayed. Four regions can be delineated: (1.) a core region, with smooth, stable vertical profiles of salinity and temperature (i.e., not double-diffusive). (2.) The upper surface of the lens, with strong diffusive-sense stratification. (3.) The lower surface, with strong finger sense stratification. (4.) An outer "intrusive" region, in which quasi-horizontal exchange of lens water with the relatively fresher Atlantic water is evidenced by the "wiggles" in the salinity profiles. The salinity amplitude of these wiggles is about 0.05 PSU (0.14 PSU peak-to-peak), and the 0.7 PSU salinity contrast between the core water and the Atlantic water of the same density outside occurs over the intrusive region, approximately $12km$ wide.

In this (second) survey, June 1985, the boundary between the unmixed core and the intrusive region was at an average radius of $17km$. The boundary was at $30km$ in the previous survey in October 1984. Figure 3 of the paper by Hebert (1988a) shows salinity profiles at the centre of the Meddy from each of the four surveys. The salinity stratification is stable and smooth in surveys I and II. The appearance of salinity inversions in survey III, in October 1985, demonstrates that the intrusions have reached the centre of the Meddy by this time. Thus, in the year between October 1984 and October 1985, intrusions worked into the Meddy core from a radius of $30km$, causing the core to become mixed with Atlantic water. This is the first observation that intrusive interleaving motions, whatever their dynamics, caused the mixing of a substantial mass of water. A full description of the experiment, including the detailed structure and size as revealed by the four surveys, will appear elsewhere (Armi *et al.*, 1988, in prep.).

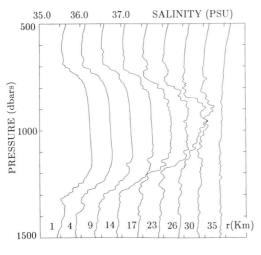

Figure 1. A sequence of vertical profiles of salinity, ranging from near the lens center to the edge. The radius of each station is marked at the bottom of each trace. Successive traces have been offset to the right by 0.5 PSU.

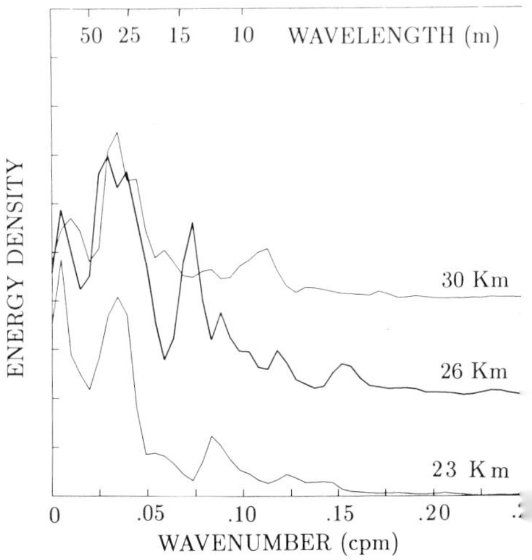

Figure 2. Temperature gradient autospectr from the intrusive zone, computed from the firs difference of temperature at 2 db intervals, wit 25 degrees of freedom per spectral estimate The radius of each station is marked to th right of each trace. Successive traces have bee displaced upwards for clarity.

The vertical scale of these intrusions can be seen from figure 2, which shows the autospectrum of temperature gradient at several stations from the interleaving region. The spectra show distinct peaks at a wavelength of about $30m$. The wavenumber of the spectral peak is variable, usually in the range $25 - 75m$. A search for systematic spatial variations of the peak wavelength has not yet been done.

The fact that the intrusive region advanced from $30km$ to the center in a year is extremely important, giving an estimate of the speed of advance of the intrusive motions of $(30km/1$ year$) = 1mm/s$. A velocity this small is not directly observable with present observational techniques. We now assume that the intrusions were driven by vertical mixing or friction (Stern, 1967; McIntyre, 1970). Models of such processes find that the inverse growth rates of such instabilities are typically a few days or weeks. The Meddy is at least eight months old by the time of the second survey, and so we expect that the intrusions have attained a steady, finite amplitude by then. Indeed, we found that the amplitude of the intrusions did not change markedly between the first, second, and third surveys.

We therefore expect that, in the centre of the intrusive region, a steady state balance will have been achieved in the salinity conservation equation, in which lateral advection balances vertical mixing (c.f. Joyce, 1977):

$$u\frac{\partial \overline{S}}{\partial x} = \frac{\partial}{\partial z}\left(K_S \frac{\partial S}{\partial z}\right) \tag{1.}$$

Allowance for the time rate of change would decrease the estimate of K_S afforded by eq. (1.). We take the following values from the observations above:

Table 1

Symbol	Value	Description
ΔS	$0.7 PSU$	(salinity contrast)
$2a$	$12 km$	(frontal zone width)
u	$1 mm/s$	(velocity of advance of intrusions)
$\partial \overline{S}/\partial x$	$0.7 PSU/12 km$	(salinity gradient)
$H = 2\pi/m$	$30 m$	(vertical wavelength)
δS	$0.05 PSU$	(salinity amplitude of intrusions)

In addition, we take the buoyancy frequency for the region to be $3 \times 10^{-3} s^{-1}$, an average value obtained from the hydrographic observations. Substituting the above values into (1) gives an estimate for the intrusion-scale vertical eddy diffusivity of salt:

$$K_S = \left(\frac{H}{2\pi}\right)^2 \frac{u\Delta S}{2a\delta S} = 3 \times 10^{-5} m^2/s.$$

The horizontal salt flux due to intrusive motions is the depth average of $u(z)S(z)$, and this translates to a horizontal diffusivity (in the radial direction) of

$$K_{HS} = (u\delta S)/2(\partial \overline{S}/\partial x) = 0.4 m^2/s.$$

Hebert's (1988a) figure 3 also shows that the Meddy decreased in thickness at the centre, primarily by erosion from below of roughly $200m$ in the year October '84 – October '85. Hebert (1988b) finds that this erosion rate is comparable with that expected by vertical salt finger transports, but that the net rate of loss of salt due to this mechanism is insignificant when compared to the loss by intrusions.

3 REVIEW OF INTRUSION MODELS

3.1 Laboratory Models

(i) Effects Introduced at a Boundary. The earliest laboratory observations of lateral intrusions dealt with the motions produced when a tank containing a salinity stratification is heated at a sidewall. Thorpe, Hutt, and Soulsby (1969) describe the series of layers formed, in which the temperature, salinity, and lateral velocity have a periodic structure which allows the lateral transport of the heat put in at the tank wall. Hart (1971) developed a theory for the formation of such layers, which predicted the formation of cells of both signs (rotating both clockwise and anti-clockwise). Comparison of theory and experiment showed that cells of one sign (the ones with upflow at the wall) rapidly entrained the other. Chen, Briggs, and Wirtz (1971), and others have extended the investigations of Thorpe, Hutt, and Soulsby (1969) to other configurations and geometries.

Turner and Chen (1974) give a visually striking review of many other phenomena involving layers which form when horizontal fluxes of heat or salt are introduced at a boundary (In many of these and later experiments, the analogue sugar/salt system was used, but we will discuss them in terms of heat and salt). Some of these were later investigated systematically. For example, Linden and Weber (1977) studied the layers formed when a sloping sidewall produces a diffusively driven lateral flux into the interior of a stratified tank. Tsinober (1983) studied the layers formed when a point source of heat was introduced at the center of a stratified tank. Huppert and Turner (1980) studied the layers produced by a vertical boundary of melting ice. In all these experiments, quasi-horizontal layers formed with T, S, and velocity structure such that the heat and salt fluxes introduced at the boundary tended to be carried away. The characteristic vertical scale, H of these layers was approximately

$$H = g(\Delta\rho/\rho)_{\text{lateral}}/N^2 \tag{2.}$$

where $(\Delta\rho/\rho)_{\text{lateral}}$ is the density contrast which would be produced by the boundary influence. For example, in the case of the heated sidewall, $(\Delta\rho/\rho)_{\text{lateral}} = \alpha\Delta T$, and the layer scale is close to the amount a fluid parcel would rise in the stratification if it were heated an amount ΔT. What has not been investigated in these experiments, which could be relevant to horizontal heat and salt transport in oceanic fronts, is the speed of advance of the layered region into undisturbed fluid.

(ii) Interior Fronts. Turner (1978) described several experiments in which interior property gradients were produced, and the resulting formation of intrusive layers which tended to spread, transporting heat and salt horizontally. In discussing the observed slope of the intrusive features, he described much of their physics. We reproduce that discussion below and elaborate on it in the light of present understanding.

Figure 3 depicts a thermohaline front after intrusions have formed, showing the salinity structure (temperature is similar), velocity profile, and the resulting double-diffusive structure, with alternating regions suitable for salt fingering (warm and salty on top) and diffusive (cool and fresh on top) convection. We now consider the process in stages, although they all occur simultaneously. (1.) As a parcel of warm, salty water is advected in from the left to the frontal region, it finds itself sitting between layers of cooler, fresher water above and below, so that salt fingers can form below the parcel, and diffusive convection can occur above it. Both the fingers and the diffusive convection will decrease the T and S of the water parcel, but it will gain or lose density according to whether the diffusive or finger buoyancy flux is the larger. Let us assume that the finger buoyancy flux is the larger, as the laboratory observations suggest that this is so. (2.) The finger fluxes cause the parcel of water to become cooler, fresher, and lighter as it continues to be advected to the right across the front. (3.) As the density of the parcel changes, it rises so as to stay at (almost) the level corresponding to its current density. Thus the layers must slope slightly relative to isopycnals so as to allow water parcels to stay at the appropriate density level as they advect and change density. The cold fresh water parcels must correspondingly sink as they move across in the opposite direction.

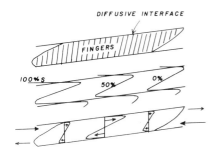

Figure 3. Structure of the laboratory frontal intrusions, showing salinity and velocity profiles.

What drives the intrusive motions? Small density differences exist between each fluid parcel and the average density for that level, $\bar{\rho}(z)$, such that the warm, salty intrusions are slightly lighter than average, and the cold fresh intrusions are slightly heavier. These cause baroclinic pressure gradients, and the resulting buoyancy forces tend to drive the warm, salty layers "uphill" to the right, and the cold, fresh ones "downhill" to the left against inertia and frictional forces. If the diffusive fluxes dominate (as they do in the heated sidewall experiments), the slope and density perturbations are in the opposite sense, so the buoyancy forces still drive the intrusions.

Thus, these intrusive motions are driven by a feedback loop in which lateral advection enables vertical double-diffusive mixing, which changes the density of water parcels. The resulting buoyancy forces drive the advection.

One of the experiments described by Turner (1979) was made more controlled and quantitative by Ruddick and Turner (1979). In these experiments, a long tank with a removable central barrier was stratified, on the one side with sugar solution, and on the other with salt, such that the density was the same on either side. When the barrier was removed, a sharp thermohaline front resulted, and intrusive features formed after a few minutes, which extended laterally. By assuming that the redistribution of salt and heat by the advection and finger fluxes must result in a net release of potential energy, Ruddick and Turner derived a formula for the vertical scale, H, of the intrusions:

$$H = 1.5(1 - \gamma)g\beta\Delta S/N^2 \qquad (3.)$$

where $\beta\Delta S$ is the salinity contrast, in density terms, across the front, and γ is the salt finger flux ratio. This compared well with the experimental observations, and also with the few published observations of frontal intrusions that were available at the time. Note also that this scaling is of the same form as eq. (2).

Substituting the values from table 1 appropriate to the Meddy observations gives a predicted intrusion scale of $300m$, an order of magnitude larger than observed. A possible reason for such strong disagreement will be discussed later, in the context of a theory by Niino (1986), for fronts of finite width.

3.2 Linear Theories of Frontal Intrusions

There is a class of intrusion models which are in essence linear instability theories. These assume a basic state with constant vertical gradients of temperature $\overline{T}(x, z)$ and salinity $\overline{S}(x, z)$. A thermohaline front is represented by horizontal gradients of temperature and salinity which compensate ($\alpha\overline{T}_x = \beta\overline{S}_x$), so that isopycnals are horizontal, and there is no geostrophic shear. (McIntyre (1970) treated the case of no $T - S$ variation, geostrophic shear, and different diffusivities for momentum and density. The intrusive features predicted by his theory tend to be much smaller than thermohaline intrusions, about $3m$, and so will not be considered further here.) The equations of motion are then linearised about this basic state, resulting in equations of motion for the (assumed infinitesimal) perturbation velocities, salinity, and temperature of the form (McDougall, 1985a);

$$u_t - fv = -\rho^{-1}p_x - \nabla \cdot \mathbf{F}^x \qquad (4.)$$

$$v_t + fu = -\rho^{-1}p_y - \nabla \cdot \mathbf{F}^y \qquad (5.)$$

$$w_t + -\rho^{-1}p_z - g(\beta S - \alpha T) - \nabla \cdot \mathbf{F}^Z \qquad (6.)$$

$$u_x + v_y + w_z = 0 \qquad (7.)$$

$$\beta S_t + u\beta\overline{S}_x + w\beta\overline{S}_z = -\beta\nabla \cdot \mathbf{F}^S \qquad (8.)$$

$$\alpha T_t + u\alpha\overline{T}_x + w\alpha\overline{T}_z = -\alpha\nabla \cdot \mathbf{F}^T \qquad (9.)$$

where $F^{x,y,z}$ are the fluxes of x, y, z momentum, and F^S and F^T are the fluxes of salt and heat. The essential physics of intrusions is contained in the conservation equations for heat and salt: production of thermohaline finestructure by horizontal advection, $u\beta S_x$, and it's dissipation by vertical mixing, $\beta \, \partial F^S/\partial z$.

The advantages of linear models are that they can always be solved, they give predictions of wavelengths and growth rates, and they elucidate the physics. However, to be completely useful, one needs realistic parameterizations for the fluxes. Although we can guess at their form (Schmitt, 1981, Kelley, 1984), present parameterizations are too uncertain to be useful. More importantly, linear models can only predict exponential growth or decay of the perturbation amplitude, and give no information about the limiting finite amplitude. Hence the most important aspect of fronts, the lateral transport of heat and salt, is not addressed by linear models.

(i) Infinitely wide fronts. The important physics of intrusion models lies in the parameterization of the fluxes of salt, heat, and momentum. Standard turbulent diffusivities, all equal to each other, will lead to a stable system in which all perturbations decay. Stern (1967) devised a parameterization which brilliantly and simply captured the physics of salt finger buoyancy fluxes:

$$F^S = K_S \partial S / \partial z \tag{10.}$$

$$\alpha F^T = \gamma \beta F^S. \tag{11.}$$

By linking the heat flux to the salt flux via the flux ratio, γ, Stern ensured that the buoyancy flux would be positive, representing a release of potential energy. For such a parameterization to be physically valid, the basic vertical stratification must be unstable to salt fingers (salty on top), and the perturbations must be small enough that the vertical stratification is not significantly altered by the perturbations. Stern assumed an infinitely wide front (constant $\overline{S}_x, \overline{T}_x$), and no momentum fluxes. He found that *any* planewave perturbation can grow, so long as the slope of the wavecrests is such that warm salty intrusions rose as they advected across (thus predating the laboratory observations by many years). He noted that the growth timescale was approximately the diffusion timescale H^2/K_S, and that the growth rate became infinite at small vertical wavelength. In Stern's (1967) and all other linear models, the form of the motion is alternating cells of clockwise and anticlockwise motion, of equal height. As in the case of the heated sidewall experiments, it is expected that nonlinear effects rapidly alter the relative sizes of the cells, so that the "fingering" cells come to dominate the flow. However, no systematic observations of the relative thickness of fingering and diffusive regions have been made.

Toole and Georgi (1981) added friction to the infinitely wide front model of Stern (1967) in the form:

$$F^z = A \partial u / \partial z, \qquad F^x = F^y = 0 \tag{12.}$$

and found that the growth rate was maximised at a particular vertical wavelength. If H was too large, the finger flux divergence was too small to produce significant buoyancy forces. If H was too small, friction tended to limit the growth rate. The scale predicted by the Toole and Georgi theory is

$$H = 2\pi \left[\frac{4(K_S A)^{1/2} N}{g(1-\gamma)\beta \overline{S}_x} \right]^{1/2}. \tag{13.}$$

If we define a kind of Rayleigh number as $(1-\gamma)g\beta \overline{S}_x H^4/(K_S A)$, and an aspect ratio (thermoclinicity?) of $(1 - \gamma)g\beta\overline{S}_x/N^2$, then Toole and Georgi's scale is given by:

$$(\text{Rayleigh No.}) \cdot (\text{aspect ratio}) = 0(1). \tag{14.}$$

Because Toole and Georgi assume an infinitely wide front in their theory, there is no externally imposed length scale, and so the predicted scale must involve diffusivities and some form of Rayleigh Number. Toole and Georgi also display several graphs of growth rate, and from those it can be shown that the growth rate is approximately $m^2 K_S$, the inverse diffusion timescale for salinity.

McDougall (1985a) repeated Toole and Georgi's analysis for a "slab" layer model, in which essentially the same parameterization was used, except that the fluxes depended on interlayer differences in salinity rather than gradients. His results paralleled Toole and Georgi's, except that H^2 is replaced by H in their formulae.

Substituting the values from table 1 appropriate to the Meddy observations, using $K_S = 3 \times 10^{-5} m^2 s^{-1}$ as deduced from the advective–diffusive balance, and choosing $\gamma = 0.7$ in eq. (13) results in a predicted wavelength of $10m \cdot (Pr)^{1/4}$. If we select $Pr \simeq 40$ (Ruddick, 1985), then the predicted wavelength is $25m$, in good agreement with the observed $30m$. Note also that this prediction is not very sensitive to the poorly known Prandtl number.

(ii) The physics of intrusion slope. Posmentier and Hibbard (1982) re-examined Stern's model, and defined three types of T-S changes, or "flux ratios" which one could compute:

(1.) Eulerian, as would be seen by an observer watching the time development of intrusions at a point: $r_E = \alpha \frac{\partial T}{\partial t} / \beta \frac{\partial S}{\partial t}$.

(2.) Lagrangian, which must of course equal the finger flux ratio if only salt fingers change the fluid properties: $r_L = \alpha \frac{DT}{Dt} / \beta \frac{DS}{Dt}$.

(3.) Synoptic, the gradients an observer would observe with a quick hydrographic survey: $r_s = \alpha \frac{\partial T}{\partial \ell} / \beta \frac{\partial S}{\partial \ell}$, where ℓ is the lateral distance measured along an intrusion. This ratio is related to the slope ϑ of intrusive features relative to isopycnals, and is equal to 1 for zero slope.

He found that all three flux ratios were different. McDougall (1985b) explained this difference in physically simple terms, and we expand on his explanation here.

We consider a planewave disturbance, sloped such that warm intrusions slope upward as they extend across the front, as in figure 3. In coordinates rotated to match the slope ϑ of the intrusions, the lateral momentum equation is:

$$\frac{\partial u}{\partial t} = -g \sin \vartheta (\rho - \overline{\rho})/\rho_o + \frac{\partial}{\partial z}\left(A \frac{\partial u}{\partial z}\right) \tag{15.}$$

In the upper (warm, salty) half of each intrusion, the density must be less than the average $\overline{\rho}(z)$, so that the buoyancy force "pushes" the warm, salty water "uphill" against the retarding effects of inertia and friction. Thus the warm, salty water parcels must also be lighter than average, and conversely the cool, fresh parcels must be more dense than average.

We will now consider the Eulerian evolution at a single point in the fluid which will become a warm, salty intrusion. The temperature and salinity at this point will increase with time, but the density must decrease. In figure (4a.) we show this point on the $(\alpha T, \beta S)$ plane, and show the effect of the salt finger fluxes as a tendency to decrease the temperature and salinity in a ratio set by the flux ratio. The advective terms in the conservation equations (8 and 9) will tend to increase T and S in a ratio set by the synoptic flux ratio, which is in turn determined by the layer slope. In (4a.) we assume the synoptic flux ratio is equal to γ, as has been assumed in observational studies of intrusions (Joyce, Zenk, and Toole, 1978). The resultant of the finger flux and advective changes is the Eulerian rate of change (the $\partial/\partial t$ terms in 8 and 9). If the finger flux and advective terms lie along a line of slope γ, the Eulerian rate of change must also lie along that line. Thus the intrusion must either get saltier and more dense, as shown, or fresher and less dense. Neither is consistent with a growing intrusion.

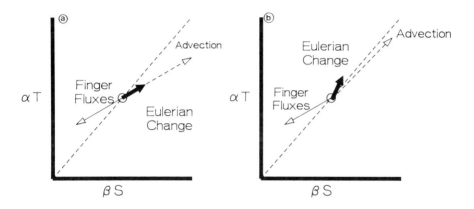

Figure 4. Eulerian evolution in the $(\beta S, \alpha T)$ plane of a warm, salty intrusion, showing the effects of advection, salt finger fluxes, and the resultant Eulerian change. The dashed line represents constant density. (a) Synoptic flux ratio = finger flux ratio. (b) Synoptic flux ratio > finger flux ratio.

In figure (4b) is shown the same situation, except that the slope of the intrusions relative to isopycnals has been decreased slightly, so that the synoptic flux ratio is closer to 1. Now the resultant Eulerian $T - S$ change lies in a different direction, such that the layer can become warmer, saltier, and less dense. McDougall (1985a) finds that the synoptic flux ratio in growing intrusions is typically 0.9 rather than the value of 0.56 that had been previously assumed. Such small slopes relative to isopycnals are virtually impossible to observe with confidence. Gregg (1980) mapped out a thermohaline intrusion in three dimensions, and found that it moved across isopycnals with a synoptic flux ratio of 0.2 (a very large slope). Schmitt et al. (1988) found that the thermohaline steps East of Barbados exhibited lateral gradients of αT and βS in the ratio 0.85 ± 0.02. They interpreted this as an "advective" flux ratio, due to a salt finger flux ratio of 0.85. However an alternate possibility is that the steps are thermohaline intrusions produced at the broad front between the North and South Atlantic Central Waters, with a synoptic flux ratio of 0.85. We note parenthetically that double-diffusively driven intrusions with a synoptic flux ratio of 1 are not possible, since these would have zero slope relative to isopycnals, and so ($g \sin \vartheta$) in eq. 15 would be zero, resulting in no buoyancy force.

Holyer (1983) considered the linear instability problem (4-9), but with the salt and heat fluxes parameterized by molecular diffusion, including horizontal fluxes associated with horizontal gradients. The horizontal flux divergences caused her analysis to be dominated by sloping salt fingers – she in essence studied the salt fingering instability in the presence of isopycnal salinity and temperature gradients. In the case where the vertical stratification is stable to salt fingers, Holyer found an unstable mode which grew more slowly than salt fingers, but which had similar scale (governed by diffusion), and had small slope relative to isopycnals in the "wrong" sense, such that warm salty intrusions sank slightly. This sense of slope can be explained in terms of the greater diffusivity of heat than salt in her flux parameterization, which caused warm salty intrusions to cool off and become more dense. One presumes that oceanic intrusions, with scales of tens of meters or more, feel the effects of fluxes on similar scales, and that parameterizations like that of Stern (1967) are more appropriate.

(iii) <u>Effects of Rotation</u>. When the Coriolis force on the intrusion velocities is considered (but no large-scale velocity shear is allowed) in linear instability models, the dynamics of thermohaline intrusions is not altered much at all. Stern (1967), Toole and Georgi (1981), and Posmentier and Hibbard (1982) considered such effects in their models. Kerr and Holyer (1986), using Holyer's molecular parameterization, and McDougall (1985a), using what amounts physically to the Toole and Georgi parameterization, showed that the Coriolis force on the cross-frontal velocity is largely balanced by an alongfront slope of order f/N which generates baroclinic pressure gradients. The cross-frontal momentum balance, growth rate, and slope are unaffected by rotation, a result reminiscent of the dynamics of a Kelvin wave. Gregg (1980) found that intrusive features tended in practice to form tongues of width about $2N/f$ times the thickness, rather than broad sloping sheets.

(iv) <u>Finite Width Fronts</u>. There is a basic incompatibility between the intrusion scale predicted by Ruddick and Turner (1979), and that predicted by Toole and Georgi (1981), which is all the more perplexing because both scales claim agreement with oceanic observations. Ruddick and Turner's scale, eq. (3) predicts $H \propto (\beta \Delta S) \cdot N^{-2}$, while Toole and Georgi's scale, eq. (13), predicts $H \propto (N/\beta \overline{S}_x)^{1/2}$. The two formulae are almost inverse to each other, so that as N increases, eq. (3) predicts decreasing H but eq. (13) predicts increasing H. Van Aken (1982) tried to resolve this difference by investigating the N–dependence of intrusions in the Rockall Trough, and found an intermediate result: $H \propto N^{-1/2}$. However, his comparison did not take into account the variation of ΔS or $\overline{S}_x$.

Niino (1986) considered a linear instability model with the same flux parameterization as Toole and Georgi (1981), adding the important difference of a finite width front. He allowed a central frontal zone, of width 2a, with a constant horizontal salinity gradient, plus semi-infinite zones on either side which had zero horizontal salinity gradient. The resulting eigenvalue problem could be scaled so that only three parameters emerged. The first parameter characterizes the stability of the basic vertical profiles to double-diffusive mixing: $\mu = (1 - \gamma)/(R_\rho - 1)$, where the "density ratio" $R_\rho = \alpha \overline{T}_z/\beta \overline{S}_z$. The second is the turbulent Prandtl number, A/K_S. Niino's results depended only weakly on these two parameters. The most important parameter was

$$G = \frac{[(1 - \gamma)g\beta\Delta S]^6}{K_S a^2 N^{10}} = \frac{[(1 - \gamma)g\beta\overline{S}_x]^6 a^4}{K_S N^{10}} \tag{16.}$$

If G is small, say less than 100, the front can be considered narrow, and the scale of the fastest growing intrusion was close to that given by Ruddick and Turner (1979). For G larger than about 10^5, the front can be considered wide, and the Toole and Georgi (1981) model is appropriate. For intermediate values of G, Niino's results should be used.

Substituting the values from table 1 into (16) gives $G \approx 10^{10}$, so that the Meddy front is clearly a wide front, and the Toole and Georgi theory should be appropriate. Physically, this means that the front is wide enough that the region outside the frontal zone is of little importance to the dynamics of the intrusions.

3.3 Finite Amplitude Models of Frontal Intrusions

The linear theories discussed in the previous section predict exponential growth, with no limits on the amplitude of the salinity or velocity perturbation, and hence no information about the cross-frontal flux to be expected.

Joyce (1977) worked out a diagnostic model that is the analogue of the Osborne-Cox (1972) model applied to frontal-scale motions. Assuming that the advective-diffusive balance (1) holds, multiply by S on both sides, and average vertically over the intrusive region, performing the *rhs* integration by parts. Denoting this average by an overbar, the result is

$$\overline{(uS)}\overline{S}_x = -K_S \overline{\left(\frac{\partial S}{\partial z}\right)^2} \tag{17.}$$

which states that the rate of production of finestructure salinity variance by horizontal advection in a horizontal salinity gradient is balanced by the dissipation of that variance via vertical mixing. Of course, that mixing produces microscale fluctuations which must be dissipated at the same rate by molecular diffusion. Joyce notes that (17) need hold in a statistical sense only for the model to be valid. The model is not generally useful unless K_S is known. It effectively uses knowledge of K_S to infer u, and hence the horizontal flux. In our observations of the Meddy, we have used our knowledge of u to infer K_S, and a horizontal diffusivity of $0.4m^2/s$. Application of the Joyce (1977) method should give a slightly larger result due to the variance of $\partial S/\partial z$ at wavelengths outside the $30m$ peak which was not considered.

The experiments of Ruddick and Turner (1979) showed that the thicker intrusions spread the fastest, and an unpublished result from those experiments is that the intrusions spread at a constant speed $u \approx 0.005NH$. No mechanism was put forth for this result. Substituting the observed thickness and buoyancy frequency gives a predicted intrusion speed of $0.5mm/s$, in reasonable agreement with the observations.

McDougall (1985b) considered the finite amplitude evolution of the "wide" front instability of Toole and Georgi (1981) and McDougall (1985a), for the case of large Prandtl number. He found that the momentum balance is that buoyancy forces balance frictional stress (the acceleration terms in eq. 15 are negligible), and thus that the evolution of the system is governed by the evolution of the temperature/salinity perturbations. He assumed that the finite amplitude disturbances maintain the slopes and wavenumber set by the fastest growing disturbances. He found that the salinity perturbations eventually grew so large that the vertical salinity gradient reversed, producing a region above each warm salty intrusion which was diffusively stratified. He argued that a steady state is reached when a three-way balance is achieved between the advective transport, the salt finger fluxes, and the diffusive sense fluxes of salt and heat. In figure (4b), this balance corresponds to a diffusive flux attempting to decrease T and S, but increase the density, of the warm, salty intrusion, and a balance being achieved such that the Eulerian rate of change is zero. McDougall argued that this steady state is a stable equilibrium of the equations, and that u is proportional to NH. However, application of the model in practical oceanic situations requires detailed knowledge of both the finger and diffusive fluxes. Also, quantitative predictions depend critically on the unproven assumption that the slope is unchanged from that of the linear disturbance. The laboratory observations of Ruddick and Turner (1979) show the intrusion slope decreasing with time.

4 SUMMARY

We have suggested that thermohaline intrusions were the dominant mode of mixing of a particular lens of Mediterranean water, moving inward $30km$ in a year. A vertical diffusivity for salt of $3 \times 10^{-5} m^2/s$, and a horizontal salt diffusivity of $0.4 m^2/s$ were estimated for this mechanism.

Models of thermohaline intrusions were reviewed, and their predictions compared to the lens observations. It was found that the Meddy front can be considered to be a wide front in the sense that the dynamics depend on the local value of the horizontal salinity gradient. The linear theory of Toole and Georgi (1981) adequately predicts the scale of the observed intrusions. Not enough is known about the fluxes in a diffusive sense stratification to test or apply the finite amplitude model of McDougall (1985b).

ACKNOWLEDGEMENTS

We would like to thank Larry Armi, Tom Rossby, Phil Richardson, Jim Price, and Neil Oakey for the collaborative spirit which has made this project such a pleasure.

REFERENCES

Chen, C.G., Briggs, D.G. and Wirtz, R.A., 1971. Stability of thermal convection in a salinity gradient due to lateral heating. Intl. J. Heat Mass Transfer 14, 57-65.
Gregg, , 1980. Three-dimensional mapping of a small thermohaline intrusion. J. Phys. Oceanogr., 10, 1468-1492.
Hart, J.E., 1971. On sideways diffusive instability. J. Fluid Mech. 49, 279-288.

Hebert, D., 1988a. Mediterranean Salt Lenses (this book).

Hebert, D., 1988b. A Mediterranean Salt Lens, Ph.D. thesis, Dalhousie University, Halifax, NS, Canada.

Holyer, J.Y., 1983. Double–diffusive interleaving due to horizontal gradients. J. Fluid Mech., 137, 347-362.

Huppert, H.E. and J.S. Turner, 1980. Ice blocks melting into a salinity gradient, J. Fluid Mech., 100, 367-384.

Joyce, T.M., 1977. A note on the lateral mixing of water masses. J. Phys. Oceanogr., 7, 626-629.

Joyce, T.M., W. Zenk and J.M. Toole, 1978. An anatomy of the Antarctic polar front in the Drake Passage. J. Geophys. Res., 83, 6093-6113.

Kelley, D., 1984. Effective diffusivities within oceanic thermohaline staircases. J. Geophys. Res., 89, 484-488.

Kerr, O.S. and J.Y. Holyer, 1986 The effect of rotation on double–diffusive interleaving. J. Fluid Mech., 162, 23-33.

Linden, P.F. and J.E. Weber, 1977. The formation of layers in a double–diffusive system with a sloping boundary. Journal of Fluid Mechanics, 81, 757-773.

McDougall, T.J., 1985a. Double–diffusive interleaving. Part I: Linear stability analysis. J. Phys. Oceanogr., 15, 1532-1541.

McDougall, T.J., 1985b. Double–diffusive interleaving. Part II: Finite–amplitude, steady-state interleaving. J. Phys. Oceanogr., 15, 1542-1556.

McIntyre, M.E., 1970. Diffusive destabilization of the baroclinic circular vortex. Geophys. Fluid Dyn., 1, 19-57.

Niino, H., 1986. A linear stability theory of double–diffusive horizontal intrusions in a temperature–salinity front, J. Fluid Mech., 171, 71-100.

Osborne, T.R. and C.S. Cox, 1972. Oceanic fine structure. Geophys. Fluid Dyn., 3, 321-345.

Posmentier, E.S. and C.B. Hibbard, 1982. The role of tilt in double–diffusive interleaving. J. Geophys. Res., 87, 518-524.

Ruddick, B.R. and J.S. Turner, 1979. The vertical length scale of double diffusive intrusions. Deep–Sea Res., 26A, 903-913.

Ruddick, B.R., 1985. Momentum transport in sheared thermohaline staircases. J. Geophys. Res., 90, 895-902.

Schmitt, R.W., 1981. Form of the temperature–salinity relationship in the central water: evidence for double–diffusive mixing. J. Phys. Oceanogr., 11, 1015-1026.

Schmitt, R.W., H. Perkins, J.D. Boyd, M.C. Stalup, 1988. C–SALT: An investigation of the thermohaline staircases in the Western tropical North Atlantic. Deep Sea Research, in press.

Stern, M.E., 1967. Lateral mixing of water masses. Deep Sea Res., 14, 747-753.

Thorpe, S.A., P.K. Hutt and R. Soulsby, 1969. The effects of horizontal gradients on thermohaline convection. J. Fluid Mech., 38, 375-400.

Toole, J.M. and D.T. Georgi, 1981. On the dynamics and effects of double–diffusively driven intrusions. Prog. Oceanogr., 10, 121-145.

Tsinober, A.B., Y. Yaholom, and D.J. Schlien, 1983. A point source of heat in a stable salinity gradient. J. Fluid Mech., 135, 199-217.

Turner, J.S., 1978. Double–diffusive intrusions into a density gradient. J. Geophys. Res., 83, 2887-2901.

Turner, J.S. and C.F. Chen, 1974. Two–dimensional effects in double–diffusive convection. Journal of Fluid Mechanics, 63, 577-592.

van Aken, H.M., 1982. The density ratio of frontal inversions in the North Rockall Trough. J. Phys. Oceanogr., 12, 1318-1323.

MIXING AND LARGE-SCALE OCEAN DYNAMICS

PETER B. RHINES
School of Oceanography, WB-10, University of Washington,
Seattle, Washington

ABSTRACT

We summarize various connections between small-scale ocean mixing and larger scale motions, with particular attention to potential vorticity dynamics. The driving of larger scale circulation by buoyancy flux is described, emphasizing three-dimensional signals driven by localized mixing, in addition to the classic one-dimensional estimation of vertical velocity driven by quasi-uniform vertical mixing. For the sake of understanding the general circulation, improved knowledge of mixing processes is particularly needed at the mixed layer base, in boundary currents and at great depths. Coupling of horizontal and vertical diffusion processes by velocity shear is an essential part of ocean mixing. The small values of vertical diffusivity implied by microstructure measurements continue to disagree with inference from large-scale tracer balance. Yet, our knowledge of the mid-depth and deep circulation is sufficiently poor that reliable mid-ocean density and vorticity balances have not yet been calculated.

1 THE DYNAMICAL CONSEQUENCES OF MIXING

Over the entire range of scales from a few meters to thousands of kilometers, the oceanic velocity and density fields are sensitive to density mixing processes, both along ('isopycnal') and across ('diapycnal') neutral surfaces . The aspect ratio of the great water masses is typically very large, with vertical scales of a km or less corresponding to a horizontal extent of many thousands of km. The plume-like upper salinity minimum of the North Pacific, which is high in tritium, oxygen and Freons, curves into the subtropical gyre from the north. It has a depth scale of order 200m, yet a downstream lateral scale of perhaps 4000 km (and cross-stream scale of order 2000 km). There is a competition between vertical and horizontal mixing processes, eroding such water masses as they circulate. Estimates of lateral and diapycnal diffusivities in mid-ocean are, respectively, $\kappa_h \sim 5\times10^2$ m^2 sec^{-1} and $\kappa_v \sim 10^{-5}$ m^2 sec^{-1}. Their ratio, 5×10^7, is comparable with the square of the plume's aspect ratio. Thus we cannot on the basis of crude scale analysis favor vertical mixing over horizontal, as a plume broadens and thickens vertically. The range of variability of the κ's allows their ratio to vary by at least a factor of 10^3.

If water mass properties were to diffuse in a fluid without velocity gradients, one could simply examine the vertical and horizontal spreading of concentration to understand the respective mechanisms. But shear-induced dispersion is an interaction between vertical and horizontal diffusion mechanisms, which greatly changes the picture. Persistent vertical shear of the horizontal velocity, for example, stretches isopleths of tracer concentration and enhances vertical diffusion. But meanwhile the shear is tending to reduce the local vertical thickness of the marked water mass, so that vertical profiles of the tracer might be taken to show little diffusion. The net result is an accelerated diffusion along isopycnal surfaces.

The interesting quantities being diffused include passive chemical tracers, and dynamically active scalars: the temperature, salinity and potential vorticity. The potential vorticity of a somewhat idealized Boussinesq fluid is, at large scale, simply - fN^2/g (i.e., the product of Coriolis frequency and squared buoyancy frequency, $-(g/\rho)d\rho/dz$ where ρ is potential density and g is gravity). Thus in mapping potential vorticity one is tracing inversions and pycnostads about the ocean, with the modulating contribution of β. The problem of mixing and large-scale dynamics is very interactive: the quantities being mixed can be dynamically active, and through shear dispersion and flow destabilization the dynamics can alter mixing rates.

The circulation generated by mixing events passes through a number of distinct regimes, unfolding toward larger length- and time-scale: gravitational collapse, f-plane geostrophic adjustment, and then beta-plane adjustment. Below we describe some simple models of these events.

2 SMALL-SCALE DYNAMICS

2.1 Gravitational collapse

The first two of these stages are most relevant to individual mixing events with lateral scale less than a km. At the smallest

time- and length scales Coriolis effects are negligible, and one
has a production of internal gravity waves and vortical modes,
which are discussed elsewhere in this symposium. In keeping with
the emphasis on larger scale circulation, it is worth rememember-
ing that radiating waves also produce an extreme type of vortical
mode: a rectified large-scale nearly horizontal flow [It is almost
paradoxical that the 'irrotational' (i.e., vanishing vertical
vorticity) gravity wave emanating from collapse should generate
larger scale flow which itself possesses vertical vorticity.] The
atmospheric dynamics community has produced an enormous literature
on the subject. Thus, the dramatic appearance of a quasi-biennial
oscillation in the tropical atmosphere has been clearly identified
with the wave momentum transport. Typically critical-layer absorption or normal dissipation will cause shelves of momentum to
move toward the wave source, without developing particularly
localized jets. The elegant laboratory experiment of Plumb and
McEwan (1978) attests to the robustness of the process, and modern
atmospheric general circulation models are beginning to incorporate the momentum flux due to mountain waves.

2.2 Geostrophic adjustment

Internal waves and potential vorticity bearing motions are
sorted out at this stage. Studies of the geostrophic adjustment
of initially resting blocks of fluid (e.g., Gill (1982)) suggest
that the fraction of the liberated potential energy, PE, (that is,
δPE, the difference between the initial total PE and the PE of the
time-mean end state) that appears as mean kinetic energy, KE, is
typically 1/3. However this fraction clearly depends upon the
scale of the initial PE distribution (Killworth (1986), Middleton
(1987)). A one-layer fluid with a free surface initally in the
form of a top-hat will have KE/δPE → 1/3 for large lateral scale
(relative to the Rossby radius), yet KE/δPE → 0 for small initial
scale. In some cases the full evolution is very simple; if, for
example, the wind stirs the upper ocean leaving a horizontal
density gradient in the mixed layer, we might set the density ρ =
αx at time t=0. This leads to a value of KE/δPE = 1/2. The problem
is interesting, because the full time history of the adjustment is
simply

$$(u,v,w) = (g\alpha z/f\rho_0)(\sin ft, 1-\cos ft, 0)$$
$$\rho = \alpha x + (g\alpha z/f^2\rho_0)(1-\cos ft)$$

where z=0 is halfway between the boundaries, at z=±H and g,f and ρ_0 have their usual meanings. Here the adjustment of the initial vertically mixed layer rebounds in inertial oscillations, about a geostrophically adjusted time-mean state. As there is no vertical stratification at the outset, the slumping of particles through one Rossby radius, $\alpha gH/f^2\rho_0$, sets that stratification.

In such problems the geostrophically adjusted field more usually fails to be in exact steady balance, and must evolve over longer time scales: the inertial time-scale of two-dimensional tur-bulence, and the time-scales of quasigeostrophic Rossby and topo-graphic waves. Vorticity dynamics of the geostrophically adjusted patch then takes hold.

In the presence of lateral boundaries, these adjustment problems take on a whole new aspect: the propagation of fast boundary waves can provide drainage pathways reaching far to the distance (Gill, (1976)). With linear damping these Kelvin or topographic waves can be arrested, and become a steady circulation. But, in addi-tion, the potential vorticity of the draining water-mass may be very poorly matched to the wave-pathways laid out by the geostrophic adjustment, and may choose to follow a different course.

A slumping water mass experiences overwhelming Coriolis ac-celeration acting upon the outward motion of the fluid. This causes interesting energetic problems. If the flow is driven by a cylindrical source of mass at the origin, the kinetic energy at large radial distance is infinite, and growing like t^2 (in two dimensions). In three dimensions the energy is finite yet growing like t^2. But, mixing, heating and cooling do not supply fluid in this way, and they have a limited capacity to provide mechanical energy. W.R. Young (private communication) suggests that, when blobs are created by deep convection, their finite store of poten-tial energy provides an upper bound on the mechanical energy of the resulting geostrophically adjusted blobs, and that this may provide a natural determination of the preferred size of sub-mesoscale eddies in the oceans.

2.2 The resulting quasi-geostrophic circulation

The predominant effect of mixing on the large scale is thus the collective alteration of the density field over a scale of tens of km and greater. The effect of mixing on the quasigeostrophic potential vorticity is traditionally written by linearizing about the mean stratification, $\rho^*(z)$. The Boussinesq density equation gives the vertical velocity,

$$w\rho^*_z = -D\rho'/Dt - \nabla \cdot H,$$

where $\rho = \rho^\wedge + \rho'$ and H is the density flux due to mixing or explicit sources (e.g., radiative heating). The potential vorticity equation sees this vertical velocity as a source term,

$$DQ/Dt = -f(\nabla \cdot H/\rho^*_z)_z$$

where $Q = \nabla^2\psi + (f^2/N^2\psi_z)_z$ (the quasi-geostrophic Q is essentially the Ertel potential vorticity, $(\nabla \times u + 2\Omega) \cdot \nabla\rho/\rho$, multiplied by the mean vertical separation of isopycnal surfaces). Thus an isolated region of heating or cooling produces a dipole (with respect to the vertical coordinate) of forcing of potential vorticity, owing to the vertical derivative. An isolated region of mixing, on the other hand, produces a triplet of Q.

As a linearized example of the response, take cooling to be:

$$\nabla \cdot H = -2A/r'(1+r')^3$$

where r' is the spherical radius with aspect ratio flattened by f/N, $r'^2 = x^2 + y^2 + (zN/f)^2$, and $(N^2 = (-g/\rho^*)d\rho^*/dz$. The origin is in the middle of the fluid. The volume integral of $\nabla \cdot H$ is $-4\pi A$. The solution for ψ to the linearized, f-plane version of the potential vorticity equation has the form

$$\psi = [-Af/\rho^*_z]zt/r'(1+r')^2.$$

Viewed in a vertical plane the circulation is dipolar, with convergence above the cooling and divergence below, and the swirling horizontal flow is a simple stacked vortex pair, cyclonic above and anticyclonic below. The linear solution accelerates with time, reminding us of the energetic paradox mentioned above.

Mixing, as opposed to explicit heating or cooling, produces a region of weakened stratification bounded above and below by intensified stratification. This triplet of potential vorticity forcing causes outflow at the level of most intense mixing, bounded above and below by inflow toward the mixing region. The triple vortex is readily observed in the laboratory, by stirring a region of the interior of a spun-up stratified fluid. The more common experiment (e.g., Stommel and Frazel (1968)) of injecting a blob through a hypodermic needle is of less geophysical interest, for the fluid enters with zero absolute angular momentum, and the subsequent events are dominated by its spin-up.

2.3 Kelvin's theorem with mixing

We can look more generally at the production of vorticity and potential vorticity by mixing. A 'microscopic' view of the effects of mixing can be had from Kelvin's circulation theorem viewed in isopycnal coordinates. Normally, in a stratified fluid one considers the vorticity normal to a waving isopycnal surface, which feels no buoyancy twisting. For the nondiffusive problem with a simple, Boussinesq fluid this isopycnal surface is also a material surface. Suppose instead that fluid particles may move through the surface due to mixing, or perhaps an imposed heat source. Let Γ be the 'circulation', $\int \mathbf{u} \cdot d\boldsymbol{\ell}$, about C, $d\boldsymbol{\ell}$ being an increment of arc length. I use quotation marks because it is not reliably equal to either the Eulerian circulation about a mean contour, or a Lagrangian average motion of fluid particles initially lying on C. It is best thought of as the average vorticity contained within C. Then we find

$$\frac{d}{dt}(\int \mathbf{u} \cdot d\boldsymbol{\ell} + fA) = \int (F + \nabla \cdot H(\frac{\partial \mathbf{u}}{\partial \rho})) \cdot ds$$

Here $-\nabla \cdot H$ is $D\rho/Dt$, the time rate-of-change of density, following a fluid particle, due to mixing or other diabatic processes. f is Coriolis frequency and A is the area enclosed by C (more exactly, the term is $2\Omega A^*$, where A^* is the projection of A normal to the rotation vector, Ω). F is an external body force, or possibly the divergence of the turbulent momentum flux, if our velocity u is meant to be an average at larger scale. The last term thus represents a source of average vorticity due to mixing. Haynes and McIntyre (1987) emphasize that this source term is a boundary contribution at the outer rim of the domain. They show that, formally, the potential vorticity changes as a result of the divergence of a flux, a flux which is parallel to isopycnal surfaces. Thus, in their words, 'there can be no net transport of Rossby-Ertel potential vorticity across any isentropic surface, whether or not diabatic heating, mixing, and frictional or other forces are present.' Written in isentropic coordinates (x,y,ρ), their formula is

$$d(\sigma Q)/dt + \nabla \cdot J = 0$$

where σ is the mass density in $xy\rho$ space and

$$J = u\sigma Q + J_\rho + J_f$$

The two contributions J_ρ and J_f being given (to the accuracy of the hydrostatic approximation) by

$$J_\rho = (\rho dv/d\rho, -\rho du/d\rho, 0)$$

and

$$J_f = (-G, F, 0)$$

where F and G are the horizontal components of the local frictional or other force per unit mass. This suggests that the net potential vorticity, integrated between two potential density surfaces, is a very robust quantity, provided one integrates out to a region where 'nothing is happening'. Mixing, diffusion, and even external forces change this integrated quantity only if they exist at the outer rim, C. But the result can also be very misleading: if one exerts an external force on the fluid somewhere in its interior, there is no production of integrated absolute vorticity in a domain that completely encompasses the region of forcing, strong plusses and minuses of absolute vorticity, and interesting circulation do occur. In the present context, a small region of cooling causes potential vorticity anomalies above and below (as in the above example). Even after integrating vertically between isopyncal sheets, we find new potential vorticity above and below the source. Yet integrating laterally over a region encompassing the cooling, we see inward advection of planetary vorticity, f, feeding the convergence, and this is equal and cancelling in the net budget of p.v. So the robustness of the absolute circulation masks the strong generation and destruction (or concentration and dilution) of p.v. by mixing.

3 LARGE-SCALE CIRCULATIONS

3.1 Beta effects

At later time and at larger scale than these local events, the flow generated by mixing activity comes under the influence of the beta effect, topography and oceanic boundaries. Though beta is often weak, compared to other terms in the mean potential vorticity field, over large times and distances it is very persistent. A useful prototype solution is the steady Green's function, driven by intense forcing in a small region near the origin. Suppose we have a linear Rayleigh damping of the potential vorticity, meant to represent mixing effects. The solution with free slip boundaries at z= 0,-D to

$$R(\nabla^2 \Psi + (F\Psi_z)_z) + \beta\Psi_x = \delta(x,y)\cos mz$$

is just

$$\Psi = \cos mz\ e^{-\alpha x}\ K_0(\gamma r)$$

where $\alpha = \beta/2R$, $\gamma^2 = \alpha^2 + Fm^2$, $F = f^2/N^2$, $r^2 = x^2 + y^2$, and m = (integer)π/D. R is a linear damping coefficient, and N has been taken to be uniform. Ψ is the geostrophic stream-function. This flow pattern is a set of long, cigar-shaped gyres, 'β-plumes' extending westward from the forcing. They are essentially arrested Rossby waves. The higher vertical modes are less effectively carried westward by β, and resemble f-plane circulations calculated above, with Prandtl-ratio scaling, NH/fL ~ 1. The barotropic mode, m=0, behaves like $(8\pi R^2\alpha r)^{-1/2} \exp(-\alpha r(1+\cos\theta))$ for $\alpha r \gg 1$. Significant circulation reaches far westward (where $1+\cos\theta=0$) yet only a frictional western boundary current width, α^{-1}, to the east. In the more complete setting of the nonlinear wind-driven circulation, the potential vorticity field, and ray paths for β-plane adjustment, are severely distorted. Plumes of circulation like those above are guided along characteristics which may be set by shape of the wind-driven gyres. What does the δ-function represent physically? It could be a 'tweak' of mechanical stress curl. For a homogeneous density with barotropic response, the δ-function forcing represents a mass-source, say at the upper boundary. The Green function appropriate to cooling a small region is simply $\partial(\text{above }\Psi)/\partial z = -m(\tan(mz))\Psi$. That appropriate to mixing in

a small region (yet still varying like cosmz in the vertical) is
∂^2(above ψ)$/\partial z^2 = -m^2\Psi$.

The above Green function selects vertical modes, due to the
rigid top and bottom boundary. It is interesting also to write
down Green's function for the unbounded problem

$$R(\nabla^2\Psi + (F\psi_z)_z) + \beta\Psi_x = \delta(x,y,z); \quad \psi \to 0 \quad \text{at large distance.}$$

For uniform N this may be written

$$\Delta\Phi - \alpha^2\Phi = \delta(r')$$

{or $\partial\delta(r')/\partial z$ for localized cooling, or $\partial^2\delta(r')/\partial z^2$ for localized
mixing}, where $\Psi = \Phi\exp(-\alpha x)$, Δ is the 3-dimensional Laplacian in
stretched coordinates, $r'^2=x^2+y^2+z'^2$, $z'=F^{-1/2}z$. The solutions are

$$\Psi = \frac{\partial^n}{\partial z^n} \frac{1}{r'} e^{-\alpha(x+r')}$$

where n=0 for a $\delta(r')$ righthandside forcing, n=1 for 'cooling' and
n=2 for 'mixing'. $\alpha = \beta/2R$ as above. The surfaces of constant Ψ
form a single family of 'stretched' ellipsoids, or rather doubly
stretched, first flattened horizontally by a factor f/N, and then
elongated westward by β. Think of the exponential as having
isosurfaces in the form of paraboloids of revolution (about the x-
axis, in x,y,z' space, with focus at the origin), and then the
1/r' factor simply closes these off to the west, producing the
distorted ellipsoids. For small R/βL this elongation is very
great. Ψ encodes both the dominant horizontal velocity field, and
through $\partial\Psi/\partial z$, the perturbation density field. For n=1 it gives
the double vortex structure of the earlier example of cooling
response, yet extended to the β-plane and made steady with linear
damping. For n=2 it gives the triple gyre structure induced by
localized mixing. Rhines (1983) discusses how one can add image δ-
functions to account for boundaries, and thereby produce a particularly simple picture of closure of the circulation through a
linear western boundary current.

The sequence of reactions of the large-scale flow to localized
forcing depends very much on the source strength; heating, cooling
and mixing have a mechanical effect that scales inversely to the
stratification. Thus a given air-sea heat flux will produce more

intense currents, for example, in the weakly stratified subpolar
regions than in the tropics, and a given mixing-induced flux of
buoyancy will cause more to happen in the deep water than in the
thermocline. This discussion has been most relevant to the
cumulative response to the large regions of cooling and mixing at
the poleward edge of subtropical gyres, and in subpolar circulations.

3.2 Gyres

The developing dynamical picture of the wind-driven gyres has
had much discussion in the literature. Some theories emphasize
direct ventilation and potential vorticity conservation following
fluid driven downward by Ekman pumping from the mixed layer into
the geostrophic interior. Other theories emphasize the way in
which recirculation and nearly closed streamlines can lead to a
more global determination of the circulation. Mixing processes are
crucial to these models, even though it may seem that the
'laminar' ventilated thermocline does not need them. Rhines
(1986a,b) emphasizes that, owing to the large recirculation index
of such gyres, particularly for fluid originating near the
poleward edge of the gyre, there is a blending of ventilated
waters with a large component of recirculating flow from the western boundary. This flow tends to emanate in a jet, lying along a
frontal boundary separating vastly differing water masses. The
exposure of the recirculating flow to the atmosphere enhances
ventilation; the degree of enhancement is very sensitive to the
penetration of mixing. Tracer budgets suggest the strength of
this effect; they show a vastly greater ventilation than the product of Ekman pumping velocity and surface concentration would imply. [The basin-budget calculation of Sarmiento (1983), and the
local β-triangle measurements of Jenkins (1987) both give an effective ventilation of about 3 times that 'laminar ventilation'
rate, which itself totals about 8 Sv. (1 Sv.= 10^6 $m^3 sec^{-1}$) in the
density range 27.4 < σ_θ < 26.2.]. This enhancement increases
downward, even as the direct Ekman pumping contribution decreases.
There is further statistical support in the scatter plot of ^{3}He
and tritium (Jenkins (1980)); ^{3}He is 'dead tritium', and its concentration in the Sargasso Sea never rises to the level that it
would in a laminar flow; mixing again is implicated. The detailed
sites of the mixing are obscure, however. It must be diapycnal,
and circulation integrals (discussed below) tend to emphasize

regions of high shear. In advection-diffusion simulations of passive tracers in gyres, Musgrave (1985) indeed finds the greatest
tracer influx in the poleward-western regions of the flow, where
advection enhances their gradient. We are still greatly in need of
information on diapycnal mixing rates in regions of fronts and
jets, to progress further with the matter (the modelling studies
usually assume uniform κ_v). The largeness of the recirculation
index is reflected also in the mean thickening and thinning of
respective isopynal prisms, as one moves downstream (Rhines,
1986b).

The principal limitation in using tracer data to infer mixing
is imperfect knowledge of both the velocity fields and the boundary conditions for the tracer. It is dismaying that even today we
have such uncertainties, for example, over the gross transport of
the circulation. Wunsch and Roemmich (1985) argue that something
like 15 Sv. of the ~ 30 Sv. passing north through the Florida
Straits is participating in the meridional thermohaline circulation, whose strength is becoming increasingly well-known through
the global heat balance. The estimated Sverdrup transport is itself only about 15 Sv. east of the eddy-rich western Atlantic, and
one can invoke nonlinear dynamics to explain the ten-fold transport increase over this value, in the inertial recirculation.
Given the extensive new surveys of hydrography and chemical
tracers, the detailed North Atlantic circulation will soon be far
clearer.

Clever use of chemical tracer fields has particular promise;
Sarmiento (private communication) provides a picture of mixing
into the gyres from the 'side', by looking at ^{228}Ra, whose source
is the seafloor sediments. A plume of ^{228}Ra is seen entering the
No. Atlantic subtropical gyre in the northeast Atlantic, almost as
if it had entered from the atmosphere. Since its source is probably the western continental rise, and its half-life is 5.75 years,
this tracer must be carried quickly across the ocean by the separated Gulf Stream. This supports the notion that midocean ventilation has a very strong 'sideways' diapycnal mixing component.

The behavior of mixed-layer fluid, moving round the circuit
while the seasons pass, has been stressed by Federiuk and Price
(1984) and Woods (1985). Fluid ventilated in regions of deep

mixing might move equatorward beneath the shoaling mixed layer base, entering the geostrophic interior from the side rather than from above. Owing to the great amount of shear in the mixed layer, and the uncertain response of surface drifters in strong winds, we are far from a quantitative understanding of this Lagrangian process.

The large-scale balance of tracer advection and diffusion is a promising complement to microstructure measurements in inferring the level of mixing, and indeed, its importance. Well-defined tongues of salinity, Freons and tritium in the upper few hundred m. of the North Pacific are examples, where vertically thin anomalies are traceable as they spiral downward in the subtropical gyre.

   Hogg (1987) uses an inverse model to give estimates of the circulation and mixing coefficients (typically $\kappa_V \sim 5 \times 10^{-4}$ at 700m, $\sim 2 \times 10^{-4}$ m^{2}sec^{-1} at 1200m) in the eastern North Atlantic. This estimate of the diapycnal mixing coefficient in the thermocline is much larger than has been suggested by microstructure measurements. This, and the similar estimates from Olbers **et al.** and others, continue to disagree with more direct inference from microstructure velocity and temperature measurement. This awkward situation needs attention.

Global property-property diagrams on isopycnal surfaces provide the classical global-scale diagnosis of mixing. Kawase et al, (1985), for example, use nutrient data in the Atlantic to infer both mixing and regeneration in the interior. Isolated diapycnal mixing regions are sometimes implied, for example by silica data near the equator. Silica is deficient there, compared to simple end-point mixing of salinities, at $\sigma_\theta = 27.4$.

   Beyond the evaluation of mixing coefficients there are structural questions. Jenkins' β-triangle work (1987) is able to follow an isopycnally descending water mass which exhibits systematic ageing, in addition to eddy noise. The decay of tritium to ^{3}He provides a measure of travel time from the mixed layer. This clock verifies the timescales of the ventilation process anticipated from dynamics. In addition, combined with other tracers, the information gives estimates of both vertical and lateral mixing rates, and of oxygen consumption. In the spirit of Iselin's

orginal theory for the origin of the temperature-salinity relation, Jenkins finds the eddy variance of a tracer on an isopycnal
surface to vary according to its character in the mixed layer:
helium and oxygen are far more uniform in the mixed layer than
salinity and tritium. The atmospheric boundary condition is, in
effect, far smoother for helium and oxygen which also equilibrate
rapidly. Salinity reflects the more structured precipi-
tion/evaporation function, and tritium contrasts can build up by
nonuniform mixed layer deepening. The 'inherited variance' of
salinity and tritium show can easily be mistaken for isopycnal
eddy stirring at depth. Jenkins also points out that time-
dependent surface boundary conditions for the tracer can seriously
hamper inference based on steady advection-diffusion modelling;
salinity in the North Atlantic is the case in point.

In addition to the quantitative questions about of diapycnal
mixing are the qualitative changes that they produce.   Layered
circulation models (e.g. Luyten and Stommel (1986),   Rhines et al.
(1985),) respond to buoyancy forcing which is imposed as a pumping of fluid between layers. The   production of potential
vorticity by mixing, described above, can develop counter-rotating
gyres at the basin scale. If less strong, mixing will still lead
to veering of the horizontal current with depth, which itself can
steer water masses on a diversity of pathways. A central diagnostic equation due to Bryden (1976) relates two 'unmeasurable'
quanitities, the geostrophic vertical velocity and the convergence
of buoyancy flux, to the (measureable) vertical distribution of
horizontal velocity. This is

$$(\rho_0 f/g)\,|\mathbf{u}|^2 \partial\Phi/\delta z \;=\; -\mathbf{u}_h\cdot\nabla\rho$$
$$= \rho_t + \nabla\cdot\mathbf{H} + w\rho_z.$$

Here Φ is the angle of the horizontal velocity vector, $\mathbf{u}_h$ with
respect to east. ρ_t may be measured, or may vanish after sufficient time averaging. Veering of the current with depth is thus a
central diagnostic indicating a combination of vertical velocity
and buoyancy flux convergence. When combined with potential
vorticity dynamics one has β-spiral analysis, in which the observed geometry of tilted isopycnal surfaces constrains the three-
dimensional mean velocity field. Because of the overdetermined nature of the β-spiral inversion one can begin to estimate cooling

and mixing (i.e., ∇·H) in addition to the three-dimensional
velocity field (e.g., Bigg (1983)). The signature of mixing is
particularly evident in regions of deep wintertime convection, for
example Station Juliette in the northeastern Atlantic (Stommel et
al. (1985)).

Olbers et al. (1985) do a β-spiral calculation at a grid of
points in the North Atlantic, using Levitus data, and giving
estimates of residual mixing terms. Though one can criticize the
use of planetary geostrophy (i.e., neglect of relative vorticity)
in the vicinity of jets, nevertheless the results have appealing
structure. They find a large vertical momentum mixing coeffient,
which may support theory and models of the 'form drag' process of
transporting momentum vertically; there is an equivalence between
the 'thickness' component of lateral potential vorticity flux and
vertical momentum flux. Their estimate for κ_V is typically a few
times 10^{-5} cm^{2}sec^{-1} in the upper 800m, yet increasing by an order
of magnitude in regions of large eddy energy, and decreasing
downward to $\kappa_V \sim 10^{-5}$ in the layer 800-2000m deep.

3.3 Deep circulation

The inference of circulation and mixing rates from global
balances is traditional, both by 'pattern recognition' of tracers,
which are compared with simple advection diffusion models, and by
quantitative estimates of integrals of the circulation: for example, the vertical profile of heat-flux, fresh-water flux, and mass
flux across long hydrographic sections. While these calculations
reinforce the traditional picture of the great meridional cells of
circulation in the Atlantic, the picture elsewhere is incomplete.
The classical pattern of meridional circulation, with quasi-
uniform upwelling into the thermocline balancing downward diffusion of heat and tracers, has been supported largely by inference
from observed deep western boundary currents. As to the magnitude
and sense of the circulation in the quieter interior, there are
many uncertainties. Given the crucial relation between diapycnal
mixing and vertical velocity, it is sobering that we are so in
doubt over the <u>sign</u> of the large-scale vertical velocity in many
areas. The meridional circulation of the Pacific, for example, as
it is generally accepted, involves a northward flow of deep water,
which is returned at mid-depth. (e.g., Broecker et al. 1985) Meanwhile Bryden et al. (1987), on the basis of a hydrographic section

at 24°N, show a southward zonally averaged flow at all depths
greater than 3000m. Their profile requires that the average vertical velocity north of 24°N be downward at all depths below 1000m,
in conflict with traditional ideas about mid-ocean balances. It
is valuable to have such widely divergent estimates of the circulation, if only to prevent us from thinking that the task is
merely a quantitative one, in which we improve measurements of κ_V
to feed into a diffusive thermohaline circulation model.

Given the suspicions that vertical mixing rates may rise something like N^{-1} in the deep water (Gargett, 1984), which have some
recent support from Moum and Osborne (1986) in the western Pacific, it seems that the deep circulation is a crucial area in which
to look more closely at mixing processes. Even a crude interpretation of deep tracer balances implicates mixing as a dominant
process.

The canonical vertical mixing rate of $10^{-4} cm^2 sec^{-1}$ yields a
scale velocity of less than a mm sec^{-1},

$$v \sim \frac{f}{\beta}\left(\frac{(\kappa_V \rho_z)_z}{\rho_z}\right)_z \sim \frac{a\kappa_V}{D^2}$$

where a is the Earth's radius and D the vertical depth scale. This
assumes that the mean potential vorticity gradient is dominated by
β (subscripts indicate differentiation except in κ_V). The crude
estimates of vertical and horizontal velocity induced by diapycnal
mixing are accordingly small, particularly when one applies the
thermocline estimates of κ_V based on direct measurement. The gyre
circulation, however, by necessity alters Q away from βy. It will
be important to explore mixing- and convection driven circulations
in the presence of realistic deep potential vorticity fields, and
to map these fields.

κ_V and w are difficult enough to measure, but what controls the
circulation is $\partial w/\partial z$. This is the crux of Gargett's argument for
the sensitivity of the circulation to rather subtle variations in
κ_V. Let $\kappa_V = AN^\alpha$, and express $\rho(z)$ in terms of N. A one-dimensional
vertical advection diffusion balance,

$w\rho_z = (\kappa_v \rho_z)_z$, or with crude Boussinesq reasoning
$wN^2 = (\kappa_v N^2)_z$, gives

$$w = A(2+\alpha)N^{\alpha-1}N_z = A((2+\alpha)/\alpha)(N^\alpha)_z$$
$$w_z = A((2+\alpha)/\alpha)(N^\alpha)_{zz}.$$

The extreme dependence of w_z on the stratification is noteworthy.
Taken literally, it would yield a multiplicity of reversing gyres
in the deep lateral circulation. Reversals in w itself would be
present. Gargett's case is $\alpha=-1$, whence $w = (A/N^2)N_z$. Thus the
vertical velocity is upward only in regions where N increases upward. For a typical exponential profile of N, $N=\exp(\gamma z)$, we find

$$w = A\gamma\exp(-\gamma z),$$
$$w_z = -A\gamma^2\exp(-\gamma z).$$

A mid-ocean planetary geostrophic vorticity balance, $fw_z=\beta v$, thus
would have equatorward midocean flow rather than poleward, and upward vertical velocity that <u>increases</u> toward the seafloor. Such
parameterizations, with meridional velocity depending on $\partial^3\rho/\partial z^3$,
will yield a vertical motion field with complex structure. Match-
ing such interior fields to the solid boundaries and connecting
boundary currents remains a challenge. Such a regime could form
over a sloping bottom, but, if the seafloor is level, has
eventually to give way to abyssal poleward flow.

Parameterizations may be tested with numerical general circulation models. With uniform diffusion F. Bryan (1987) finds both the
transport of the meridional circulation and also thermocline depth
vary like $\kappa_v^{1/3}$, for a variety of surface boundary conditions.
This occurs in steady, diffusive circulations, where an essentially fixed meridional temperature contrast penetrates deeply, and
drives thermal wind circulation, when κ_v is large; other dependen-
cies on κ_v may be expected in more inertial, eddy-resolving
models. The effect of non-uniform κ_v has not yet been carefully
examined. The result may well depend on the accuracy of the interior dynamic regime of the model. As tantalyzing as these results
and their eddy-resolved counterparts are (Cox, 1986), crucial effects (western boundary currents, mixed layer and abys-
sal/topographic circulation) are difficult to model at present.
And, there is no hope in the next decade of running an eddy-

resolving primitive equation model for the $O(10^3)$ model years required to equilibrate the circulation. One cannot imagine fully
resolving these regions without carrying out corresponding, in-
tensely sampled observational programs.

In apposition to the uniform-upwelling model, Kawase (1987) sup-
poses vertical motion and mixing to be determined by the flow
structure, rather than being prescribed. He discusses a source-
driven model of the deep circulation, showing how the self-damping
of water masses can prevent the Stommel-Arons circulation from
filling out. The calculation uses a linear Rayleigh damping of
the height field in a simplified free-surface model of the circulation. Thus one has a stark contrast between the disposition of
water masses in an ocean with uniform diffusion driving upwelling
into the thermocline, and one in which the vertical mixing and
vertical velocity depend on the amount and 'intensity' of the
water mass. The calculation, and its stratified GCM counterpart
(F.Bryan op cit.) show the amazing rapidity with which boundary-
and equatorial waves carry the circulation meridionally (in a few
months), the slow penetration of the interior ocean by baroclinic
Rossby waves, and the yet slower final equilibration of the
temperature and salinity fields. Nonlinearity is likely to affect
these conclusions, and important fast waveguides are created by
bottom topographic variations. It remains to be seen how great
these effects will be.

The morphology of simple models like Kawase's begins to provide
tests of mixing rates. For instance, the more remote western
boundary currents in Kawase's calculation are the last part of the
circulation finally to fill out. Southward flow of North Atlantic
Deep Water along the western boundary south of the Equator is one
of these dynamically remote boundary currents. It may help to discriminate between deep circulation driven by broadly distributed
upwelling, and that driven by self-damping of the water masses.

As we have remarked, the deep circulation is a particularly important region in which to investigate diapycnal mixing. Diagnostic studies of a variety of locations in the abyssal Atlantic
have been made, for example by Whitehead and Worthington (1982),
Hogg et al. (1982) and Saunders (1987). These tend to show the im-

portance of geothermal heating on abyssal heat balance, and to require diapyncal diffusivity in the range (1 to 4)x10^{-4} m^2 sec^{-1}.

3.4 <u>Enhancement of mixing by the circulation, with fixed dif-
fusivity</u>

Ordinary laminar flows often contrive to increase the gradients
of velocity and tracer concentration near a boundary which acts as
source or sink of the tracer. This is necessary to match the
tracer flux through the diffusive sublayer to the much larger advective transport of the interior. The flux can be doubly enhanced if there is an augmentation of mixing by shear. The importance of this dependence of mixing on large-scale shear is evident
in diagnositic integral expressions which summarize in 'natural
coordinates' the global balance of a tracer. Suppose there is a
steady gyre-like flow with closed streamlines. For large Peclet
number, UL/κ, the tracer homogenizes through shear dispersion and
simple diffusion (Rhines and Young, (1983)). The value of the
tracer in this region is found by multiplying the advection-
diffusion equation by ψ and integrating:

$$\iint \psi [\partial\theta/\partial t + J(\psi,\theta) = \nabla\cdot\kappa\nabla\theta]dxdy$$

which yields for $<\theta> \equiv \iint\theta \, dxdy/\iint dxdy$

$$<\theta> = \int\theta\kappa\mathbf{u}\cdot\mathbf{d\ell}/\int\kappa\mathbf{u}\cdot\mathbf{d\ell},$$

(W.R.Young, private communication). The integrals are taken around
streamlines, at which curves θ, κ and u are evaluated. The stream-
line in question may lie at a free-slip boundary, or in the interior. [The strong assumption is that closed streamlines adequately
model the actual oceanic gyres, which are a combination of closed
recirculation and 'open' ventilation and water mass transformation.] This is the way in which a sourceless tracer equilibrates,
to the point that $\int\kappa\nabla\theta\cdot\mathbf{n}d\ell = 0$. Boundary values of θ are weighted
by κu. Is the weight function large or small in boundary currents
where the flow itself is large? Laminar oceanic general circulation models, by fixing their lateral diffusivity, are guaranteed
to have enhanced flux in boundary currents and jets. Conversely,
present eddy-resolving general circulation models (e.g. Holland et
al. 1984) fail to generate eddy fluxes in the western boundary

currents, and hence are likely to deemphasize these regions in the tracer flux budget or the potential vorticity budget.

Shear dispersion in two-dimensional (x,y) circulations is described in more detail by Rhines and Young (1983). The principal effect is to speed the mixing of a tracer about mean streamlines. The mixing time is $\sim Pe^{-2/3}(L^2/\kappa_h)$, for large Pe. In gyres that pass through boundary currents the effect can in addition lead to mixing across streamlines, and hence throughout the gyre. Given the strong vertical shear of the oceans, one also has the possibility of folding vertical mixing onto the horizontal. Young et al. (1983) show that shear dispersion due to the vertical shear (provided either by mean circulation or inertial oscillations) causes diapycnal mixing to enhance lateral mixing. The effective lateral diffusion coefficient, κ_{eff}, for inertial oscillations is given by

$$\kappa_{eff} = \kappa_h + \kappa_v(\alpha^2/2f^2)$$

where α is the rms vertical shear. For typical internal wave spectra, the effective lateral diffusivity is $\sim 1.3 \times 10^3 \kappa_v$.

3.5 Possible enhancement of the diffusivity itself by the circulation.

The other side of the coin is the enhanced mixing that can occur owing to stretching of the isopleths of an active tracer like potential vorticity. A fine example is the rotation of Q-contours by the upper wind-gyre circulation, seen in many numerical studies (e.g., Cox 1985, Holland et al., 1984). In the subtropics a tongue of low potential vorticity is carried anticyclonically around the gyre, leading to a change in sign of dQ/dψ. Conditions favorable for baroclinic instability are established, particularly in two regions, the tight inertial recirculation and the broader North Equatorial Current. The models show that lateral mixing by mesoscale eddies in these regions is important in the shaping of the wind-driven gyre. More generally, the mean circulation provides critical surfaces at which wave-breaking may occur, enhancing mixing. This is largely unexplored territory in modelling studies.

Many recent theories of the large scale circulation exhibit
frontal discontinuities and mid-ocean jets. These occur due to
the conditioning of the environment for Rossby wave propagation by
the circulation itself. If there is a feedback on mixing rates, as
there obviously is in the case of mesoscale ring formation, then
this link will be important to future modelling. During the Liege
meeting, another explicit example (Rhines (1987)) involving deep
circulation over topography was given, in which shear dispersion
occurs owing to a gradient in the group velocity, and this generates greatly enhanced velocity, and enhanced tracer gradient and
mixing.

4 CONCLUSION

Through an improved sense of the deep ocean circulation, based
on large-scale property distributions and direct dynamical
measurements, we should be able to create a 'geography' of mixing
which is more satisfactory than what we have today. This process
has already begun to give productive picture of mixing in the
thermocline. Yet, one is still faced with the familiar discrepancy
between diapycnal mixing rates inferred from microscale and macroscale measurements.

There are many more areas where mixing and large-scale dynamics
meet: basic processes like isopycnal mixing by eddies, which are
not yet precisely defined (because the mixing occurs on layers of
varying thickness, and not on 'sheets'); the many relations between particle motions and stirring and mixing of tracers; the
connections between tracer transport as seen by an Eulerian observer, and Lagrangian mean circulation (indeed, a useful definition of Lagrangian mean circulation is difficult to find in any
but the simplest circumstances); the quantitative discussion of
mesoscale eddy shear- and strain fields, and their conditioning of
mixing effects; the development of specialized mixing sites
through wave focussing at critical layers and elsewhere; boundary
mixing; classic property-property diagrams on neutral surfaces;
the way in which inverse models may relate tracer patterns and
circulation, and finally, detailed intercomparison of active and
passive tracers (salinity, temperature and potential vorticity on
the one hand, and the many passive chemical/biological tracers in
the oceans, on the other). Indeed, the future will see models of
shear dispersion of mobile biological tracers, which may be at

work in shaping the global distributions of nutrients and productivity.

Acknowledgment. The National Science Foundation (Grant OCE 84-45194, OCE 86-13725) generously supported this work.

5 REFERENCES

Armi, L. and H.Stommel, 1983, Four views of a portion of the North Atlantic subtropical gyre. J.Phys.Oceanogr. 13:828-857.

Bigg, G.R., 1983, An analysis of the β-spiral method, Ocean Modelling 54, 7-12.

Broecker, W.S., T.Takahashi, and Timothy Takahashi, 1985, Sources and flow patterns of deep-ocean waters as deduced from potential termperature, salinity, and initial phosphate concentration, J.Geophys. Res. 90:6925-6939.

Bryan, F., 1987, Parameter sensitivity of primitive equation ocean general circulation models, J.Phys.Oceanogr. 17:970-985.

Bryden, H., 1976, Horizontal advection of temperature for low-frequency motions, Deep-Sea Res. 23:1165-1174.

Bryden, H., D.Roemmich and J.A. Church, 1987, Ocean heat transport across 24°N in the Pacific, Deep-Sea Res., submitted.

Cox, M., 1985, An eddy-resolving numerical model of the ventilated thermocline, J.Phys.Oceanogr.15:1312-1324.

Federiuk, J.M. and J.F. Price, 1984, Subduction mechanisms in the eastern North Atlantic, Eos 65:943.

Gargett, A.E., 1984, Vertical eddy diffusivity in the ocean interior, J.Mar.Res. 42: 359-393.

Gill, A.E., 1976, Adjustment under gravity in a rotating channel, J.Fluid Mech. 77:603-621.

Gill, A.E., 1982, Atmosphere-Ocean Dynamics, Academic Press.

Haynes, P.H. and M.E.McIntyre, 1986, On the evolution of vorticity and potential vorticity in the presence of diabatic heating and frictional or other forces, J.Atmos.Sci. 44:828-841.

Hogg, N., 1987, A least-squares fit of the advective diffusive equations to Levitus Atlas data, J.Mar.Res. 45:347-375.

Hogg, N., P.Biscaye, W.Gardner and W.J.Schmitz, 1982, On the transport and modification of Antarctic bottom water in the Vema Channel, J.Mar.Res. 40(Suppl.):231-263.

Holland, W.R., Keffer.T. and P.Rhines, 1984, Dynamics of the oceanic general circulation: the potential vorticity field, Nature 308:698-705.

Hoskins, B.J., M.E.McIntyre and A.W.Robertson, 1985, On the use and significance of isentropic potential vorticity maps, Q.Jour.Royal Met.Soc. 111:877-944.

Jenkins, W.J., 1980, Tritium and ^{3}He in the Sargasso Sea, J.Mar. Res.38:533-569.

Jenkins, W.J. 1987, ^{3}H and ^{3}He in the beta triangle: observations of gyre ventialtion and oxygen utilization rates, J.Phys. Oceanogr.17: 763-783.

Kawase, M. 1987, Establishment of mass-driven abyssal circulation, to appear in J.Phys.Oceanogr.

Kawase, M. and J. Sarmiento, 1985, Nutrients in the Atlantic thermocline, J.Geophys.Res. 90:8961-8979.

Killworth, P.D., 1986, A note on Van Heijst and Smeed, Ocean Modelling 69:7-9.

Luyten, J.R. and H.Stommel, 1986, Gyres driven by combined wind and buoyancy flux, J.Phys.Oceanogr. 16:1551-1560.

Moum, J.N. and T.R. Osborne, 1986, Mixing in the main thermocline, J.Phys. Oceanogr. 16:1250-1259.

Musgrave, D.L., 1985, A numerical study of the roles of subgyre-scale mixing and the western boundary current on homogenization of a passive tracer, J.Geophys.Res 90:7037-7043.

Olbers, D.J., M.Wenzel and J.Willebrand, 1985, The inference of North Atlantic circulation patterns from climatological hydrographic data, Rev. Geophys. 23:313-356.

Middleton, J., 1987, Energetics of linear geostrophic adjustment, J.Phys.Oceanogr. 17:735-740.

Plumb, R.A. and A.D.McEwan, 1978, The instability of a forced standing wave in a viscous stratified fluid: a laboratory analogue of the quasi-biennial oscillation, J.Atmos.Sci. 35:1827-1839.

Rago, T.A. and H.T. Rossby, 1987, Heat transport into the north Atlantic ocean north of 32°N latitude, J.Phys.Oceanogr. 17:854-871.

Rhines, P.B., 1983, Lectures in geophysical fluid dynamics, in **Lectures in Applied Mathematics**, 20, American Math. Soc.:3-58.

Rhines, P.B. and W.R. Young, 1983, How quickly is a passive tracer mixed within closed streamlines? J.Fluid Mech. 133:135-148.

Rhines, P.B., W.R.Holland and J.C.Chow, 1985, Experiments with buoyancy forced ocean circulation, Tech.Rep. Tn-260+Str, 108pp.

Rhines, P.B., 1986a, Vorticity dynamics of the oceanic general circulation, Ann.Revs.Fluid Mech.18:433-497.

Rhines, P.B., 1986b, Lectures on ocean circulation dynamics, in <u>Large-Scale Transport Processes in Oceans and Atmosphere</u>, J.Willebrand and D.Anderson (eds.), D.Reidel:105-161.

Rhines, P.B., 1987, Deep planetary circulation and topography: simple models of forced zonal flows, submitted to J.Phys.Oceanogr.

Sarmiento, J.L., 1983, A tritium box model of the North Atlantic thermocline, J.Phys.Oceanogr. 13:1269-1274.

Saunders, P.M., 1987, Flow through Discovery Gap, J.Phys.Oceanogr. 17:631-643.

Stommel, H. and R.Frazel, 1968, Hidaka's onions (Tamanegi), Records of Oceanographic Works in Japan 9:229-281.

Stommel, H., J.R. Luyten and C. Wunsch, 1985, A diagnostic study of the northern Atlantic subpolar gyre, J.Phys.Oceanogr. 15:1344-1348.

Whitehead, J.A. and L.V. Worthington 1982, THe flux and mixing rates of Antarctic Bottom Water within the N.Atlantic, J.Geophys.Res, C87:7903-7924.

Woods, J.D., 1985, The physics of thermocline ventilation, in <u>Coupled Ocean-Atmosphere Models</u>, J.C.J.Nihoul, Ed., Elsevier.

Wunsch, C. and D.Roemmich, 1985, Is the North Atlantic in Sverdrup balance? J.Phys.Oceanogr. 15:1876-1880.

Young, W.R., P.B. Rhines and C.J.R. Garrett, 1982, Shear-flow dispersion, internal waves and horizontal mixing in the ocean, J.Phys.Oceanogr. 12: 515-527.

VORTICAL MOTIONS

MÜLLER, P.
University of Hawaii, Department of Oceanography and Hawaii Institute of Geophysics, 1000 Pope Road, Honolulu, Hawaii

ABSTRACT
    In the linear case, the incompressible Boussinesq equations on an f–plane support two types of motion: steady geostrophic flows and internal gravity waves. One of the distinguishing properties is that steady geostrophic flows carry (linear perturbation) potential vorticity while internal gravity waves do not. This difference has led to the idea that a complex nonlinear flow might be viewed as consisting of two components: a "vortical" component that carries potential vorticity and is stagnant, and a "gravity" component that does not carry potential vorticity and propagates. This paper reviews the underlying concepts and algebra of such a decomposition, describes some of the attempts, and discusses various obstacles. The intrinsic theoretical problem is that potential vorticity is a nonlinear quantity whereas any decomposition is an essentially linear operation, requiring the validity of the superposition principle.

1 INTRODUCTION

Motions in the scale range bounded at large scales by synoptic–scale quasi–geostrophic eddies and at small scales by three–dimensional turbulence exhibit a complex kinematic structure and dynamic evolution. They represent a superposition of internal gravity waves, current finestructure, instabilities and other kinds of motion that interact in a complicated manner. To understand and diagnose these small–scale motions and their counterparts in numerical models a distinction is often attempted between "waves" and "turbulence" based on notions that waves have a dispersion relation, propagate, and superpose whereas turbulence is strongly nonlinear, spreads slowly and causes mixing. Here we review attempts that are based on the concept of potential vorticity.

Small–scale motions are adequately described by the incompressible Boussinesq equations on an f–plane (if temperature–salinity effects are neglected). In the linear limit, these equations support two types of motions: internal gravity waves and steady geostrophic flows. A distinguishing property is that steady geostrophic flows carry potential vorticity whereas internal gravity waves do not. Since potential vorticity is conserved for nonlinear inviscid and adiabatic flows (Ertel, 1942) one might attempt, in the general nonlinear case, to distinguish between a "vortical" component that carries potential vorticity and is "stagnant" (potential vorticity can only be transported by advection) and a "gravity" component that does not carry potential vorticity and propagates. This distinction is a well–observed phenomenon in laboratory experiments (e.g., Lin and Pao, 1979). Objects moving through a stratified quiescent fluid first create a three–dimensional wake. The internal gravity wave component rapidly propagates away and the wake collapses into two-dimensional horizontal vortices or "pancakes" which carry the potential vorticity of the wake.

The incompressible Boussinesq equations, the linear eigenmodes, and Ertel's potential vorticity theorem are reviewed in the first three sections. Then, some obvious and well known examples of nonlinear vortical and gravity flows are given. The complete decomposition of an arbitrary flow into vortical and gravity components using the linear eigenmodes is described in Section 8. The major theoretical obstacle is discussed in the summary section.

2 EQUATIONS OF MOTION

Small-scale motions in the ocean are governed by the incompressible Boussinesq equations on an f-plane. Neglecting viscosity and diffusion they take the form

$$\partial_t \underset{\sim}{u} + \underset{\sim}{u} \cdot \underset{\sim}{\nabla} \underset{\sim}{u} - f \underset{\sim}{u} \times \hat{\underset{\sim}{z}} + \rho_0^{-1} g \, \delta\rho \hat{\underset{\sim}{z}} = -\rho_0^{-1} \underset{\sim}{\nabla} \, \delta p \qquad (1.a)$$

$$\partial_t \, \delta\rho + \underset{\sim}{u} \cdot \underset{\sim}{\nabla} \, \delta\rho + \underset{\sim}{u} \cdot \underset{\sim}{\nabla} \, \tilde{\rho} = 0 \qquad (1.b)$$

$$\underset{\sim}{\nabla} \cdot \underset{\sim}{u} = 0 \qquad (1.c)$$

where $\underset{\sim}{u} = (u, v, w)$ is the current vector, f the Coriolis parameter, $\hat{\underset{\sim}{z}}$ the vertical unit vector, $\delta\rho$ the deviation of the density from a stably stratified background state $\tilde{\rho}(z)$, δp the deviation of the pressure from the background state , ρ_0 a constant reference value of density, and $N^2 = -g \, \rho_0^{-1} \, \partial\tilde{\rho}/\partial z$ the square of the Brunt-Väisälä frequency. Often the Eulerian vertical displacement ξ, defined by $\delta\rho = \tilde{\rho}(z - \xi) - \tilde{\rho}(z) = -\xi \, \partial\tilde{\rho}/\partial z + ...$, is used instead of $\delta\rho$. The density equation must then be replaced by

$$\partial_t \, \xi + \underset{\sim}{u} \cdot \underset{\sim}{\nabla} \, \xi = \underset{\sim}{u} \cdot \hat{\underset{\sim}{z}} \qquad (1.b')$$

Note that only the variables u, v, and $\delta\rho$ or u, v and ξ are prognostic. The equations of motion determine the time rate of change of these variables. The vertical velocity w and the pressure δp are diagnostic variables and can be inferred from the incompressibility condition (1.c) and the divergence of the momentum balance

$$\frac{1}{\rho_0} \nabla^2 \, \delta p = -\rho_0^{-1} g \, \partial_z \, (\delta\rho) + f \, \hat{\underset{\sim}{z}} \cdot (\underset{\sim}{\nabla} \times \underset{\sim}{u}) - \underset{\sim}{\nabla} \cdot (\underset{\sim}{u} \cdot \underset{\sim}{\nabla} \, \underset{\sim}{u}) \qquad (2)$$

without any time integration. The horizontal momentum balance, the density (or displacement) equation (1.b), the incompressibility condition (1.c), and the pressure equation (2) constitute a better-conditioned form of the equations of motion.

3 LINEAR EIGENMODES

For infinitesimally small perturbations the equations of motion can be linearized. Assume an unbounded ocean, constant Brunt–Väisälä frequency and solutions of the form

$$
\begin{bmatrix} u \\ v \\ \xi \end{bmatrix} = \text{Re} \left\{ a\,(\underset{\sim}{k}) \begin{bmatrix} U\,(\underset{\sim}{k}) \\ V\,(\underset{\sim}{k}) \\ Z\,(\underset{\sim}{k}) \end{bmatrix} \exp\left[i\,(\underset{\sim}{k} \cdot \underset{\sim}{x} - \omega t) \right] \right\}
\tag{3}
$$

where $a\,(\underset{\sim}{k})$ is the amplitude, $\underset{\sim}{k} = (k_x, k_y, k_z)$ the wavenumber vector, ω the frequency and $U\,(\underset{\sim}{k})$, $V\,(\underset{\sim}{k})$ and $Z\,(\underset{\sim}{k})$ the amplitude factors, forming the three components of the polarization vector. The linearized equations of motion are then converted to a set of coupled linear algebraic equations.

$$
-i\omega\,U - fV + \rho_0^{-1}\,ik_x\,P = 0 \tag{4.a}
$$
$$
-i\omega\,V + fU + \rho_0^{-1}\,ik_y\,P = 0 \tag{4.b}
$$
$$
-i\omega\,Z - W \qquad\qquad = 0 \tag{4.c}
$$

with

$$
W = -\frac{1}{k_z}\,(k_x\,U + k_y\,V) \tag{5.a}
$$

$$
P = i\,\frac{\rho_0}{k^2}\,\{N^2\,k_z\,Z - f(k_x V - k_y\,U)\} \tag{5.b}
$$

A nontrivial solution requires the determinant to be zero. Because there are three prognostic variables this determinant is a cubic polynomial in ω. The existence of three roots corresponds to there being three independent modes of motion. Two of the roots give the dispersion relation for internal gravity waves

$$
\omega_{\underset{\sim}{k}}^{\,s} = s \left[\frac{N^2\,(k_x^2 + k_y^2) + f^2\,k_z^2}{k^2} \right]^{1/2} \qquad s = +, - \tag{6}
$$

The two possible signs correspond to waves propagating in the direction of $\underset{\sim}{k}$ and $-\underset{\sim}{k}$. The third root is $\omega_{\underset{\sim}{k}}^{0} = 0$. These three roots are schematically sketched in Figure 1.

288

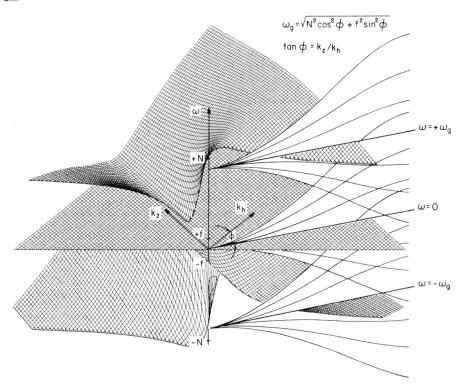

Fig. 1. Dispersion surfaces of the internal gravity and vortical mode. The hatched surfaces depict the three dispersion relations $\omega_{\mathbf{k}}^{+}$, $\omega_{\mathbf{k}}^{-}$ and $\omega_{\mathbf{k}}^{0}$ in a space spanned by the horizontal wavenumber $k_h = (k_x^2 + k_y^2)^{1/2}$, the vertical wavenumber k_z and frequency ω. For a particular aspect ratio $k_z/k_h = \tan \phi$ the eigenfrequencies are given by the intersection of the (hatched) dispersion surfaces and the plane $k_z/k_h = \tan \phi$. These intersections are straight lines and labelled $\omega = +\omega_g$, $\omega = 0$ and $\omega = -\omega_g$. Because of nonlinear interactions, energy will be distributed about the dispersion curves. The relative distribution of energy is sketched as fan–like curves, which may be regarded as contours of relative energy density (From Holloway, 1983).

The polarization vector of the internal gravity mode is given by

$$
\begin{bmatrix} U^s (\mathbf{k}) \\ V^s (\mathbf{k}) \\ Z^s (\mathbf{k}) \end{bmatrix} = \begin{bmatrix} (\omega_{\mathbf{k}}^s \, k_x + i \, f \, k_y) \, k_z \\ (\omega_{\mathbf{k}}^s \, k_y - i \, f \, k_x) \, k_z \\ -i \, (k_x^2 + k_y^2) \end{bmatrix} \qquad s = +, - \qquad (7)
$$

and the one for the $\omega_{\underline{k}}^0 = 0$ mode by

$$
\begin{bmatrix}
U^o\,(\underline{k}) \\[1ex]
V^o\,(\underline{k}) \\[1ex]
Z^o\,(\underline{k})
\end{bmatrix}
=
\begin{bmatrix}
-k_y \\[1ex]
k_x \\[1ex]
-\dfrac{f}{N^2}\,k_z
\end{bmatrix}
\tag{8}
$$

The $\omega_{\underline{k}}^0 = 0$ mode is in geostrophic ($-fv + \rho_0^{-1}\,\partial_x \delta p = 0$, $fu + \rho_0^{-1}\,\partial_y \delta p = 0$) and hydrostatic ($g\delta\rho + \partial_z\,\delta p = 0$) balance, it is horizontally nondivergent ($\partial_x u + \partial_y v = 0$) and satisfies the thermal wind equations ($f\partial_z u = \rho_o^{-1}\,g\,\partial_y\delta\rho$, $f\partial_z v = -\rho_o^{-1}\,g\partial_x\delta\rho$).

Note that only the direction of the polarization vectors is uniquely defined. The length can be arbitrarily normalized, defining the amplitude in (3). General solutions to the linear equations can be constructed by superposing the solutions (3).

4 POTENTIAL VORTICITY

In 1942 Ertel proved the following hydrodynamic theorem: If a scalar quantity satisfies $d\lambda/dt = 0$ then

$$
\frac{d}{dt}\left\{ \frac{1}{\rho}\,(\nabla \times \underline{u} + 2\,\underline{\Omega}) \cdot \underline{\nabla}\,\lambda \right\} = \rho^{-3}\,(\underline{\nabla}\rho \times \underline{\nabla}p) \cdot \underline{\nabla}\,\lambda
\tag{9}
$$

in a rotating inviscid fluid. Here $\underline{\Omega}$ is the rotation vector. The proof requires the Euler equations (inviscid momentum balance) and the continuity equation (mass balance). In many circumstances the energy equation for such fluids reduces to $d\rho/dt = 0$. The density ρ can then be used as the scalar λ and the theorem reduces to the conservation statement

$$
\frac{d}{dt}\left\{ \frac{1}{\rho}\,(\underline{\nabla} \times \underline{u} + 2\,\underline{\Omega}) \cdot \underline{\nabla}\,\rho \right\} = 0
\tag{10}
$$

The conserved quantity

$$
\pi = \frac{1}{\rho}\,(\underline{\nabla} \times \underline{u} + 2\,\underline{\Omega}) \cdot \underline{\nabla}\,\rho
\tag{11}
$$

is called potential vorticity. Its conservation is an extremely powerful constraint in a wide range of applications. Conservation of potential vorticity is the backbone of the theory of large– and synoptic–scale motions in the ocean and atmosphere.

For small–scale motions governed by the incompressible Boussinesq equations (1) conservation of potential vorticity can be expressed as

$$
\frac{d}{dt}\,\pi = \frac{d}{dt}\left\{ (\underline{\nabla} \times \underline{u} + f\,\hat{z}) \cdot \underline{\nabla}\,(\,\tilde{\rho} + \delta\rho) \right\} = 0
\tag{12a}
$$

or

$$
\frac{d}{dt}\,\hat{\pi} = \frac{d}{dt}\left\{ (\underline{\nabla} \times \underline{u} + f\,\hat{z}) \cdot \underline{\nabla}\,(\,z - \xi) \right\} = 0
\tag{12b}
$$

Note that the potential vorticity is a nonlinear quantity with respect to the prognostic variables u, v, and $\delta\rho$ or u, v, and ξ.

The linear perturbation potential vorticity is defined by

$$PV = \underset{\sim}{z} \cdot (\underset{\sim}{\nabla} \times \underset{\sim}{u}) - \int \partial_z \xi \tag{13}$$

It vanishes for linear internal gravity waves $(s = +, -)$

$$PV^s \, (\underset{\sim}{k}) = i \, \{k_x \, V^s \, (\underset{\sim}{k}) - k_y \, U^s \, (\underset{\sim}{k}) - \int k_z \, Z^s \, (\underset{\sim}{k})\}$$
$$\equiv 0 \tag{14}$$

and is given by

$$PV^o \, (\underset{\sim}{k}) = i\{k_x \, V^o \, (\underset{\sim}{k}) - k_y \, U^o \, (\underset{\sim}{k}) - fk_z \, Z^o \, (\underset{\sim}{k})\}$$
$$= i \, (k_x^2 + k_y^2 + \frac{f^2}{N^2} k_z^2) \tag{15}$$

for the $\omega_{\underset{\sim}{k}}^o = 0$ mode. The internal gravity wave mode is hence characterized by the property that it does not carry linear perturbation potential vorticity and the $\omega_{\underset{\sim}{k}}^o = 0$ mode by the property that it does. To emphasize this property Müller et al. (1986) refer to the $\omega_{\underset{\sim}{k}}^o = 0$ mode as the vortical mode.

5 SPECIAL NONLINEAR SOLUTIONS

There exist special solutions to the nonlinear equations (1) that are easily identified as either vortical, i.e., potential vorticity carrying motion, or gravity motions. An example is the steady single vortex, conveniently described in cylindrical coordinates (r, φ, z). For any pressure field $\delta p = \delta p \, (r, z)$ the density is given by $\delta\rho = -g^{-1} \, \partial \delta p / \partial z$ and the azimuthal velocity by the gradient flow formula

$$v_\varphi = -\frac{f}{2} \, r \pm \left[\frac{f^2 \, r^2}{4} + \frac{r}{\rho_0} \, \frac{\partial \delta p}{\partial r} \right]^{1/2} \tag{16}$$

The vertical and radial velocity vanish. This flow is stagnant and carries perturbation potential vorticity, given by

$$\pi - f \, \frac{\partial \tilde{\rho}}{\partial z} = \frac{1}{r} \, \frac{\partial}{\partial r} \, (rv_\varphi) \, \frac{\partial \tilde{\rho}}{\partial z} + f \, \frac{\partial \delta\rho}{\partial z} + \frac{1}{r} \, \frac{\partial}{\partial r} \, (rv_\varphi) \, \frac{\partial \delta\rho}{\partial z} - \frac{\partial v_\varphi}{\partial z} \, \frac{\partial \delta\rho}{\partial r} \neq 0 \tag{17}$$

and must hence be identified with the vortical mode.

A second example is a <u>single</u> finite amplitude internal gravity wave which is a solution of the nonlinear equations, since the incompressibility condition $\underset{\sim}{\nabla} \cdot \underset{\sim}{u} = 0$ implies $\underset{\sim}{u} \cdot \underset{\sim}{\nabla} = 0$ for a single wave. The single finite amplitude internal gravity wave does not carry any perturbation potential vorticity, $\pi - \int \partial \tilde{\rho} / \partial z = 0$, since $(\underset{\sim}{\nabla} \times \underset{\sim}{u}) \cdot \underset{\sim}{\nabla} = 0$ for a single wave.

In general, any solution with $\pi - \int \partial\bar{\rho}/\partial z = 0$ must be identified with the gravity mode. The problem begins when $\pi - \int \partial\bar{\rho}/\partial z \neq 0$. The flow might consist of a mixture of vortical and gravity motions and an algorithm is needed to determine which part of the flow is associated with the potential vorticity field.

6 VORTICAL MOTIONS

Solutions to the incompressible Boussinesq equations (1) are characterized by various dimensionless parameters. The important ones for the present purpose are the

$$\text{aspect ratio} \qquad \delta = \frac{H}{L}$$

$$\text{Rossby number} \qquad Ro = \frac{U}{fL}$$

$$\text{and} \qquad \text{Burger number} \qquad B = \frac{N^2 \, H^2}{f^2 \, L^2}$$

Here U denotes a typical horizontal velocity scale, L a horizontal length scale, and H a vertical length scale. In general solutions must be expected to consist of both vortical and gravity components. However, in certain parts of the 3-dimensional parameter space spanned by δ, Ro and B the solutions reduce to pure vortical motions. The best known examples are geostrophic and quasi-geostrophic flows. If $\delta \ll 1$, Ro $\ll 1$ and B $\ll$ Ro the equations (1) reduce to the geostrophic equations on an f-plane with the flow being characterized by carrying and conserving "geostrophic potential vorticity"

$$\pi_g = \int \partial_z (\bar{\rho} + \delta\rho) \tag{18}$$

Similarily, if $\delta \ll 1$, Ro $\ll 1$ but B $\gg$ Ro the equations (1) reduce to the quasi-geostrophic equations on an f-plane with "quasi-geostrophic potential vorticity"

$$\pi_{qg} = \rho_0^{-1} \, f^{-1}\{\partial_x\partial_x + \partial_y\partial_y + \partial_z \, f^2 \, N^{-2} \, \partial_z\} \, \delta p \tag{19}$$

being carried and conserved.

The limit which might be most relevant to small-scale motions is $\delta \ll 1$, Ro $\gg 1$ and B $\gg$ Ro^2 (or equivalently Froude number F = Ro B$^{-1/2}$ = U/NH $\ll 1$ or Richardson number Ri = F^{-2} = N^{2}H^2/U$^2 \gg$1). In this limit the equations (1) reduce to (Riley et al., 1981; Lilly, 1983; Müller, 1984)

$$\partial_t u + u\partial_x u + v\partial_y u + \rho_0^{-1} \partial_x \delta p = 0 \tag{20a}$$

$$\partial_t v + u\partial_x v + v\partial_y u + \rho_0^{-1} \partial_y \delta p = 0 \tag{20b}$$

$$g\delta\rho + \partial_z \delta p = 0 \tag{20c}$$

$$w = 0 \tag{20d}$$

$$\partial_x u + \partial_y v = 0 \tag{20e}$$

Since the flow is horizontally non–divergent there is only one prognostic variable and hence only one mode of motion. By taking the curl and divergence of the horizontal momentum balance one obtains a diagnostic equation for the pressure

$$\rho_0^{-1} \left(\partial_x\partial_x + \partial_y\partial_y\right) \delta p = -\partial_x u \, \partial_x u - \partial_x v \, \partial_y u - \partial_y u \, \partial_x v - \partial_y v \, \partial_y v \tag{21}$$

and a prognostic equation for the relative vorticity

$$\left(\partial_t + u\partial_x + v\partial_y\right) \left(\partial_x v - \partial_y u\right) = 0 \tag{22}$$

When the streamfunction Ψ of the horizontal velocity field is introduced, the prognostic equation reduces to:

$$\partial_t \left(\nabla_h^2 \, \Psi\right) + J(\Psi, \nabla_h^2 \, \Psi) = 0 \tag{23}$$

where ∇_h^2 is the horizontal Laplacian and J the Jacobian. The motion is vortical motion with the potential vorticity given by the relative vorticity, i.e.,

$$\pi = \partial_x v - \partial_y u = \nabla_h^2 \, \Psi \tag{24}$$

The motions described by these equations are purely horizontal, but they can vary in the vertical. The horizontal momentum balance is ageostrophic. Pressure gradients are balanced by acceleration. The motions are, however, decoupled in the vertical (at least to the order considered here). Equation (23) applies to each horizontal plane separately. Vertical variations are hence determined by the distribution of sources and sinks. No scaling laws with z or N are known.

In each horizontal plane equation (23) describes two–dimensional turbulence. This is why Lilly (1983) coined the term "stratified two–dimensional turbulence" for these motions. Two–dimensional turbulence has been studied extensively since it describes barotropic quasi–geostrophic motions with scales smaller than the Rossby radius. The principal features of two–dimensional turbulence are the conservation of energy and enstrophy and the transfer of energy to low wavenumbers (red energy cascade) and of enstrophy to high wavenumbers (blue enstrophy cascade). Kraichnan (1967) suggested that these cascades can occur in statistically steady inertial ranges. If energy and enstrophy are steadily generated at wavenumber $\hat{k}$, an inertial range

$$E(k) \sim k^{-5/3} \quad \text{for } k < \hat{k} \tag{25}$$

cascades energy (but no enstrophy) to small wavenumbers and an inertial range

$$E(k) \sim k^{-3} \quad \text{for } k > \hat{k} \tag{26}$$

cascades enstrophy (but no energy) to large wavenumbers. These inertial ranges require energy dissipation at small wavenumbers and enstrophy dissipation at high wavenumbers.

Observations of the kinetic energy spectrum of the atmosphere (e.g., Nastrom et al., 1984) show a –3 power law dependence on wavenumber at synoptic wavelengths, from about 3,000 km to 1,000 km, and a –5/3 power law in the mesoscale range, between 500 km and 5 km. These power laws have been considered evidence of enstrophy and energy cascading inertial ranges as sketched in Figure 2 (Charney, 1971; Gage, 1979). For the mesoscale range, this interpretation has been challenged by Van Zandt (1982), who interprets mesoscale fluctuations as internal gravity waves. It is not clear to what extent two-dimensional turbulence inertial ranges exist in the ocean.

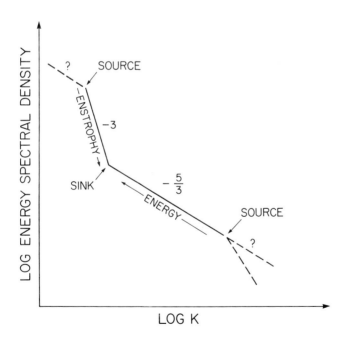

Fig. 2. Schematic representation of two-dimensional turbulence inertial ranges in the atmosphere. The enstrophy cascading inertial range reflects the spectrum of synoptic-scale fluctuations with a source at a wavenumber corresponding to the internal Rossby radius of deformation. The energy cascading inertial range reflects the spectrum of mesoscale fluctuations with an unknown source at high wavenumbers (From Larsen et al., 1982).

7 GRAVITY MOTIONS

Pure gravity motions and their nonlinear interactions can be studied by considering the special cases of weak resonant interactions and motions in a nonrotating vertical plane. In the weak resonant interaction approach the flow field is regarded at every time instant as a superposition of infinitesimal linear internal gravity waves. Resonant interactions only lead to a slow exchange of energy and momentum between various waves, resulting in a slow change of the amplitude but in no change of the polarization vector. Resonant weak interaction theory predicts the existence of an inertial range that cascades energy from low to high vertical wavenumbers (Figure 3). The inertial range has a vertical wavenumber spectrum

$$E\ (k_z) \sim k_z^{-2} \tag{27}$$

as observed. The slope cannot be obtained from simple dimensional arguments since the dominant transfer mechanisms, induced diffusion and parametric subharmonic instability, involve waves with largely different wavenumbers and are hence nonlocal in wavenumber space. Dimensional arguments would lead to $E\ (k_z) \sim k_z^{-3}$.

Motions in a nonrotating vertical plane also represent pure gravity motion since the perturbation potential vorticity, $\pi - \int \partial \tilde{\rho} / \partial z$, vanishes. Numerical simulations of these two–dimensional flows have been used to study strongly interacting internal gravity waves and to explore the limitations of the weak resonant interaction approach (e.g., Shen and Holloway, 1986).

8 NORMAL MODE DECOMPOSITION

Infinitesimal perturbations satisfy the dispersion relations. Energy is concentrated on the dispersion curves in Figure 1. In the actual nonlinear case energy is not confined to the dispersion curves but spread about them, as indicated by the fan–like curves in Figure 1. The broadening is due to nonlinear interactions. At large wavenumbers the fans become very broad and the different modes start to overlap. In this case, the polarization vectors can be used to define gravity and vortical motions. The three polarization vectors (7) and (8) represent a complete set of basis vectors in $\{u\ (\underline{k}),\ v\ (\underline{k}),\ \xi\ (\underline{k})\}$ –space and at any instant in time an arbitrary $\{u,\ v,\ \xi\}$ –field can uniquely and completely be decomposed into a gravity and a vortical mode component.

In such a modal decomposition the vortical mode is defined as that part of the motion which carries linear perturbation potential vorticity $PV = \partial_x v - \partial_y u - f \partial_z \xi$. The amplitude and the space– and time–scales of PV at small scales have been estimated by Müller et al. (1988) using the current meter array of the Internal Wave Experiment. This array resolves horizontal scales from about 5 m to 1 km. The principal findings are: The linear perturbation potential vorticity has a variance of about $10^{-6} s^{-2}$, implying a Rossby number and vertical strain of order 10. The associated total energy is $2 \cdot 10^{-4}\ m^2\ s^{-2}$ and the associated inverse Richardson number 0.7. The vortical mode might hence be the major

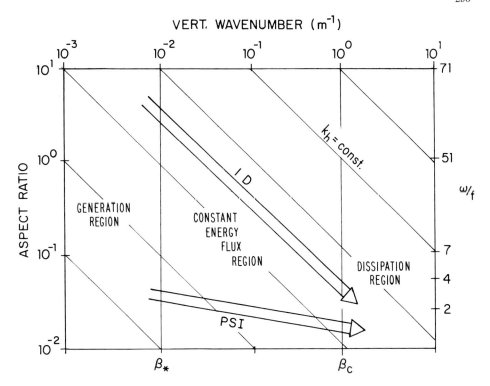

Fig. 3. Schematic representation of the dynamical balance of internal gravity waves. Energy is generated at low vertical wavenumbers $\beta < \beta_*$. Between β_* and β_c there is an inertial range where the induced diffusion (ID) mechanism at high frequencies and the parametric subharmonic instability (PSI) mechanism at low frequencies provide a constant energy flux to high wavenumbers. At $\beta > \beta_c$ energy is dissipated (From McComas and Müller, 1981).

contributor to the observed shear in the ocean. The observed frequencies (Figure 4) of the vortical mode are interpreted as Doppler frequencies. A low frequency "steppy" vortical mode field is advected vertically past the sensors by internal gravity waves. The horizontal wavenumber spectrum of the advected vortical mode has a +2/3 power law (Figure 5).

Though Müller et al. (1988) did not attempt a complete normal mode decomposition they compared their results to the Garrett and Munk spectrum that is supposed to describe the gravity mode. Figure 6 shows that the vortical mode dominates both the total energy and inverse Richardson number spectrum in the resolved horizontal wavenumber range. The comparison might not be entirely fair since the Garrett and Munk spectrum is based on data that do not adequately resolve the small scales. A complete decomposition is needed.

Nonlinear interactions change the modal amplitudes in time. Linear perturbation potential vorticity PV is not conserved. The vortical mode in the normal mode decomposition is not necessarily the "stagnant" part of the motion. To follow the dynamical evolution of PV the equations of motion need to be projected onto the normal modes. This is an extremely cumbersome algebraic exercise.

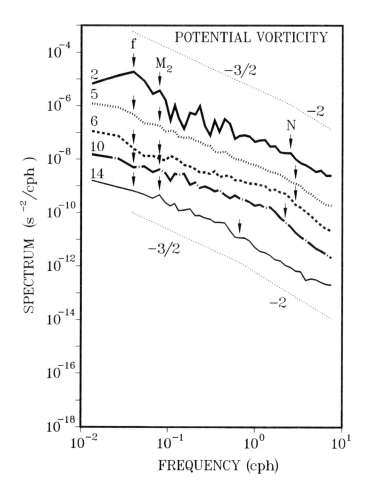

Fig. 4. Frequency spectrum of linear perturbation potential vorticity PV at five levels in the vertical. The systematic decrease in amplitude from level to level is due to the fact that the potential vorticity signal is averaged over a larger and larger area with increasing depth. (From Müller et al., 1988).

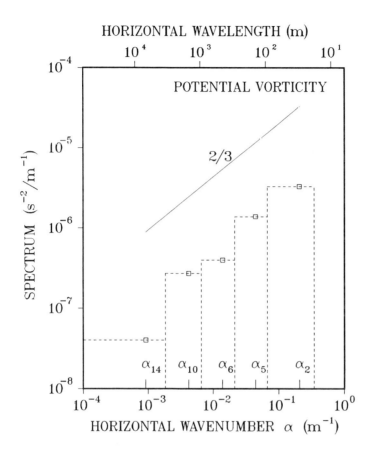

Fig. 5. Horizontal wavenumber spectrum of linear perturbation potential vorticity. (From Müller et al., 1988).

298

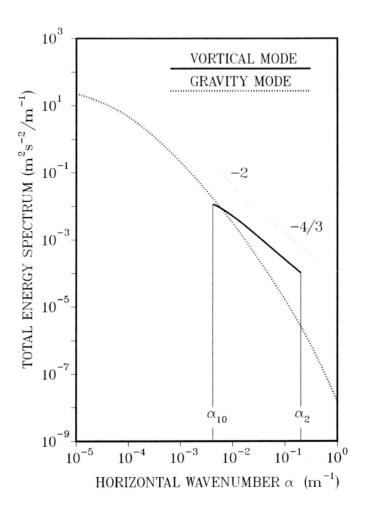

Fig. 6a. Horizontal wavenumber spectra of total energy for the vortical and gravity mode. The vortical mode spectra are calculated from Müller et al.'s 1988 model spectrum and the gravity mode spectra from the Garrett and Munk 1976 (Cairns and Williams, 1976) model spectrum (From Müller et al., 1988).

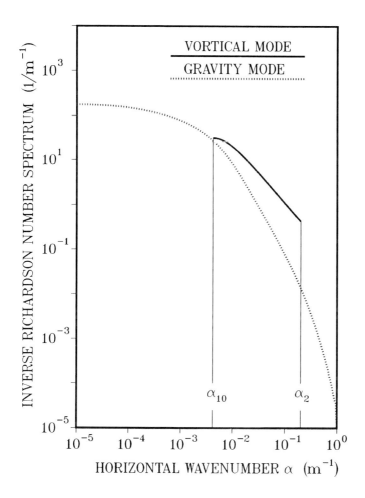

Fig. 6b. Inverse Richardson number for the vortical and gravity mode. The vortical mode spectra are calculated from Müller et al.'s 1988 model spectrum and the gravity mode spectra from the Garrett and Munk 1976 (Cairns and Williams, 1976) model spectrum (From Müller et al., 1988).

9 SUMMARY

Various attempts have been described that distinguish between or isolate the "vortical" and "gravity" component in an incompressible Boussinesq fluid. These attempts are based on the fact that the complete solution to the linearized equations consists of a superposition of propagating internal gravity waves and steady stagnant geostrophic flows. The steady geostrophic flow component carries the linear perturbation potential vorticity of the flow. Linear perturbation potential vorticity is the linear limit of Ertel's potential vorticity. Ertel's potential vorticity is conserved under nonlinear interactions. These facts suggest that an arbitrary complex nonlinear flow can be thought of as consisting of two components: a vortical component that carries the potential vorticity and is stagnant, and a gravity component that does not carry potential vorticity and propagates.

The decomposition into vortical and gravity modes is clear and unambiguous in the linear case. In the nonlinear case, there exist a few special solutions that are obviously vortical or gravity motions. Vortical and gravity motions are also isolated if special parameter limits or geometries are considered. In the general nonlinear case, however, no concept or algorithm is yet known to determine which part of the flow field is associated with Ertel's potential vorticity. The normal mode decomposition with respect to the linear polarization vectors, which is unique and complete at every time instant, yields a vortical component that is characterized by linear perturbation potential vorticity, which is not conserved under nonlinear interaction and hence does not necessarily reflect the stagnant part of the motion.

The intrinsic theoretical problem is that Ertel's potential vorticity is a nonlinear quantity whereas any decomposition is an essentially linear operation, requiring the validity of the superposition principle. An optimal representation of the flow field would be one in which Ertel's potential vorticity is one of the components. It is not known whether or not such a representation exists.

Nevertheless, vortical and gravity motions are the two distinctly different ingredients that comprise small-scale oceanic motions. Distinguishing between a vortical and a gravity component is therefore an extremely useful approach to use in untangling and understanding the complex kinematic structure and dynamic behavior of small-scale motions.

ACKNOWLEDGEMENTS:

I would like to thank F. Henyey, G. Holloway, E. Kunze and R. Lien for stimulating and helpful discussions and K. Chong Morgan, L. Takimoto and C. Miles for expert technical assistance. This work was supported by the Office of Naval Research and is Contribution 1952 of the Hawaii Institute of Geophysics.

References

Cairns, J. L. and Williams G. O., 1976. Internal wave observations from a
  midwater float, Part II. J. Geophys. Res. 81: 1943–1950.
Charney, J. G., 1971. Geostrophic turbulence. J. Atmos Sci., 28: 1087–1095.
Ertel, H., 1942. Ein neuer hydrodynamischer, Wirbelsatz, Meteorol, Z.
  59: 277–281.
Gage, K.S., 1979. Evidence for a $k^{-5/3}$ law inertial range in mesoscale
  two-dimensional turbulence. J. Atmos. Sci., 36: 1950–1954.
Holloway, G., 1983. A conjecture relating oceanic internal waves and small-
  scale processes. Atmosphere–Ocean, 21: 107–122.
Kraichnan, R. H., 1967. Inertial ranges in two-dimensional turbulence. Phys.
  Fluid., 10: 1417–1423
Larsen, M. F., Kelley M. C. and Gage K. S., 1982. Turbulence spectra in the
  upper troposphere and lower stratosphere at periods between 2 hours and 40 days.
  J. Atmos. Sci., 39: 1035–1041.
Lilly, D. K., 1983. Stratified turbulence and the mesoscale variability of
  the atmosphere. J. Atmos. Sci., 40: 749–761.
Lin, J. and Pao Y., 1979. Wakes in stratified fluids. Ann. Rev. Fluid Mech., 11:
  317–338.
McComas, C. H. and Müller P., 1981. The dynamic balance of internal waves.
  J. Phys. Oceanogr. 11: 970–986.
Müller, P. 1984. Small-scale vortical motions. In: P. Müller and R. Pujalet
  (Editors), Internal Gravity Waves and Small-Scale Turbulence, Proceedings,
  'Aha Huliko'a Hawaiian Winter Workshop. Hawaii Institute of Geophysics,
  Honolulu, pp. 249–261.
Müller, P., Holloway, G. Henyey F. and Pomphrey N., 1986. Nonlinear
  interactions among internal gravity waves. Rev. of Geophys., 24:
  No. 3, 493–536.
Müller P., Lien R. C. and Williams R., 1988. Estimates of potential vorticity
  at small scales in the ocean. J. of Phys. Oceanogr. (In press).
Nastrom, G. D., Gage K. S. and Jaspersion W. H., 1984. Kinetic energy spectrum of
  large and mesoscale atmospheric processes. Nature, 310: 36–38.
Riley, J. J., Metcalfe R. W. and Weissman M. A., 1981. Direct numerical
  simulations of homogeneous turbulence in density–stratified fluids. In: B. J. West
  (Editor), Nonlinear properties of internal waves. American Inst. of Physics, 76:
  79–112.
Shen, C. and Holloway G., 1986. A numerical study of the frequency and the
  energetics of nonlinear internal gravity waves, J. Geophys. Res., 91: 953–973.
Van Zandt, T. E., 1982. A Universal spectrum of buoyancy waves in the atmosphere.
  Geophys. Res. Lett., 9: 575–578.

THE RELATION BETWEEN LAGRANGIAN AND EULERIAN SPECTRA: A CONNECTION
BETWEEN LARGE-SCALE WAVES AND SMALL-SCALE TURBULENCE

K. R. Allen and R. I. Joseph*
The Johns Hopkins University/Applied Physics Laboratory, Laurel, Maryland 20707
(U.S.A.)

ABSTRACT
    We obtain an exact relation between Lagrangian and Eulerian spectra for the
case in which the canonically conjugate Lagrangian variables are distributed in
accordance with the canonical distribution from statistical mechanics.   By
considering a simplified example with one spatial dimension we obtain an exact
closed form expression for the Eulerian frequency wavenumber spectrum.   We find
that even though the Lagrangian frequency wavenumber spectrum is proportional
to a delta function which confines the system to the dispersion surface, the
Eulerian frequency wavenumber spectrum is not so confined and exhibits turbulence properties at small wavenumbers.

1 INTRODUCTION

    Problems in fluid dynamics can be formulated either in terms of Lagrangian

or Eulerian variables. Which is preferable depends upon the specific objec-

tives of the formulation. If, for example, we are interested in solving a flow

problem for which the specifics of boundary conditions and the details of

forcing are important, then there is little question that the formulation in

terms of Eulerian variables is the more tractable. On the other hand, if our

interest is in the statistical treatment of a large stochastic system, then

consideration should be given to a formulation in terms of Lagrangian vari-

ables. This is because the usual methods of statistical mechanics are formu-

lated in terms of Lagrangian variables. However, most observations and empir-

ical studies are in terms of Eulerian variables and we must, therefore, obtain

some method of relating results which are given in terms of the two different

sets of variables.

    We shall denote the Lagrangian displacement and velocity by $\vec{\eta}_L(\vec{\chi},t)$ and

$\vec{v}_L(\vec{\chi},t)$ where t is time. The Lagrangian label $\vec{\chi}$ corresponds to the position of

---

*Permanent Address: Department of Electrical and Computer Engineering, The
Johns Hopkins University, Baltimore, Maryland 21218 (U.S.A.)

a fluid parcel under some reference condition often taken to be the undisturbed condition. Once selected, a specific value for $\vec{\chi}$ remains with the fluid parcel and does not change throughout the dynamic evolution of the system. The Eulerian displacement and velocity will be denoted by $\vec{\eta}_E(\vec{x},t)$ and $\vec{v}_E(\vec{x},t)$ where a given value for the Eulerian label $\vec{x}$ corresponds to a specific point in space and refers to the fluid parcel which happens to be at that point at the time t. Thus, a given value for the Eulerian label $\vec{x}$ does not always refer to the same fluid parcel.

A formulation in terms of Lagrangian variables retains a faithful correspondence with the particles of Newtonian mechanics. Such a formulation can be given in terms of a Hamiltonian and Hamilton's canonical equations, or the Lagrangian variables can be used directly in Hamilton's principle to obtain the well-known variational form of continuum mechanics. In the case of Eulerian variables no such straightforward Hamiltonian formulation is possible. While the Eulerian equations of motion have been cast into a variational form, which is sometimes referred to as canonical, such a formulation requires the introduction of additional variables and constraints, and is significantly different from the usual well-known Hamiltonian dynamics (Seliger and Whitham, 1968).

We can expand the Lagrangian displacement and velocity in terms of the spatial eigenfunctions associated with writing the quadratic part of the Hamiltonian in separated form. In a complete treatment this would require, in addition to the internal wave modes, the inclusion of surfaces waves, the geostrophic modes, and perhaps even sound. While it is straightforward to include these, our focus in this paper is upon fundamentals so that we are not, at this time, interested in results which can be used for detailed comparison to experiment. We shall, therefore, make as many simplifying assumptions as seem prudent. It will be clear that the formulation can be generalized to include details, but to do so now would serve only to increase the already considerable mathematical complexity, obscure the underlying physics, and add nothing of fundamental importance.

In keeping with this we shall, for now, consider only the contributions from the internal wave modes. We shall also neglect the Coriolis effect, set the Väsiälä profile equal to the constant N, and employ periodic boundary conditions in the vertical as well as in the horizontal directions. Under these conditions we can write the Lagrangian displacement as

$$\vec{\eta}_L(\vec{\chi},t) = \sum_{j=1}^{M} \sqrt{2/V\rho\Omega_j} \, [(\ell_{jo}\hat{\epsilon}_3 - \ell_{j3}\hat{\ell}_{jo})/\ell_j][\cos(\hat{\ell}_j\cdot\vec{\chi})q_{j1}(t)$$

$$+\sin(\hat{\ell}_j\cdot\vec{\chi})q_{j2}(t)] = \sum_{j=1}^{M}\sum_{s=1}^{2}\sum_{\alpha=1}^{3} f_{js\alpha}(\vec{\chi})q_{js}(t)\hat{\epsilon}_\alpha \tag{1}$$

where ρ is the density, V is the volume of the ocean, and 2M is the number of degrees of freedom (later V and M will be taken arbitrarily large). In (1) $\vec{\ell}_j$ is a three-dimensional wavevector which, because we have chosen to use real eigenfunctions, is constrained to lie in a hemisphere of wavenumber space. The $\ell_{j\alpha}$, $1\leq\alpha\leq3$, are, respectively, the cartesian components of $\vec{\ell}_j$ in the directions x, y, and z with z vertical. The vector $\vec{\ell}_{jo}$ is the horizontal component of $\vec{\ell}_j$, ℓ_j is the magnitude of $\vec{\ell}_j$, ℓ_{jo} is the magnitude of $\vec{\ell}_{jo}$, and $\hat{\ell}_{jo}$ is a unit vector in the direction $\vec{\ell}_{jo}$. The ε_α, $1\leq\alpha\leq3$, are, respectively, unit vectors in the directions x, y, and z. The $q_{js}(\tau)$, $1\leq s\leq2$, are real independent generalized displacements and the dispersion relation Ω_j is given by

$$\Omega_j = N\ell_{jo}/\ell_j \tag{2}$$

We shall also write the Lagrangian velocity as

$$\vec{v}_L(\vec{\chi},t) = \sum_{j=1}^{M} \sum_{s=1}^{2} \sum_{\alpha=1}^{3} f_{js\alpha}(\vec{\chi})\Omega_j p_{js}(t)\hat{\varepsilon}_\alpha \tag{3}$$

where $p_{js}(t)$ is the canonically conjugate momentum associated with $q_{js}(t)$. The Hamiltonian H(p,q) can thus be written

$$H(p,q) = H_o(p,q) + \lambda V_I(p,q) = \sum_{j=1}^{M} \sum_{s=1}^{2} (\Omega_j/2)[p_{js}^2(t) + q_{js}^2(t)] + \lambda V_I(p,q) \tag{4}$$

where $H_o(p,q)$ is the quadratic part of the Hamiltonian and (p,q) denotes the set $(\cdots,p_{js}(t),q_{js}(t),\cdots)$ of canonically conjugate dynamical variables. The interaction potential $\lambda V_I(p,q)$ is of cubic and higher order in the dynamical variables and gives rise to the nonlinear interactions.

The Lagrangian equations of motion for the $p_{js}(t)$ and $q_{js}(t)$ are obtained by using the Hamiltonian given by (4) in Hamilton's canonical equations. In a general case they are nonlinear and cannot be solved exactly. The linear approximation is obtained by setting $\lambda V_I(p,q)=0$ in which case it is easy to show that

$$q_{js}(t) = q_{js}(0)\cos(\Omega_j t) + p_{js}(0)\sin(\Omega_j t) \tag{5}$$

and

$$p_{js}(t) = p_{js}(0)\cos(\Omega_j t) - q_{js}(0)\sin(\Omega_j t) \tag{6}$$

where the $p_{js}(0)$ and $q_{js}(0)$ are initial values. For large many body systems the precise specification of the initial values is out of the question and we must resort to the use of statistical methods. In order to deal with the nonlinear contributions it is necessary to use perturbation or other approximation techniques. We shall refer to these nonlinearities as dynamic nonlinearities, and even though they can sometimes be treated as weak they play an important role in the time evolution of the relevant statistical distributions.

The Eulerian displacement and velocity can also be written in the form

$$\vec{\eta}_E(\vec{x},t)= \sum_{j=1}^{M} \sum_{s=1}^{2} \sum_{\alpha=1}^{3} f_{js\alpha}(\vec{x})a_{js}(t)\hat{\epsilon}_\alpha \tag{7}$$

and

$$\vec{v}_E(\vec{x},t)= \sum_{j=1}^{M} \sum_{s=1}^{2} \sum_{\alpha=1}^{3} f_{js\alpha}(\vec{x})\Omega_j b_{js}(t)\hat{\epsilon}_\alpha \tag{8}$$

While the form of (7) and (8) is identical to that of (1) and (3), in a general case the $a_{js}(t)$ and $b_{js}(t)$ are not related in any simple way to the canonically conjugate variables $p_{js}(t)$ and $q_{js}(t)$. The linearized equations of motion for the two sets of variables are the same, so that for small enough amplitude disturbances we may write $a_{js}(t)=q_{js}(t)$ and $b_{js}(t)=p_{js}(t)$. However, the nonlinear terms associated with the two sets of variables are different. The dynamic nonlinearities also contribute to the Eulerian equations of motion, but because individual fluid parcels are constantly flowing into and out of the region of interest there is an additional nonlinear flow term, $(\vec{v}_E\cdot\nabla)\vec{v}_E$, which we shall call the advective nonlinearity. Thus, for larger amplitude disturbances the exact transformation between the variables rapidly becomes intractable. It is important to realize that the two types of nonlinearity are fundamentally different. The dynamic nonlinearities are associated with the details of the forces between collections of fluid parcels. The advective nonlinearity is associated with the flow of fluid parcels into and out of a fixed region of space, and is strictly an Eulerian frame concept. From a Lagrangian frame point of view, the advective nonlinearity is a kinematic effect.

The usual formulations of statistical mechanics are in terms of Hamiltonian dynamics and, in a strict sense, require the use of Lagrangian variables. A typical formulation utilizes the phase space density function $g(p,q,t)$ so that if $F(p,q,t)$ is any function of the dynamical variables and time, then the expectation value $E[F(p,q,t)]$ is given by

$$E[F(p,q,t)]= \int F(p,q,t)g(p,q,t) \prod_{j=1}^{M} \prod_{s=1}^{2} dp_{js}dq_{js} \tag{9}$$

where from this point on, unless otherwise noted, it is to be understood that the various p_{js} and q_{js} are evaluated at the time t. Thus, in order to obtain a statistical theory in terms of Lagrangian variables we have only to specify the phase space density function and to use (9) to compute the various expectation values. For example, the steady-state phase space density function associated with canonical equilibrium is given by

$$g(p,q)=(1/Z)\exp\{-H(p,q)/\Lambda\} \tag{10}$$

where Z is the partition function and Λ is the average energy per degree of freedom. The expression given by (10) describes the joint distribution of the canonically conjugate dynamical variables p_{js} and q_{js}. In a general case, however, this does not provide us with direct information about the joint distribution of the Eulerian amplitudes a_{js} and b_{js} where again we have suppressed explicit display of the time t. It is only for the case of weakly excited systems (i.e., small Λ) that (10) is also the expression for the joint distribution of the Eulerian amplitudes.

Even in the case of Lagrangian variables, the use of the full expression for the Hamiltonian given by (4) in the phase space density function given by (10) is often too complicated to be of practical value. The difficulty is that the interaction potential $\lambda V_I(p,q)$ is a complicated function of the dynamical variables and its use in (10) leads to intractable expressions. The weak interaction approximation consists of neglecting the term $\lambda V_I(p,q)$ when (4) is used in (10). This results in a phase space density function which is of a Gaussian form and leads to many interesting calculations which are mathematically tractable. The weak interaction approximation has been studied by Prigogine (1962) and his coworkers who, by using the Liouville equation and Hamiltonian dynamics, obtain a master equation which describes the time evolution of $g(p,q,t)$ and show that its long time solution corresponds to (10) with $H(p,q)$ replaced by $H_o(p,q)$. This treatment includes only internal interactions. That is, contributions from external sources of energy and external dissipation are neglected. This is done on the assumption that even though the internal interactions are weak they are still larger than the external interactions and dominate the time evolution of the phase space density function. This must, of course, be verified for any particular physical system, but it is often valid and if it is the phase space density function is given by (10) with $H(p,q)$ replaced by $H_o(p,q)$.

The Prigogine treatment of the weak interaction approximation is in terms of Lagrangian variables. It is also possible to develop such a theory in terms of Eulerian variables (e.g., Benney and Saffman, 1966). In the case of Eulerian variables, however, contributions from the advective as well as the dynamic nonlinearities must be included. Thus, it is clear that the conditions for the validity of the weak interaction approximation within the Eulerian frame are different from those within the Lagrangian frame. If the contributions from the advective as well as the dynamic nonlinearities can be treated as weak, then the weak interaction approximation is valid within the Eulerian frame. In this case the Lagrangian and Eulerian variables are equivalent, and the weak interaction approximation is also valid within the Lagrangian frame. The converse, however, is not true. It is clear that there exists a class of systems for which the weak interaction approximation is valid within the Lagrangian frame, but for which the contributions from the advective nonlinearity are not necessarily small. We shall refer to the case for which the advective nonlinearity as well as the dynamic nonlinearities can be treated as weak as the absolutely weak interaction approximation, and we shall refer to the case for which only the dynamic nonlinearities can be treated as weak as the dynamically weak interaction approximation.

One major result of this paper is to show that in the dynamically weak interaction approximation we can still obtain exact expressions for the Eulerian frequency wavenumber spectrum. By considering an example system with one spatial dimension we obtain an exact closed form expression for the Eulerian frequency wavenumber spectrum. We find that even though the Lagrangian frequency wavenumber spectrum is proportional to a delta function which confines the system to the dispersion surface, the Eulerian frequency wavenumber spectrum is not so confined. Specifically, at small wavenumbers there is a remnant of the dispersion surface which becomes more and more smeared as the wavenumber increases until finally at large wavenumbers the system is completely turbulent. The higher the level of excitation the smaller the wavenumber at which turbulent behavior is evident.

2 THE WEAK INTERACTION APPROXIMATION

2.1 The absolutely weak interaction approximation

Before considering the more general case we will first briefly explore the consequence of the absolutely weak interaction approximation. The results can serve as a base line for future discussion. In this case the Lagrangian and Eulerian variables are equivalent so that for now we can consider Lagrangian spectra. We need first to point out that the sum over j is truncated at $j=M$. For example, M might be chosen to correspond to the very small scales at which molecular viscosity is important (cf. Holloway, 1986). If we are interested in details at these small length scales, then, of course, a more adequate treatment must be provided. At larger length scales, however, a simple truncation is expected to be adequate. We shall find it convenient to implement this truncation by introducing the phase space density function

$$g(p,q) = \prod_{0<j} \prod_{s=1}^{2} (\Omega_j/2\pi\Lambda_j) \exp\{-(\Omega_j/2\Lambda_j)(p_{js}^2+q_{js}^2)\} \qquad (11)$$

as a slight generalization of the canonical distribution. In (11) we have allowed Λ_j, the average energy of the jth mode, to depend upon the mode index, while for the canonical distribution Λ_j is independent of j. If Λ_j is set approximately equal to the constant E_o for j<M and decreases rapidly to zero for M<j, then the above truncation is achieved, and we may drop the explicit display of the upper bound M.

The two space point two time point vertical displacement correlation function $C_{L\eta33}(\vec{X},\tau)$ is given by

$$C_{L\eta33}(\vec{X},\tau)=E[\eta_{L3}(\vec{X}+\vec{X},t+\tau)\eta_{L3}(\vec{X},t)]=(1/8\pi^3\rho N^2)\int \Lambda(\vec{\ell})\cos(\vec{\ell}\cdot\vec{X})\cos(\Omega(\ell)\tau)d^3\ell \quad (12)$$

where we have allowed the volume of the system to become arbitrarily large and, both here and in subsequent expression, integrals will be extended to cover all of wavenumber space. In writing (12) we have used (1) for the definition of the $f_{js3}(\vec{X})$, (5) for the explicit time dependence of the $q_{js}(t+\tau)$, and (9) and (11) to show that

$$E[q_{js}(t+\tau)q_{j's'}(t)]=\delta_{jj'}\delta_{ss'}(\Lambda_j/\Omega_j)\cos(\Omega_j\tau) \quad (13)$$

By using (12) we find that the frequency wavenumber spectrum is given by

$$S_{L\eta33}(\vec{k},\omega)=\int d^3X \int d\tau\, C_{\eta33}(\vec{X},\tau)\exp\{-i(\vec{k}\cdot\vec{X}-\omega\tau)\}$$

$$=(\pi/\rho N^2)\Lambda(\vec{k})[\delta(\omega-\Omega(\vec{k}))+\delta(\omega+\Omega(\vec{k}))] \quad (14)$$

The delta functions in (14) confine the system to the dispersion surface described by $\Omega(\vec{k})$ and the system is, therefore, wave-like. The three-dimensional wavenumber spectrum $\hat{S}_{L\eta33}(\vec{k})$, which is proportional to the 'so called' wave action spectrum, is given by

$$\hat{S}_{L\eta33}(\vec{k})=(1/2\pi)\int S_{L\eta33}(\vec{k},\omega)d\omega=(1/\rho N^2)\Lambda(\vec{k}) \quad (15)$$

We note that the wavenumber spectrum given by (15) is not similar to that which has been proposed by Garrett and Munk (GM) (1972, 1975). This is sometimes taken as evidence that the ocean is not near canonical equilibrium (e.g., McComas and Muller, 1981). However, some caution should be exercised here since the GM model spectrum is not unique. This is because experiments do not actually measure the three-dimensional wavenumber spectrum, but instead some incomplete collection of marginal spectra obtained from it. Depending upon how $\Lambda(\vec{k})$ is chosen, (14) can yield results for some of the marginal spectra which compare favorably with experiment. In fact, a more detailed study shows that virtually all of the usual marginal spectra and coherencies associated with moored measurements can be well described by the methods which lead to (14).

However, this also appears to require values for the wavenumber cutoffs which are much larger than those associated with molecular viscosity and it is difficult to see how to justify such a choice. Further, it is clear that the horizontal wavenumber spectrum obtained from (15) does not agree with experiment. For practical purposes the horizontal wavenumber spectrum obtained from (15) is white, while experiment indicates an inverse power law decay. Hence, while there is some interesting partial agreement with experiment, significant discrepancies remain. We will next show that some of these discrepancies are removed by considering the dynamically weak interaction approximation.

2.2 The dynamically weak interaction approximation

In this section we show how to write the Eulerian variables in terms of the canonically conjugate dynamical variables p_{js} and q_{js} and obtain a general relationship between Lagrangian and Eulerian spectra. We shall illustrate the procedure by considering the Eulerian velocity. It is straightforward to obtain the corresponding expressions for displacement. The components of the Eulerian velocity are defined in terms of the weighted averages

$$v_{E\alpha}(\vec{x},t)= \int v_{L\alpha}(\vec{\chi},t)\Delta(\vec{x}-\vec{u})d^3u= \int v_{L\alpha}(\vec{\chi},t)\Delta(\vec{x}-\vec{u})J[\vec{u}]d^3\chi \tag{16}$$

where

$$\vec{u}=\vec{\chi}+\vec{\eta}_L(\vec{\chi},t) \tag{17}$$

is the position of the Lagrangian fluid parcel, $J[\vec{u}]$ is the Jacobian determinant associated with the transformation given by (17), and $\Delta(\vec{x}-\vec{u})$ is a coarse-grained delta function (cf. Hardy, 1963) which can be written as

$$\Delta(\vec{x}-\vec{u})=(1/2\pi)^3 \int \exp\{i\vec{m}\cdot[\vec{x}-\vec{\chi}-\vec{\eta}_L(\vec{\chi},t)]-m^2\sigma^2/2\}d^3m \tag{18}$$

where σ defines the spatial extent of the coarse-graining.

The two space point two time point Eulerian velocity correlation functions $C_{Ev\alpha\beta}(\vec{X},\tau)$ are given by

$$C_{Ev\alpha\beta}(\vec{X},\tau)=E[v_{E\alpha}(\vec{x}+\vec{X},t+\tau)v_{E\beta}(\vec{x},t)]$$

$$=(1/2\pi)^3 \int d^3y \int d^3m \exp\{i\vec{m}\cdot(\vec{X}-\vec{y})-m^2\sigma^2\} M_{\alpha\beta}(\vec{y},\tau,\vec{m}) \tag{19}$$

where

$$M_{\alpha\beta}(\vec{y},\tau,\vec{m})=E[v_{L\alpha}(\vec{\chi},t+\tau)v_{L\beta}(\vec{\chi}',t)J[\vec{\chi}+\vec{\eta}_L(\vec{\chi},t+\tau)]$$

$$\times J[\vec{\chi}'+\vec{\eta}_L(\vec{\chi}',t)]\exp\{-i\vec{m}\cdot[\vec{\eta}_L(\vec{\chi},t+\tau)-\vec{\eta}_L(\vec{\chi}',t)]\}] \tag{20}$$

$\vec{y}=\vec{\chi}-\vec{\chi}'$, and we have used (16) and (18). In obtaining (19) we have noted that (20) is a function only of the spatial difference $\vec{y}$, so that the transformation $\vec{y}=\vec{\chi}-\vec{\chi}'$, $\vec{z}=(\vec{\chi}+\vec{\chi}')/2$ generates a factor $\delta(\vec{m}-\vec{m}')$ which produces a considerable simplification. We may then use (19) to write

$$S_{Ev\alpha\beta}(\vec{k},\omega)= \int d^3X \int d\tau\ C_{Ev\alpha\beta}(\vec{X},\tau)\ \exp\{-i(\vec{k}\cdot\vec{X}-\omega\tau)\}$$

$$= \int d^3y \int d\tau\ M_{\alpha\beta}(\vec{y},\tau,\vec{k})\ \exp\{-i(\vec{k}\cdot\vec{y}-\omega\tau)-k^2\sigma^2\} \tag{21}$$

The expressions for the frequency wavenumber spectra given by (21) are central results of this paper. In order to make practical use of them, however, we must first obtain a tractable expression for (20). The complicating feature in (20) is the complex exponential. All of the other factors simply generate products of the p_{js} and q_{js}. The complex exponential can be written

$$\exp\{-i\vec{m}\cdot[\vec{\eta}_L(\vec{\chi},t+\tau)-\vec{\eta}_L(\vec{\chi}',t)]\}=\exp\{-i\sum_{0<j}\sum_{s=1}^{2}(P_{js}p_{js}+Q_{js}q_{js})\} \tag{22}$$

where

$$P_{js}=\sum_{\alpha=1}^{3}m_\alpha f_{js\alpha}(\vec{\chi})\sin(\Omega_j\tau)\ ;\ Q_{js}=\sum_{\alpha=1}^{3}m_\alpha[f_{js\alpha}(\vec{\chi})\cos(\Omega_j\tau)-f_{js\alpha}(\vec{\chi}')] \tag{23}$$

Then by using (9), (11), and (22) it can be shown that

$$E[p_{j_1s_1}\cdots q_{j_ns_n}\ \exp\{-i\sum_{0<j}\sum_{s=1}^{2}(P_{js}p_{js}+Q_{js}q_{js})\}]$$

$$=(i)^n\frac{\partial}{\partial P_{j_1s_1}}\cdots\frac{\partial}{\partial Q_{s_ns_n}}\ \exp\{-\sum_{0<j}\sum_{s=1}^{2}(\Lambda_j/2\Omega_j)(P_{js}^2+Q_{js}^2)\} \tag{24}$$

By using (24) the expression given by (20) can be evaluated exactly. The only complication is that in the full three-dimensional case there are a substantial number of terms generated by the product of the two Jacobian determinants. While this makes calculations tedious, it is only tedium and there are no fundamental difficulties. However, because of the complexity associated with the full three-dimensional problem we shall, in this paper, confine our attention to a simplified example with one spatial dimension. We will find that the one-dimensional example is rich in physical detail and there is little doubt that the important features will carry over to three dimensions.

3. THE ONE-DIMENSIONAL EXAMPLE

In the one-dimensional case the Lagrangian displacement can be written as

$$\eta_L(\chi,t)= \sum_{0<j} \sqrt{2/L\rho\Omega_j}\,[\cos(\ell_j\chi)q_{j1}(t)+\sin(\ell_j\chi)q_{j2}(t)]$$

$$= \sum_{0<j} \sum_{s=1}^{2} f_{js}(\chi)q_{js}(t) \tag{25}$$

where L is the length of the system and $\ell_j=2\pi j/L$. The Lagrangian velocity is given by

$$v_L(\chi,t)= \sum_{0<j} \sum_{s=1}^{2} f_{js}(\chi)\Omega_j p_{js}(t) \tag{26}$$

It can be shown that the two time point two space point Lagrangian velocity correlation function is given by

$$C_{Lv}(X,\tau)=(1/2\pi\rho) \int \Lambda(\ell)\cos(\ell X)\cos(\Omega(\ell)\tau)d\ell \tag{27}$$

where we have allowed the length of the system to become arbitrarily large and recall that integrals over ℓ are extended to include both negative and positive values. In obtaining (27) we have used (25) to define the $f_{js}(\chi)$, (26), (6) to obtain the explicit τ dependence, and (9) and (11) to evaluate the various expectation values. The Lagrangian frequency wavenumber spectrum is then given by the two-dimensional Fourier transform of (27) which yields

$$S_{Lv}(k,\omega)=(\pi/\rho)\Lambda(k)[\delta(\omega-\Omega(k))+\delta(\omega+\Omega(k))] \tag{28}$$

Finally, the Lagrangian wavenumber spectrum is given by

$$\hat{S}_{Lv}(k)=(1/2\pi) \int S_{Lv}(k,\omega)d\omega=(1/\rho)\Lambda(k) \tag{29}$$

We note that (28) and (29) are analogous to (14) and (15) which were obtained for the case of three spatial dimensions.

In the one-dimensional case the Jacobian determinant is given by $J[u]=[1+\partial\eta_L(\chi,t)/\partial\chi]$ so that the Eulerian velocity is

$$v_E(x,t)=(1/2\pi) \int d\chi \int dm\, v_L(\chi,t)[1+\partial\eta_L(\chi,t)/\partial\chi]\exp\{im[x-\chi-\eta_L(\chi,t)]-m^2\sigma^2/2\}$$

$$=(1/2\pi) \int d\chi \int dm\, v_L(\chi,t)(i/m)\partial\exp\{im[x-\chi-\eta_L(\chi,t)]-m^2\sigma^2/2\}/\partial\chi$$

$$=-(i/2\pi) \int d\chi \int dm\, (1/m)\exp\{im[x-\chi-\eta_L(\chi,t)]-m^2\sigma^2/2\}\partial v_L(\chi,t)/\partial\chi \tag{30}$$

where the last step involves integration by parts. By using (30) it can be shown that

$$C_{Ev}(X,\tau)=E[V_E(x+X,t+\tau)V_E(x,t)]=(1/2\pi) \int dy \int dm\, M(y,\tau,m)\exp\{im(X-y)-m^2\sigma^2\}/m^2 \tag{31}$$

where

$$M(y,\tau,m)=E[\{\partial V_L(\chi,t+\tau)/\partial\chi\}\{\partial V_L(\chi',t)/\partial\chi'\}\exp\{-im[\eta_L(\chi,t+\tau)-\eta_L(\chi',t)]\}] \tag{32}$$

The frequency wave number spectrum is the two-dimensional Fourier transform of (31) which yields

$$S_{Ev}(k,\omega)=(1/k^2)\int dy \int d\tau\, M(y,\tau,k)\exp\{-i(ky-\omega\tau)-k^2\sigma^2\} \tag{33}$$

By using (6), (24), (26), and (32) it is tedious but straightforward to show that

$$M(y,\tau,m)=-\{[\partial^4 D(y,\tau)/\partial y^2\partial\tau^2]-m^2[\partial^2 D(y,\tau)/\partial y\partial\tau]^2\}\exp\{-m^2 D(y,\tau)\} \tag{34}$$

where

$$D(y,\tau)=(E_0/2\pi\rho)\int [h(\ell)/\Omega^2(\ell)][1-\cos(\ell y)\cos(\Omega(\ell)\tau)]d\ell \tag{35}$$

and we have set $\Lambda(\ell)=E_0 h(\ell)$. By using (33) and (34) we find

$$S_{Ev}(k,\omega)=-(1/k^2)\int dy \int d\tau\, \{[\partial^4 D(y,\tau)/\partial y^2\partial\tau^2]-k^2[\partial^2 D(y,\tau)/\partial y\partial\tau]^2\}$$

$$\times\ \exp\{-i(ky-\omega\tau)-k^2 D(y,\tau)-k^2\sigma^2\} \tag{36}$$

By integrating (36) over ω it is straight forward to show that the wavenumber spectrum is given by

$$\hat{S}_{Ev}(k)=-(1/k^2)\int [\partial^4 D(y,0)/\partial y^2\partial\tau^2]\ \exp\{-iky-k^2 D(y,0)-k^2\sigma^2\}dy \tag{37}$$

where we have used (35) to show that $\partial^2 D(y,0)/\partial y\partial\tau=0$.

4. DISCUSSION

The expressions given by (36) and (37) are central results of this paper. It is clear that in a general case they yield results that are different from the corresponding Lagrangian spectra. In this section we shall discuss the important differences between Eulerian and Lagrangian spectra. In order to facilitate the discussion we will consider the specific convergence factor $h(\ell)$ given by

$$h(\ell)=\ell\mu/\sinh(\ell\mu) \tag{38}$$

and the dispersion relation for longitudinal sound given by

$$\Omega(\ell)=c|\ell| \tag{39}$$

where c is the speed of sound. In (38) the factor μ is a Lagrangian frame length scale such that $h(\ell)$ is approximately unity if $|\ell|\mu<1$ and decays exponentially if $1<|\ell|\mu$. Our choice of convergence factor given in (38) and of the dispersion relation given by (39) are for practical reasons only. They are representative, and by using them we will be able to obtain closed form expressions for both the Eulerian frequency wavenumber spectrum and the Eulerian wavenumber spectrum. This will provide us with a convenient vehicle to facilitate the discussion. It can be shown that the important features of the results do not depend upon the specific choice for $h(\ell)$.

It is convenient to write the function $D(y,\tau)$ defined by (35) in terms of an Eulerian frame length scale ν and the dimensionless function $F(y,\tau)$ such that

$$D(y,\tau)=\nu^2 F(y,\tau) \tag{40}$$

where

$$\nu=\sqrt{E_0\mu/2\pi\rho c^2} \tag{41}$$

and

$$F(y,\tau)=4\int [1/\ell \ \sinh(\ell\mu)][1-\cos(\ell y)\cos(\ell c\tau)]d\ell$$
$$=\ell n\{\cosh[\pi(y-c\tau)/2\mu]\}+\ell n\{\cosh[\pi(y+c\tau)/2\mu]\} \tag{42}$$

Then by using (36), (40), and (42) it can be shown that

$$S_{Ev}(k,\omega)=(\nu^4 c\pi^2/2\mu^2)\exp\{-k^2\sigma^2\}P(k,\omega)P(k,-\omega)4^{k^2\nu^2}/(3+k^2\nu^2)(2+k^2\nu^2)[\Gamma(2+k^2\nu^2)]^2$$

$$\times \{Q(k,\omega)R(k,-\omega)+Q(k,-\omega)R(k,\omega)-2(3+k^2\nu^2)(2+k^2\nu^2)\} \tag{43a}$$

where

$$P(k,\omega)=|\Gamma(1+(k^2\nu^2/2)+i\mu(k+\omega c)/2\pi)|^2 \tag{43b}$$

$$Q(k,\omega)=[2+(3k^2\nu^2/2)+(k^4\nu^4/4)+(3\mu^2/2\pi^2\nu^2)(1+k^2\nu^2/6)(1+\omega/ck)] \tag{43c}$$

and

$$R(k,\omega)=k^2\nu^2(1+k^2\nu^2)/[(k+\omega/c)^2(\mu/2\pi)^2+(k^2\nu^2/2)^2] \tag{43d}$$

Finally, by using (37), (40), (41), and (42) we obtain

$$\hat{S}_{Ev}(k)=(2\nu^4 c^2\pi^3/\mu^3)\exp\{-k^2\sigma^2\}[1+k^2\nu^2+3(\mu/\pi\nu)^2]/(3+2k^2\nu^2)(1+k^2\nu^2)$$

$$\times |\Gamma(1+k^2\nu^2+ik\mu/\pi)|^2 4^{k^2\nu^2}/\Gamma(2+2k^2\nu^2) \tag{44}$$

where $\Gamma(z)$ is the gamma function of complex argument z.

While the expression for the frequency wavenumber spectrum given by (43) is somewhat cumbersome, it is given in terms of tabulated functions whose properties are well known. The factor $\exp\{-k^2\sigma^2\}$ simply accounts for the small but non-zero size of the coarse-graining and, for our purposes here, can be set equal to unity. We first note that if $\nu\ll\mu$, then (43) can be approximated by

$$S_{Ev}(k,\omega)=(2\pi^2\nu^2c^2/\mu)\,[k\mu/\sinh(k\mu)]\,[\delta(\omega-ck)+\delta(\omega+ck)] \tag{45}$$

which is the same as the Lagrangian frequency wavenumber spectrum given by (42). Thus, in this limit we recover the results of the absolutely weak interaction approximation. In obtaining (45) the factor given by (43d) can be approximated by $R(k,\pm\omega)=4\pi^2c\delta(\omega\pm ck)/\mu$. The more interesting case is that for which $\mu\ll\nu$. If $k\nu\ll\mu/\nu$, then (43d) is proportional to a delta function and (43a) can be written

$$S_{Ev}(k,\omega)=(2\pi^4\nu^4c^2/3\mu^3)\,[\delta(\omega-ck)+\delta(\delta+ck)] \tag{46}$$

There are two important differences between (45) and (46). The first is that (46) is proportional to ν^4 while (45) is proportional to ν^2. The second difference is that (46) is valid only for very small values of k. If $1\ll k\nu$, then by using (43) and well-known expansions for $\Gamma(z)$ it can be shown that

$$S_{Ev}(k,\omega)=(\pi^3\nu^4c/2\mu^2k^4\nu^4)\exp[-(k^2+\omega^2/c^2)\mu^2/2\pi^2k^2\nu^2] \tag{47}$$

We note that (47) is not confined to a dispersion surface, and thus does not correspond to a wave-like system. Hence, for small values of $k\nu$ there exists a remnant of the dispersion surface, but for large values of $k\nu$ the dispersion surface is completely smeared and the spectrum corresponds to turbulence.

The expression for the wavenumber spectrum given by (44) is somewhat more easily analyzed. If $\nu\ll\mu$, then we simply recover the Lagrangian spectrum. The more interesting case is that for which $\mu\ll\nu$ and (44) can be written

$$S_{Ev}(k)=(2\pi^3\nu^4c^2/\mu^3)\,|\Gamma(1+k^2\nu^2+ik\mu/\pi)|^2 4^{k^2\nu^2}/(3+2k^2\nu^2)\Gamma(2+2k^2\nu^2) \tag{48}$$

If $k\nu\ll1$, then (48) becomes

$$\hat{S}_{Ev}(k)=2\pi^3\nu^4c^2/3\mu^3 \tag{49}$$

which is white as is the corresponding Lagrangian spectrum. We note again, however, that (49) is proportional to ν^4, while the corresponding Lagrangian spectrum is proportional to ν^2. If $1\ll k\nu$, then it can be shown that

$$\hat{S}_{Ev}(k)=\pi^{7/2}\nu^4c^2/2\mu^3(k\nu)^3 \tag{50}$$

which is significantly different from the corresponding Lagrangian spectrum. Thus, we have found that when $\mu \ll \nu$, the Eulerian wavenumber spectrum is white until $k\nu \cong 1$ and then transitions to a $(k\nu)^{-3}$ decay.

Our proposal is that, in fact, the ocean is at (or near) canonical equilibrium. However, we must be careful to distinguish between Lagrangian and Eulerian variables. The canonical distribution yields the wavenumber spectrum given by (20), but this is in terms of Lagrangian variables only. The empirical GM action spectrum is given in terms of Eulerian variables and, as we have shown, if the contributions from the advective nonlinearity are large, then the two types of spectra are different. Thus, it is the kinematic distortion ㄱaused by the transformation from Lagrangian to Eulerian variables which converts a white Lagrangian wavenumber spectrum into an Eulerian wavenumber spectrum which exhibits an inverse power law decay. This potentially removes the discrepancy between the computed and the observed horizontal wavenumber spectrum which was discussed in the second section. This also provides us with a possible explanation for the required values of the cutoff parameters in computing the one-dimensional frequency spectra and coherencies. That is, the appropriate cutoffs correspond to the Eulerian frame cutoff ν rather than the Lagrangian frame cutoff μ.

An equally important point is that the Eulerian wavenumber spectrum is not confined to a dispersion surface, and hence is not wave-like. This would mean that the observed towed spectra, which are modeled by GM as due to internal waves, actually correspond to turbulence from an Eulerian frame point of view. Even though the system is entirely wave-like from the Lagrangian frame point of view, strong contributions from the advective nonlinearity result in turbulent behavior in the observation (i.e., Eulerian) frame.

In a more complete treatment the geostrophic modes would also have to be included. They will contribute directly to measured velocities, participate in advecting both themselves and the internal waves, and will be advected by the internal waves. The end result will be a complicated mixture of all of the above. The methods we have illustrated for the case of internal waves are equally applicable to the geostrophic modes as well as to a combination of both internal wave and geostrophic modes. The level of tedium, of course, increases as the complexity of the system increases, but no fundamental difficulties arise. The advantage of this approach is that the dynamics is treated within the Lagrangian frame, where it is nearly linear, and advection is treated as a kinematic effect for which statistical averages can often be computed exactly. The alternative is to deal with the strongly nonlinear Eulerian equations of motion and the problem of finding adequate approximate solutions. This later approach will surely require infinite order perturbation methods as well as consideration of the possiblity that wave-like perturbation expansions (i.e., transitions that begin and end on the dispersion surface) will not converge.

In order to establish that this approach is appropriate for the oceanic internal wave system it must be shown that the dynamically weak interaction approximation is valid while the absolutely weak interaction approximation is not valid. This will require a detailed study of the dynamic nonlinearities within the Lagrangian frame for the full three-dimensional oceanic internal wave system. While it is ultimately important to complete this study, the results of the one-dimensional example are clear cut and provide a compelling motivation to pursue these methods.

5 REFERENCES

Benney, D. J., and Saffman, P. G., 1966. Nonlinear interactions of random waves in a dispersive medium. Proc. Roy. Soc., Ser. A, 289: 301-320.
Garrett, C. J. R., and Munk, W. H., 1972. Space-time scales of internal waves. Geophys. Fluid Dyn., 3: 225-264.
Garrett, C. J. R., and Munk, W. H., 1975. Space-time scales of internal waves: a progress report. J. Geophys. Res., 80(3): 291-297.
Hardy, R. J., 1963. Energy-flux operator for a lattice. Phys. Rev. 132: 168-177.
Holloway, G., 1986. Eddies, waves, circulation, and mixing: Statistical geofluid mechanics. Ann. Rev. Fluid Mech., 18: 91-147.
McComas, C. H., and Müller, P., 1981. The dynamic balance of internal waves. J. Phys. Oceanogr., 11: 970-986.
Prigogine, I., 1962. Non-Equilibrium Statistical Mechanics. Interscience Pub., New York.
Seliger, R. L., and Whitham, G. B., 1968. Variational principles in continuum mechanics. Proc. Roy. Soc., Ser. A. 305: 1-25.

EVIDENCE AND CONSEQUENCES OF FOSSIL TURBULENCE IN THE OCEAN

C. H. GIBSON
Departments of Applied Mechanics and Engineering Sciences and Scripps Institution of Oceanography University of California, San Diego La Jolla, CA 92093, USA.

ABSTRACT

Turbulence in stably stratified fluids such as the ocean appears intermittently in isolated patches. Inertial forces characteristic of turbulence are rapidly overcome by buoyancy forces of the entrained fluid, converting the turbulent kinetic energy to internal waves at the intrinsic frequency N of the ambient fluid. The scrambled temperature, salinity and density fluctuations persist in the bobbing fluid as fossil turbulence remnants and preserve information about the previous turbulence such as its overturning scale $L_{P_o} \approx 0.6(\varepsilon_o/N^3)^{1/2}$ and dissipation rate ε_o at the beginning of fossilization. A fossil turbulence model developed in previous work allows microstructure to be classified according to hydrodynamic state using hydrodynamic phase diagrams. These diagrams compare indicated Froude numbers and Reynolds numbers of the microstructure with critical values established by theory and experiment. Four states are possible: active turbulence, fossil turbulence, mixed active-fossil turbulence and non-turbulence. Most microstructure patches in the ocean are fossil turbulence for the largest scales and active turbulence for the smallest scales. "Active" patches which dominate averages in most data sets are found to be fossil, indicating the turbulence processes are undersampled and the average values subject to undersampling errors. Dissipation rates of velocity and temperature variance ε and χ are distributed as lognormals with large intermittency factors $\sigma^2_{\ln\varepsilon}$ and $\sigma^2_{\ln\chi}$, where σ^2 is the variance of either $\ln\varepsilon$ or $\ln\chi$. Values of σ^2 in the ocean range from 1 to 7 depending on the layer, with 1-3 confined to the surface and the extremely intermittent range 3-7 comprising the interior. Very large dissipation rates implied by extrapolating the measured distributions are not found in the data, but are implied by the fossil turbulence interpretation of the dominant patches. Egregious quantitative undersampling errors may arise in the interpretation of microstructure data if such large intermittency factors and the fossil hydrodynamic states are ignored, and if the data set is small. Qualitative undersampling errors are also possible; for example, false minimum average ε and χ values may actually reflect maximum quantitative undersampling errors in layers of maximum dissipation and maximum intermittency. The preceding fossil turbulence description is contrasted with a recent model of oceanic billow turbulence in which the billow forms and begins to collapse before the turbulence begins, and both the turbulence and other microstructure collapse together leaving no fossil turbulence remnants.

1. INTRODUCTION

Fossil turbulence in this paper is defined as a remnant fluctuation of some hydrophysical field, usually temperature, produced by active turbulence which persists after the fluid is no longer actively turbulent at the scale of the fluctuation. This is the definition used in a series of papers by Gibson (1980, 1981ab, 1982abc, 1983, 1986, 1987ab), referred to in the following as the Gibson (1980-1987) fossil turbulence model. Active turbulence is defined below. The term "fossil turbulence" arose in the late 1960's to describe persistent refractive index fluctuations detected by radar returns from atmospheric regions where the vertical components of the original active turbulence were thought to be suppressed by buoyancy, or from measured oceanic temperature microstructure patches with imperceptible velocity microstructure. The data and concept are described by a workshop report, Woods (1969), on fossil turbulence for a Colloquium on Spectra of Meteorological Variables. Attempts were made at the Colloquium to distinguish turbulence from internal waves, and to define these terms. Stewart (1969) chose the "syndrome" type definition of turbulence (such as used for diseases) by listing commonly recognized properties, but expressed doubts whether turbulence could be distinguished from internal waves except in extreme cases. The recommended criterion for the existence of turbulence was that the vertical Reynolds flux of density $\overline{\rho'w'}$ be positive; that is, $\overline{\rho'w'} \geq 0$ where ρ is density, w is vertical velocity and primes indicate fluctuations about the mean. Laboratory tests of Stillinger et al. (1983) and Itsweire et al. (1986) use this criterion to determine the stratified turbulence inertial-viscous transition, and confirm the proposed expression of Gibson (1980) in (1) below within experimental uncertainty.

More precise velocity microstructure measurements such as Crawford (1976) and Osborn (1978) show that all sampled patches of strong temperature fluctuations in the ocean are accompanied by velocity fluctuations substantially above ambient. Using the Stewart definition of turbulence without the dominant inertial force criteria and the Nasmyth (1970) definition of fossil turbulence that assumes all velocity fluctuations of any kind have decayed at small scales, many oceanographers conclude that fossil turbulence does not exist in the ocean, as pointed out by Gregg (1987). However, Gibson (1986) shows that viscous dissipation rates ε in most detected microstructure patches are much less than needed to create patches with the large vertical length scales and high values of temperature dissipation rate χ observed. Furthermore, the stratified turbulence criterion in (1) implies at least an order of magnitude greater ε value in an active-fossil patch, with $\varepsilon \geq 30\nu N^2$, than in the ambient internal wave field, with $\varepsilon \approx \nu N^2$ according to the Garrett and Munk spectrum. Gregg (1987) adopts a turbulence definition attributed to Phillips (1977) that includes all random motions without a dispersion relation, but the definition is taken out of context since Phillips clearly includes Reynolds number (inertial/viscous) and Richardson number (buoyancy/inertial) force criteria in his discussion of the properties of turbulence, as does Stewart (1969).

Assuming "anything that wiggles is turbulence", to paraphrase the Gregg (1987) definition, eliminates the need for the fossil turbulence concept. However, this assumption lumps random motions dominated by buoyancy forces together with turbulence (as defined here) even though the two modes of motion have very different properties. For example, buoyancy dominated (wave) motions dissipate very slowly compared to inertially dominated (turbulent) motions and cannot entrain external fluid even when they are near saturation. They should be classed as a form of internal wave rather than turbulence, and are termed "fossil vorticity turbulence" by Gibson (1980) to emphasize that they represent a remnant in the vorticity field of previous active turbulence. An interesting unanswered question is whether such waves are a unique class which *must* be formed by damping turbulence, or whether they can also arise by driving linear internal waves toward saturation. It seems unlikely that large scale saturated internal waves can be produced without the development of internal smaller scale turbulence.

Gregg (1987) introduces a model for billow turbulence which deviates strongly from the Gibson (1980-1987) model in almost every respect. The two models are contrasted in the cartoon of Figure 1, and will be discussed in detail below. According to the Gregg (1987) model, turbulence begins when the vertical microstructure overturn scale of the billow L_P is maximum rather than minimum as assumed by the Gibson (1980-1987) model, and thereafter L_P collapses along with, and proportional to, the turbulence overturning scale L_R so that no fossil remnant of previous more active turbulence activity ever exists. In Fig. 1, vertical overturn scales of the velocity microstructure in the patches are indicated by solid lines and vertical overturn scales of the temperature microstructure are shown by shaded lines. According to the Gibson (1980-1987) model, turbulence begins when the thickness of a shear layer exceeds the local inertial- viscous scale $11 \times L_K$ and grows by a process of entrainment till it reaches the inertial-buoyancy scale $0.6 \times L_R$ where fossilization begins. Although the turbulence collapses, meaning that L_R and ε monotonically decrease with time in the absence of any sources of turbulent kinetic energy, the temperature microstructure overturn scales remain relatively constant, preserving information about the previous higher turbulence dissipation rate as a fossil remnant. The model is based on the observed behavior of boundary layers, wakes and jets in stratified fluid, all of which follow the same pattern and scaling, as discussed by Gibson (1987b).

In the present paper, the fossil turbulence model of Gibson (1980-1987) is contrasted with the always-active (WYSIWYG) turbulence model used by Gregg (1987) and others. Quite different conclusions result regarding the interpretation of oceanic microstructure depending on the choice of turbulence model. Possible consequences of misinterpretation are discussed. The principal hazard is that egregious undersampling errors, both quantitative and qualitative, may result in gross underestimates of mean

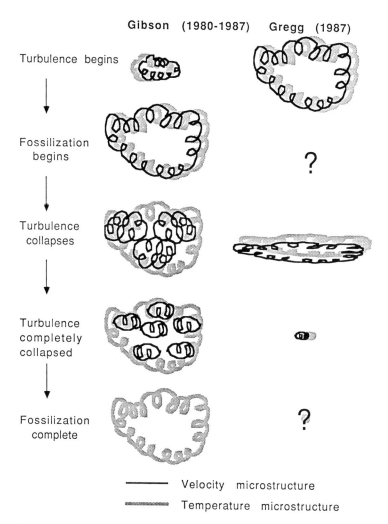

Figure 1. Schematic of Gibson (1980-1987) and Gregg (1987) models of billow turbulence. According to the Gibson model (Fig. 2), turbulence begins at small scales and grows by entrainment till buoyancy forces overcome inertial forces and fossilization begins. Velocity overturn scales (dark lines) become smaller than temperature overturn scales (shaded lines) which persist as "fossil-turbulence" remnants of the turbulence at the point of fossilization. According to the Gregg model, turbulence is caused by the collapse of a larger billow (Fig. 3). The velocity microstructure and temperature microstructure collapse together and vanish without a trace.

dissipation rates and vertical diffusivities in some layers of the ocean. A major finding of oceanic microstructure research of the last decade, Gregg and Briscoe (1979), is that canonical estimates such as Munk (1966) of turbulence mixing and vertical diffusivities in the interior of the ocean are orders of magnitude too high. This finding is questionable when the data is examined statistically, Baker and Gibson (1987), and probably incorrect when subjected to tests of hydrodynamic state using the Gibson (1980-1987) fossil turbulence model. Alternative explanations for the observed mean flow properties of the ocean other than the canonical interpretation of vertical turbulence diffusion, for example boundary mixing with horizontal diffusion, have not been convincing for horizontally homogeneous regions far from boundaries.

Baker and Gibson (1987) suggest that the deviations claimed lack statistical significance when the extreme intermittency of oceanic dissipation rates is considered.

2. Fossil Turbulence Model of Gibson (1980-1987)

Gibson (1980) defines active turbulence as a nearly isotropic, eddy-like state of fluid motion where the inertial forces of the eddies are larger than either the buoyancy or viscous forces which tend to damp them out. This definition includes Reynolds number and Froude number force criteria, and isotropy, as necessary components. Classification according to hydrodynamic state is accomplished using hydrodynamic phase diagrams, as described by Gibson (1986, 1987b, 1988). Two-dimensional turbulence is similarly defined, but isotropy is required only in the horizontal. Forced viscous eddies (Moffatt vortices) or random motions (viscous sublayers) dominated by viscous forces are excluded. Nonrealizable flows (Euler flows and vortices) are excluded. Nearly isotropic forced internal waves are random, rotational and dissipative, but are not turbulence by this definition because they are dominated by buoyancy forces. Assuming a universal critical Richardson number of 1/4 at the buoyancy-inertial-viscous transition point to turbulence in a stratified medium, Gibson (1980) proposed the following criterion for the existence of stratified turbulence

$$\varepsilon \geq 30 \nu N^2 \equiv \varepsilon_F \tag{1}$$

and using the universal turbulent velocity spectrum, Gibson (1981b) proposed a criterion for the existence of turbulence with wavelength λ:

$$1.2 L_R \geq \lambda \geq 15 L_K \tag{2}$$

where L_R is the Ozmidov scale $(\varepsilon/N^3)^{1/2}$, L_K is the Kolmogoroff scale $(\nu^3/\varepsilon)^{1/4}$, ν is the kinematic viscosity, $N \equiv (g \bar{\rho}_{,z}/\rho)^{1/2}$ is the intrinsic frequency of the stratified fluid, ε is the viscous dissipation rate $\varepsilon \equiv 2\nu \overset{\leftrightarrow}{e} : \overset{\leftrightarrow}{e}$, $e_{ij} \equiv (v_{,i} + v_{,j})/2$ is the ij component of the rate-of-strain tensor $\overset{\leftrightarrow}{e}$, $\mathbf{v}$ is the velocity and commas denote partial differentiation. These criteria now have rather firm laboratory support and are consistent with available field measurements in the atmosphere, ocean and lakes, as discussed by Gibson (1986, 1987b).

Criteria for classifying microstructure according to hydrodynamic state were first proposed by Gibson (1980) based on the evolution of an isolated patch of strong turbulence as a model. The model is illustrated schematically in Figure 2 and has several characteristics which differ from the model of Gregg (1987) and others. The patch of strong turbulence grows till buoyancy forces prevent further large scale overturns. The turbulence then "collapses" in the sense that the Ozmidov scale L_R, and therefore the maximum velocity overturning scale $0.6 \times L_R$, collapses as the dissipation rate ε decreases below the value at the beginning of fossilization ε_o and approaches the value at complete fossilization ε_F from (1). The patch has now been converted from turbulence to saturated internal waves, using the definitions discussed above. The kinetic energy is nearly the same as it was just before fossilization begins, but the dissipation rate is much less, so the patch of waves, termed fossil vorticity turbulence by Gibson (1980), will be very persistent. Since the frequency of the waves is N, their propagation velocity is zero so the patch will twist and bob for many N^{-1} periods if the Reynolds number at fossilization Re_o is large. Density fluctuations within the patch cannot collapse because the twisting motions continuously change the "collapse" direction. These fossil density turbulence fluctuations continue to be at least locally isotropic as long as the horizontal twist angles are of order 180 degrees, and will continue to "mix" because their gradient scale will be maintained at the Batchelor scale $L_B \equiv (D/\gamma)^{1/2}$ whatever the rate-of-strain γ may be, even though γ may be less than the minimum turbulence value of 5N, from (1), where $\gamma \equiv (\varepsilon/\nu)^{1/2}$ and D is the molecular diffusivity of the density. Several properties of the microstucture patch preserve information about the previous higher turbulence dissipation rate ε_o and turbulent mixing rate $\chi_o \equiv 2D(\nabla \rho)^2$, where o-subscripts indicate turbulence at the beginning of fossilization, as discussed by Gibson (1986, 1987b).

3. Billow Turbulence Model of Gregg (1987)

Figure 3 shows the evolution of length scales L_P, L_R and L_K and the associated mixing structures according to the Gregg (1987) model of billow turbulence. In the initial stage 1 the Kelvin-Helmholz billow begins to roll up following a sequence of schematic diagrams of the process by Thorpe (1969), but the temperature and viscous dissipation rates are assumed to be negligibly small rather than maximum as in Fig.

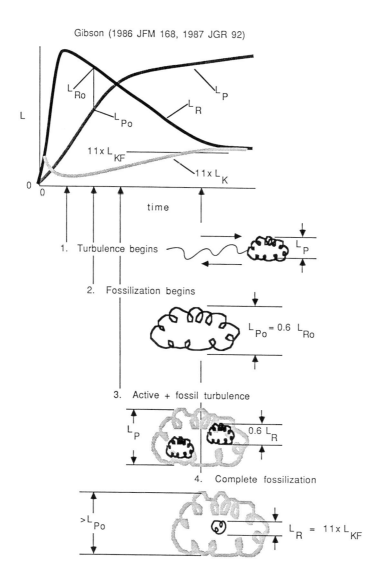

Figure 2. The Gibson (1980-1987) billow turbulence model. Length scales versus time are shown at the top, and relative overturn scales at four stages are shown below. Ozmidov scales L_R (dark line) are maximum when turbulence begins, stage 1, and decrease toward $L_P/0.6$ where fossilization begins, stage 2. L_P (dark shading) is the patch size, or temperature microstructure overturn scale, assumed to be equivalent, and remains constant or slightly increases during the collapse of the turbulence. Kolmogoroff scales L_K (light shading) decrease and then increase, reflecting their inverse dependence on ε. Stage 3 shows mixed active and fossil turbulence, with $L_P \geq 0.6 \times L_R \geq 11 \times L_K$. Stage 4 shows complete fossilization, where viscous, inertial and viscous forces of the eddies converge, leaving only the fossil-turbulence remnant in the temperature field.

324

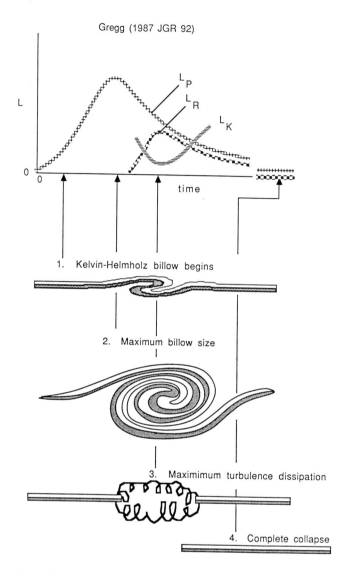

Figure 3. The Gregg (1987) billow turbulence model. Kelvin-Helmholz billow begins at stage 1 and grows to some large (unspecified) size, stage 2. The billow collapses and produces turbulence, stage 3. The turbulence and temperature microstructure collapse together, vanishing completely at stage 4. Length scales L_P (plusses), L_R (checkers), and L_K (horizontal shading) evolve as shown at the top, for comparison with Fig. 2.

2. Thorpe's diagrams are based on flow visualizations in a laboratory study in which the actual dissipation rates were not measured. They give a clear qualitative picture of the rollup of vortices on an interface with constant velocity and density increments and fixed thickness, leading to buoyant damping when the scale of the rollup increases the Richardson number above a critical value. However, it is not clear that the schematic is representative of actual transition events in stratified media such as the ocean since such natural events

probably take place where the shears are unusually large due to prolonged vorticity production (where the term $\nabla\rho\times\nabla p$ is maximum) and where the density gradient layer is thinner than either the velocity gradient layer or other surrounding density gradient layers, rather than thicker as shown in Thorpe's schematic. The shearing is likely to amplify the density gradient even further, delaying transition due to an increased Richardson number and leading to large viscous and temperature dissipation rates at the interface rather than small values inferred by Gregg (1987). If the enhanced shear is compensated by enhanced density gradient, turbulence formation may be prevented altogether. However, if the vorticity production is prolonged, local kinetic energy may accumulate and may be dissipated in a burst of turbulence.

The initial condition of such a burst would be a vortex sheet stabilized by a strong density gradient across the sheet. The first turbulent eddies to develop would be at the interface buoyant-inertial-viscous scale (see Gibson (1987a)) $L_{KF_{int}}=(\nu/N_{int.})^{1/2}$, where $N_{int.}\gg N_{ambient}$. These will mix away the stabilizing gradient so that the turbulent burst can develop in the usual way by a cascade from small scales to large scales by a process of vortex pairing and entrainment, drawing energy from the shear layer surrounding the burst. Dissipation rates within the burst will first increase and then decrease during this growth period, as shown by Fig. 2, depending on whether the velocity increases as $v\sim z^n$ with $n>1/3$, since $\varepsilon\sim v^3/z$ where z is the vertical burst thickness. Eventually the burst thickness will approach the inertial-buoyancy scale $0.6\times L_R$ and the fossilization process illustrated by Fig. 2 will begin.

The notion that a maximum billow size can develop before turbulence appears, shown as stage 2 of Gregg's model in Fig. 3 and by the Thorpe diagram g, is contrary to experience and theoretical expectations. Vortex sheets are unstable to perturbations on all scales, but the time scale for growth $z_{pert.}/v_{pert.}$ is smaller for perturbations of small scales unless the perturbation velocity $v_{pert.}$ increases with scale $\sim z_{pert.}^m$ where the

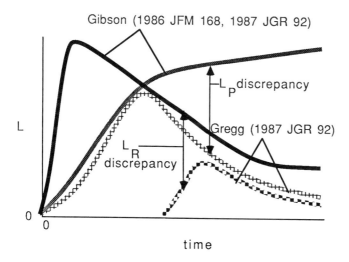

Figure 4. Comparison of length scales L_P and L_R of the Gibson (1980-1987) and Gregg (1987) models, as a function of time from Fig. 2 and Fig. 3, same shading. Discrepancies are shown by vertical double arrows.

exponent $m > 1$. High amplitude large scale perturbations are possible, such as a shear layer emerging from a flapping splitter plate in a laboratory flow, but are unlikely in the ocean interior.

Even if a shear layer should roll up to maximum size before L_K scale eddies have time to develop, the viscous dissipation rate ε would still decrease monotonically rather than reaching a maximum due to a "collapse" as proposed by Gregg (1987), stage 3 in Fig. 3. The idea that the vertical overturn scales of the temperature microstructure mixed by the turbulence L_P can collapse and vanish without a trace, as shown by stage 4 of Gregg's model in Fig. 3, seems equally unlikely. The most commonly observed oceanic fossil of previous more active turbulence is the ocean surface mixed layer thickness, which reflects the dissipation rate of the season's strongest storm. Interior mixed layer thicknesses, and other overturn scales tabulated by Gibson (1987b) will not collapse as rapidly as the turbulence overturn scale L_R, as shown in Fig. 3 according to the Gregg (1987) model.

Figure 4 compares the predictions of the Gibson (1980-1987) and Gregg (1987) models for inertial-buoyancy, or Ozmidov, length scales L_R and temperature overturn scales L_P of billow turbulence in the ocean. The discrepancies are very large, and result in large discrepancies in the interpretation of microstructure data sets, as discussed below. Similar large discrepancies exist between a model of "near-inertial" turbulence put forth by Gregg (1987, Fig. 45), where the overturn scales grow without bound in a layer with constant ε, contrary to the "secondary turbulence" model of Gibson (1987a) which shows that large overturn scales for fossil microstructure patches of strong events will be *reduced*, not increased, by parasitic turbulence events which form at the boundaries, not in the interior, of the patch and erode it back toward the original uniform vertical temperature gradient without microstructure. However, further discussion is beyond the scope of the present paper.

4. Quantitative Undersampling Errors in Microstructure Interpretation

Turbulence dissipation rates in most stratified layers of the ocean are generally recognized to be intermittent in time and space, but the consequences of this intermittency to sampling strategies and the inference of mean values and their confidence intervals has only begun to be appreciated. Baker and Gibson (1987) have compared available data sets in various ocean layers to normal and lognormal distributions, and find that for most layers the data are indistinguishable from lognormal once noise levels are taken into account. The intermittency of the random variables ε and χ are measured by the variances of their natural logarithms σ^2_{ln}, termed "intermittency factors". Intermittency factors are estimated from lognormal probability plots which form straight lines with intercept μ at the 50% probability point and slope $1/\sigma$, where μ and σ^2 are the mean and variance, respectively, of the random variable, either $ln\varepsilon$ or $ln\chi$. Values range from 1-3 for mixed layers, but from 3-7 for seasonal thermocline, main thermocline and equatorial undercurrent layers. Serious undersampling errors are likely for the latter layers unless the intermittency and lognormality of the data are considered. Examining the hydrodynamic state of the data is a test for time intermittency over periods as long as fossil turbulence persists, but little is known of how long this may be. Gibson (1982a) estimates periods as long as days or weeks may be possible.

Undersampling errors are likely because the mean to mode ratio of lognormal random variables is very sensitive to the value of the intermittency factor

$$\frac{\text{mean}}{\text{mode}} = \exp(3\sigma^2/2) \approx G \qquad (3)$$

where the mode is the most likely value to be observed. G is the quantitative undersampling error factor, defined as

$$G \equiv \frac{\text{true value}}{\text{erroneous value}} \qquad (4)$$

The factor G may be used to correct erroneous underestimates of mean values due to undersampling and intermittency from (4). The approximate value of G to be expected if only one sample of a lognormal random variable is taken to be representative of the mean is the mean to mode ratio in (3). Figure 5 shows log G plotted versus the intermittency factor σ^2. For the oceanic range of intermittency factor 3-7, the quantitative undersampling error factor ranges from about 2-4 orders of magnitude. More than one sample is usually taken in a microstructure data set, but how many are independent is not clear without examination of the hydrodynamic state of the samples. If none of the microstructure patches in the data set is in the fully active turbulence state, as is commonly observed, then the conclusion can be drawn that the space average inferred from that set will be less than the space-time average for the layer in the preceding time period. More data must be taken over a longer time period. If one wants to estimate the strength of storms in an unfamiliar region one should check the local depth of the seasonal thermocline. Similarly, if one wants to infer mixing rates in an interior layer of the ocean, one should look for the thickness of internal mixed layers;

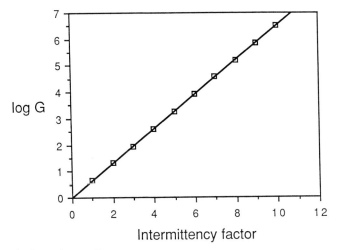

log G

Intermittency factor

Figure 5. Quantitative undersampling error factor G as a function of intermittency factor σ^2 for lognormal random variables, where G≈mean/mode for a small data set.

that is, the fossil turbulence parameters of the microstructure patches.

5. Hydropaleontology or WYSIWYG Turbulence?

Determining space time averages for a layer from the meager amount of microstructure data likely to be available is an art in its infancy. The term "hydropaleontology" is suggested by Gibson (1980) to characterize a potentially efficient strategy for deriving reliable averages of turbulent mixing parameters from minimal data taking advantage of the properties and parameters of fossil turbulence. Gregg (1987) rejects this term as more colorful than illuminating and prefers to assume, along with Caldwell (1983), Dillon (1984), Gregg (1984) and Gregg et al. (1986) that "what you see is what you get", that their data sets are representative (p. 5281) of the space time process, and to define away or deny the need for the fossil turbulence concept. As discussed above, this "WYSIWYG" approach is based on highly questionable fluid mechanical models of stratified turbulence, distortions of generally accepted definitions of turbulence in fluid mechanics, and an unfortunate tendency to presume that laboratory studies are somehow irrelevant to oceanic processes. Laboratory studies clearly show that scalar microstructure scales do persist long after turbulence overturning scales have collapsed and disappeared, providing a fossil turbulence remnant of the previous dissipation rates. Gibson (1987) shows that dissipation rates at the beginning of fossilization for stratified grid turbulence measurements of Stillinger et al. (1983) and Itsweire et al. (1986) are correlated at the 92% level with estimated previous dissipation rates from the fossil turbulence overturn scales at the point of complete fossilization. The Gregg (1987) model of billow turbulence shown in Fig. 3 may be necessary to reject fossil turbulence in ocean data, but it is contradicted by all available laboratory evidence and seems contrary to theoretical expectations.

6. Qualitative Undersampling Errors

Layers and regions of maximum dissipation rates in the ocean also tend to be layers and regions of maximum intermittency and thus maximum quantitative undersampling error. The magnitude of the quantitative undersampling error may be so large that a qualitative undersampling error may result; that is, a layer or region of maximum dissipation rate might be erroneously identified as a layer of minimum dissipation rate.

The criterion for a qualitative undersampling error in $\bar\varepsilon$ is that

$$\frac{\bar{\varepsilon}_{1\,est.}}{\bar{\varepsilon}_{2\,est.}} \leq \left[\frac{G_2}{G_1}\right] \times \left[\frac{\bar{\varepsilon}_1}{\bar{\varepsilon}_2}\right] ; \quad \frac{\bar{\varepsilon}_1}{\bar{\varepsilon}_2} \geq 1 \tag{5}$$

where the error factors G depend on the intermittency factors of the respective layers and the degree of undersampling. If the layer is not undersampled then the values and ratio of G factors are 1.0. However, since G depends strongly on σ^2, and since σ^2 may be large in layers of large $\bar{\varepsilon}$ then the criterion of (5) may be satisfied.

The scenario envisaged leading to possible quantitative and qualitative undersampling errors is illustrated schematically by Figure 6, which shows the profile of temperature mixing rate χ_T, intermittency factor σ^2, error factor G and estimated χ_{mean} observed in the the upper ocean during the MILE mixed layer experiment, as described by Gibson (1983). Individual dropsonde profiles, as show on the left, should be representative of the most probabable values, or mode values. They usually gave minimum χ_T values close to the mode of the distribution at the seasonal thermocline depth, where the mode may be minimum but the average χ_T is actually maximum, as indicated by repeated vertical profiles and towed body measurements which accumulated enough data to form meaningful $\bar{\chi}_T$ averages. The quantitative and qualitative undersampling errors are shown by the middle diagram. The reason for the qualitative undersampling error suggested by Gibson (1983) is that the energy scale of the turbulence in the mixed layer is the mixed layer depth of 35 m, which is much smaller than the scale of the turbulence, in this case two-dimensional horizontal turbulence with scales 5-50 km, which dominates the mixing in the seasonal thermocline. Gibson (1981a) proposes that the intermittency and lognormal character of stratified turbulence is attributable to the same sort of nonlinear cascade model used by Gurvich and Yaglom (1967) to explain the intermittent lognormality of high Reynolds number turbulence in Kolmogoroff's third universal similarity hypothesis for unstratified turbulence. The fluid mechanical, statistical model for undersampling errors is summarized by the cartoon at the bottom of Fig. 6.

A similar possible undersampling error has been suggested by Gibson (1983) for the core depth of the equatorial undercurrent, where the horizontal mixing scales may be several hundred kilometers and the periods of temporal intermittency may be several months or several years. Large discrepancies between towed body dissipation rates, Williams and Gibson (1974) in the Pacific and Belyaev et al. (1973) in the Atlantic, and dropsonde dissipation rates at core depths are attributed by Gregg (1987) to vibrational noise in the towed body signals. However, the large reported values from towed bodies were indicated by redundant sensors operating well above noise levels of the instruments for repeated measurements. No evidence other than the discrepancy with dropsonde data has been put forth to indicate the towed body measurements are invalid. Gregg (1987) argues that the Belyaev et al. (1973) velocity sensors are sensitive to temperature fluctuations, but the huge, high frequency temperature fluctuations at core depths necessary to produce the "noise" are even more inconsistent with the dropsonde data than the velocity interpretation. Core velocities were higher for the towed measurements in the Pacific by a factor of two than for the Crawford (1982) measurements. When both intermittency corrections and the higher core velocity are taken into account, no significant difference exists in the estimated mean dissipation rates, as shown by Gibson (1983). Large dissipation rates were also detected at core depths by Williams and Gibson (1974) at 1 degree north, consistent with the broad maximum inferred by Caldwell (1987) for the meridional profile of the vertical average. Much lower dissipation rates were measured with the same instruments (unreported) near the Hawaiian Islands.

Figure 7a shows profiles of estimates of mean $\bar{\varepsilon}$ and $\bar{\chi}$ from Gregg (1987, Fig. 12, p. 5261) from 385 profiles over a 4.5 day period of the TROPIC HEAT expedition. Figure 7b shows the temperature, salinity, velocity, N, velocity gradient and Ri, from Gregg et al. (1985, Fig. 1). Both $\bar{\varepsilon}$ and $\bar{\chi}$ profiles show minima at the high velocity core depth of the undercurrent, with the $\bar{\varepsilon}$ minimum below the inertial-viscous transition value $\varepsilon_{tr} = \varepsilon_F$ given by (1) suggesting that no turbulence at all can exist at core depths. However, the vertical eddy diffusivity values reported are an order of magnitude greater than the molecular diffusivity according to Gregg (1987, caption Fig. 12) suggesting the χ_T values are also an order of magnitude above that of the ambient mean gradient. Such microstructure in nonturbulent fluid is presumably fossil turbulence. Large

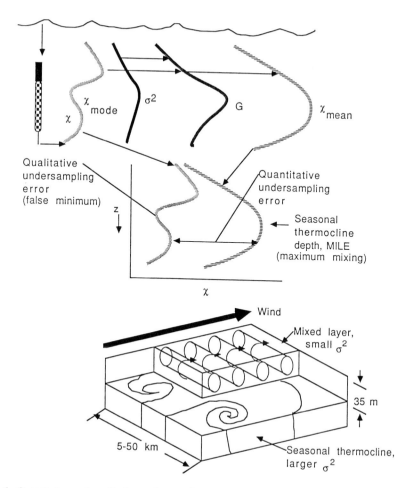

Figure 6. Quantitative and qualitative undersampling errors in an individual dropsonde profile of χ (lightly shaded line), from the Gibson (1983) description of the MILE mixed layer experiment. Multiple χ profiles by Dillon and Caldwell showed that σ^2 (black line), and therefore G, was maximum at seasonal thermocline depths. Towed body measurements gave enough data for the mean $\overline{\chi}_{mean}$ to converge and to determine a precise distribution function for χ. The mode agreed well with the individual profiles, as might be expected. The upper profiles show how the G profile (dark shading) might be used to estimate the χ_{mean} profile (horizontal shading). The schematic at the bottom shows the mixing model leading to undersampling errors.

---

values of χ_T below the core in Fig. 7a are in the range 10^{-5} to 10^{-4} K^2/s reported by Williams and Gibson (1984). They indicate values of ε_o of 0.01-0.1 cm^2/s^3 using the expression $\varepsilon_o = 13DCN^2$ from Gibson (1980), where C is the Cox number $C \equiv (\nabla T)^2/(\overline{\nabla T})^2$, in good agreement with tow body values. Figure 7c shows an individual profile reported by Gregg et al. (1985, Fig. 5) with large χ and with ε values at this depth of 0.01 cm^2/s^3: less than the active turbulence value, but more than two orders of magnitude greater than the mean values shown in Fig. 7a. Gregg et al. (1985) attribute this high level of activity to lateral intrusions and possible double diffusive effects, and suppress the large vertical heat flux that would be implied in Fig. 7a. However, neither of these effects render the Osborn-Cox model invalid if a sufficiently wide horizontal

330

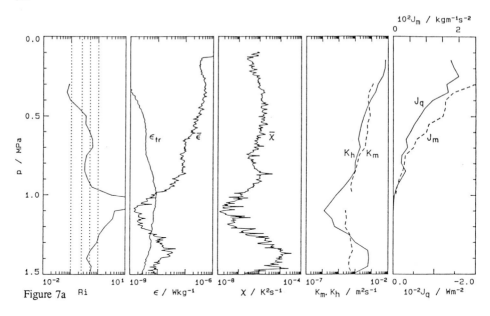

Figure 7a

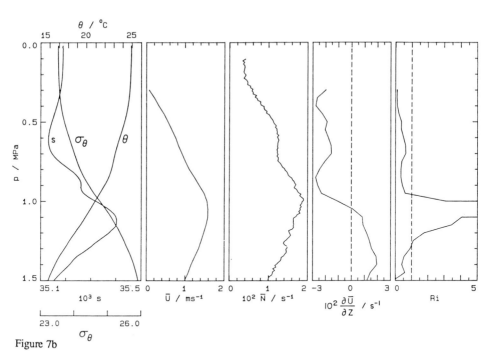

Figure 7b

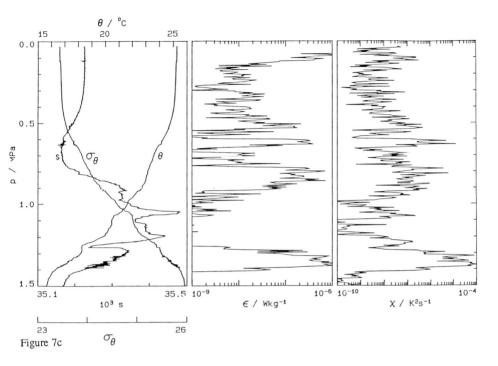

Figure 7c

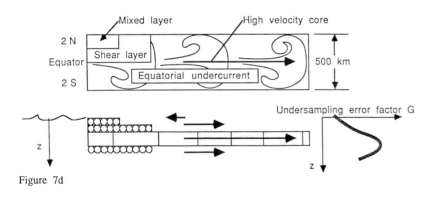

Figure 7d

Figure 7. a. Profiles of dissipation rates estimated from dropsonde measurements in the equatorial undercurrent, from Gregg (1987, Fig. 12). Large vertical heat fluxes J_q below the core, corresponding to indicated large K_h values, are not shown. Such large fluxes would contradict the hypothesis of a nonturbulent undercurrent core. b. Profiles of ambient properties for 7a., from Gregg et al. (1985, Fig.1). c. An individual profile, with large ε and χ values below the undercurrent core, from Gregg et al. (1985, Fig. 5). The salinity profile S at the left shows evidence of lateral mixing processes at core depths, possibly leading to maximum intermittency and maximum undersampling errors. d. Schematic of the equatorial undercurrent which might lead to quantitative and qualitative undersampling errors similar to those found for the seasonal thermocline during MILE, Gibson (1983), shown in Fig. 6. Note that horizontal scales of eddies at the equatorial undercurrent core may be much larger than those at seasonal thermocline depths at high latitudes giving larger intermittency and undersampling errors.

control volume is considered and it seems that horizontal eddies would produce the observed profiles just as well as the postulated intrusions and seem conceptually identical. Clearly such a large vertical heat flux below the core would imply that the Gregg et al. (1985) hypothesis of a nondiffusive barrier at the undercurrent core is invalid. Figure 7d shows the postulated scenario of undersampling error in the equatorial undercurrent, similar to that shown in Fig. 6 for the seasonal thermocline, except that the scale of horizontal eddies which dominate mixing at the undercurrent core are much larger because the velocity is larger and the coriolis force near the equator is smaller.

If the average vertical diffusivity at the core were actually 4 orders of magnitude less than values in the layers above and below, as shown by Fig. 7a, one might expect large vertical gradients in the mean temperature and oxygen concentrations at core depths, reflecting the vertical flux downward through the undercurrent of these quantities. Similar strong mean gradients of the nutrient concentrations diffusing up should also be found at core depths, but no such maximum vertical mean gradients are observed. Judging by the well known high biological productivity of equatorial regions, the equatorial undercurrent appears to be very well ventilated with heat and oxygen diffusing down and nutrients diffusing up, consistent with strong turbulent mixing throughout its entire vertical profile and inconsistent with the nonturbulent barrier at the core inferred from dropsonde measurements. This inference may be a qualitative undersampling error.

Caldwell (1987) reports a qualitative result similar to that of Gregg (1987) in Fig. 7a based on 12 days of sampling yielding 1749 profiles. This is a time period sufficient for only one 1000 km feature of the undercurrent to be convected past the ship at the core velocity, and hardly justifies the suggestion by Caldwell (1987, p. 184) that dissipation rates of order 0.1 cm^2/s^3 detected by towed bodies are never present in the core. Caldwell (1987) reports that a peak in the mean dissipation rate $\bar{\varepsilon}$ less than a fraction of a degree wide reported by Crawford (1982) is actually subject to both quantitative and qualitative undersampling errors as defined above since the peak falls below a broader, higher peak in the horizontal profile, a few meridional degrees wide, when the data set increases. It seems quite possible that the vertical mean dissipation rate profiles of Fig. 7a will also change substantially in both quantitative and qualitative form in the future as more data from different years and seasons are included in the averages.

7. Summary and Conclusions

The models of Gibson (1980-1987) and Gregg (1987) for the evolution of billow turbulence in stratified fluids have been compared and found to be widely divergent. The Gregg model indicates no fossil turbulence stage, but is contrary to laboratory observations which fail to show the assumed collapse of scalar microstructure after the decay of turbulence and which do not support the assumed evolution of dissipation rates and overturn scales. In contrast, substantial evidence exists in support of the Gibson (1980-1987) fossil turbulence model, as reviewed by Gibson (1986, 1987b, 1988). Such major differences between models of the basic properties of stratified turbulence render agreement over interpretation of oceanic microstructure data impossible.

Based on the Gibson (1980-1987, 1988) method of classifying stratified microstructure according to hydrodynamic state, the data sets available appear to be primarily remnants of previous more active mixing periods. Consequently, the probability of substantial quantitative and even qualitative undersampling errors in the interpretation of oceanic microstructure exists. It is urged that future data sets be examined carefully as to the hydrodynamic state of the most "active" patches which dominate average values of ε, χ and inferred vertical diffusivities. If these patches are fossil, then it may be that more data is needed to characterize the mixing processes of the layer. Claims that existing microstructure data in the ocean show that canonical values of vertical eddy viscosities and eddy diffusivities are orders of magnitude too large or that minimum values exist at the equatorial undercurrent core are probably incorrect and certainly premature.

8. Acknowledgements

Support for this paper was provided by ONR Contract N00014-85-C-0104.

9. References

Baker, M. A. and C. H. Gibson, 1987. Sampling turbulence in the stratified ocean: Statistical consequences of strong intermittency. J. of Phys. Oceanogr., in press.

Belyaev, V. S., M. M. Lubimtzev and V. V. Ozmidov, 1974. The rate of dissipation of turbulent energy in the upper layer of the ocean. J. Phys. Oceanogr., 5:499-505.

Caldwell, D. R., 1983. Oceanic turbulence: Big bangs or continuous creation? J. Geophys. Res., 88:7543-7550.

Caldwell, D. R., 1987. Small-scale physics of the ocean. Rev. Geophys., 25:2:183-192.

Crawford, W. B., 1976. Turbulent energy dissipation in the Atlantic equatorial undercurrent. Thesis, The University of British Columbia, Canada.

Crawford, W. B., 1982. Pacific equatorial turbulence. J. Phys. Oceanogr., 16:1847-1854.

Dillon, T. M., 1984. The energetics of overturning structures: Implications for the theory of fossil turbulence. J. Phys. Oceanogr., 14:541-549.

Gibson, C. H., 1980. Fossil temperature, salinity, and vorticity turbulence in the ocean. in Marine Turbulence, J. Nihoul (Ed.), Elsevier Oceanography Series, Elsevier Publishing Co., Amsterdam, 221-257.

Gibson, C. H. , 1981a. Buoyancy effects in turbulent mixing: Sampling turbulence in the stratified ocean. AIAA J., 19:1394-1400.

Gibson, C. H., 1981b. Fossil turbulence and internal waves. in American Institute of Physics Conference Proceedings No 76: Nonlinear Properties of Internal Waves, Bruce West (Ed.), American Institute of Physics, 159-179.

Gibson, C. H., 1982a. Alternative interpretations for microstructure patches in the thermocline. J. Phys. Oceanogr., 12:374-383.

Gibson, C. H., 1982b. On the scaling of vertical temperature gradient spectra. J. Geophys. Res., 87: C10:8031-8038.

Gibson, C. H., 1982c. Fossil turbulence in the Denmark Strait. J. Geophys. Res., 87:C10:8039-8046.

Gibson, C. H., 1983. Turbulence in the equatorial undercurrent core. in Hydrodynamics of the Equatorial Ocean, J. C. H. Nihoul (Ed.), Elsevier Oceanography Series, Elsevier Publishing Company, Amsterdam, 131-154.

Gibson, C. H. , 1986. Internal Waves, fossil turbulence, and composite ocean microstructure spectra. J. Fluid Mech., 168:89-117.

Gibson, C. H., 1987a. Oceanic turbulence: Big bangs and continuous creation. J. Physicochem. Hydrodyn., 8:1:1-22.

334

Gibson, C. H., 1987b. Fossil turbulence and intermittency in sampling oceanic mixing processes. J. Geophys. Res., 92:C5:5383-5404.

Gibson, C. H., 1988. Hydrodynamic phase diagrams for microstructure in stratified flows. Proceedings,Third International Symposium on Stratified Flows, Pasadena, Feb. 3-5, in press.

Gregg, M. C., 1984. Persistent turbulent mixing and near-inertial internal waves. in Internal Gravity waves and Small-Scale Turbulence, Proceedings, Hawaiian Winter Workshop, edited by P. Muller and R. Pujalet, 1-24.

Gregg, M. C., 1987. Diapycnal mixing in the thermocline: A review. J. Geophys. Res., 92:C5:5249-5286.

Gregg, M. C., and M. G. Briscoe, 1979. Internal waves, finestructure, microstructure and mixing in the ocean. Rev. Geophys. Space Phys., 17:1524-1548.

Gregg, M. C., H. Peters, J. C. Wesson, N. S. Oakey and T. J. Shay, 1985. Intensive measurements of turbulence and shear in the equatorial undercurrent. Nature, 318:140-144.

Gregg, M. C., E. A. D'Asaro, T. J. Shay, and N. Larson, 1986. Observations of persistent mixing and near-inertial internal waves. J. Phys Oceanogr., 16:856-885.

Gurvich, A. S., and A. M. Yaglom, 1967. Breakdown of eddies and probability distributions for small scale turbulence. Phys. Fluids, 10:59-65.

Itsweire, E. C., K. N. Helland, and C. W. Van Atta, 1986. The evolution of grid-generated turbulence in a stably stratified fluid. J. Fluid Mech., 155:299-338.

Munk, W., 1966. Abyssal recipes. Deep Sea Research, 13:707-730.

Nasmyth, P. W., 1970. Ocean turbulence. Ph. D. dissertation, Univ. of British Columbia, Vancouver, Canada, 69pp.

Osborn, T. R., 1978. Measurements of energy dissipation adjacent to an island. J. Geophys. Res., 83:C6:2939-2957.

Phillips, O. M. 1977. The dynamics of the upper ocean. Cambridge University Press, New York, 336pp.

Stewart, R. W., 1969. Turbulence and waves in a stratified atmosphere. Radio Science, 4:12:1269-1278.

Stillinger, D., K. Helland and C. Van Atta, 1983. Experiments on the transition of homogeneous turbulence to internal waves in a stratified fluid. J. Fluid Mech., 131:91-122.

Williams, R. B., and C. H. Gibson, 1974. Direct measurements of turbulence in the Pacific Equatorial Undercurrent, J. Phys. Oceanogr., 4:104-108.

Woods, J. D. Ed., 1969. Report of working group (V. Hogstrom, P. Misme, H. Ottersten and O. M. Phillips): fossil turbulence. Radio Science 4:1365-1367.

MEASURING TURBULENT TRANSPORTS IN STRATIFIED FLOWS

H.W.H.E. GODEFROY and M. KARELSE, Delft Hydraulics, P.O. Box 177, 2600 MH Delft
(The Netherlands)
B. BARCZEWSKI, University of Stuttgart, Stuttgart (FRG)
R. SPANHOFF, Tidal Waters Division, Ministry of transport and public works, The
Hague (The Netherlands)

ABSTRACT

The possibility of combining in one instrument optical velocity- and concentration-measuring techniques for stratified flows is investigated. An immersible Laser Doppler Anemometer (ILDA) developed at Delft Hydraulics is shown to
measure shear stresses in strongly stratified flows where conventional LDA
systems fail because of refraction-index variations. Experiments under various
flow conditions with an existing Laser Fluorometer (LF) of the University of
Stuttgart reveal a high sensitivity for concentration measurements with uranine
as tracer. The performance is also compared with that of in-situ conductivity
sensors. The most recent literature data on the uranine absorption and emission
spectra are erroneous and better-shaped spectra are given. An optimized LF
system turns out to need only a low-powered Argon laser, so it can be rather
compact. In fact, it is proposed to base the desired immersible instrument on
the easily transportable ILDA, with an extra fluorescent-light detector in
backscatter mode. The blue line of the Argon laser is then used for laser fluo-
rometry and the green line for anemometry. This new instrument will be developed as a laboratory device; by technical modifications it will be possible to
use such device also in the ocean.

1 INTRODUCTION

Correlations between velocity- and concentration fluctuations are paramount

in theoretical descriptions of turbulent mass transport in stratified flows. An

instrument that measures these correlations, that is, that measures simultane-

ously at the same place the instantaneous velocity and composition of the flow,

would be invaluable in laboratory and field experiments performed to further

our theoretical understanding of mixing processes. Especially so when such an

instrument can probe the turbulent mixing layer, with its refraction-index

variations. Up to now such an instrument is lacking.

The present paper suggests to combine into one instrument optical techniques

for velocity and concentration detection. Preliminary tests are presented that

demonstrate the feasibility of the concepts. They concern an immersible two-

dimensional Laser Doppler Anemometer (ILDA), developed at Delft Hydraulics

(Godefroy and Vegter, 1984 and Godefroy, 1986) and a Laser Fluorometer (LF), built for concentration measurements in the Institut für Wasserbau of the University of Stuttgart (Barczewski 1985, 1986a,b).

Section 2 summarizes an investigation into the possibilities and limitations of the ILDA for measurements of the turbulent characteristics of stratified flows, especially at the turbulent mixing layer.

For flume experiments with the LF method under similar conditions as for the ILDA, the Stuttgart instrument was transported to Delft Hydraulics. In section 3 these experiments and their results are presented. The first part of that study is directed to the sensitivity of the method, with emphasis on the minimum laser power needed. The ultimate combined instrument aimed at should namely be compact and easily transportable, therefore a small, air-cooled low-power Argon laser would be preferred as light source, when sufficient. This part of the study was performed with homogeneous flows.

The possibilities of the LF method in turbulent mixing layers are the subject of the second part. Here stratified-flow conditions were used.

The LF instrument was developed by Barczewski for other applications and it turned out that several parts of it, notably the optical device, can be significantly improved for the present goal. Still, the results of the present study indicate that the LF method is well suited for concentration measurements in turbulent mixing layers, and that it might be rather straightforwardly integrated with the ILDA, especially since the optics of both components are of about the same construction. As a first stage the research will be directed towards the development of a new, integrated instrument for measurements of turbulent transports in laboratory circumstances. In a latter stage such type of device can be developed also for measurement in the ocean.

2 MEASURING TURBULENT VELOCITY FLUCTUATIONS

For measurements of turbulent intensities and correlations between velocity components in flumes two-dimensional Laser Doppler Anemometers (LDA) are very useful. A LDA system can be kept completely outside the flume, so there will be no disturbance of the flow at all. However, in general LDA cannot be used in turbulent stratified flows since salinity and thus refraction-index variations exclude a long light path under these conditions.

An immersible probe has, therefore, been developed which combines the advantages of the LDA method, such as high accuracy and small measuring volume and linear response to velocity variations, with a short light path. The shape of the probe of this immersible instrument (ILDA) has been optimized to give minimum disturbance of the flow field. The immersible part of the total ILDA body consists of three stainless steel tubes, see figure 1. The largest tube contains some lenses of the laser transmitter part. Each of the smaller ones contains an ordinary optical fiber to guide the light to the photodiodes. The dis-

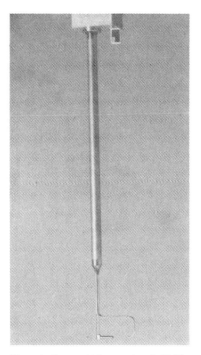 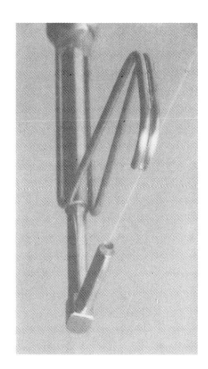

Fig. 1 Immersible part of ILDA.

tance between the tube end of the "transmitter" and the other two tube ends
with the fibers is 58 mm. This means that the point of intersection of the
three beams, which determines the position of the measurement volume, lies just
in between, at about 3 cm from both tube ends. The relative disturbance of the

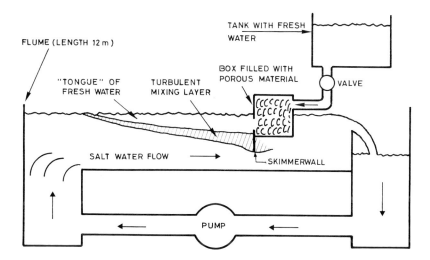

Fig. 2. Sketch of experimental facility used for stratified flows.

338

flow velocity at the measurement volume is small (an increase of ~ 4%) and
nearly independent from flow conditions (a maximum variation of ~ 2%).

Detailed information on the ILDA used at Delft Hydraulics were published by
Godefroy and Vegter (1984) and Godefroy (1986).

To investigate the possibilities and limitations of the ILDA for measurements in stratified flows, especially for obtaining information of the turbulence structure at the mixing layer, measurements were performed in a fresh-
water wedge type of flow (see figure 2). In a transparent flume with a length
of 12 m, a width of 0.5 m, and a maximum depth of 0.5 m a skimmer wall was

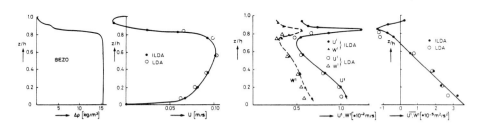

Fig. 3. Measured density (Δρ), velocity (u), turbulence intensities
(u' and w') and shear stress (u̅'̅ ̅w̅'̅) profiles in flume of figure 2.

placed with under it a salt-water underflow and with above it a small fresh-
water flow in the opposite direction. This type of flow creates a strong vertical stratification at the interface and a particular shape of the shear-stress
profile with opposite signs for bottom and interfacial shear stresses. To study
the influence of the immersible body of the ILDA, a conventional LDA was placed
close to the ILDA, with its measuring volume 0.115 m upstream of that of the
latter, at the same vertical and lateral position in the flume.

In figure 3 the results of this test (for more information see Karelse et
al. (1986)) are shown for both detectors. The figure gives
- the vertical density distribution (resulting from conductivity measurements)
  for reference, as well as
- the vertical mean-velocity distribution,
- the vertical distribution of the turbulence intensities, and
- the vertical distribution of the shear stress, as given by both laser systems.

The ILDA appears to give reliable information of turbulent intensities and
shear stresses over the whole vertical, including the mixing layer, while the
LDA system only gives information in the homogeneous underlayer. The small
shifts in the profiles of LDA and ILDA can be explained by the difference in
longitudinal position.

3 EXPERIENCES WITH THE LASER FLUOROMETRY METHOD

Laser-induced fluorescence of tracer materials dissolved in small quantities in the optically highly transparent medium water provides, in principle at least, an easy method to monitor the movement and mixing of thus labeled water bodies and to study the transport and dilution of therein dissolved materials of interest, that usually are much harder to detect. Assuming a known absolute calibration of a laser-fluorescensing detection apparatus, and known initial concentrations of the tracer and of the material(s) of interest, the concentrations of the latter can be studied quantitatively, e.g. during turbulent mixing. In this section we describe some experimental tests of this measuring principle, performed to see whether a compact concentration-measuring instrument for inhomogeneous-flow conditions, analogous to and integrated with the above-discussed ILDA, is feasible.

3.1 Uranine as fluorescent dye and its optical properties

(i) The choice of uranine. Uranine has several advantages above other fluorescent dyes, that make it a natural choice for our purposes:
- Uranine has a relatively narrow absorption spectrum, that peaks at about 490 nm. So the spectrum matches nicely the 488 nm blue laser line of the easy-to-operate and relatively cheap Argon-ion lasers. In addition, the spectrum drops off thus rapidly that the absorption for the green laser beam of 514.5 nm is about 5 times less effective (fig. 4), which is rather convenient since the latter might be used for LDA in the desired combined instrument.
- Uranine's fluorescence efficiency is high: the quantum efficiency (ratio of emitted and absorbed photons) is 71%.
- The change of this efficiency with temperature is small. Only about -0.3% per degree Celsius.
- The fluorescence efficiency does not decrease under influence of chemical substances dissolved in the water such as oxygen: there is no so-called quenching effect for uranine.
- Uranine gives no environmental pollution. It is for example used as a bath salt. So large quantities of water, such as needed in flume experiments, can be labeled without creating a waste problem.
- It dissolves well and fast in water.
- It does not attach to dust and silt particles.
- It is chemically stable.

(ii) Self absorption by uranine. Due to absorption along its path in the water, the intensity of the blue laser light at the measurement volume will be reduced with respect to the original laser beam itself, our reference. The fluorescent-light intensity is thus reduced as well, and it will also be partially absorbed before being detected. This is called the shadow effect, and it might cause an error in the detector signal if not accounted for. In practice this problem can easily be circumvented with an immersible system with only a short optical path in the water.

We have measured the absorption coefficient α of uranine at 488 nm, and found about $5 \times 10^5 \mathrm{cm}^{-1}$ (see figure 4). Thus for a concentration c = 10 ppb, implying $\alpha c \approx 0.005 \ \mathrm{cm}^{-1}$, and a path length l = 3 cm, the Lambert-Beer formula

$$I = I_0 \cdot \exp l(-\alpha c l)$$

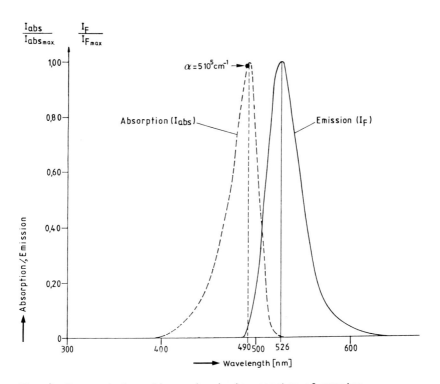

Fig. 4. Measured absorption and emission spectra of uranine

gives the stimulating light to be reduced with only 1.5%, which can often be neglected in practice.

The absorption by uranine of the fluorescent light can safely be neglected with respect to that of the stimulating light (see fig. 4) as can the absorption by the water itself ($\alpha c \approx 0.0002 \ \mathrm{cm}^{-1}$ for 450 nm < λ < 550 nm).

(iii) Total absorption in the water. In the flume experiments described below, we have checked the light absorption in the water for the various distances of the measurement volume from the optically transparent wall. These optical paths in the water, in the following called depth, were obtained by moving the optical apparatus to and from this wall, which leaves all angles with respect to the optical axis, and thus the detection efficiency constant. The Lambert-Beer formula can therefore be applied to check the absorption coefficient α.

The deduced values turned out to be larger than expected for uranine (see above), up to a factor 10 in unfavourable cases. Dust and/or tiny air bubbles in the water might very well have caused this: during several experiments the water in the flume displayed some turbidity detectable with the naked eye. This calls for an opticle path in the water as short as possible in an operational instrument. Even then it will probably be desirable to perform some calibrations before and after actual measurements. These have the additional advantage that possible changes in the contamination of optical components and drift in e.g. the photomultiplier response can be accounted for.

(iv) The detection wavelength. Originally, Barczewski chose the wavelength 589 nm for the detection of the fluorescent light (Barczewski, 1985), based on the uranine absorption and emission spectra given by Früngel and Koch (1976). He used a relatively narrow-band interference filter in front of the photomultiplier.

During the course of the present study it became apparent that the uranine data presented by Früngel and Koch were wrong, notably the positions of the maxima in their absorption and emission spectra. In addition, these spectra did not show the mirror symmetry (compare fig. 4) dictated by physical arguments. Therefore we have remeasured these spectra (fig. 4), both in Delft and in Stuttgart, confirming the older literature data of Lewschin (1927, 1931) and replacing the emission maximum at 520 nm.

A more broad-band interference filter than originally used by Barczewski, and now centered around ca 530 nm should give a much stronger output signal of the detector. Of course, this filter should block the 514.5-nm green laser light of the Argon-ion laser.

3.2 Instrumental and experimental details

A complete LF system is built up of an optical and an electronic device. The optical device has an emitting and a receiving part. The emitting part consists mainly of a laser as a light source, an optical filter which is passed only by the blue laser light exciting the fluorescence, and a lens that focusses the light to a tiny measurement volume. The receiving part consists mainly of a

342

lens that collects fluorescent light emitted from the measurement volume, an aperture (pinhole) that filters the light spatially, so that virtually only light from the measurement volume can pass through, an optical filter that eliminates all light other than in a wavelength band near the maximum of the fluorescence spectrum, and a photomultiplier as a light detector.

In the Stuttgart device both parts are "fixed" together as a confocal back-scatter system. In such a system the optical configuration is about the same as for a LDA backscatter system. That is why Barczewski could use an optical device that was originally designed as a two-colour LDA backscatter system. For LF measurement only the blue light beam is used (see fig. 5). This system has the advantage that no further optical alignments need to be made, and that the stimulating blue light hits the optical axis only at the place of the measurement volume, which is focussed to the photomultiplier, so this system has a good spatial resolution. Its disadvantage is that the light efficiency is poor. Only about 25% of the incident blue light reaches the measurement volume and the total efficiency is only about 10%. In an optical device especially designed for a LF system a much higher efficiency can be realized.

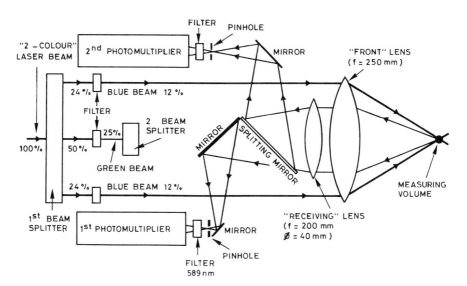

Fig. 5. Sketch of the optical device of the Stuttgart laser-fluorometer (only the blue laser line and the first photomultipier were used).

To eliminate the influence of background light, the technique of light chopping and synchronous detection is used to separate electronically the signal due to the background light from the alternating signal produced by the chopped light from the laser.

For measurements of turbulent concentration fluctuations the chopping frequency has to be more than twice the highest desired fluctuation frequency. As chopper a rotating disk with 120 tiny holes is used to realize chopping frequencies up to more than 5 kHz.

The electronic device has to measure the value of the output signal of the photomultiplier during each on and off period. These values correspond with the fluorescence intensity stimulated by the blue laser light plus the background light respectively with the background light only. By substracting these two values the resulting signal should be proportional to the fluorescent light intensity stimulated by the laser light only.

The device from Stuttgart uses two peak detectors for the measurement of the maximum and minimum values. This design has the advantage that it is simple and cheap. Its disadvantage is that, because of the noise in the signal of the photomultiplier, the output signal is rather noisy also (the difference between maximum and minimum instantaneous value is not the same for each chopped period). Besides, the value of the output signal overestimates the correct value, so there is e.g. already a signal if there is no chopped light at all (see fig.

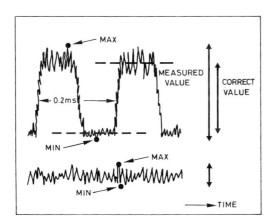

Fig. 6. Effect of noise on output signal for Stuttgart device as determined by peak detectors. top: with chopped laserlight
bottom: without laserlight

6). A better design, which should give a much less noisy signal, would be to measure the mean value of the photomultiplier signal during the time the laser beam passes the chopper and also during the time the beam is blocked.

Such improvements of the electronics have not been systematically investigated yet in this preliminary feasibility study. Only various relatively simple measures have been taken. For instance we have low-pass filtered (fourth-order linear phase filter) the detector net-output signal, which can be considered as a mere first step to the proposed averaging; only the noisy behaviour is partly improved, not e.g. the ambient light correction.

3.3 Results of the LF experiments

The LF method was tested on dilute uranine solutions under homogeneous as well as under inhomogeneous conditions using the optical and electronic device from Stuttgart with a "medium" power air cooled Argon laser.

(i) Homogeneous-flow conditions. A small 640-liter flume with a measuring section (glass walls) of 0.3 m width and 1.0 m length was used for the homogeneous conditions. Uranine was dissolved to 10, 20 and 40 ppb in the water, respectively. At each of these concentrations the detected output signal was recorded for three power settings of the laser (about 15, 20 and 25 mW of the 488-nm blue laser beam) and for three optical path lengths in the solution.

From the following observations it is concluded that the method is well suited for concentration measurements.

(a) At the concentration of 10 ppb uranine the output signal amounted about ten times the value of the ambient light, that is, without laser light. This ratio was another order of magnitude better with dimmed lights in the experimental hall. So the signal-to-noise ratio is convenient, even at this low concentration. The optical path in the fluid was 36 mm long, a realistic value for an immersible system. The output signal was stable within 2% during 1-s measuring periods. Superior S/N ratios might be expected with the above-indicated optical (enhanced efficiency, optimized transmission-filter wavelength) and electronic (signal averaging) improvements.

If the values of the output signal for the chopped laser-off periods are subtracted from the laser-on ones, the resulting values are almost linearly proportional with the uranine concentration.

For a system with a higher light efficiency or a more powerful laser, the relative effect of the background signal is less and a better linearity would be obtained, as found by Barczewski (1985). In an earlier study with the present system and with a 300-mW laser he observed excellent linearity in the concentration range 0.5 - 200 ppb. Then, a calibration at one concentration only suffices.

(b) The detector signal is linear with respect to the laser power, that is

    . the ratio of the signals for the various depths (36, 135 and 278 mm, respectively) was independent of the laser power, and

    . the signals for each depth followed linearly the laser intensity as measured independently with an optical laser-power meter.

The only noticed possible complication in the method is the above-mentioned (sec. 3.1) higher absorption than expected for pure uranine in water. This calls for the shortest optical path as possible, for the use of filtered water, and, if necessary, for calibrations before and after actual measurements.

(ii) <u>Inhomogeneous-flow conditions</u>. Experiments under inhomogeneous-flow conditions were performed in the same flume (see fig. 2) as used for the ILDA tests. Now small amounts of uranine, up to 25 ppb, were added to the salt water. Several light-path lengths in the fluid were used, as well as various density differences, up to 26 kg/m³.

In these tests the performance of the LF was compared with that of other concentration meters, namely conductivity sensors that give the salt content. The first, more general tests were made with a so-called BEZO, a continuously registrating instrument developed at Delft Hydraulics, with a measurement volume of about 0.25 cm³; a tiny suction pipe brings the fluid to be probed to a measuring cell. For the present experiments a standard in-use BEZO was modified to an increased frequency response of more than 20 Hz.

The later, more refined comparisons of the LF were made with a miniature but harder to handle probe obtained from Dr. Head, who developed and constructed it (Head, 1983). Basically it consists of four 0.25-mm short, 38-μm diameter wires protruding in the flow from a 1 mm diameter rod; the measurement volume is circa one mere mm³.

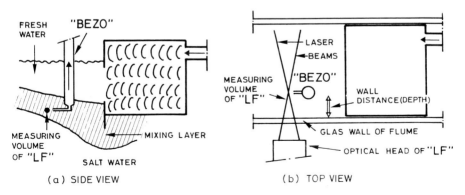

Fig. 7 Detailed sketch of positions of measurement volumes in the flume of figure 2.

The output signals of the conductivity instrument in use passed a low-pass fourth-order linear phase filter identical to that on the LF. This allowed a direct comparison between the two types of instruments. For the present purposes a visual inspection suffices, so the used analogue registration with a fast two-channel recorder is efficient.

Comparison with BEZO. Figures 8 and 9 give examples of the recordings of the concentration fluctuations in the mixing layer as measured simultaneously by LF and BEZO (see fig. 7). The salt-water flow velocity amounted circa 0.1 m/s, and the density difference with the fresh water was 26 kg/m^3. Under these flow conditions no fast fluctuations occurred in the mixing layer, therefore the frequency of the low-pass filters could be set at 5 Hz without loss of information. Each figure gives two different parts of the registrations, with different recorder speeds.

The general behaviour of the concentration fluctuations is seen to be reproduced consistently by the two sensors. The LF signal reveals more details than the BEZO, which are largely filtered out by the relatively large volume of the latter. The correspondence between the two signals is much better in fig. 9 than in fig. 8, as a consequence of the difference in the optical paths in the fluid, respectively 33 mm and 66 mm. With larger paths the light beams are more subject to bending by concentration differences, while the BEZO remains unaffected.

Comparison with Head conductivity probe. Figures 10 and 11 show parts of the registrations of the LF compared with the Head sensor. The measuring volumes were now placed only 25 mm from the wall (fig. 7); laser power in the blue Argon line was 25 mW, flow velocity of the salt water again was circa 0.1 m/s in the case of fig. 12, and the density difference with the fresh water 26 kg/m^3.

In order to profit from the higher frequency response of this miniature conductivity sensor, the frequencies of both low-pass filters, on LF and Head signals, were raised to 60 Hz. This clearly enhances the noise level of the LF as compared with figs. 8 and 9.

The concentration fluctuations as measured in the turbulent mixing layer (fig. 7) by both sensors are given in fig. 10. Discarding the noise on the LF signal and some minor hum on the Head probe (see 1-s registration), the correspondence between both detection systems is excellent. The finer details in the concentration fluctuations are now about equally well reproduced, maybe with a slight preference for the Head probe, partly by a masking effect of the noise on the LF signal.

Despite the increased frequencies (60 Hz) of the low-pass filters and the expected frequency responses of the same order, no concentration fluctuations

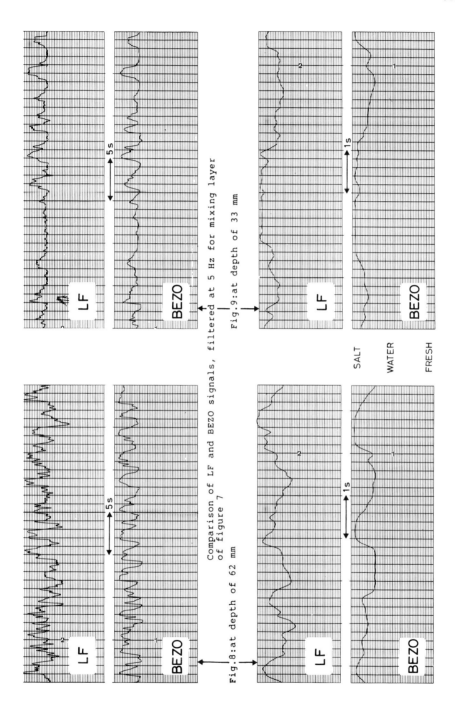

Comparison of LF and BEZO signals, filtered at 5 Hz for mixing layer
of figure 7

Fig.8:at depth of 62 mm

Fig.9:at depth of 33 mm

348

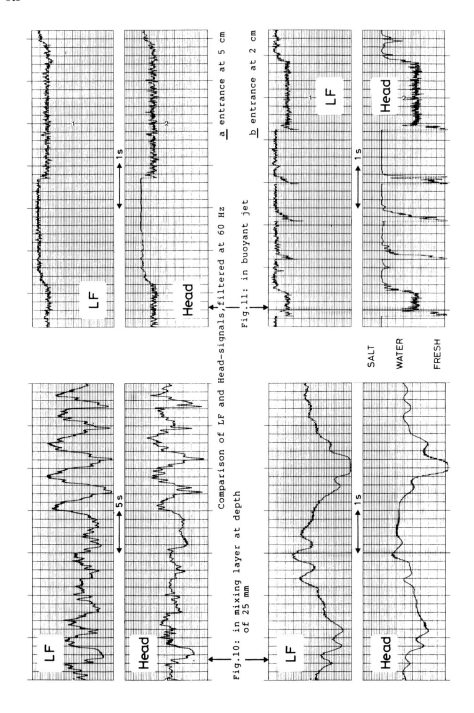

LF

Head

a entrance at 5 cm

b entrance at 2 cm

Fig.11: in buoyant jet

Comparison of LF and Head-signals, filtered at 60 Hz

LF

Head

1 s

5 s

Fig.10: in mixing layer at depth
of 25 mm

1 s

LF

Head

LF

Head

SALT
WATER
FRESH

of more than 5 Hz, say, are observed in fig. 10, simply because they did not occur in the mixing layer.

Therefore, we have experimented in the flume with a small buoyant jet of fresh water. Switching on and off the jet generated fast fluctuations. They also occurred during the jet-on periods. The pipe of the jet was only 2 mm in diameter and it was placed 5 cm (fig. 11a) respectively 2 cm (fig.11b) upstream of the measuring volumes. The jet was switched manually, with various switching times.

The jet-on and jet-off periods are well reproduced by both detector systems; especially in fig. 11b the high frequency nature of the responses can be seen (fast drops in the signals).

Without a jet stream (maximum values in fig. 11) turbulence is virtually absent and the recorded signal reflects the noise level of the instruments. The Head detector is almost free of noise, while the noise on the LF signal appears to be of the same order as the turbulent fluctuations during the jet-on periods (lower values in fig. 11a). Therefore only the strongest fluctuations in the Head signal can also be identified in the LF registration; in these cases the agreement is good.

The Head recording for 2 cm distance between jet and probes (fig. 11b) clearly shows the turbulence only to develop a fraction of a second after the start of the jet, provided the latter lasts long enough; then the Head signal fluctuates roughly around the middle of the fresh- and salt-water values.

The LF reproduces this behaviour also, the time of the signals agreeing excellently with those of the Head. However, now the amplitude is reduced: it by far does not approach the fresh-water value. Probably the measuring volume of the LF exceeds the jet diameter (2 mm), always mixing in a salt-water contribution in the signal.

During the above experiments the precise paths of the laser beams in the flume could be clearly seen thanks to the strong fluorescence, much better than with immersible LDA systems. The dancing of the beams, that is the variations in their positions due to refraction-index variations, could now well be observed.

Strong dancing was seen in the mixing layer. Displacements of 0.2 to 0.3 mm, and sometimes 0.5 mm, in directions perpendicular to the beams were quite common at 35 mm depth. Both laser beams moved quite independently, in different directions. Deeper in the fluid the effect got worse; at the other wall of the flume (50 cm) a large blurred light beam was observed.

Roughly speaking, the displacements perpendicular to the beams were proportional to the square of the path in the fluid. In practice, concentration measurements can be made with a LF system up to 3 to 4 cm. A larger pinhole (fig. 5), such as 1.0 mm stead of 0.3 mm in diameter, will give larger output signals, thanks to a larger measuring volume, but will hardly extend the applicable path lengths.

4 CONCLUSIONS

(i) Longitudinal and vertical instantaneous local velocities and the related turbulent characteristics in stratified flows, including the mixing layer, can be measured with the ILDA.

(ii) Already with the present LF equipment and with a small 10-mW blue laser line and as little as 10 ppb uranine, concentration fluctuations can well be measured.

- Up to 3 cm in the fluid the shadow effect is virtually absent.
- With an optical path not exceeding 4 cm, the method works also in turbulent mixing layers despite their refractive index variations.
- Absolute concentration determinations require a calibration, preferentially before and after the measurements. One uranine concentration suffices.

(iii) Combining the present velocity- and concentration-measuring techniques into one immersible instrument appears feasible, according to these preliminary investigations. It might be based on a small air-cooled two-colour beam Argon-ion laser. The beams are separated, allowing only the blue line to be chopped for synchronous detection (LF), and combined again, whereupon they traverse the same optical system. The latter is of the same concept as the ILDA, but now with an additional part for the detection of backscattered fluorescent light.

5 REFERENCES

Barczewski, B., 1985. Development of a Laser-Fluorometer for Sensitive, Local Concentration Measurements in Flows. In: Proceedings of the 21st IAHR Congress, vol. 5, Melbourne, Australia, August 19-23, pp. 1-6.

Barczewski, B., 1986a. Tracer measurements in flows. In: A.C.E. Wessels (Editor), Proceedings Symposium Measuring Techniques in Hydraulic Research, Delft, The Netherlands, April 22-24, 1985. A.A. Balkema, Rotterdam, pp. 77-96.

Barczewski, B., 1986b. A new probeless technique for concentration measurements in turbulent flows. In: Proceedings International Symposium on Buoyant Flows, Athens, Greece, September 1-5, pp. 130-132.

Früngel, F. and Koch, C., 1976. Practical Experience with the variosens equipment in measuring chlorophyll concentrations and fluorescent tracer substances, like rhodamine, fluorescine, and some new substances. IEEE J. Oceanic Eng., OE-1: 21-32.

Godefroy, H.W.H.E., 1986. Measurements with a new type immersible Laser Doppler Anemometer. In: A.C.E. Wessels (Editor), Proceedings Symposium Measuring Techniques in Hydraulic Research, Delft, The Netherlands, April 22-24, 1985. A.A. Balkema, Rotterdam pp. 37-56.

Godefroy, H.W.H.E. and Vegter, D., 1984. A new type immersible Laser Doppler Anemometer. In: Proceedings Transducer Tempcon Conference 84, Harrogate, Yorkshire, U.K., November 27-29, pp. 384-403.

Head, M.J., 1983. The use of miniature four-electrode conductivity probes for high resolution measurement of turbulent density or temperature variations in salt-stratified water flows. Ph.D. Diss., University of California, San Diego.

Karelse, M., Godefroy, H.W.H.E., Moser, G.M. and Van Leussen, W., 1986. Turbulence measurements in a stratified flow. In: Proceedings International Symposium on Buoyant Flows, Athens, Greece, September 1-5, pp. 383-393.

Lewschin, W.L., 1927. Die Auslöschung der Fluoreszenz in festen und flüssigen Farbstofflösungen. Z. Phys., 43: 230-253.

Lewschin, W.L., 1931. Das Gesetz der Spiegelkorrespondenz der Absorptions- und Fluoreszenzspektren. Z. Phys., 72: 368-381.

NEW APPROACH TO COX NUMBER IN QUASI-HOMOGENEOUS LAYERS

A. ICHA and R. SIWECKI
Institute of Oceanology, Polish Academy of Sciences, P.O. Box 68,
ul. Powstańców Warszawy 55, 81-967 Sopot (Poland)

ABSTRACT

The paper presents a proposed modification to the definition of the Cox number, which would allow finite values of this number when the mean temperature gradient equals zero. On the basis of experimental data collected during an expedition to Spitsbergen in 1984-1985, the new Cox number was verified with the help of a series of measurements of temperature fine structure. It was found that in non-homogeneous layers, the values of Cox number calculated on the basis of a formula elaborated in this paper are of the same order as the values obtained in the traditional manner. In homogeneous or quasi-homogeneous layers the proposed Cox number assumes values corresponding to the level of molecular exchange or to the level of turbulence of small intensity, which differs substantially from the results obtained so far, based on a classical Cox number.

1 INTRODUCTION

Among approaches to the problem of turbulence most popular were transport models resulting from a consideration of simplified models of turbulent transport based on physical premises. In transport models, only averaged fields of turbulent fluctuations are dynamic variables; the dynamics of interaction between various scales of movement is given little attention. The number of transport models created so far is very large, but they mostly refer to the description of velocity pulsation turbulent fields (Laundner and Spalding, 1972). Less attention has been paid to the behaviour of mean-square temperature pulsations $\overline{T'^2}$.

In oceanographic practice, the analysis of the equation describing the spatial-temporal evolution $\overline{T'^2}$ made it possible to determine the criterion of the intensity of turbulent processes in the sea, known as Cox number (Osborn and Cox, 1972 ; Gregg, 1976 ; Fedorov, 1978). The physical interpretation of results obtained on the basis of this criterion should, however, be made with caution, since with mean temperature gradients $d\overline{T}/dz$ approaching zero, Cox number tends to assume artificially large values. Physically this could mean a strong predominance of turbulent processes over the processes of molecular exchange and, as a consequence, strong turbulent mixing of the medium. Nevertheless, experimental investigations do not confirm this assertion: in layers characterized by zero temperature gradients low turbulence intensity is very often observed; it is comparable with the intensity of molecular exchange processes (Ozmidov, 1980).

The aim of this paper is to propose a modification of the currently used operational definition of Cox number, which would obtain finite values of this number at $d\overline{T}/dz \rightarrow 0$. Moreover, some verification is presented here; it is based on a series of measurements of the fine structure of the temperature field made by one of the authors in 1984-1985.

2 BASIC EQUATIONS

Let D be the volume occupied in R^3 with incompressible fluid with density ρ and viscosity ν. The state of the moving fluid is described by velocity field $\overline{u}(x,t) = (u_1, u_2, u_3)$ (pressure field is a square functional of the velocity field) , where $x = (x_1, x_2, x_3)$ $t \in (t_0, \infty)$. Let us designate sea water temperature in area D by $T(x,t)$ and assume it is a hydrodynamically passive substance. This assumption makes it possible to write the equation for $T(x,t)$ in following form (Monin and Yaglom, 1971)

$$\frac{dT}{dt} + \frac{d}{dx_i}(u_i T) = k_t \frac{d^2 T}{dx_i dx_i} \qquad \forall\ D \qquad (1)$$

where k_t is a coefficient of molecular temperature conductivity. Einstein's summation convention was used in equation (1).

We shall use a classical Reynolds averaging procedure in order to obtain an equation for $\overline{T'^2}$. Let $T' = T - \overline{T}$, $u'_i = u_i - \overline{u}_i$, where line — designate an averaging operator in which the weight function is Lebesque integrable and fulfils Reynolds axioms (Frost and Moulden, 1977) . It is not difficult to show that $\overline{T'^2}$ fulfils the following equation (Fedorov, 1978)

$$\frac{d\overline{T'^2}}{dt} + \frac{d}{dx_i}(\overline{u}_i \overline{T'^2} + \overline{u'_i T'^2}) + 2\overline{u'_i T'}\frac{d\overline{T}}{dx_i} = \frac{d}{dx_i}(k_t \frac{d\overline{T'^2}}{dx_i}) - 2\overline{e}_t \qquad (2)$$

where

$$\overline{e}_t = k_t \overline{\frac{dT'}{dx_i}\frac{dT'}{dx_i}} \qquad (3)$$

is the temperature dissipation rate.

3 COX NUMBER

By analysing equation(2)one may obtain some practical results which are experimentally verifiable. The traditional method assumed the following simplification(Fedorov, 1978): a)states examined are stationary, b)influence of temperature non-homogeneities molecular diffusion is negligible, c)advection processes do not occur, d)layers are homogeneous on an x_1, x_2 plane. These assumptions make it possible to present equation(2)in following form(u_3=w, x_3=z)

$$\overline{w'\,T'}\,\frac{d\overline{T}}{dz} = -k_t(\overline{\frac{dT'}{dz}})^2 \qquad (4)$$

We shall apply parametrization of the vertical heat flux in expression(4)by means of a thermal conductivity turbulent coefficient K_T, according to which $\overline{w'\,T'} = -K_T\,\frac{d\overline{T}}{dz}$ (Schmidt's hypothesis). As a result, we will obtain(Osborn and Cox, 1972 ; Gregg, 1976 ; Fedorov, 1978 ; Ozmidov, 1980)

$$\frac{K_T}{k_t} = C_z^2 = \overline{(\frac{dT'}{dz})^2}/(\frac{d\overline{T}}{dz})^2 \qquad (5)$$

Expression(5)is known as Cox number and makes it possible to estimate the order of magnitude of coefficient K_T on the basis of direct measurements.

Using this criterion when $d\overline{T}/dz \longrightarrow 0$ is - as may be seen from expression(5)and the analysis of experimental data made so far - devoid of physical sense. We will present a method leading to the obtaining of finite values of Cox number at $d\overline{T}/dz \longrightarrow 0$. According to formula(5), obtained from equation(2)under assumptions a)- d), the physical sense of Cox number consists in assuming in the theoretical model that the generation of temperature fluctuations by vertical turbulent movements in the field of mean temperature vertical gradient is balanced by the dissipation of these pulsations. In our model we will take into account a greater number of terms in equation(2): simplifications a), b), d)will still be valid, but we will take into consideration the vertical turbulent transport of temperature fluctuations(second term in brackets, on the left side of equation(2)). We thus obtain the following expression

$$\frac{d}{dz}\,\overline{w'\,T'^2} + 2\overline{w'\,T'}\,\frac{d\overline{T}}{dz} = -2k_t(\overline{\frac{dT'}{dz}})^2 \qquad (6)$$

Several authors(cf. Monin and Yaglom, 1971)emphasized the analogy between equation(2)and turbulent energy balance equation. We will apply ideas used when closing this equation for determining the form of the first term in(6) , on the left. The simplest approximation of the gradient type (close Boussinesq type approximation)has the form

$$\overline{w' T'^2} = -K' \frac{\overline{dT'^2}}{dz} \tag{7}$$

where K' has the sense of turbulent exchange coefficient$[m^2/s]$; its structure may be determined on the basis of the following considerations.

Coefficient K' should be proportional to turbulent thermal conductivity coefficient K_T ; it should also depend on parameters characterizing microstructural layers. Since processes of turbulent transport in those layers are of a local character, these parameters should include thickness of layer H and the parameters defining the internal structure of the layer, among which we will include Brunt-Väisälä frequency N, temperature dissipation rate defined by formula(3)and buoyancy parameter ag, where $a= -\frac{1}{\rho}(\frac{d\rho}{dT})_{s,p}$ is the coefficient of thermal expansion. These three parameters may form a combination with a dimension of length $L_T = a\overline{g e}_t^{1/2}N^{-5/2}$ which we idenify with the mixing length, at which temperature non-homogeneities maintain their individual character. According to the above we should have 1) $K'=BK_T$; 2) $K'=f(K_T,H,L_T)$ where $L_T = f_1(N,\overline{e}_t,ag) = a\overline{g e}_t^{1/2}N^{-5/2}$.

Using Buckingham theorem(Barenblatt, 1978)and assumption 1), we will obtain

$$K' = K_T F(L_T/H)$$
$$F(L_T/H) = L_T/H \tag{8}$$

Substituting dependence(8)in equation(6)we will obtain a new expression for Cox number in the form

$$\frac{K_T}{k_t} = C^2 = \frac{2\overline{(dT'/dz)^2}}{2(d\overline{T}/dz)^2 + B(\overline{dT'^2}/dz^2)} \tag{9}$$

where

$$B = a\overline{g e}_t^{1/2}N^{-5/2} /H \tag{10}$$

Dependence(9)is a generalization of the previous, widely used formula(5)and leads to finite values of Cox number at $d\overline{T}/dz \rightarrow 0$. Let us emphasize that formula(9)allows for a direct experimental verification.

4 CHARACTERISTIC OF EXPERIMENTAL DATA

The experimental data presented in this paper were collected
during an expedition of the Institute of Oceanology and Institute
of Geophysics of the Polish Academy of Sciences to Spitsbergen
between August 1984 and end of July 1985. Oceanological investigations concentrated in internal waters of the Hornsund (Fiord Fig. 1)

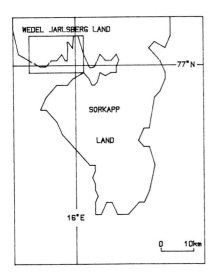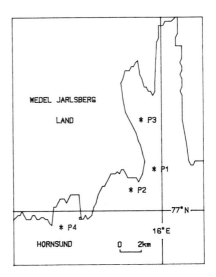

Fig. 1. Situation of the investigated area and area of oceanographic investigations in Hornsund.

Information about the vertical structure of temperature fields,
salinity and density was obtained from multiple profiles with a
specially-designed small-sized STD probe. The number of profiles
in a series varied between 25 and 30 with an interval between measurements of $\sim$ 2.5 minutes. The measurements were made in the surface 10-m layer with a mean velocity of $\bar{v}_s$ = 10 cm/s. Recording of
signals from the probe was made with a sampling step Δt = 1s. The
STD probe contained pressure and conductivity sensors produced by
Plessey Environmental Systems (time constant τ = 0.1s) and a thermistor temperature sensor made at the Institute of Oceanology (time
constant τ = 0.3s) (Siwecki and Kućmierz, 1985).

Information about the velocity field was obtained by means of a
Kelvin-Hughes currentmeter with a direct display of the velocity
modulus and direction of flow. The vertical velocity and flow

direction profiles were made with a sampling step 0.5 m before and directly after a series of STD measurements.

Vertical distribution of Cox number obtained from formulas(5) and(9)and distribution of mean temperature gradient for four series of measurements made in different hydrological conditions and seasons are presented in this paper. Sampling step of distributions with depth was $\Delta z = H = 0.2$ m. Information about each measurement series is given in Table 1.

TABLE 1

Description of measurements series.

Fig. No.	Date	Location see Fig.1	No. of measurements	Mean current velocity	Comments
2	1984-08-31	P1	30	~6 cm/s	measurements in drifting ice field;wind velocity 2-3 m/s
3	1984-10-24	P2	29	<than instrument sensitivity	
4	1985-04-12	P3	25	<than instrument sensitivity	measurement from stationary ice cover
5	1985-04-30	P4	25	~4 cm/s	no wind

5 DISCUSSION

The results of calculations obtained are presented in Figs 2 - 5. The also include distributions of Cox number calculated from formulas(5)and(9). Distributions of mean temperature gradients in the layers investigated are presented as well. The characteristic feature of all distributions is the same order of magnitude of Cox number calculated on the basis of(5)and(9)in non-homogeneous layers with a mean temperature gradient of $d\overline{T}/dz \neq 0$. On the other hand in homogeneous or quasi-homogeneous layers, the values of Cox number calculated from the new formula(9)are much lower than those calculated from formula(5)assuming values corresponding to the level of molecular exchange(Figs 3, 4)or corresponding to the level of small turbulence intensity(Fig. 5).

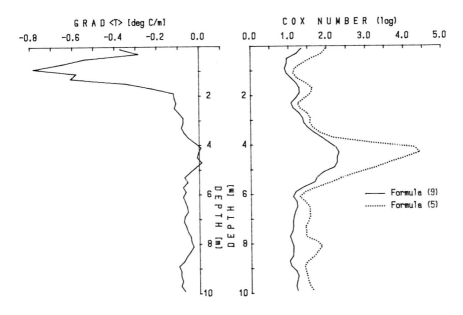

Fig. 2. Vertical distributions of temperature gradient and Cox number (Station P1).

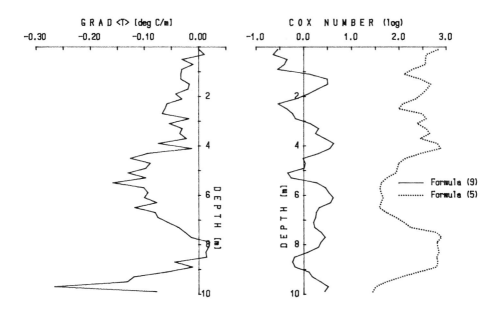

Fig. 3. Vertical distributions of temperature gradient and Cox number (Station P2).

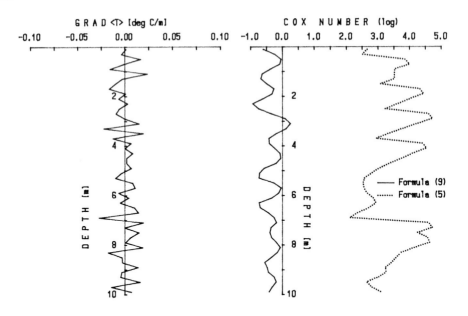

Fig. 4. Vertical distributions of temperature gradient and Cox number (Station P3).

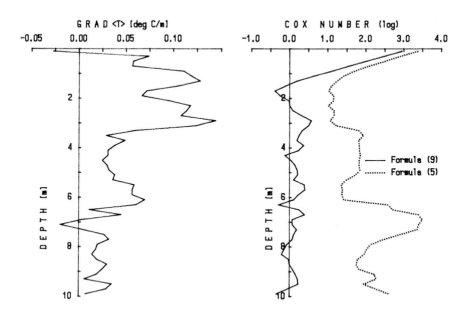

Fig. 5. Vertical distributions of temperature gradient and Cox number (Station P4).

It is thus the opinion of the authors that Cox number calculated
from expression(9)is a better indicator for identifying processes
which determine the character of microstructure in the ocean than
classic Cox number which leads, at best, to incorrect interpretation of experimental data. An additional confirmation of conclusions concerning the microstructure of the temperature field, resulting from Figs 2 - 5, may be the absence of measurable currents
in the study area, except for small currents with a velocity of
6 - 1 cm/s observed during recording of data from Figs 2 and 5.
Thus, there were no causes which could result in considerable turbulence of the layers and the expected level of turbulence could
approach the intensity of molecular exchange processes cf.(Table 1
; fig. 4). Let us also pay attention to an interesting fact connected with the distribution presented in Fig. 5. In the 0-0.5 m
layer one may observe a variable-sign temperature gradient which
according to relationship(5)has to lead to a great value of Cox
number(10^2 - 10^3). Cox number calculated from formula(9)gives the
same order of magnitude for this layer. It thus appears that the
near-surface layer is characterized by a considerable turbulence
intensity which might be connected with a local break-down of surface wave.

ACKNOWLEDGEMENTS

The authors would like to thank prof. dr Czesław Druet for
discussing problems connected with the issues presented in the
paper.
    The study was made within the framework of Research Problem
M.R. I. 15.

REFERENCES

Barenblatt, G. I., 1978. Similarity, self-similarity, intermediate
    asymptotic. Leningrad, Gidrometeoizdat(in Russian), 206 pp.
Fedorov, K. N., 1978. Fine structure of hydrophysical fields in
    the ocean. In:Oceanology. 1 :Ocean physics. Moscow, Publish.
    House Nauka: 113-147(in Russian).
Frost, W. and Moulden, T.H., 1977. Handbook of turbulence. Vol. 1.
    Plenum Press, New York and London, 535 pp.
Gregg, M. C., 1976. Microstructure:signature of mixing in the
    ocean. Naval Research Reviev, vol. 29, no. 11, 1-21.
Laundner, B. E. and Spalding, D.B., 1972. Lectures in mathematical
    models of turbulence. Academic Press, New York, 210 pp.
Monin, A. S. and Yaglom, A.M., 1971. Statistical fluid mechanics,
    Vol. 1. MIT Press, 769 pp.
Osborn, T. R. and Cox, C. S., 1972. Oceanic fine structure. Geop-
    hys. Fluid Dynamics, no. 3, 321-345.

Ozmidov, R. V., 1980. Oceanic turbulence. Studia i Materiały
    Oceanol. no. 29, 21-84 (in Polish, English abstract).
Siwecki, R. and Kućmierz, H., 1985. Application of the STD recor-
    der in the measurements carried out during the cruise of R/V
    Akademik Mstislav Keldysh. Studia i Materiały Oceanol. no. 47
    272-280 (in Polish, English abstract).

THEORIES OF INTERNAL WAVE INTERACTION AND STABLY STRATIFIED TURBULENCE:
TESTING AGAINST DIRECT NUMERICAL EXPERIMENTATION

GREG HOLLOWAY AND DAVE RAMSDEN
Institute of Ocean Sciences, P.O. Box 6000, Sidney, B.C., Canada and
Daleth Research, 1426 Pembroke Street, Victoria, B.C., Canada

ABSTRACT

Direct numerical experiments, both in 2D (vertical plane) and in 3D geometries, reveal the roles of nonlinear energy transfer and of vertical buoyancy (heat) flux in governing the wavenumber spectra of stably stratified turbulence. Buoyancy flux is typically negative (downwards) at large scales and positive (upwards, "restratifying") at small scales where "small scale" may include the overturning scale. Numerical experiments provide a basis for testing some aspects of theoretical research. A practical concern arises because energy spectra and dissipation rates do not clearly imply the magnitude nor even the sign of the overall buoyancy flux.

1 INTRODUCTION

Over decades of research, very considerable efforts have been directed toward the understanding of turbulence in stably stratified environments and of energy transfer by interactions among internal gravity waves. Yet major uncertainties remain. Because of a number of dimensional parameters that occur, the problem area is ambiguous to simpler dimensional analysis. Approaches from the view of turbulence have depended upon uncertain scaling and plausibility arguments. From the view of waves, resonant interaction theories have been explored on the assumption of sufficiently weakly wavelike motions. Tentative efforts have been made to incorporate higher amplitude effects in renormalized wave-wave closure theories. Simulation of ray trajectories in large scale random shear flows are another approach toward understanding how energy is redistributed both in physical space and in wavenumber space. Finally, various saturation hypotheses have been suggested that may relate internal wave energetics to the occurrence of small scale turbulence.

A major impediment to progress has been the unavailability of observations that are adequate to test differences among the various theories. For the most part theories predict characteristics of variance spectra that are broadly consistent with observations. However, theories based upon very

different, sometimes contradictory, hypotheses may yield nearly the same
predictions for variance spectra. Limited observations are unable to
discriminate among the theories. If all one wanted was an account, "right" or
"wrong", for the observed variance spectra, then differences on theoretical
issues might be viewed as "academic". However, matters of great practical
importance, such as the prediction of transport and mixing of heat, chemical
substance and momentum, are sensitive to theoretical differences. These
concerns urge us to attempt to resolve some theoretical issues. It's a
daunting challenge; the present paper should be read more as a progress report
than as a resolution.

We report results from two kinds of numerical experimental studies. First
we restrict fields to two-dimensionality by requiring all variations to be
independent of one horizontal coordinate. While all three components of
velocity as well as earth's rotation are permitted, the restriction to 2D
variation allows us to employ computing resource to resolve interactions among
widely disparate scales of motion. Second, we lift the 2D restriction and
perform experiments with fully 3D variation but with much more limited range
of resolved scales. We discuss also some new results from waves-turbulence
renormalized closure that may be compared with numerical experimental results.
Our concerns are directed mainly toward the wavenumber distribution of
vertical buoyancey (heat) transport and the transfers across wavenumber of
kinetic and available potential energies.

2 2D (VERTICAL PLANAR) EXPERIMENTS

All motion fields are restricted to be independent of one horizontal
coordinate, x say. Evolution is then defined by the x-directed component of
vorticity, ζ, the x-directed component of velocity, u, and buoyancy
fluctuation, b, defined by $b(y,z,t) = -\rho(y,z,t) - z$ where $\rho(y,z,t)$ is density
scaled by a reference vertical gradient so that b is buoyancy anomaly about a
reference state $\rho = \rho_0 - z$ where ρ_0 is a (large) constant reference density.
The Boussinesq equations of motion are

$$\frac{\partial}{\partial t} \zeta + J(\psi, \zeta) - f\frac{\partial}{\partial z} u - \frac{\partial}{\partial y} b - \nu\nabla^2 \zeta = F_\zeta \tag{1}$$

$$\frac{\partial}{\partial t} u + J(\psi, u) + f\frac{\partial}{\partial z} \psi \qquad - \nu\nabla^2 u = F_u \tag{2}$$

$$\frac{\partial}{\partial t} b + J(\psi, b) \qquad + \frac{\partial}{\partial y} \psi - k\nabla^2 b = F_b \tag{3}$$

where ψ, given by $\nabla^2\psi = \zeta$, is streamfunction in the y-z plane,
$J(A,B) \equiv \partial_y A\partial_z B - \partial_z A\partial_y B$, f/2 is the vertical component of earth's rotation, ν
is kinematic viscosity, κ is the effective diffusivity for buoyancy, and $F\zeta$,
F_u and F_b are fields of external forcing applied to the ζ, u and b fields.
Time is scaled so that a stability frequency is set to unity.

Our motivation for experiments in 2D is that this permits us to deploy
computational resource to resolve interactions among widely disparate scales
of motion. A spectral truncation wavenumber can be more than 100 times a
gravest wavenumber. This is important to theories that distinguish wavenumber
local and nonlocal interactions. One may remark also that Equations (1-3) are
completely faithful to the 3D Boussinesq equations; only we choose to consider
a restricted class of solutions of those 3D equations. However, our immediate
goal is to develop a numerical experimental dataset against which to test
statistical theory; later we turn to 3D experimentation.

Figure 1 provides pictorial "snapshots" of some of the experimental output,
here showing total buoyancy b(y,z,t) + z. In each of these cases, random
excitations $F\zeta$ and F_b are such as to inject both planar kinetic energy $\frac{1}{2}\overline{|\nabla\psi|}^2$
and available potential energy $\frac{1}{2}\overline{b^2}$ at equal rates with wavenumber spectra of
forcing given by $k^3/(k^2 + k_o^2)^6$ with $k_o = 7$. Overbars denote domain average.
For the cases shown, f = 0 and no forcing is applied to the transverse kinetic
energy $\frac{1}{2}u^2$. By changing the r.m.s. amplitude of random excitation,
statistically stationary states were achieved at turbulent Froude numbers of
approximately F = 0.1, 0.3, 1.0, 3.0 where $F^2 \equiv \overline{(\partial v/\partial z)^2}/N^2 = \overline{(\partial^2\psi/\partial z^2)^2}$.
These are the cases shown in Figure 1. One sees that the onset of manifest
overturning occurs near F = 1. In fact we've observed overturning from about
F = 0.7 over a variety of forcing conditions.

A wealth of numerical experimental results from such 2D experiments have
been reported by Shen and Holloway (1986) and Ramsden and Holloway (1987), the
latter including f $\neq$ 0 and $\nu \neq \kappa$. In Figure 2a we illustrate certain features
of the numerical experiments near F = 1 with f = 0 and $\nu = \kappa$.

Kinetic (KE) and available potential energy (APE) spectra are graphed
against magnitude wavenumber k = $|k|$. At smaller F the spectra are more
nearly in equipartition. At larger F the KE spectrum steepens while the APE
spectrum becomes more shallow, reflecting a tendency toward 2D turbulence with
buoyancy advected relatively passively.

KE and APE transfer functions are graphed against k. Positive transfer
indicates that nonlinearities (i.e. triple correlations) are tending to
increase KE or APE at any k. Negative transfer indicates nonlinearity
decreasing KE or APE at any k. The shapes of the transfer functions are found
to be relatively insensitive to turbulent intensity F. Only the overall
amplitude of transfers increases with increasing F.

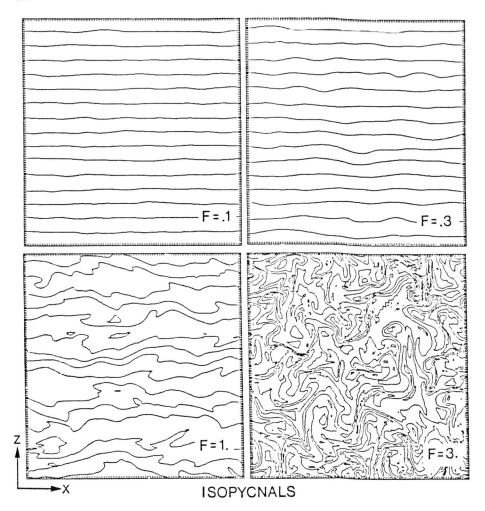

ISOPYCNALS

Fig. 1. "Snapshots" of total buoyancy at turbulent Froude number 0.1, 0.3, 1. and 3.

Vertical buoyancy flux $b\partial_y\psi$ is graphed in the third panel of Figure 2a. Flux cross-spectra are shown as functions of vertical wavenumber k_z, horizontal wavenumber k_h and total wavenumber k. In each case the flux is negative (i.e. downward transport of buoyancy, conversion of KE to APE) at low wavenumbers with positive flux (i.e. upward or counter-gradient buoyancy transport, conversion of APE to KE, "restratification") at high wavenumbers. Interestingly, an overturning scale (r.m.s. Thorpe displacement) occurs in the higher wavenumber, restratifying portion of spectrum. This illustrates that

(a) waves of length scale considerably larger than an overturning scale are able to sustain systematic mean buoyancy flux, and

(b) overturning events do not necessarily imply down-gradient transport. Indeed. overturning may be systematic of counter-gradient restratification!

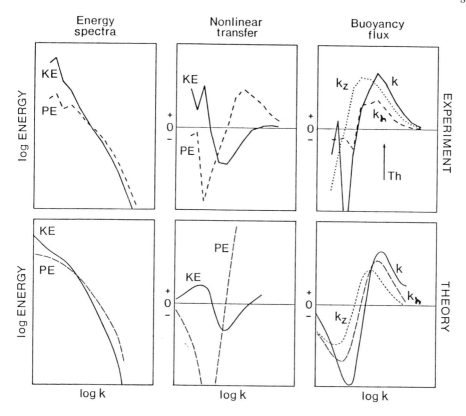

Fig. 2. Numerical experiment and statistical theory are compared.
Top (a): KE and APE spectra as functions of total wavenumber. KE and APE
transfer functions in total wavenumber. Buoyancy flux cross-spectra as
functions of total (solid), horizontal (dashed) and vertical (dash-dot)
wavenumber. Bottom (b): The same quantities are evaluated from statistical
closure theory.

The general shape of the buoyancy flux cross-spectrum, i.e. negative at low
wavenumbers and positive at higher wavenumbers, is seen at both larger and
smaller F. At smaller F, however, buoyancy flux is dominated by temporal
oscillations so that long averaging times over many stability periods are
required in order to isolate the (relatively) weak mean flux.

The shapes of the KE and APE transfer functions, and the resultant shapes
of the buyoyancy flux cross-spectra, can be attributed to the imposed
restriction to two-dimensionality. (It should be recalled though that these
are possible solutions of the fully 3D equations.) Restricted to 2D, the KE
transfer is positive toward low wavenumber in order that nonlinear
interactions conserve enstrophy ζ^2. Unless direct dissipation at low
wavenumbers is adequate to dispose of the transferred KE, a stationary energy

balance can only be obtained by exchanging KE to low wavenumber APE by means of negative buoyancy flux. APE is then efficiently transferred to high wavenumber where a build-up of APE induces reconversion to KE by means of positive buoyancy flux, a scenario which has been further elaborated in Holloway (1984).

While 2D numerical experiments are not typical of 3D reality, these experiments do provide a precise data set against which to test waves/ turbulence theories which may be posed either in 2D or in 3D. We examine the 2D testing of theories before turning to some preliminary results in 3D.

3 THEORY TESTING IN 2D

A goodly part of extant theory, especially that which addresses weak wave-wave interactions (Muller and Olbers, 1975; Olbers, 1976; McComas, 1977; Pomphrey, Meiss and Watson, 1980) or wave transports in qusei-random shear flows (Henyey and Pomphrey, 1983; Flatté, Henyey and Wright, 1985), are posed as readily in 2D as in 3D. Direct testing in 3D reality is not yet feasible. Computer capacity is not yet so great as to admit 3D numerical testing. For the present we employ 2D numerical testing.

From the outset it appears that the greater part of extant theory must fail! Both the weak wave-wave interaction and the wave-shear flow interaction theories adopt as their bases the free wave eigenfunction. Energy is slowly exchanged among free waves by weak (resonant) interactions and/or the waves are refracted by background shear flows. Only when wave energy is transported to some high wavenumber is it "removed" from a calculation by assumed "breaking". Since each free wave is unable to sustain net vertical buoyancy flux, it would appear that all such theories are systematically unfaithful to the 2D Boussinesq equations that seem to require negative buoyancy flux at low wavenumbers to avoid indefinite accumulation of low wavenumber KE.

An effort that goes beyond weak wave interaction is seen in Carnevale and Frederiksen (1983, hereafter CF) who investigate renormalized closure theory for interaction among wave modes given by complex amplitude coefficients $a_{\underline{k}}^{\pm} = k\psi_{\underline{k}} \pm b_{\underline{k}}$ where the motion is considered to lie in a vertical plane without earth's rotation. Formally the closure investigated by CF could include non-zero buoyancy flux expressed by complex correlations $\langle a_{\underline{k}}^{+} a_{-\underline{k}}^{-} \rangle$; in practice the full theory would seem so tedious of calculation that CF employ a simplifying assumption (diagonalization) which retains only second moments $\langle |a_{\underline{k}}^{+}|^2 \rangle$ and $\langle |a_{\underline{k}}^{-}|^2 \rangle$. If however, as we've argued above, buoyancy flux is required to satisfy wavenumber KE and APE budgets in 2D flows, then diagonalization is not permitted even for motion fields of very small amplitude.

We have pursued somewhat an alternative closure approach, based upon the primitive variables $\psi_{\underline{k}}$ and $b_{\underline{k}}$ to obtain equations for evolution of the second moments (a) velocity variance $V_k = k^2 \langle|\psi_{\underline{k}}|^2\rangle$, (b) buoyancy variance $R_{\underline{k}} = \langle|b_{\underline{k}}|^2\rangle$ and (c) buoyancy flux $F_{\underline{k}} = \text{Im}\langle k_y\psi_{\underline{k}}b_{-\underline{k}}\rangle$ where we are considering 2D flow without earth's rotation. Following Holloway (1979) the result is

$$\tfrac{1}{2}\partial_t V_k - F_k + T_{v,k} = E_{v,k} - D_{v,k} \tag{4}$$

$$\tfrac{1}{2}\partial_t R_k + F_k + T_{r,k} = E_{r,k} - D_{r,k} \tag{5}$$

$$\partial_t F_k + \omega^2(V_k - R_k) + 2\mu F_k = E_{f,k} \tag{6}$$

$$\omega_k^2 = k_y^2/k^2 \tag{7}$$

$$T_{v,k} = \overset{\Delta}{\underset{p,q}{\Sigma}} \theta\, b_{kpq} \cdot V_q(-V_p + V_k) \tag{8}$$

$$T_{r,k} = \overset{\Delta}{\underset{p,q}{\Sigma}} \theta\, c_{kpq} \cdot V_q(-R_p + R_k) \tag{9}$$

$$b_{kpq} = \frac{|p\times q|^2}{k^2 p^2 q^2}(p^2 - q^2)\cdot(k^2 - q^2) \tag{10}$$

$$c_{kpq} = |p\times q|^2/|q|^2 \tag{11}$$

$$\theta = \theta_{kpq} = \mu_{kpq}/(\omega^2_{kpq} + \mu^2_{kpq}) \tag{12}$$

$$\mu_{kpq} = \mu_k + \mu_p + \mu_q \tag{13}$$

$$\omega^2_{kpq} = k_y^2/|k|^2 + p_y^2/|p|^2 + q_y^2/|q|^2 \tag{14}$$

$$\mu^2 = \lambda \underset{p}{\Sigma} k^2 p^2/(p^2 + k^2)\, V_p \tag{15}$$

While the underlying derivation is no more tedious than that of CF, we retain non-zero buoyancy flux. This gain is achieved at the cost of another

assumption which occurs in equations (12, 14). ω^2_{kpq} ought to refer to square
of the sum of three free wave frequencies, each frequency signal $\pm$ according
to $a^{\pm}$. Insofar as we deal with V_k and R_k this $a^{\pm}$ sign information is not
available and we "approximate" ω^2_{kpq} by the sum of squared frequencies. As
well, equation (15) is a compromise between a more difficult formulation by
Kraichnan (1971) and a simpler heuristic suggested by Pouquet et al. (1975).

Evaluations of equations (4-15) are shown in Figure 2b for F near unity.
Theoretical and numerical experimental results are broadly consistent although
precise testing of quantitative skill is not implied. (The figures are
unlabelled.) To the extent that we've pursued such testing, we've found that
quantitative agreement improves at higher F but appears to deteriorate rapidly
for F smaller than unity. We guess that this deterioration is due to
inadequacy of equations (12, 14) in weaker, more wavelike flows. However, we
leave this speculation in order to move on to fully 3D numerical experiments
and theory.

4 NUMERICAL EXPERIMENTS IN 3D

Employing spectral transform methods with filtered leapfrog timestepping,
we've integrated the Boussinesq equations

$$\partial_t \underline{u} + \underline{u} \cdot \nabla \underline{u} + \underline{f} \times \underline{u} + \nabla_p - b\hat{z} - \nu\nabla^2 \underline{u} = \underline{F}_u \tag{16}$$

$$\partial_t b + \underline{u} \cdot \nabla b + W + \kappa\nabla^2 b = F_b \tag{17}$$

$$\nabla \cdot \underline{u} = 0 \tag{18}$$

on a collocation grid of 32 x 32 x 32 prints. $\underline{F}_u$ and F_b are random
excitations of momentum and of buoyancy which are included so that we can
achieve forced/dissipative statistically stationary states. A report on these
experiments is available in Ramsden (1987). Here we make only a few brief
remarks.

By means of random excitation $\underline{F}_u$ and F_b with variance spectra
$k^3/(k^2 + k_o^2)^6$ as employed in the 2D experiments, we've developed statistic-
ally stationary Froude number (F) cases where $F \equiv ((\partial u/\partial z)_{rms} + (\partial v/\partial z)_{rms})/N$.
At values of F near F = .1, F = .3, F = 1. and F = 3., Figure 3 shows vertical
buoyancy flux cross-spectra as functions of total wavenumber k, magnitude
horizontal wavenumber k_h and vertical wavenumber k_z. In these experiments
we've caused KE and APE to be injected at a ratio KE:APE of 2:1.

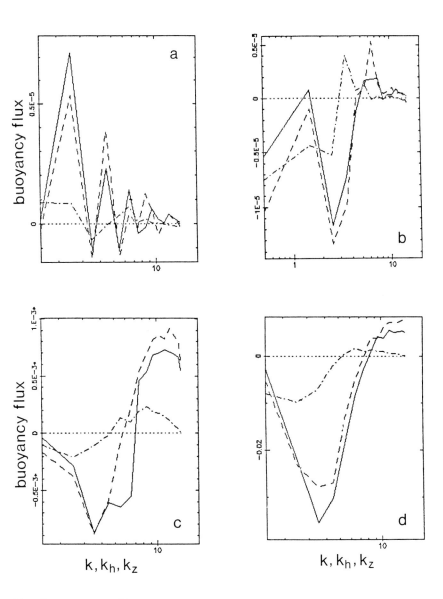

Fig. 3. Buoyancy flux cross-spectra are shown as functions of total (solid), horizontal (dashed) and vertical (dash-dot) wavenumber for flows with random excitation of KE and APE in ratio 2:1. Amplitude of excitation is varied to achieve turbulent Froude numbers (a) F = 0.1, (b) F = 0.3, (c) F = 1.0, and (d) F = 3.0.

When F is small, the buoyancy flux is quite oscillatory in time and averaging over many stability periods is required. Even after long averaging, results at F = .1 are not significantly different from zero at any scale. At F = .3 a positive (counter gradient) tail arises at high wavenumber. By F = .7 (not shown) negative buoyancy flux is established at low wavenumbers with positive flux at high wavenumbers. This is seen at F = 1., F = 3. and at F = 10. (not shown) where the positive tail persists but at reduced amplitude.

An immediate question is the extent to which these results depend upon the method of forcing. A variety of alternative forcings have been explored with some examples near F = 1. illustrated in Figure 4. Previously we forced KE and APE in ratio 2:1, motivated by a desire to make a "least prejudiced" choice among internal gravity waves and geostrophic modes. Employing the broad spectrum of forcing used previously, we explore the dependence on KE:APE forcing ratio with an example in which only KE is forced and an example where KE and APE are forced equally (ratio unity). The results are nearly opposite to each other. With forcing of KE only, buoyancy flux is negative (i.e. downwards, converting KE to APE) at nearly all scales although a very slight positive tail is seen. With equal forcing of KE and APE, buoyancy flux is decidedly positive (i.e. upwards, converting APE to KE) at all scales.

Because of the limited range of scales that we are able to resolve in 3D experiments, the spectra of external forcing and of dissipation are substantially coincident. Seeking to separate these scales, we've performed experiments with random forcing limited to the range k ≤ 2. An example with forcing of KE only is given in Figure 4. While buoyancy flux is negative at low wavenumbers where KE dominates, the flux is positive ("restratifying" or reconverting APE back to KE) over middle and high wavenumbers.

We complete Figure 4 with another example which is forced only over k ≤ 2. In this case we seek to avoid possible spurious effects due to the randomizing character of the external excitations. Instead we specify that, for k ≤ 2, internal wave modes will satisfy identically their linear, inviscid dynamics. All other wave modes with k > 2 and all geostrophic modes are fully nonlinearly coupled to the prescribed large scale waves which are the sole energy source. For the example shown, f = 0.1 and the large scale waves are in approximate KE:APE equipartition. The resulting buoyancy flux shows a distinctly positive tendency over most of the available wavenumber space.

We explored somewhat the role of Prandtl number although this is quite limited by available wavenumber range. Prandtl number 3 flows are found to exhibit a greater tendency toward positive buoyancy flux than corresponding flows at Prandtl number unity.

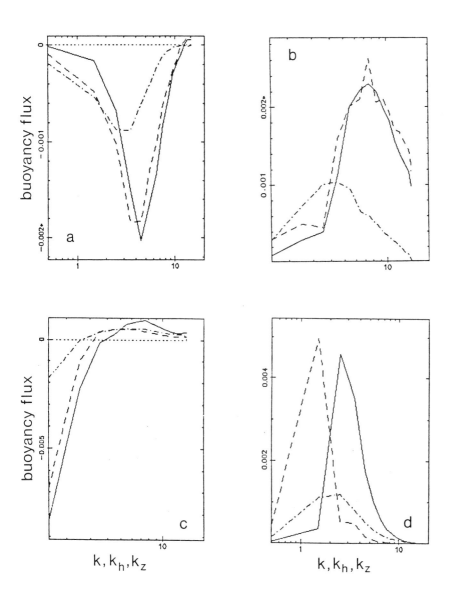

Fig. 4. Buoyancy flux as in Fig. 3, here near F = 1.0. Panel a: only random KE excitation. Panel b: equal rates KE and APE excitation. Panel c: only KE excitation, restricted to scales k ≤ 2. Panel d: free internal waves at scales k ≤ 2 provide energy source.

Here we have not shown KE and APE transfer functions, which may be seen in Ramsden (1987). Despite the considerable variation among buoyancy flux cross-spectra seen in Figures 3 and 4, the transfer functions are remarkably similar. Both the KE and APE transfers are negative in low wavenumbers and positive in high wavenumbers. In particular the KE transfer shows none of the backward cascade tendency resulting in positive transfer to low wavenumber such as characterizes 2D experiments (c.f. Figure 2a).

It may seem curious that buoyancy flux in 3D has so much in common with buoyancy flux in 2D, especially given that the KE transfer functions are so different in the two geometries. We postpone comment on this until Section 6.

5 THEORY IN 3D

Detailed second moment closure theory such as equations (4-15) for 2D poses a daunting challenge in 3D, although an important beginning can be seen in Sanderson, Hill and Herring (1987). Classical scaling ideas dating to Bolgiano (1959), Monin (1962), Shur (1962), Lumley (1964, 1965) or Phillips (1965) are largely concerned with the role that buoyancy flux cross-spectra may play in governing energy spectra of stratified turbulence. Weinstock (1985) has further observed that a threory such as Lumley (1964) can be extended to predict a reversed transfer of APE toward low wavenumbers. Unfortunately, the limited range of wavenumbers available to our numerical experiments prevent meaningful tests of these ideas.

An alternative approach in 3D, which borrows from second moment closure theory but differs sharply from the above-mentioned scaling arguments, has been suggested by Holloway (1983, 1986). A feature of Holloway's approach is to suppose that buoyancy flux is energetically insignificant relative to the KE and APE nonlinear transfers. Both KE and APE make simple forward transfers from low to high wavenumber. In fact all of our numerical experiments, which ranged up to F = 10., are consistent with this view. The shapes of the KE and APE spectra are presumed by Holloway (1983, 1986) to be controlled by the functional form of θ_{kpq} from equation (12). However, the shape of the buoyancy flux cross-spectrum is not clearly determined from theory and Holloway (1986) speculated on a simple, everywhere negative shape. Clearly this speculation is systematically contradicted by all of our numerical experiments.

6 ACONCLUSIVE DISCUSSION

In terms of an objective to understand stratified turbulence well enough to calculate its role inmaintaining oceanic distributions, direct numerical experiments are not yet applicable. 3D experiments are severely limited in their available resolution, certainly on the computers available to this study but also on larger machines elsewhere. Higher resolution is available in 2D experiments (a good deal higher resolution than we have employed here); however, applicability is limited by the 2D constraint on nonlinear transfer terms.

On the other side, statistical theory for this difficult waves/vortex modes/turbulence problem is not yet so well developed and so well tested that one may be confident in application. Should one despair? We think not and suggest that progress can be seen in two areas.

First, theories both in 2D and in 3D can be tested against numerical experiments without requiring very high resolution. In Figure 2 we saw some encouraging correspondence between 2D theory and experiment, especially that theory produced buoyancy flux cross-spectra in some agreement with numerical experiment. Theory in 3D will require more development, perhaps along lines seen in Sanderson, Hill and Herring (1987), before careful testing against numerical experiment can be realized. Still we can now see these items of theoretical progress.

Second, both the experiments and theory help to reveal errors in the ways that stratified turbulence is usually understood. It is thought that density overturning flow structures are the mechanisms that mix buoyancy downwards (or mass upwards) while larger scale flow sustains no average vertical buoyancy flux. We seem to find nearly the opposite: that large scale motion supports negative (downwards) buoyancy flux whereas motion on the scale of overturning often coincides with positive ("restratifying") buoyancy flux. When this was first reported (Holloway, 1984) from 2D experiments, it could still be supposed that the "conventional wisdom" was intact and that the "strange" results were only artifacts of 2D. Now we see that this "strangeness" afflicts 3D numerical experiments as well. How can that be? A sketch, given in Figure 5, conveys our effort to provide more intuitive explanation. Although the sketch is 2D, we have 3D in mind. The key point is that nonlinear interaction is more effective at scattering APE than KE from large scales down to small scales. This imbalance results in tendency towards excess KE at large scales and excess APE at small scales which, in turn, drives oppositely signed buoyancy flux at large and small scales in order to maintain KE:APE ratios. The outstanding question in 3D is why APE transfer to small scales is more efficient than KE transfer. Here we remark tentatively, hoping later to quantify: the constraint to incompressibility in the velocity field impedes triad-wise transfer of velocity variance whereas transfer of buoyancy variance is relatively unrestricted. Such an effect is seen, perhaps, in theoretical calculation of eddy Prandtl number for passive scalar advection (Herring et al., 1982).

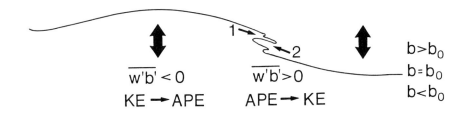

Fig. 5. The sketch depicts a deformed isopycnal surface. At very large scales KE and APE are exchanged in a quasi-oscillatory way. When large scale KE has mostly gone to APE, inducing large scale tilts to isopycnals, the isopycnal surface is most readily deformed by interaction with smaller scale waves. This selectively bleeds off large scale APE. At smaller scales, including overturning, buoyancy flux is positive (counter-gradient) while denser fluid (arrow 1) overrides lighter fluid (arrow 2). Overturning is here seen as a release of APE to KE.

As a very practical matter we are concerned for widely different possible forms of buoyancy flux that appear to be consistent with very similar KE and APE spectra and with dissipation thereof. We are brought to the question: Do turbulence dissipation measurements imply ocean mixing?

7 ACKNOWLEDGEMENT

This work has been supported in parts by the U.S. Office of Naval Research under contract N00014-85-C-0440 and by the Canada Department of Supply and Services under contract FP941-6-0202.

8 REFERENCES

Bolgiano, R., 1959. Turbulent spectra in a stably stratified atmosphere. J. Geophys. Res. **64**, 2226-2229.
Carnevale, G.F. and Frederiksen, J.S., 1983. A statistical dynamical theory of strongly nonlinear internal gravity waves. Geophys. Astrophys. Fluid Dyn., **23**, 175-207.
Flatte, S.M., Henyey, F.S. and Wright, J.A., 1985. Eikonal calculations of short-wavelength internal-wave spectra. J. Geophys. Res., **90**, 7265-7272.
Henyey, F.S. and Pomphrey, N., 1983. Eikonal description of internal wave interactions: A non-diffusive picture of induced diffusion. Dyn. Atmos. Oceans, **7**, 189-208.
Herring, J.R., Schertzer, D., Lesieur, M., Newman, G.R., Chollet, J.P., and Larcheveque, M., 1982. A comparative assessment of spectral closures as applied to passive scalar diffusion. J. Fluid Mech., **124**, 411-437.
Holloway, G., 1979. On the spectral evolution of strongly interfacing waves. Geophys. Astrophys. Fluid Dyn., **11**, 271-287.
Holloway, G., 1984. Probing the internal wave strong interaction regime by numerical experimentation, in Inernal Gravity Waves and Small-Scale Turbulence, Proceedings, 'Aha Huliko'a Hawaiian Winter Workshop, edited by P. Muller and R. Pujalet, pp. 221-248, Hawaii Institutue of Geophysics, Honolulu.

Holloway, G., 1986. Considerations on the theory of temperature spectra in stably stratified turbulence:. J. Phys. Oceanog. **16**, 2179-2183.

Kraichnan, R.H., 1971. An almost-Markovian, Galilean-invariant turbulence model. J. Fluid Mech. **47**, 513-524.

Lumley, J.L., 1964. The spectrum of nearly inertial turbulence in a stably stratified fluid. J. Atmos. Sci., **21**, 99-102.

Lumley, J.L., 1965. Theoretical aspects of research on turbulence in stratified flows. Atmospheric Turbulence and Radio Wave Propogation, A.M. Yaglom and V. I. Tatarsky Eds., Moscow, 105-110.

McComas, C.H., 1977. Equilibrium mechanisms within the oceanic internal wave field. J. Phys. Oceanogr., **7**, 836-845.

Monin, A.S., 1962. Turbulence spectrum in thermally stratified atmosphere. Izv. Akad. SSSR, Ser. Geofis., **3**, 397-407.

Muller, P., and Olbers, D., 1975. On the dynamics of internal waves. J. Geophys. Res., **80**, 3848-3860.

Olbers, D.J., 1976. Nonlinear energy transfer and the energy balance of the internal wave field in the deep ocean. J. Fluid Mech., **74**, 375-399.

Phillips, O.M., 1965. On the Bolgiano and Lumley-Shur theories of the buoyancy subrange. Atmospheric Turbulence and Radio Wave Propogation, A.M. Yaglom and V.I. Tatarsky Eds., Moscow, 121-128.

Pomphrey, J., Meiss, J.D., and Watson, K.M., 1980. Description of nonlinear internal wave interactions using Langevin methods. J. Geophys. Res., **85**, 1085-1094.

Pouquet, A., Lesieur, M., Andre, J.C., and Basdevant, C., 1975. Evolution of high Reynolds number two-dimensional turbulence. J. Fluid Mech. **72**, 305-319.

Ramsden, D., 1987. Three dimensional stratified turbulence, Daleth Research, Victoria, BC Canada (unpublished report).

Ramsden, D., and Holloway, G., 1987. Direct simulation and turbulence closure evaluation for large amplitude internal wave interaction in the vertical plane. Third International Symposium on Stratified Flows, Calif. Inst. Tech., Pasadena, Calif.

Sanderson, R.C., Hill, J.C., and Herring, J.R. Transient behavior of a stably stratified homogenous turbulent flow, in Advances in Turbulence, G. Comte-Bellot and J. Mathieu, eds. Springer-Verlag.

Shen, C., and Holloway, G., 1986. A numerical study of the frequency and the enegetics of nonlinear internal gravity waves. J. Geophys. Res., **91**, 953-973.

Shur, G.H., 1962. Experimental studies of the energy spectrum of atmospheric turbulence. Proc. Cent. Aerolog. Observ. USSR, **43**, 97-90.

Weinstock, J., 1985. On the theory of temperature spectra in a stably stratified fluid. J. Phys. Oceanogr., **15**, 475-477.

THE INFLUENCE OF BOTTOM DISSIPATION ON WAVE GROWTH

S.L. WEBER
Royal Netherlands Meteorological Institute, P.O. Box 201, 3730 AE De Bilt, The Netherlands.

ABSTRACT

Wave growth in shallow water is studied numerically, assuming certain expressions for wind input, dissipation by 'white-capping' and bottom friction, and by explicitly calculating the resonant four-wave interactions. The results are explained from the balance between the different source terms at the various stages of growth. Some basic results are presented here. A fuller account will be published elsewhere (Weber, 1987).

1 INTRODUCTION

It has long been known from measurements that wave growth is restricted by water depth when the wave length to depth ratio becomes larger than about three. The waves 'feel' the bottom and energy is lost through some kind of interaction mechanism. There are many possible mechanisms; the most likely for continental shelf conditions are percolation (coarse sand) and friction in the turbulent bottom boundary layer (fine sand, or sand with ripples), see Shemdin et al. (1978).

2 POSING THE PROBLEM

Wave growth is determined by the balance between wind input, dissipation by 'white-capping', the resonant wave-wave interactions and bottom dissipation (see e.g. Hasselmann et al., 1973):

$$\frac{\partial F}{\partial t} + c_g \cdot \frac{\partial F}{\partial x} = S_{tot} \quad \text{(flat bottom, no currents)} \tag{1}$$

with $F = F(f,\theta,x,t)$ the two dimensional frequency spectrum, f the frequency and θ the direction of a wave component, x the space coordinate, t time. Wave energy propagates with the group velocity c_g, $S_{tot} = S_{tot}(f,\theta,x,t)$ is the sum of the source terms.

To get more insight in the relative importance of the different source terms and their role in wave evolution, equation (1) was solved numerically for a fetch-limited situation ($\partial F/\partial t = 0$) for different water depths. The source terms were taken as follows:

2.1 Windinput

The windinput is parametrized in terms of the turbulent friction velocity u^*:

$$S_{in}(f,\theta) = \max\left[0, 0.25 \frac{\rho_a}{\rho_w} \left(\frac{28u^*}{c_p} \cos\theta -1\right)\right]\omega F(f,\theta) \qquad (2)$$

with ρ_a and ρ_w densities of air and water respectively, c_p the phase velocity and ω the radian frequency. u^* was taken to be .71 m/s. This expression is based on the measurements by Snyder et al. (1981).

2.2 The non-linear interactions

The non-linear interactions are given by Hasselmann (1961) :

$$S_{nl}(\underline{k}_4) = \omega \int A (\underline{k}_1, \underline{k}_2, \underline{k}_3, \underline{k}_4) * [N_1N_2 (N_3 + N_4) -$$

$$N_3N_4 (N_1 + N_2)]\, \delta (\underline{k}_1 + \underline{k}_2 - \underline{k}_3 - \underline{k}_4)\, \delta (\omega_1 + \omega_2 - \omega_3 - \omega_4)\, d\underline{k}_1 d\underline{k}_2 d\underline{k}_3 \qquad (3)$$

$N_i = N(\underline{k}_i) = F(\underline{k}_i)/\omega$ is the number density, A the interaction coefficient.

The method of calculating this expression is described in Hasselmann and Hasselmann (1981).

2.3 White-capping

Komen et al. (1984) obtained an equilibrium expression for deep water wave growth assuming the following expression for the dissipation by 'white-capping':

$$S_{dis}(f, \theta) = - 1.59\, \bar{\omega}\, k/\bar{k}\, (E\, \bar{k}^2)^2\, F (f, \theta)$$

$$\qquad (4)$$

with $\bar{\omega} = \frac{1}{E} \int F(f, \theta)\, \omega\, df\, d\theta$ and $\bar{k} = \{ \frac{1}{E} \int F (f, \theta)\, k^{1/2}\, df\, d\theta \}^2$

2.4 Bottom dissipation

The bottom dissipation is given by an empirical expression, originally derived from some North Sea swell data, which is used in the Dutch operational wave model GONO (see e.g. the SWIM study, 1985):

$$S_{bot}(f,\theta) = - \frac{C}{g} \frac{k}{\sinh 2kh} F(f,\theta), \qquad (5)$$

where C is a friction coefficient, taken to be 0.038 m^2/sec^3.

A simple, empirical expression is used, rather than a more complicated one like the Hasselmann and Collins (1968) expression for bottom friction, because it is in general hard to say which physical process is responsible for the bottom dissipation during the different stages of wave growth, as bottom conditions change with the changing wave conditions. The general idea that especially the longer waves in the wave spectrum should 'feel' the bottom is well represented by this expression. One of the aims of the present study was to find out if the input of energy to the longer waves by the wave-wave interactions can compensate the energy loss due to the bottom dissipation.

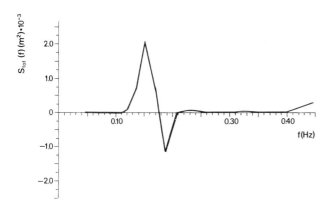

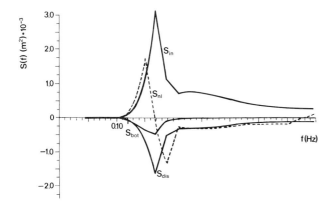

Fig. 1 Total source term S_{tot} and separate source terms S as functions of frequency f (integrated over angles). Depth is 15m, peak frequency is .17 Hz.

3 THE RESULTS

At an early stage of wave growth bottom dissipation can be neglected. The
wave-wave interactions dominate the form of the total source term, the net
input of energy is positive and the growth is on the forward face of the
spectrum (fig.1). As the spectrum continues to grow and to shift to the left,
so that the waves grow longer, the bottom dissipation becomes a sink of energy
on the forward face of the spectrum, which eventually balances the transfer of
energy from the shorter waves to the longer waves by the wave-wave
interactions (fig.2).

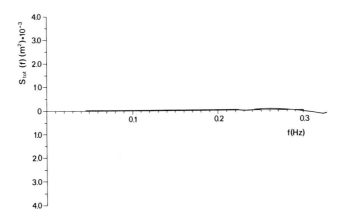

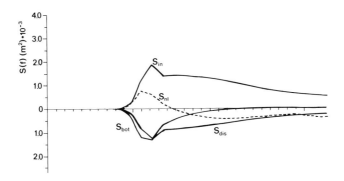

Fig. 2. Total source term S_{tot} and separate source terms S as functions of
frequency f (integrated over angles). Depth is 15 m, peak frequency is .12 Hz.

Growth is halted and the asymptotic values for both the peak frequency and the total energy are depth dependent. In deep water equilibrium is only reached when there is no more input of energy by the wind for waves with frequency around the peak frequency, whereas for shallow water the balance between the non-linear transfer and the bottom term causes the equilibrium to be reached in an earlier stage. So the asymptotic values for the peak frequency are higher for lower water depths (where the bottom dissipation is stronger). As the total amount of energy gained by the waves is clearly reduced by the bottom dissipation, the energy values are lower for lower water depths (fig.3).

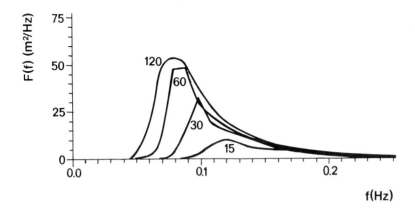

Fig. 3. Fully grown wave energy spectra for depths 15 m (energy E = .8 m^2, peak frequency f_p = .12 Hz), 30 m (E = 1.6 m^2, f_p = .098 Hz), 60 m (E = 2.6 m^2, f_p = .083 Hz) and 120 m (E = 3.0 m^2, f_p = .076 Hz).

4 COMPARING THE RESULTS WITH SWIM

In SWIM three operational shallow water wave forecasting models were intercompared for several cases, the present study corresponds to case I of SWIM (SWIM study, 1985). All three SWIM-models use parametrizations of the non-linear transfer, whereas the present model calculates the non-linear transfer explicitly from expression (3). One SWIM-model (BMO) uses Hasselmann and Collins' expression for the bottom dissipation, one model (GONO) uses (5) and one model (HYPAS) has no explicit bottom dissipation term, but prescribes a depth dependent spectral shape.

Qualitatively the equilibrium conditions of the present model agree best
with the BMO and the GONO results. In BMO and GONO the explicit bottom
dissipation term reduces the total energy, the peak frequency is then
automatically increased through the use of a deep water diagnostic relation
between the energy and the peak frequency. Explicitly calculating the non-
linear interactions also gives depth dependent values for both the total
energy and the peak frequency, but now through the balance between the
interactions and the bottom dissipation on the forward face of the spectrum.

5 ACKNOWLEDGEMENTS

This work was supported by the Netherlands Organisation for the Advancement
of Pure Research (ZWO). The non-linear transfer routines were developed by S.
and K. Hasselmann from the Max Planck Institut für Meteorologie in Hamburg and
were modified for shallow water applications by S.L. Weber. One simulation run
with the explicit non-linear transfer calculations takes about 30-60 minutes
CPU-time on a Cray XMP 48.

6 REFERENCES

Hasselmann, K., 1961. On the non-linear energy transfer in a gravity-wave
    spectrum, part 1: general theory, J. Fluid Mech., 12, 481-500.
Hasselmann, K., Barnett, T.P., Bouws, E., Carlson, H., Cartwright, D.E., Enke,
    K., Ewing, J.A., Gienapp, H., Hasselmann, D.E., Kruseman, P., Meerburg, A.,
    Müller, P., Olbers, D.J., Richter, K., Sell, W. and Walden, H., 1973.
    Measurements of wind-wave growth and swell decay during the Joint North Sea
    Wave Project (JONSWAP), Dtsch. Hydrogr. Z., A8, No.12, 95 pp.
Hasselmann, K. and Collins, J.I., 1968. Spectral dissipation of finite-depth
    gravity waves due to bottom friction, J. Mar. Res. 26, 1-12.
Hasselmann, S. and Hasselmann, K., 1981. A symmetrical method of computing the
    non-linear transfer in a gravity-wave spectrum, Hamburger Geophysikalische
    Einzelschriften, Reihe A: Wiss Abh., 52, 163 pp.
Komen, G.J., Hasselmann, S. and Hasselmann, K., 1984. On the existence of a
    fully developed wind-sea spectrum, J. Phys. Oceanogr., 14, 1271-1285.
Shemdin, O., Hasselmann, K., Hsiao, S.V. and Herterich, K., 1978. Non-linear
    and linear bottom interaction effects in shallow water: Turbulent fluxes
    through the sea surface, wave dynamics and prediction, NATO Conference Ser.
    V, Vol. 1, Plenum Press, 647-665.
Snyder, R.L., Dobson, F.W., Elliot, J.A. and Long, R.B., 1981. Array
    measurements of atmospheric pressure fluctuations above surface gravity
    waves, J. Fluid Mech., 102, 1-59.
The SWIM-group: Bouws, E., Ewing, J.A., Ephraums, J., Francis, P., Günther,
    H., Janssen, P.A.E.M., Komen, G.J., Rosenthal, W. and de Voogt, W.J.P.,
    1985. Shallow water intercomparison of wave prediction models (SWIM),
    Quart. J. Royal Met. Soc. 111 (470), 1087-1115.

Van Vledder, G.Ph. and Weber, S.L., 1987. Guide for the Program EXACT-NL,
    Royal Neth. Met. Inst., unpubl., 23 pp.
Weber, S.L., 1987. The energy balance of finite depth gravity waves,
    calculated numerically with an empirical bottom dissipation term and the
    exact non-linear transfer, accepted for publication in J. Geophys. Res.

INTERNAL TIDAL OSCILLATIONS AND WATER COLUMN INSTABILITY IN THE
UPPER SLOPE REGION OF THE BAY OF BISCAY

R.D. PINGREE
Institute of Oceanographic Sciences, Wormley, Surrey, GU8 5UB

ABSTRACT
    Internal tidal oscillations are examined on the upper slopes of
the Bay of Biscay where the M_2 barotropic currents reach maximum
values. Instabilities in the water column were observed near the
200 m contour in association with the troughs of large amplitude
internal waves propagating on-shelf. At one position near the
500 m contour, the on-shelf and off-shelf tidal flow caused the
isotherms to move vertically by 400 m and strong tidal currents
occurred near the bottom. Estimates of Richardson number
indicate instability in the lower half of the water column. In
water depths of 1000 m and 1500 m, the isotherms at about 800 m
depth were observed to oscillate semidiurnally through a depth
range of 200 m and showed an advance of phase with increasing
depth indicating ray-like propagation of internal energy into the
ocean interior. The mean slope currents showed a tendency for a
down-slope near-bottom Eulerian residual near the 500 m contour
which might result from the strong baroclinic oscillations on the
slopes.

1   INTRODUCTION

    The barotropic tides force internal tides where water depth

changes take place in the presence of stratification (Baines,

1982). The generation of internal tides can be particularly

marked in the shelf-break slope region where changes in topography

are abrupt. Previous studies in the region where the semidiurnal

barotropic tidal currents reach maximum values (75 cms^{-1} at

spring tides) near the shelf-break in the Celtic Sea have shown

that internal tides and waves propagate both oceanward and

on-shelf (Pingree et al., 1986). In summer, on the shelf, the

density structure is characterised by a two layer distribution

with a sharp thermocline separating the upper and lower mixed

layers. The internal tide is then often dominated by an

interfacial mode and two layer models can be used to represent the

on-shelf propagation of long internal tides (Mazé, 1983). The

ocean situation is however more complex since stratification is

present throughout the water column. Higher modes may be excited

and rays may propagate both into the ocean interior and onto the

shelf from critical places on the slopes (New, 1987). In this
paper, observations of internal tidal oscillations in the
shelf-break slope region are presented with some examples showing
evidence of instability and mixing. These results have been
largely drawn from a programme aimed at examining the slope
currents in the region and it is clear that more comprehensive
measurements of internal tidal oscillations on the slopes are
still required. Evidence is also accumulating indicating that the
forcing of internal tides by the barotropic tide affects the
near-bottom residual currents on the upper slopes.

2 ON-SHELF PROPAGATING INTERNAL TIDES AND WAVES

Measurements revealing the structure of the on-shelf
propagating internal tide were made close to the shelf-break near
position A (Fig. 1) in June 1986 from the G.A. Reay whilst
maintaining position alongside a dahn buoy drogued at a depth of
80 m (Fig. 2). Making measurements following a dahn buoy rather
than maintaining a fixed geographical position gives results that
minimise the severe distorting effects of the strong barotropic

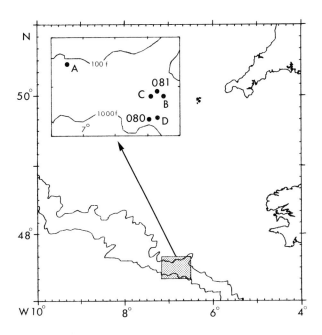

Fig. 1. Map showing slope area and station positions where the
temperature structure in the water column was examined.

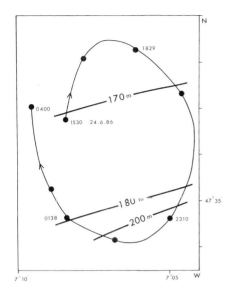

Fig. 2. Track of a dahn buoy
drogued at a depth of 80 m.

tidal currents. The vertical section illustrated in Fig. 3 was
contoured from 100 CTD profiles and shows a deeply penetrating
internal tidal trough where the 13°C contour has been forced from
a minimum depth of 20 m to below 100 m. After the passage of the
wave trough, the isotherms are observed to broaden in the lower
part of the thermocline, suggestive of internal wave mixing and
perhaps also indicating higher modal oscillations, and numerous
temperature inversions indicating instabilities in the water
column occur in this general region. The trailing edge of the
internal tidal trough was associated with the largest vertical
extent over which the water column was unstable, and the
inversions imply overturning on a 20 m scale.

One tidal period later the thermocline descended dramatically
again, with the 12.5°C contour falling from a depth of 25 m to 120
m in 1.5 hours, and vertical profiles of temperature, salinity, σ_t
and chlorophyll 'a' through this internal tidal trough are shown
in Fig. 4. Since the salinity is essentially uniform with depth
in this region the temperature profile mirrors the σ_t profile and
so temperature inversions indicate unstable water. The major
instability at a depth of about 100 m has a vertical scale of 50 m
and the temperature inversion amounts to about 0.1°C. An
instability of this vertical scale is also apparent in the
chlorophyll 'a' profile as would be expected.

The main criticism to the interpretation that inversions of
temperature represent overturning is that near-breaking internal

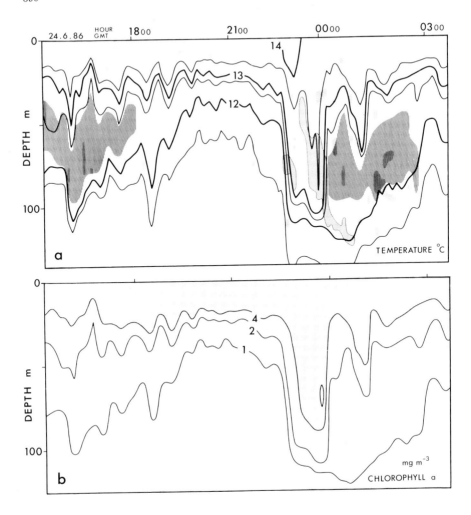

Fig. 3. (a) Isotherm displacement over a tidal cycle following the drogued dahn (Fig. 2) showing large amplitude internal tidal trough at 2300 GMT (24.6.86). The vertical shading indicates a region of reduced vertical stability where the change of temperature was only 0.1°C over the vertical extent indicated. The stippled shading gives a vertical scale for overturning events based on temperature inversions recorded in the individual temperature profiles from which the temperature section was contoured. (b) Corresponding levels of chlorophyll 'a' showing evidence for mixing of phytoplankton associated with the internal tidal trough.

waves may have very steep slopes. Since the profiles are not instantaneous (with typical drop rates of 1 ms^{-1}), the CTD does not in fact measure the vertical profile of a line of fluid particles that is initially vertical. Evidence for mixing was

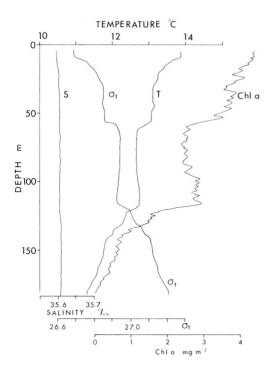

Fig. 4. Profiles of temperature (T), salinity (S), σ_t and chlorophyll 'a' through the internal tidal trough that followed the trough illustrated in Fig. 3. The station position corresponds to A of Fig. 1 and the start time of this station was 1207 GMT 25.6.86.

however also provided by measurements of inorganic nutrients. During the spring tide period from 10 to 25 June 1986 the surface levels of inorganic nitrate increased from 0.2 μM to 1.0 μM in this shelf-break region. This upward flux of inorganic nutrients across the thermocline sustains higher surface levels of chlorophyll 'a' in the shelf-break slope region (Pingree et al., 1982; Le Fèvre, 1986).

Assuming linear theory and a predominantly mode 1 oscillation gives a phase speed of about 55 cms^{-1} and a corresponding wavelength of about 25 km for the internal tide using the density structure corresponding to Fig. 3. The decay scale of the on-shelf propagating internal tide is typically about 2 wavelengths and the on-shelf structure may take on a form which tends to preserve the asymmetric character of more deeply penetrating troughs and more rounded crests. Sometimes however the waveform is a series of solitary waves or solitons (Osborne and Burch, 1980). Evidence for some initial development into a series of secondary waves can be seen in the trailing edge of the tidal trough illustrated in Fig. 3 where a further conspicuous trough penetrating to 50 m is apparent. Assuming an on-shelf

speed of 55 cms^{-1} gives a width of about 2-3 km for the dominant
internal trough shown in Fig. 3.

3 MIXING IN THE GENERATING REGION ON THE UPPER SLOPES

3.1 Temperature oscillations near the 500 m depth contour

Internal tidal oscillations of the water column on the upper
slopes in a water depth of 548 m (mooring position 081, Fig. 1)
showed marked (300 m) temperature oscillations (Fig. 5(a)) as the
tide flows on and off-shelf. These isotherms were contoured from
a thermistor chain covering the depth range 25 m to 75 m and
calibrated thermistors on Aanderaa current meters at depths of 92
m, 148 m, 294 m and 515 m and represent spring tide conditions.
The smoothed contours are the average of 40 individual tides over
six spring tide periods falling between 11.6.84 and 29.8.84. The
maximum oscillation on an individual tide was about 400 m and the
minimum and maximum temperatures recorded 33 m off bottom ranged

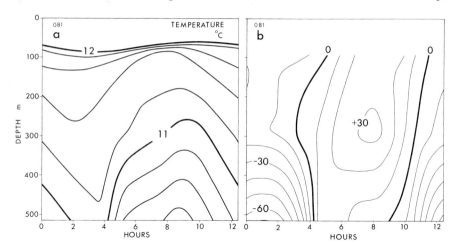

Fig. 5. (a) Mean isotherm displacements against time in hours with
respect to HW Plymouth in a water depth of 548 m at mooring
position 081 shown in Fig. 1. Isotherms above 12°C are not shown.
(b) Corresponding mean contours of up-slope current (direction
020°T) in cms^{-1}. (c) Resulting mean values of gradient Richardson
number between 294 m and 515 m depth showing minimum values about
2 hours after maximum near-bottom down-slope currents.

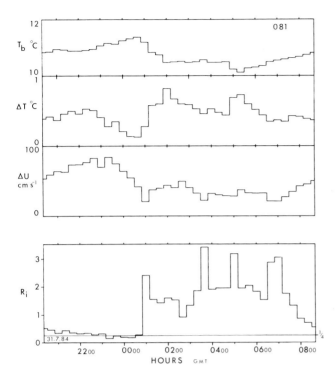

Fig. 6. Temperature 33 m off bottom at 515 m depth, T_b, temperature difference, ΔT, and the modulus of the vector current difference, ΔU, between 294 m and 515 m depth and the corresponding gradient Richardson number, Ri, over a tidal cycle. HW Plymouth corresponds to the start of the record. The maximum bottom current was 95 cms^{-1} at 2240 GMT.

from $9.8^\circ C$ to $11.6^\circ C$ representative of bottom water from about 850 m (Mediterranean water) to water with a temperature that is characteristic of the bottom mixed layer at the shelf-break and these water types are separated horizontally by about 5 to 10 km.

The corresponding component of current in the on-shelf direction ($020^\circ T$), similarly smoothed, is illustrated in Fig. 5(b). Maximum values (70 cms^{-1}) occur near the bottom and maximum and minimum bottom temperatures result from the on-shelf off-shelf tidal flow. The maximum down-slope current recorded 33 m off bottom (515 m depth) was 95 cms^{-1} in a direction $210^\circ T$ at 2240 GMT on 31.7.84.

The gradient Richardson number, Ri, is defined as
$$Ri = [(g/\rho) \Delta\rho / \Delta z]/(\Delta U/ \Delta z)^2 \tag{1}$$
where g is the acceleration due to gravity and $\Delta\rho$ and ΔU are the differences in density and current component occurring over a depth interval of Δz. Values of Ri over the lower half of the water column (i.e. between 294 m and 515 m, Δz = 221 m) are shown against time in Fig. 5(c) and values as low as 0.4 occur as the off-shelf currents decrease and the isotherms descend and broaden.

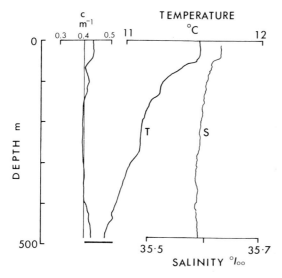

Fig. 7. Temperature, T, salinity, S, and beam attentuation coefficient, C, profiles at position 47° 28'N, 06° 35'W (Station B, Fig. 1) on 3.3.87 showing increases of beam attentuation coefficient near the bottom (bold horizontal line).

Lower values of Ri occur using the unsmoothed data and for the measurements made on 31.7.84 values of Ri < ¼ persist for as long as 1 hour (Fig. 6). Lower values of Ri would be anticipated with a closer spacing of current meters and lower mean values of Ri would be expected below the current meter 33 m off bottom.

Evidence for bottom mixing due to the strong near-bottom currents does not manifest itself as a persistent bottom mixed layer such as that which occurs on the shelf. However beam attenuation coefficient profiles generally show increased levels near the bottom (Fig. 7) indicating active erosion of bottom

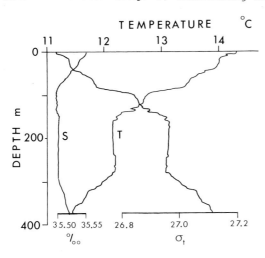

Fig. 8. Temperature, T, salinity, S, and σ_t profiles obtained at position 47° 28'N, 06° 39'W (Station C, Fig. 1) on 13.11.85 showing a region of relatively mixed water extending over a depth interval of 120 m with a temperature inversion of 0.05°C.

sediments by the bottom currents (Dickson and McCave, 1986). An unusual winter temperature profile was recorded during peak spring tides on 13.11.85 at mid-depths. This profile (Fig. 8) showed mixed water over a vertical scale of 120 m at mid-depths. The mixed water may have originated from the shelf-break region or it might have been generated locally. The observations correspond to about 2.5 hours after high water Plymouth when low Richardson numbers are anticipated at this position. The density inversion amounts to about 0.01 σ_t and is suggestive of local mixing.

3.2 Temperature oscillations near the 1000 m and 1500 m depth contour

Hourly temperature profiles were made keeping station on a dahn buoy tethered to the bottom in a water depth of 1050 m at station D (Fig. 1) on 4.3.87 and 5.3.87, 3 days after spring tides. The results (Fig. 9) show a 200 m peak to trough semidiurnal oscillation at about 800-900 depth. There is also a tendency for an advance of phase with increasing water depth with oscillations at 900 m occurring about 2 to 3 hours ahead of the oscillations at 500 m. Increases in beam attenuation coefficient similar to those

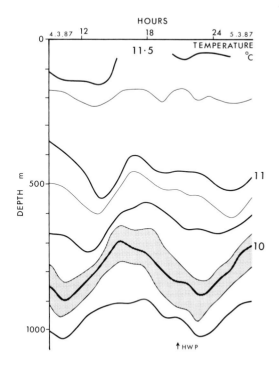

Fig. 9. Isotherm displacements against time (GMT) at position 47° 34.5'N, 7° 05'W (Station D, Fig. 1) in a water depth of 1050 m showing a 200 m peak to trough semidiurnal oscillation at a depth of about 800 m with a tendency for an advance of phase with increasing water depth.

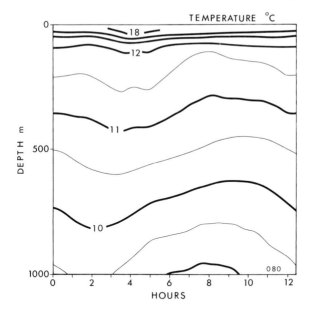

Fig. 10. Mean isotherm displacement over 8 tidal cycles at mooring position 080 (Fig. 1) in a depth of 1505 m against time with respect to HW Plymouth.

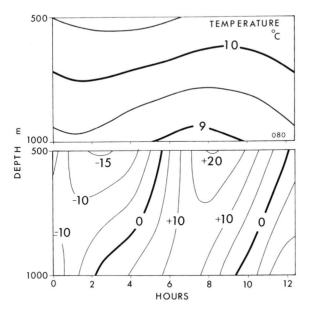

Fig. 11. Smoothed contours of temperature and on-shelf (020°T) component of current (in cms⁻¹) between 500 m and 1000 m depth at mooring 080 showing an advance of phase with increasing depth.

shown in Fig. 7 were also evident at this station.

Mooring 080 placed on the slopes in a depth of 1503 m (Fig. 1) had Aanderaa current meters at depths of 50 m, 106 m, 500 m and 1000 m and a thermistor chain covering the upper 100 m.

Temperature oscillations over a tidal period are illustrated in
Fig. 10 where the isotherms represent the average of 8 tides over
springs starting on 26.8.84. The peak to trough amplitude around
1000 m depth is about 200 m and again the phase at 1000 m is
advanced by about 2 hours with respect to conditions at about 500
m. Smoothed temperature contours representing the average of 43
tides from six spring tide periods between 11.6.84 and 31.8.84 are
shown together with the corresponding on-shelf (direction 020°C)
component of current in Fig. 11.

4 NUMERICAL SIMULATION OF INTERNAL TIDES FROM THE BISCAY SLOPE
REGION

A useful conceptual framework for the interpretation of the
observational programme which inevitably contains much variability
is provided by a model of internal tides by New (1987) using a
modal theory similar to that of Prinsenberg and Rattray (1975).
Fig. 12 gives the contours of minimum Richardson number that were
derived under spring tide conditions with summer stratification and

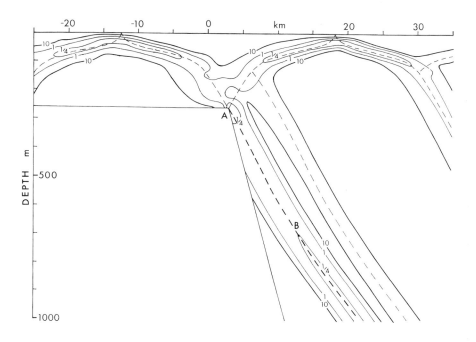

Fig. 12. Contours of minimum Richardson number derived by New
(1987 and 1988) from a linear modal model under spring tide
conditions and with summer stratification. The broken lines
represent rays emanating from the point A at the shelf-break.

shows that Richardson numbers of < $\frac{1}{4}$ are likely to be encountered in this region. The model also shows rays emanating from the point A at the shelf-break along paths which have slope c, given by

$$c = \pm \ [(\sigma^2 - f^2)/(N^2(z) - \sigma^2)]^{\frac{1}{2}} \tag{2}$$

where σ is the semidiurnal tidal frequency, f, is the coriolis parameter and N(z) is the buoyancy frequency at the depth z.

For the summer situation with $N = 2.2 \times 10^{-3} s^{-1}$ over the upper slopes the internal wave characteristic, c = 0.04, matches the bottom slope at a depth of about 300-400 m. Below this depth the bottom slope gradient is greater than the characteristic slope to depths of about 3000 m. A beam of internal energy propagating downwards from 300-400 m depth along the path B (Fig. 12) would be at a depth of about 800 m at position 080 where large amplitude temperature oscillations were encountered (Fig. 10). In the winter situation the buoyancy frequency is reduced over the upper slopes and the generating region where c matches the bottom slopes moves to deeper depths where the bottom slope is steeper. For the winter situation (Station D, Fig. 9) with $N = 1.3 \times 10^{-3} s^{-1}$ over the upper slopes c matches the bottom slope at a depth of about 500 m, with c = 0.07. A beam of internal energy propagating downwards from these depths would arrive at station D (a distance of about 9 km from the 500 m contour near station C) at a depth of about 850 m using the buoyancy frequency derived from the results of this station. The downward beam is consistent with the advance of phase with respect to depth since the wavenumber vector is directed upwards across the beam at right angles.

5   CURRENT STRUCTURE ALONG THE UPPER SLOPES

The tidal analysis for the current meter data from mooring 081 is presented in Table 1. Of particular interest is the record from 33 m off the bottom which shows increased semidiurnal currents near the bottom. This effect cannot readily be attributed to internal tidal oscillations and possibly bottom-trapped waves (Rhines, 1970) phasing favourably with the barotropic currents (Huthnance and Baines, 1982) since the near rectilinear nature of the tidal current is not satisfactorily explained. Local undetermined along-slope topographic variations are likely to be important. The importance of nonlinear aspects is indicated by the relatively large values of the MSf and quarter diurnal components (Petrie, 1975), the latter have an internal wave characteristic c = 0.1

TABLE 1

Tidal analysis of the current records on Mooring 081 (depth 548 m, position 47°29'N, 06°37'W). All records start on 7.6.84 and are 90 days long.

Depth off bottom (m)	Tidal Constituent	Semi-major axis (cm s^{-1})	Ratio of semi-minor to semi-major axis (+ anti-clockwise)	Phase °$_g$	Orientation °$_T$
	Residual	6.5	–	–	314
456	MSf	1.7	+0.43	272	154
	μ_2+2MS$_2$	2.0	−0.78	188	167
	M$_2$	18.8	−0.44	55	023
	S$_2$	8.1	−0.39	80	015
	M$_4$	2.9	−0.28	101	019
	MS$_4$	1.7	−0.75	339	177
	Residual	6.9	–	–	327
400	MSf	2.6	+0.58	266	158
	μ_2+2MS$_2$	1.0	−0.58	132	056
	M$_2$	17.3	−0.48	44	021
	S$_2$	6.5	−0.42	59	006
	M$_4$	2.1	−0.29	107	008
	MS$_4$	1.7	−0.65	268	087
	Residual	8.0	–	–	347
254	MSf	2.6	+0.40	54	021
	μ_2+2MS$_2$	1.5	+0.15	330	113
	M$_2$	18.2	−0.64	30	033
	S$_2$	7.8	−0.86	38	002
	M$_4$	2.7	−0.19	111	047
	MS$_4$	2.1	−0.54	188	044
	Residual	15.0	–	–	210
33	MSf	7.5	−0.10	246	023
	μ_2+2MS$_2$	4.1	−0.17	3	026
	M$_2$	31.8	−0.13	14	042
	S$_2$	14.9	−0.11	77	039
	M$_4$	9.0	−0.84	27	176
	MS$_4$	6.6	−0.90	6	069

which may locally match the bottom slope (Gordon, 1979) near this position. The MSf phases with the M$_2$ and S$_2$ currents to result in a stronger down-slope current component at spring tides and the down-slope residual current of 15 cms^{-1} may also be of tidal origin.

The residual currents from the moorings 080 and 081 are shown in Fig. 13 and further residuals from other moorings along the slopes are presented in Table 2 and Fig. 14. The mean slope current at mid-depth near the 500 m contour is about 6 cms^{-1} and the transport along the upper 1000 m of the slopes is estimated at

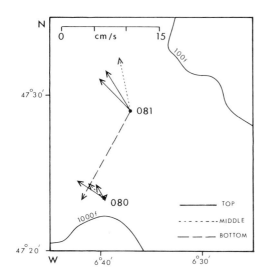

Fig. 13. Residual currents at mooring positions 081 and 080 showing the slope current and the marked near-bottom down-slope residual current at position 081.

about 0.4 x 10^6 m^3s^{-1} which is rather less than the corresponding transport along the upper part of the Hebridean slopes (about 1.5 x 10^6 m^3s^{-1} at 58°N) (Huthnance, 1986).

A noteworthy feature of the time averaged flow is the tendency for a down-slope component of the near-bottom currents with respect to conditions at mid-depth which is particularly marked at moorings 081 and 095. The surface currents also show the same tendency but to a lesser extent and exhibit more variability with shorter time scale averaging of the flow. Wunsch (1971) showed that Eulerian residual currents of this form could be expected for internal waves propagating into a shoaling region. The down-slope residuals measured here might result from compensating transports due to Stokes and Euler generated near the bottom in the general region where the internal tides are forced or result from near-bottom internal tidal surges. In an interesting experimental study Thorpe and Haines (Thorpe, 1987) showed that up-slope motions are likely to produce more boundary mixing than down-slope motions. Differences in up- and down-slope currents may result in Eulerian residual currents and these effects will be magnified where the near bottom semidiurnal oscillations are large. In this connection it is worth noting that the down-slope tendency is perhaps relatively less marked at moorings 093 and 094. At both these

TABLE 2
Residual currents from the slope moorings.

Mooring No and position	Water Depth (m)	Height of current meter off bottom (m)	Start date (length of record in days)	Residual current	
				Speed cms^{-1}	Direction $^{\circ}$T
067 47° 31.7'N 06° 36.7'W	206	116	20.6.82 (25)	10.6	278
080 47° 23.35'N 06° 39.6'W	1503	1453	7.6.84 (90)	4.9	303
		1397	7.6.84 (90)	2.4	329
		1003	7.6.84 (90)	3.2	313
		503	7.6.84 (90)	0.4	040
081 47° 29.0'N 06° 37.0'W	548	456	7.6.84 (90)	6.5	314
		400	7.6.84 (90)	6.9	327
		254	7.6.84 (90)	8.0	347
		33	7.6.84 (90)	15.0	210
085 47° 30.0'N 06° 43.8'W	456	348	9.6.84 (88)	7.3	290
		195	9.6.84 (88)	8.2	292
092 47° 43.7'N 07° 34.7'W	505	464	10.9.85 (62)	8.8	337
		237	10.9.85 (62)	3.3	005
		38	10.9.85 (62)	7.8	298
093 47° 32.9'N 07° 16.7'W	505	464	20.9.85 (52)	13.8	291
		39	20.9.85 (52)	2.3	290
		6	20.9.85 (52)	4.1	273
094 47° 29.4'N 07° 20.1'W	980	930	10.9.85 (62)	7.9	283
		470	10.9.85 (62)	4.7	301
		94	10.9.85 (62)	2.0	211
		7	10.9.85 (62)	1.8	274
095 47° 28.2'N 06° 40.4'W	475	446	16.10.85 (15)	8.0	310
		226	10.9.85 (62)	7.2	340
		7	10.9.85 (12)	15.4	208
096 47° 01.0'N 05° 26.8'W	495	463	10.9.85 (62)	4.0	293
		235	10.9.85 (62)	3.3	264
		7	10.9.85 (15)	4.7	225

locations the near bottom semidiurnal currents were remarkably reduced. At mooring 093 the M_2 currents at 6 m and 39 m off the bottom were 3 cms^{-1} and 5 cms^{-1} respectively and at mooring 094 the M_2 currents were only 1 cms^{-1} and 2 cms^{-1} at 7 m and 94 m off bottom. By contrast at mooring 081 the M_2 currents were 32 cms^{-1} at 33 m off bottom (Table 1) and 26 cms^{-1} at 38 m off bottom at mooring 092.

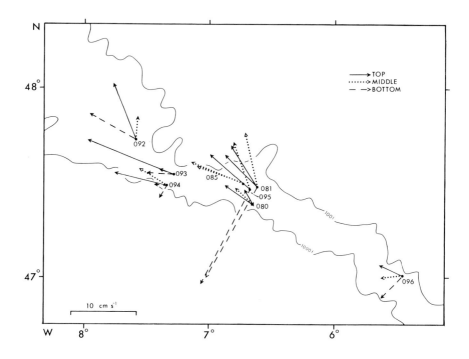

Fig. 14. Residual currents from the slope moorings showing the along-slope residual currents and the down-slope tendency for the near-bottom currents.

Additional down-slope contributions may arise from Ekman transports in the slope current boundary layer. Alternatively, as Wunsch (1970) has pointed out, no state of rest is possible in a stratified fluid with sloping walls with the condition of no flux of heat or salt at the sloping boundary and mixing on the bottom may induce local down-slope density driven circulations. Clearly further studies are required.

6   SUMMARY

Internal tidal oscillations are examined within the water column over the upper slopes of the Bay of Biscay in depths ranging from 170 m to 1503 m at spring tides in the region where the semidiurnal currents reach maximum values. At the shelf-break (200 m) the internal tide may manifest itself as a marked depression (60 m) of the thermocline. In late June near the shelf-break, this

narrow on-shelf propagating internal tidal trough has a width of
about 2-3 km and exhibits instabilities over vertical scales of
about 20 m. Relatively mixed water appeared at the depth of the
thermocline after the passage of this wave. In a water depth of
548 m the isotherms at mid-depth at one site moved semidiurnally
through a depth interval of 400 m and current measurements 33 m
off the bottom showed a marked intensification of the tidal
currents in the down-slope direction. The maximum down-slope
current was 95 cms^{-1} and the Eulerian mean down-slope current was
15 cms^{-1}. The Richardson number over the lower half of the water
column approached the value of $\frac{1}{4}$. Near the 1000 m and 1500 m
contours in the same general region isotherm oscillations near
the bottom and at mid-depths showed peak to trough amplitudes of
200 m with an advance of phase with increasing depth indicating
internal tidal energy propagating downwards into the ocean
interior. Current measurements made at other sites near the
500 m contour also showed a tendency for a down-slope Eulerian
residual current.

7 REFERENCES

Baines, P.G., 1982. On internal tide generation models. Deep-Sea
  Res., 29: 307-338.
Dickson, R.R. and McCave, I.N., 1986. Nepheloid layers on the
  continental slope west of Porcupine Bank. Deep-Sea Res., 33:
  791-818.
Le Fèvre, J., 1986. Aspects of the Biology of Frontal Systems.
  Adv. Mar. Biol., 23: 163-299.
Gordon, R.L., 1979. Tidal interactions in a region of large
  bottom slope near northwest Africa during JOINT-1. Deep-Sea
  Res., 26: 199-210.
Huthnance, J.M. and Baines, P.G., 1982. Tidal currents in the
  northwest African upwelling region. Deep-Sea Res., 29:
  285-306.
Huthnance, J.M., 1986. The Rockall slope current and shelf-edge
  processes. Proc. R. Soc. Edinb., 88B: 83-101.
Mazé, R., 1983. Movements Internes induits dans un Golfe par le
  Passage d'une Dépression et par la Marée. Application au
  Golfe de Gascogne. Thèse Docteur. L'Université de Bretagne
  Occidentale, 320 pp.
New, A.L., 1987. Internal tidal currents in the Bay of Biscay.
  Proceedings of the Society for Underwater Technology. In:
  Modelling the offshore environment. Advances in Underwater
  Technology, Graham and Trotman Ltd.
New, A.L., 1988. Internal tidal mixing in the Bay of Biscay.
  Deep-Sea Res. (submitted).
Osborne, A.R. and Burch, T.L., 1980. Internal solitons in the
  Andaman Sea, Science, 208: 451-460.
Petrie, B., 1975. M_2 surface and internal tides on the Scotian
  shelf and slope. J. Mar. Res., 33: 303-323.

404

Pingree, R.D., Mardell, G.T., Holligan, P.M., Griffiths, D.K. and
    Smithers, J., 1982. Celtic Sea and Armorican current
    structure and the vertical distributions of temperature and
    chlorophyll. Contin. Shelf. Res., 1: 99-116.
Pingree, R.D., Mardell, G.T. and New, A.L., 1986. Propagation of
    internal tides from the upper slopes of the Bay of Biscay.
    Nature, 321: 154-158.
Prinsenberg, S.J. and Rattray, M., 1975. Effects of continental
    slope and variable Brunt-Vaisala frequency on the coastal
    generation of internal tides. Deep-Sea Res., 22: 251-263.
Rhines, P., 1970. Edge-, Bottom-, and Rossby waves in a rotating
    stratified fluid. Geophys. Fluid Dyn., 1: 273-302.
Thorpe, S.A., 1987. On the reflection of a train of
    finite-amplitude internal waves from a uniform slope. J.
    Fluid Mech., 178: 279-302.
Wunsch, C., 1970. On oceanic boundary mixing. Deep-Sea Res.,
    17: 293-301.
Wunsch, C., 1971. Note on some Reynolds stress effects of internal
    waves on slopes. Deep-Sea Res., 18: 583-591.

ESTIMATES OF VERTICAL MIXING BY INTERNAL WAVES REFLECTED OFF A SLOPING BOTTOM

CHRISTOPHER GARRETT and DENIS GILBERT

Department of Oceanography, Dalhousie University, Halifax, NS, B3H 4J1 Canada

ABSTRACT

Eriksen (1982, 1985) has drawn attention to the increased vertical shear due to internal wave reflection off a sloping bottom. For a typical incident spectrum we calculate a cut–off wavenumber for the reflected spectrum such that the shear from all lower wavenumbers gives a Richardson number of order 1, and we assume that the energy flux associated with higher wavenumbers is lost to dissipation and mixing. Our results appear to be significant for deep ocean mixing rates and the energy balance of the internal wave field in the ocean but the theory is presently limited by weak assumptions.

1 INTRODUCTION

Vertical, or diapycnal, mixing rates in the ocean can often be inferred from fitting observed, smoothed, distributions of ocean properties to the predictions of a model that involves some kind of advective-diffusive balance. A classical example is the simple fit by Munk (1966) of the vertical profiles of temperature, salinity and Carbon-14 in the Pacific to a simple one-dimensional model in which vertical diffusion is matched by upwelling. He obtained a value of $1.3 \times 10^{-4} m^2 s^{-1}$ for the vertical eddy diffusivity.

Similar, or somewhat larger values, have been found in more recent studies in which the assumptions (such as the neglect of lateral processes) are less severe. For example, Hogg et al (1982) measured the flux of Antarctic Bottom Water through the Vema Channel into the Brazil Basin where water with a temperature less than $1°C$ vanishes. Assuming a steady state, they attribute this to a diapycnal mixing rate of 3 to $4 \times 10^{-4} m^2 s^{-1}$. More recently Saunders (1987) has applied the same technique in the Madeira basin of the abyssal northeast Atlantic and found a diapycnal mixing rate of 1.5 to $4 \times 10^{-4} m^2 s^{-1}$.

On the other hand, values of the diapycnal diffusivity obtained directly from microstructure measurements of temperature and velocity in the thermocline have suggested values of K_v considerably less than $10^{-4} m^2 s^{-1}$ (e.g. Gregg, 1987). However, Gargett (1984) and Moum and Osborn (1986) have pointed out that microstructure measurements suggest (though not yet conclusively) an increase of K_v with depth, achieving values of the order of $10^{-4} m^2 s^{-1}$ or more in the deep ocean.

There is also the possibility, suggested by Munk (1966) and advocated more recently by Armi (1978, 1979) that an effective diapycnal mixing for the ocean arises from vigorous mixing at ocean boundaries followed by advection and stirring along isopycnals into the ocean interior. The energetic requirements of this may be discussed with reference to the schematic ocean shown in Figure 1 in which we consider the area of the bottom $A_{boundary}(z, \delta z)$ exposed

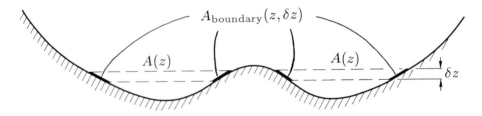

Figure 1. Schematic of an ocean basin in which boundary mixing leads to an effective vertical diffusivity in the interior.

to water between two isopycnals which, for simplicity, are assumed to be flat and lie at depths z and $z + \delta z$. If F_d is the average energy dissipation rate per unit area at this boundary, the total dissipation rate is $F_d A_{\text{boundary}}(z, \delta z)$, and we note that A_{boundary} includes the sides of topographic features such as seamounts as well as the sloping sides of ocean basins.

Meanwhile, if K_v is the effective diapycnal mixing rate in the ocean interior, with Väisälä frequency N, the rate of gain of potential energy there is $\rho K_v N^2$ per unit volume, or $\rho K_v N^2 A(z) \delta z$ for the volume of thickness δz and surface area $A(z)$. Thus if Γ is the mixing efficiency, i.e. the fraction of energy dissipated at the boundaries which is converted to interior potential energy, we have

$$K_v = \frac{A_{\text{boundary}}(z, \delta z)}{A(z)\delta z} \frac{\Gamma F_d}{\rho N^2}. \tag{1.1}$$

Armi (1979) quotes values for the ratio $A_{\text{boundary}}(z, \delta z)[A(z)\delta z]^{-1}$ for various depth ranges and different oceans. It averages to $2.2 \times 10^{-4} m^{-1}$ for the 3 to 4 km depth range and $3.5 \times 10^{-4} m^{-1}$ for 4 to 5 km. Taking $N^2 \simeq 10^{-6} s^{-2}$ as typical of the deep ocean, we see that $K_v \simeq 10^{-4} m^2 s^{-1}$ in the ocean interior requires $\Gamma F_d \simeq 0.3 \, mW.m^{-2}$.

In the stratified ocean interior Osborn (1980) and Oakey (1982) find values of Γ of about 0.2, but it seems unlikely that the efficiency of turbulence in the bottom boundary layer will be that great; even with an efficient exchange mechanism to eject mixed water from the boundary layer and replace it with stratified water from the interior, much of the dissipation probably occurs in water that is already mixed. Nonetheless, if Γ is as big as 0.1, the required dissipation rate is $3 \, mW.m^{-2}$ and equals $\rho C_D U^3$ for a drag coefficient $C_D = 3 \times 10^{-3}$ and an average bottom current $U \simeq 0.1 \, ms^{-1}$. Sub-inertial currents are typically much less than this (Dickson, 1983) as are internal waves and tidal currents over a flat bottom. Indeed D'Asaro (1982) found the bottom dissipation of internal wave energy on the flat Hatteras Abyssal Plain to be much less than $1 \, mW.m^{-2}$.

However, Eriksen (1982) drew attention to the intensification of internal waves associated with their reflection off a sloping bottom. This raises the possibility not only of increased dissipation within the benthic boundary layer, but also of shear instability and mixing in the stratified region above the bottom mixed layer, thus avoiding the problem that further mixing of water that is already homogenized can do nothing to change the interior stratification.

The purpose of this paper is to explore Eriksen's hypothesis further. In Section 2 we review the properties of internal wave reflection off a slope and discuss Eriksen's (1985) estimate of the amount of mixing that this might produce. In Section 3 we introduce an alternative hypothesis which is evaluated in Sections 4, 5 and 6. The paper concludes in Section 7 with a discussion of further work that appears to be necessary before a definitive assessment can be given of the global significance of near-boundary mixing by reflected internal waves.

2 INTERNAL WAVE REFLECTION

The magnitude of the angle to the vertical subtended by the group velocity of internal waves is uniquely determined by the frequency of the waves. Hence, as waves reflect off a fixed sloping surface, maintaining their frequency, the wave rays maintain the magnitude of this angle, rather than the angle they make with the normal to the surface. This generally leads to a change in the wavelength of the waves and, in order to preserve the energy flux normal to the surface, the wave amplitude is also changed.

In the two-dimensional situation where the direction of wave propagation is normal to the isobaths, the properties of this reflection process are easily evaluated (e.g. Phillips, 1977). Particularly large increases in wavenumber and amplitude of the reflected waves with respect to the incident waves occur if the latter propagate from deeper water and have a frequency close to the "critical frequency" at which the slope of the wave rays is equal to the bottom slope.

Eriksen (1982) has investigated the changes in wavenumber and amplitude that occur if the direction of the group velocity of the incident waves is arbitrary. Changes in wavenumber are calculable from conservation of wave frequency and of the wavenumber components on the sloping boundary. Changes in amplitude can be evaluated from the condition of zero normal flow at the boundary or, equivalently, from the requirement that the energy flux normal to the boundary be equal and opposite for the incident and reflected waves. Eriksen showed that

$$E_r/E_i = (m_r/m_i)^2 \qquad (2.1)$$

for arbitrary direction of incidence, where E, m are energy density and vertical wavenumber respectively and the subscripts i, r denote the incident and reflected waves.

The details of the results depend on the two parameters

(i) ϕ_i, the azimuth of the incident waves, defined as the angle between the horizontal component of their wavenumber or group velocity vector and the onshore normal to the isobaths (i.e. $\phi_i = 0$ for waves incident normal to the isobaths from deeper water).

(ii) $a = \tan \alpha \tan \theta_i$, where $\tan \alpha$ is the bottom slope (i.e. α is the angle between the bottom and the horizontal) and θ_i (with a range from $-\pi/2$ to $\pi/2$) is the angle to the horizontal made by the wavenumber vector. The wave frequency σ is related to θ_i by

$$\sigma^2 = N^2 \cos^2 \theta_i + f^2 \sin^2 \theta_i, \qquad \tan^2 \theta_i = (N^2 - \sigma^2)/(\sigma^2 - f^2) \qquad (2.2)$$

where N, f are the Väisälä and Coriolis frequencies. The group velocity vector is at right angles to the wavenumber vector and in the same vertical plane so that waves at the critical frequency have $\theta_i = \theta_c = \pi/2 - \alpha$ and frequency

$$\sigma_c = (N^2 \sin^2 \alpha + f^2 \cos^2 \alpha)^{1/2} \tag{2.3}$$

for which $|a| = 1$.

Incident waves with $\sigma > \sigma_c$ have $0 < a < 1$ (as $0 \leq \theta_i < \theta_c$ for the energy flux to be incident and for $\sigma > \sigma_c$) and ϕ_i can range from $-\pi$ to $+\pi$. For $\sigma < \sigma_c$ wave rays are less steep than the slope, so that the waves may impinge on the slope either from above with $a > 1$ or from below with $a < -1$. For both upward and downward subcritical ($\sigma < \sigma_c$) incidence a range of azimuths is excluded by geometry, but the sum of the two permitted ranges is 2π.

The situation is summarized in Figure 2, which also shows contours of $|m_r/m_i|$, which Eriksen (1982) showed is given by a formula equivalent to

$$m_r/m_i = (a^2 + 2Ca + 1)/(a^2 - 1) \tag{2.4}$$

with $C = \cos \phi_i$. It is clear that although there are some areas of (a, ϕ_i) space in which the energy density is decreased on reflection, the dominant tendency is for an increase. This is especially true near the critical frequency, although the frequency bandwidth for a particular amplification is less for incident azimuths ϕ_i away from zero.

Eriksen (1982) found evidence in some near bottom current meter data for an increase in the spectral energy density near the critical frequency. He also found evidence for the currents to become more aligned across isobaths near the critical frequency, as predicted by the kinematics of the reflection process, but this will not concern us further here.

The oceanographically important aspect of these changes in internal wave properties is the enhancement of the vertical shear of horizontal current, by a factor $(m_r^2 E_r)/(m_i^2 E_i) = (m_r/m_i)^4$. In fact Eriksen (1982) found evidence in one deep ocean mooring for low Richardson numbers (and hence a presumption of shear instability) over a vertical scale of up to 40 or $50m$. This suggested that internal waves breaking near the sea floor, after enhancement of their vertical shear by reflection, might be responsible for a significant amount of near-boundary mixing in the ocean, and hence of diapycnal transport averaged over ocean basins.

Eriksen (1985) pursued this hypothesis further in a later paper. Starting from a horizontally isotropic incident spectrum with a typical distribution of energy over different frequencies σ and vertical wavenumbers m_i, he computed the details of the energy spectrum of the reflected waves in (σ, m_r, ϕ_r) space. He then computed the "redistributed" bottom–normal energy flux in (σ, m_r) space, defining this as the integral over the incident azimuth of the *modulus* of the difference between the calculated reflected flux (equal to the incident flux at (σ, m_i, ϕ_i)) and what the flux would be at (σ, m_r, ϕ_r) for the typical deep-sea internal wave spectrum. The reason for doing this was that observations (Eriksen, 1982) suggest that

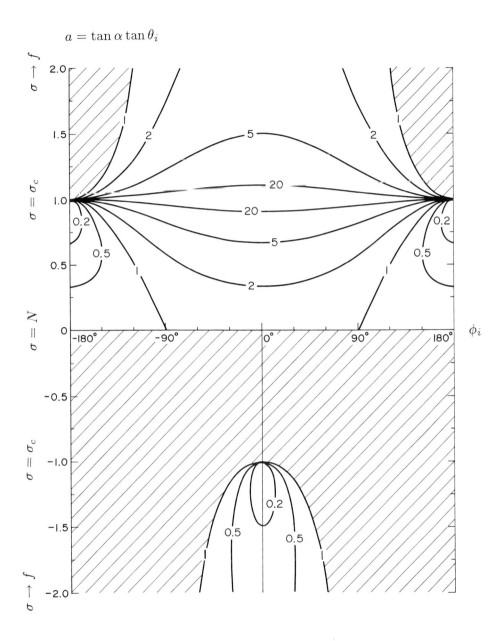

Figure 2. Wavenumber amplification, $|m_r/m_i|$, as a function of the frequency–dependent parameter a and incident azimuth ϕ_i. The regions of (a, ϕ_i) space for which incident waves are impossible are hatched. The vertical component of group velocity is down for $a > 0$ and up for $a < -1$.

the internal wave spectrum relaxes to a more typical form, without significant enhancement or horizontal anisotropy near σ_c, within about $100\,m$ of a sloping bottom.

Eriksen (1985) shows that for typical bottom slopes the redistributed energy flux is of the same order of magnitude, about $30\,mW.m^{-2}$, as the energy flux in the incident spectrum. He points out that a small fraction of this could produce mixing at a rate equivalent to a very significant basin-wide average diapycnal diffusivity.

This argument certainly draws attention to the potential importance for ocean mixing of the shear instability of internal waves reflected off a sloping sea floor. However, there are clearly difficulties in deciding how much of the reflected energy flux might be available for mixing rather than just being transferred by nonlinear interactions to other frequencies, wavenumbers and azimuths as the reflected spectrum recovers to a more horizontally isotropic and otherwise typical form. Moreover, it seems very unlikely that the energy flux lost from the internal wave spectrum can be anything like as much as $30\,mW.m^{-2}$, as this would drain the typical total internal wave energy of about $4 \times 10^3\,J.m^{-2}$ in a mere $1\frac{1}{2}$ days and lead to much less universality of the internal wave spectrum in time and space than seems to be typical (e.g. Olbers, 1983).

The purpose of this paper is to explore a more concrete hypothesis for the amount of energy flux that might be available for mixing and to evaluate its dependence on characteristics of the incident spectrum and on environmental parameters such as f, N and the bottom slope $\tan \alpha$.

3 A SCENARIO FOR WAVE BREAKING

It seems reasonable to consider an incident wave spectrum that has its energy distributed in a typical manner in (frequency, wavenumber) space. On reflection from a sloping bottom the mean square shear of a particular wave component of the spectrum changes by a factor $(m_r/m_i)^4$, which is generally considerably greater than 1 (Figure 2). Hence the mean square shear for the whole wave field is likely to increase considerably and, as suggested by Eriksen (1985), shear instability and mixing are likely. What we seek is estimates of the vertical scale on which mixing might occur and, more importantly, the rate of mixing.

The procedure to be followed in this paper is as follows:

(i) the (σ, m_r) shear spectrum is calculated for a typical incident spectrum.

(ii) the total shear spectrum of the incident and reflected waves (which is dominated by the reflected waves) is integrated over the vertical wavenumber m from low values up to some wavenumber m_p such that the Richardson number based on the mean square shear from this part of the spectrum has some critical value Ri_c (of order 1, to be discussed later).

(iii) waves with $m > m_p$ are then likely to undergo shear instability as they propagate relative to the waves with $m \leq m_p$, so that their associated energy flux can be regarded as the energy flux available for mixing. When multiplied by an appropriate efficiency factor, this gives an estimate of the rate of increase of potential energy of the basic state and hence an estimate of the diapycnal eddy diffusivity K_v.

The way in which this procedure operates is made clearer if we consider the energy in the incident spectrum to be distributed in frequency from f to N, but confined to one vertical wavenumber. On reflection the frequency of each wave in this spectrum is conserved, but the reflected wavenumber at each frequency is spread over a wide range, with the precise value depending on the incident azimuth (Figure 3). Integrating the associated shear spectrum produces the critical shear by wavenumber m_p, but it is clear that there are a significant number of waves, particularly those with frequency near σ_c and small azimuth ϕ_i, which are reflected with a higher wavenumber. It is the energy flux associated with these which is assumed to be available for mixing.

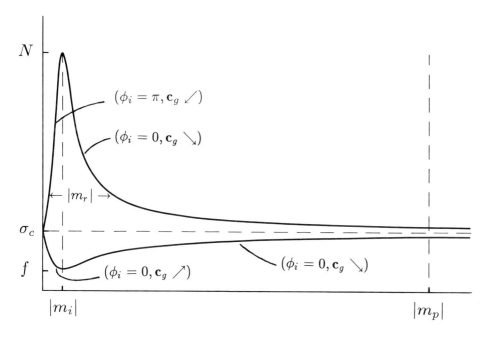

Figure 3. Schematic of the range of reflected vertical wavenumber magnitude $|m_r|$ for a given magnitude $|m_i|$ of the incident vertical wavenumber. At each frequency the range is bounded by cases with incident azimuth, ϕ_i, and direction of the vertical component of the group velocity, as shown.

412

It is thus clear that the energy flux available for mixing is distributed in a narrow frequency band around σ_c (Figure 4). The shear which leads to the critical Richardson number is also localised in frequency near σ_c (Figure 4), though with a reduced contribution very close to σ_c where the amplification is so large that, for most incident azimuths, the waves are reflected with $m_r > m_p$ and so do not contribute to the integrated shear.

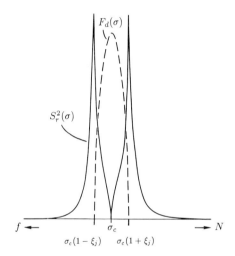

Figure 4. For a single incident vertical wavenumber, the area under the dashed line schematically represents the energy flux available for mixing, whereas the area under the solid line represents the contribution to the mean square shear of waves reflected with $|m_r| \leq m_p$.

If we now include another, higher, wavenumber in the incident spectrum, it will experience the same amplification on reflection, for a given σ and ϕ_i, as the first wavenumber. Thus there are increases in the ranges of ϕ_i and σ for which the reflected wavenumber m_r is greater than some cutoff value m_p, so that the frequency bandwidths shown in Figure 4 increase.

Of course adding more wavenumbers to the incident spectrum will affect the values of m_p and the energy flux available to mixing. The calculations involved are reasonably straightforward and will be described in the next section.

4 DETERMINATION OF THE CUTOFF WAVENUMBER

Following Eriksen (1985) we shall use Munk's (1981) representation of a typical horizontally isotropic internal wave spectrum. The energy density spectrum is $b^2 N_0 N E(\sigma, j)$ as a function of frequency σ and vertical mode number j, which we use for convenience instead of the vertical wavenumber m to which it is related by

$$m = \frac{j\pi}{b} \left(\frac{N^2 - \sigma^2}{N_0^2 - \sigma^2} \right)^{1/2}, \tag{4.1}$$

with $N_0 = 3\ cph = 5.24 \times 10^{-3} s^{-1}$ and $b = 1.3\ km$. The total energy per unit mass is $b^2 N_0 N \int_f^N \sum_{j=1}^{\infty} E(\sigma, j) d\sigma$. At this stage we need make no further assumptions about the form of $E(\sigma, j)$.

Only half of this spectrum can be incident on a sloping boundary; for $\sigma > \sigma_c$ the energy is isotropically distributed in azimuth, whereas for $\sigma < \sigma_c$ there are forbidden azimuths for both upward and downward propagating waves although the total range of angles permitted is still 2π. In (σ, j, ϕ_i) space the energy spectrum is thus $(4\pi)^{-1} b^2 N_0 N E(\sigma, j)$. Converting this to a spectrum of horizontal current introduces a factor $(1 + f^2/\sigma^2)$ and conversion to the spectrum of horizontal shear $(\overline{(\partial u/\partial z)^2} + \overline{(\partial v/\partial z)^2})$ brings in a further factor m_i^2, with m_i given by m in (4.1).

On reflection the wavenumber is changed by the factor in (2.4) and the shear of a single wave component of the spectrum by a factor $(m_r/m_i)^4$. For a given j of the incident waves, there exists a range of frequencies, say σ_{j1} to σ_{j2}, for which waves with incident azimuth ϕ_i in the range $-\phi_c(\sigma, j) < \phi < \phi_c(\sigma, j)$ have $|m_r| > m_p$ and so do not contribute to the mean square shear for waves with $|m_r| \leq m_p$. Outside this frequency range the amplification is less and all azimuths contribute (Figure 4), but at a rapidly decreasing rate away from σ_c due to the reduced amplification that is possible. The mean square shear of the reflected waves may thus be written

$$S_r^2 = (4\pi)^{-1} b^2 N_0 N (1 + f^2/\sigma^2) \sum_{j=1}^{\infty} m_i^2 (I_1 + I_2 + I_3) \tag{4.3}$$

where

$$I_1 = \int_f^{\sigma_{j1}} d\sigma \int_{-\pi}^{\pi} (m_r/m_i)^4 E(\sigma, j) d\phi \tag{4.4}$$

$$I_2 = \int_{\sigma_{j1}}^{\sigma_{j2}} d\sigma \left\{ \int_{-\pi}^{-\phi_c} + \int_{\phi_c}^{\pi} (m_r/m_i)^4 E(\sigma, j) d\phi \right\} \tag{4.5}$$

$$I_3 = \int_{\sigma_{j2}}^{N} d\sigma \int_{-\pi}^{\pi} (m_r/m_i)^4 E(\sigma, j) d\phi. \tag{4.6}$$

This development ignores the existence of forbidden azimuths for $\sigma < \sigma_c$. However, a permitted wave with azimuth ϕ propagating upwards has the same amplification as one with azimuth $\pi - \phi$ propagating downwards (as both a and C change sign in (2.4)) so that in calculating S_r^2 it is fair to ignore the forbidden zones and proceed as if all the waves were downward propagating ($a > 0$).

In principle we may now evaluate S_r^2 as a function of m_p, for particular choices of $E(\sigma, j)$, f, N and the bottom slope, by numerical evaluation of the above integrals. However, the singularity in $|m_r/m_i|$ at the critical frequency σ_c and the factor $(m_r/m_i)^4$ suggest that σ_{j1} and σ_{j2} are close to σ_c and that the integrals are dominated by contributions close to σ_c.

In that case we replace $E(\sigma, j)$ by $E(\sigma_c, j)$ in the integrals and approximate the wavenumber amplification by

$$|m_r/m_i| = \xi^{-1}(1 - f^2/\sigma_c^2)(1 + C) \qquad (4.7)$$

where $\sigma = \sigma_c(1 + \xi)$ and we have also approximated $N^2 - f^2$ and $N^2 - \sigma_c^2$ by N^2 (i.e. assumed $N^2 \gg \sigma_c^2, f^2$). This latter approximation also simplifies (4.1) to

$$m = j\pi N/(N_0 b) \qquad (4.8)$$

near σ_c. With j_p the mode number corresponding to the cutoff wavenumber m_p, we then have σ_{j1} and σ_{j2} separated from σ_c by $\sigma_c \xi_j$ where

$$\xi_j = 2Aj/j_p, \ A = 1 - f^2/\sigma_c^2, \qquad (4.9)$$

and for (4.5) we need

$$\phi_c = \phi_j = \cos^{-1}[2(\xi/\xi_j) - 1]. \qquad (4.10)$$

Hence

$$I_1 + I_3 \approx 4\sigma_c A^4 E(\sigma_c, j) \int_{\xi_j}^{\infty} d\xi \int_0^{\pi} \xi^{-4}(1 + \cos\phi)^4 d\phi \qquad (4.11)$$

$$= 4\sigma_c A^4 E(\sigma_c, j)(35\pi/24)\xi_j^{-3} \qquad (4.12)$$

and

$$I_2 \simeq 4\sigma_c A^4 E(\sigma_c, j) \int_0^{\xi_j} d\xi \int_{\phi_j}^{\pi} \xi^{-4}(1 + \cos\phi)^4 d\phi \qquad (4.13)$$

$$= 4\sigma_c A^4 E(\sigma_c, j)(29\pi/24)\xi_j^{-3} \qquad (4.14)$$

after integration by parts with respect to ξ. Hence

$$I_1 + I_2 + I_3 = (4\pi/3)\sigma_c A(j_p/j)^3 E(\sigma_c, j) \qquad (4.15)$$

so that, using $m_i \simeq j\pi N/(N_0 b)$, we have from (4.3)

$$S_r^2 = \frac{1}{3}\pi^2 N_0^{-1} N^3 \sigma_c (1 - f^4/\sigma_c^4) j_p^3 \sum_{j=1}^{\infty} j^{-1} E(\sigma_c, j). \qquad (4.16)$$

We will show later that, for typical values of the bottom slope $\tan\alpha$, S_r^2 is much larger than the mean square shear S_i^2 contributed by the incident spectrum for all frequencies and all modes less than j_p. Hence we choose j_p by equating S_r^2 to $N^2 Ri_c^{-1}$. Before evaluating this for a particular choice of $E(\sigma, j)$ we show how the energy flux available for mixing, and the residual energy spectrum, may be calculated with the same approximations near σ_c.

5 THE ENERGY FLUX AVAILABLE FOR MIXING AND THE RESIDUAL SPECTRUM

Multiplying the energy density $(4\pi)^{-1} b^2 N_0 N E(\sigma, j)$ by the component of the group velocity normal to the bottom gives the bottom-normal incident energy flux spectrum (Eriksen, 1985)

$$F_n(\sigma, j, \phi) = (4\pi)^{-1} \rho b^2 N_0 N E(\sigma, j) \cos^2\theta_i \sin^2\theta_i (\sigma m)^{-1}(N^2 - f^2) \cos\alpha(aC + 1) \qquad (5.1)$$

where, as before $a = \tan\theta_i \tan\alpha$ and $C = \cos\phi_i$. This energy flux is conserved on reflection, but we assume that waves reflected with a vertical wavenumber greater than m_p are dissipated. In the notation of Section 4, the energy flux associated with these waves is

$$F_d = \sum_{j=1}^{\infty} \int_{\sigma_{j1}}^{\sigma_{j2}} d\sigma \int_{-\phi_c}^{\phi_c} F_n(\sigma, j, \phi) d\phi \tag{5.2}$$

Since we expect, and will confirm later, that σ_{j1} and σ_{j2} are both close to σ_c (except for large values of j), we make the same approximations as before to obtain

$$F_d = (4\pi^2)^{-1}\rho b^3 N_0^2 N^2 \cos^3\alpha \sin^2\alpha \sum_{j=1}^{\infty} j^{-1} E(\sigma_c, j) \times$$

$$4 \int_0^{\xi_j} d\xi \int_0^{\phi_j} (1 + \cos\phi) d\phi \tag{5.3}$$

$$= 3(2\pi)^{-1}\rho b^3 N_0^2 N^2 \cos^3\alpha \sin^2\alpha(1 - f^2/\sigma_c^2) j_p^{-1} \sum_{j=1}^{\infty} E(\sigma_c, j) \tag{5.4}$$

We note that F_d gives the same weight to each mode as did the original spectrum $E(\sigma_c, j)$, whereas S_r^2 in (4.16) is weighted towards the low modes. We also note that while $S_r^2 \propto j_p^3$, $F_d \propto j_p^{-1}$, so that the energy flux available for mixing depends on $(Ri_c)^{1/3}$ and hence is not very sensitive to the choice of Ri_c.

It is also of interest to evaluate the residual reflected energy spectrum, i.e. the energy spectrum after multiplication of the spectral energy density by $(m_r/m_i)^2$ and removal of all waves with $|m_r| > m_p$. This may then be summed over wavenumbers less than m_p to give the residual reflected energy spectrum as a function of frequency and denoted by $1/2 b^2 N_0 N E_r(\sigma)$ for comparison with the incident spectrum $1/2 b^2 N_0 N E(\sigma)$ where $E(\sigma) = \sum_{j=1}^{\infty} E(\sigma, j)$. In particular the total expected residual spectrum after reflection from a sloping bottom is a factor $[E_r(\sigma) + E(\sigma)]/2E(\sigma)$ times that for reflection off a flat bottom. A particular illustration of this will be presented later.

If we further integrate over frequency σ we can define an increase R in total energy density as

$$R = \int_f^N [E_r(\sigma) + E(\sigma)] d\sigma / \int_f^N 2E(\sigma) d\sigma. \tag{5.5}$$

With the same approximations as before, $\int_f^N E_r(\sigma) d\sigma$ may be estimated in a manner very similar to the calculation of S_r^2 in Section 4, with the result

$$\int_f^N E_r(\sigma) d\sigma = 2\sigma_c(1 - f^2/\sigma_c^2) j_p \sum_{j=1}^{\infty} j^{-1} E(\sigma_c, j) \tag{5.6}$$

so that R may be evaluated once the form of $E(\sigma, j)$ is fixed.

6 EVALUATION FOR A MODEL SPECTRUM

We now evaluate j_p, F_d and R for the model spectrum proposed by Munk (1981) and used by Eriksen (1985):

$$E(\sigma, j) = EB(\sigma)H(j), \qquad (6.1)$$

$$B(\sigma) = 2\pi^{-1}f\sigma^{-1}(\sigma^2 - f^2)^{-1/2} \qquad (6.2)$$

$$H(j) = (j^2 + j_*^2)^{-1} / \sum_{j=1}^{\infty}(j^2 + j_*^2)^{-1}. \qquad (6.3)$$

where $\sum_{j=1}^{\infty} H(j) = 1$ and $\int_f^N B(\sigma)d\sigma = 1 - 2\pi^{-1}\sin^{-1}(f/N) \simeq 1$. This spectral shape fits internal wave observations at mid-latitudes reasonably well with $E = 6.3 \times 10^{-5}$ and the modenumber bandwidth $j_* = 3$. We shall choose this value of j_* as the results for j_p and F_d are not very sensitive to it, but we leave E unspecified for the moment. In fact F_d depends only on $\sum_{j=1}^{\infty} H(j) = 1$ independent of the form of $H(j)$, but S_r^2 and R depend upon $\sum_{j=1}^{\infty} j^{-1}H(j)$ which is equal to 0.40 for (6.3) with $j_* = 3$.

If we also note that $\sigma_c^2 = N^2 \sin^2 \alpha + f^2 \cos^2 \alpha$ and again use $N^2 - f^2 \simeq N^2$, we may write

$$S_r^2 = 0.84EfN_0^{-1}N^4\sigma_c^{-4}(\sigma_c^2 + f^2)j_p^3 \sin \alpha. \qquad (6.4)$$

Hence, if $S_r^2 = N^2 Ri_c^{-1}$, we may write

$$j_p^3 = \left(\frac{1.2N_0}{Ri_c Ef}\right) \frac{(\sin^2 \alpha + \gamma^2 \cos^2 \alpha)^2}{\sin \alpha[\sin^2 \alpha + \gamma^2(1 + \cos^2 \alpha)]}, \qquad (6.5)$$

where $\gamma = f/N$. Similarly

$$F_d = 0.30j_p^{-1}\rho b^3 EfN_0^2(\sin^2 \alpha + \gamma^2 \cos^2 \alpha)^{-3/2} \cos^3 \alpha \sin^3 \alpha \qquad (6.6)$$

and

$$R = 1 + 0.26j_p\gamma \sin \alpha(\sin^2 \alpha + \gamma^2 \cos^2 \alpha)^{-1} \qquad (6.7)$$

where we have adjusted the first term on the right hand side of the equation for R from 0.5 to 1 to give the correct limit of 1 as α tends to zero and so compensate for the neglect of higher orders in the approximation.

Following studies by Wunsch and Webb (1979) and Eriksen (1980) which suggest an internal wave spectral level rather than total energy that is independent of latitude, Munk (1981) has proposed taking $E \propto f^{-1}$ with $Ef = 4.6 \times 10^{-9}s^{-1}$. With his other canonical values, $N_0 = 5.24 \times 10^{-3}s^{-1}$ and $b = 1.3\,km$, and taking $Ri_c = 1$, we have

$$j_p = 111(\sin \alpha)^{-1/3}(\sin^2 \alpha + \gamma^2 \cos^2 \alpha)^{2/3}[\sin^2 \alpha + \gamma^2(1 + \cos^2 \alpha)]^{-1/3} \qquad (6.8)$$

$$F_d = 85j_p^{-1}(\sin^2 \alpha + \gamma^2 \cos^2 \alpha)^{-3/2} \cos^3 \alpha \sin^3 \alpha \quad mW.m^{-2} \qquad (6.9)$$

so that j_p and F_d are now, like R, functions only of the bottom slope $\tan \alpha$ and the frequency ratio $\gamma = f/N$. Figure 5 shows j_p as a function of these two parameters. We note that for

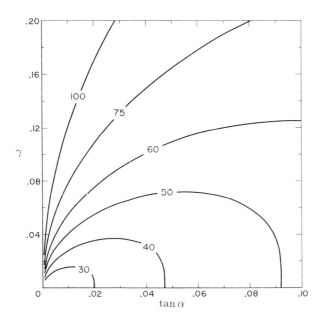

Figure 5. Cutoff mode number j_p as a function of bottom slope $\tan\alpha$ and frequency ratio $\gamma = f/N$, using (6.8).

typical bottom slopes (say 0.005 to 0.10) and for typical ratios f/N at the bottom (say 0.01 to 0.20), j_p is in the range 30 to 100. The energy flux that is lost from the internal wave field is shown in Figure 6. For typical values of the bottom slope and frequency ratio, F_d is generally less than $1\,mW.m^{-2}$. The energy density ratio R is shown in Figure 7.

The asymptotic forms of j_p, F_d and R for small and large slopes are

$$
\left.
\begin{aligned}
j_p &\approx 88\alpha^{-1/3}\gamma^{2/3}, \\
F_d &\approx 1.0\alpha^{10/3}\gamma^{-11/3}\ mW.m^{-2}, \\
R &= 1 + 23\alpha^{2/3}\gamma^{-1/3}
\end{aligned}
\right\} \quad \text{for } \tan\alpha \ll \gamma,
\tag{6.10}
$$

$$
\left.
\begin{aligned}
j_p &\approx 111(\sin\alpha)^{1/3}, \\
F_d &\approx 0.8(\sin\alpha)^{-1/3}\cos^3\alpha\ mW.m^{-2}, \\
R &= 1 + 29\gamma(\sin\alpha)^{-2/3}
\end{aligned}
\right\} \quad \text{for } \tan\alpha \gg \gamma,
\tag{6.11}
$$

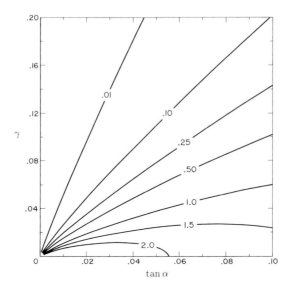

Figure 6. Energy flux F_d available for mixing as a function of bottom slope $\tan \alpha$ and frequency ratio $\gamma = f/N$, using (6.9). The units are $mW.m^{-2}$.

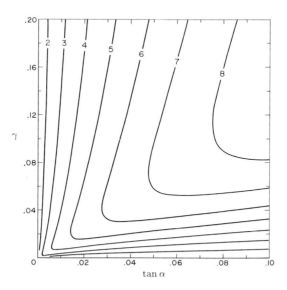

Figure 7. Energy density ratio R as a function of bottom slope $\tan \alpha$ and frequency ratio $\gamma = f/N$, using (6.7).

showing a complicated dependence on the bottom slope and f/N. In fact, for a given ratio f/N the energy flux F_d peaks when the bottom slope is about $2.9f/N$ at which point $j_p \approx 155\gamma^{1/3}$ and $F_d = 0.45\gamma^{-1/3}\,mW.m^{-2}$. For a typical $\gamma = f/N$ of 0.1 and a slope of 0.29 we have $j_p \simeq 72$ and $F_d \approx 1.0\,mW.m^{-2}$; for small slopes the energy flux F_d is considerably less.

7 DISCUSSION

7.1 Approximations

Before discussing the implications and shortcomings of the results presented in Section 6, we first review the validity of the approximations made in using (4.7) for the wavenumber amplification and in replacing $E(\sigma, j)$ by $E(\sigma_r, j)$ in the integrals over frequency. The latter approximation essentially requires the scaling frequency difference $\sigma_c \xi_j$ to be much less than $\sigma_c - f$, i.e. $j_p/j \gg 4$, whereas (4.7) is valid at $\xi = \xi_j$ if $j_p/j \gg 2$. For the values of j_p obtained in Section 6 we conclude that the approximations we have made are valid for the low modes that dominate in the expressions for j_p, F_d and R. Moreover, numerical integration of the more general formulae produces results that typically differ from our approximate values by much less than 10%.

As mentioned in Section 4, we have also ignored the contribution to the mean square shear associated with the incident spectrum. Now the model spectrum of Section 6 has a wavenumber spectrum for the shear that is white for $j \gg j_*$ and has to be integrated out to a j of a few thousand to give a shear close to N (Munk, 1981). Hence if j_p as we have calculated it is very much less than a thousand our neglect of the incident shear is a very good approximation. The incident waves with $j > j_p$ may still give up their energy flux to dissipation, but this is negligible compared to the energy flux for reflected waves with $j > j_p$ as these are so much larger. The formulae presented for j_p, F_d and R are thus mathematically accurate given the physical assumptions introduced in Section 3.

7.2 Choice of Ri_c

As mentioned in Section 5, our results are rather insensitive to Ri_c which we have equated to 1 in Section 6. We could argue that the reflected waves break when the instantaneous Richardson number equals $1/4$. For a single wave this would correspond to a Richardson number, based on the mean square shear, equal to $1/2$ rather than 1. However, we note from (6.6) that, because of the E dependence of j_p, F_d is proportional to $E^{4/3}$ and, if E varies, F_d should be increased by a factor $\overline{E^{4/3}}/\overline{E}^{4/3}$. If we assume a Rayleigh distribution for the amplitude of the important narrow frequency band near σ_c, then $p(E) = \overline{E}^{-1}\exp(-E/\overline{E})$ and $\overline{E^{4/3}} = 1.2\overline{E}^{4/3}$, i.e. F_d should be multiplied by 1.2 if we use $Ri_c = 1/2$. This is equivalent to assuming E constant but using $Ri_c = 0.86$ and gives $F_d = 0.95$ times that with E constant and $Ri_c = 1$. The issue is thus of rather minor importance.

7.3 Vertical Mixing Scale

After reflection of the internal wave spectrum from a sloping bottom we expect shear instability to be associated with all wavenumbers greater than the cut-off. The largest scale of instability might be expected after waves with $m > m_p$ have been eliminated and an event involving the rest of the spectrum occurs, starting at $Ri = 0.25$ and spreading to include the thickness over which the average shear would have given some larger value of the Richardson number, say 0.4 (see Garrett and Munk, 1972 and Thorpe, 1973).

For the white vertical wavenumber spectrum of the reflected waves (as indicated by the factor j_p^3 in S_r^2) and using the result that the mean square shear over a height h is the integral of $2(1 - \cos mh)/h^2$ times the current spectrum, we thus require $(m_p h - \sin m_p h)(m_p^3 h^3/6)^{-1} = 0.625$ for the mixed layer thickness h. This leads to $h = 3.1\, m_p^{-1}$. In view of the discussion in Section 7.2 we shall use this estimate with j_p given by (6.8) and m from (4.8) even though j_p was estimated with $Ri_c = 1$ rather than the value $1/2$ that would give a local $Ri = 1/4$. We thus have a predicted maximum mixing scale $h \simeq 1300(N_0/N)j_p^{-1}$ m. For $N \simeq 10^{-3}s^{-1}$ and $j_p \simeq 100$ this gives $h \simeq 70\,m$. We note that Eriksen (1982) did find evidence for Richardson numbers less than 0.4 over separations of tens of metres in some locations, though a detailed comparison with theory is unjustified for reasons to be discussed shortly.

7.4 The Residual Energy Spectrum

In Section 5 we discussed the residual energy frequency spectrum although we only presented there and in Section 6 the values of its integral over frequency. Here we show it, in Figure 8, as a function of frequency for the particular choices $f = 7.3 \times 10^{-5}s^{-1}$, $N = 10^{-3}s^{-1}$ and $\tan \alpha = 0.07$ so that $\sigma_c = 10^{-4}s^{-1}$. This shows the typical large enhancement of the spectrum near σ_c even after removal of the high wavenumber waves which are assumed to break. We assume that local variations in bottom slope as well as limited spectral resolution would smooth out the valley right at σ_c, leaving only a spectral bump near σ_c. This is, of course, qualitatively consistent with some observational evidence presented by Eriksen (1982) but a detailed comparison is not warranted.

7.5 Implications and Limitations

We have shown that the energy flux that is likely to be dissipated after internal wave reflection from a sloping bottom can be of the order of $1\,mW.m^{-2}$. This compares quite favourably with the estimates in Section 1 of the dissipation rate required to sustain a vertical eddy diffusivity, in the deep ocean interior, of about $10^{-4}m^2s^{-1}$. Moreover, a rate of loss of energy of $1\,mW.m^{-2}$ from the internal wave field with total energy about $4,000\,Jm^{-2}$ (Munk, 1981) corresponds to a lifetime of 46 days, which is not inconsistent with the apparent requirements of an internal wave energy level that does not vary dramatically in space and time (e.g. Olbers, 1983).

We note, however, that for F_d to be of the order of $1\,mW.m^{-2}$ we require rather steep slopes and small values of f/N. Major oceanic features such as the continental slope are barely steep enough to give large values of F_d and certainly do not cover a sufficiently large part of the total surface area of the sea floor to give a large average value of F_d. It might be argued that smaller scale features of the ocean floor are also steep, with an r.m.s. steepness that depends on the scale of resolution but is 0.07 for wavenumbers less than 0.75 cycles per km according to Eriksen (1985), quoting Bell (1975). For such features, however, the low

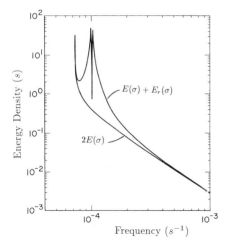

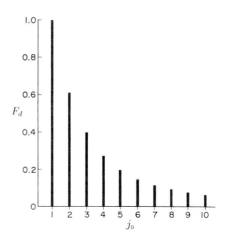

Figure 8. Total residual energy spectrum versus the model spectrum of Munk (1981), for $f = 7.3 \times 10^{-5} s^{-1}$, $N = 10^{-3} s^{-1}$ and $\tan \alpha = 0.07$.

Figure 9. Energy flux available for mixing after removal of all incident modes less than j_0, expressed as a fraction of the value for $j_0 = 1$.

modes of the incident spectrum, which carry most of the energy flux, will not be reflected as if from an infinite sloping plane. The possible consequences of this for our estimates of F_d are illustrated in Figure 9. This shows the reduction of F_d by assuming that only modes with $j \geq j_0$ are reflected as if the local slope were infinite and that lower incident modes are not amplified. Of course some shear enhancement due to diffraction of the low modes might be expected, but presumably less than for reflection from an infinite slope. Hence, if the appropriate j_0 has a corresponding wavelength comparable to the vertical height of the features, the energy flux to dissipation after reflection from the side of an abyssal hill might be reduced by an order of magnitude or so from our estimates in this paper. A thorough assessment, and more plausible estimate of F_d, might come from application of the diffraction theory developed by Baines (1971a, 1971b).

On the basis of the preceding remarks we might conclude that the immediate loss of energy through shear instability after reflection is not enough to give a significant value for K_v. However, it is clear (in Figure 8, for example) that the residual energy spectrum in both frequency and wavenumber is substantially different from that generally found in the ocean interior. Presumably, as suggested by Eriksen (1985), the reflected spectrum relaxes back to the more typical form through wave–wave interactions. It would be fruitful to explore the implications of the formalism of McComas and Müller (1981) for this relaxation and to determine the energy flux to dissipative high wavenumbers.

Further work should also include consideration of the role played by the bottom boundary layer and, perhaps, mean currents, in the reflection process at a sloping bottom. The work described in this paper explores just one relevant model that arises from Eriksen's (1982) important suggestion that internal waves cause mixing after reflection from a sloping bottom. This is almost certainly qualitatively true, but a quantitative assessment will require considerable further observational and theoretical work.

ACKNOWLEDGEMENTS

This work was partially supported by the U.S. Office of Naval Research under contract N00014-87-G-0028, and by Canada's Natural Sciences and Engineering Research Council.

REFERENCES

Armi, L., 1978. Some evidence for boundary mixing in the deep ocean. J. Geophys. Res., **83**, 1971-1979.

Armi, L., 1979. Effects of variations in eddy diffusivity on property distributions in the oceans. J. Mar. Res., **37**, 515-530.

Baines, P.G., 1971a. The reflexion of internal/inertial waves from bumpy surfaces. J. Fluid Mech., **46**, 273-291.

Baines, P.G., 1971b. The reflexion of internal/inertial waves from bumpy surfaces. Part 2. Split reflexion and diffraction. J. Fluid Mech., **49**, 113-131.

Bell, T.H., 1975. Statistical features of sea-floor topography. Deep-Sea Res., **22**, 883-892.

D'Asaro, E., 1982. Absorption of internal waves by the benthic boundary layer. J. Phys. Oceanogr., **12**, 323-336.

Dickson, R.R., 1983. Global summaries and intercomparisons: flow statistics from long-term current meter moorings. Chapter 15, 278-353 in: Eddies in Marine Science. (A.R. Robinson, ed.) Springer Verlag, 609 pp.

Eriksen, C.C., 1980. Evidence for a continuous spectrum of equatorial waves in the Indian Ocean. J. Geophys. Res., **85**, 3285-3303.

Eriksen, C.C., 1982. Observations of internal wave reflection off sloping bottoms. J. Geophys. Res., **87**, 525-538.

Eriksen, C.C., 1985. Implications of ocean bottom reflection for internal wave spectra and mixing. J. Phys. Oceanogr., **15**, 1145-1156.

Gargett, A.E., 1984. Vertical eddy diffusivity in the ocean interior. J. Mar. Res., **42**, 359-393.

Garrett, C.J.R. and W.H. Munk, 1972. Oceanic mixing by breaking internal waves. Deep Sea Res., **19**, 823-832.

Gregg, M.C., 1987. Diapycnal mixing in the thermocline: a review. J. Geophys. Res., **92**, 5249-5286.

Hogg, N.G., P. Biscaye, W. Gardner and W.J. Schmitz, 1982. On the transport and modification of Antarctic bottom water in the Vema Channel. J. Mar. Res., **40**, 231-263.

McComas, C.H. and P. Müller, 1981. Time scales of resonant interactions among oceanic internal waves. J. Phys. Oceanogr., **11**, 139-147.

Moum, J.N. and T.R. Osborn, 1986. Mixing in the main thermocline. J. Phys. Oceanogr., **16**, 1250-1259.

Munk, W.H., 1966. Abyssal recipes. Deep-Sea Res., **13**, 207-230.

Munk, W.H., 1981. Internal waves and small scale processes, in Evolution of Physical Oceanography Scientific Surveys in Honor of Henry Stommel, edited by B.A. Warren and C. Wunsch, pp. 264-291, MIT Press, Cambridge, Massachusetts.

Oakey, N.S., 1982. Determination of the rate of dissipation of turbulent energy from simultaneous temperature and velocity shear microstructure measurements. J. Phys. Oceanogr., **12**, 256-271.

Olbers, D.J., 1983. Models of the oceanic internal wave field. Rev. Geophys. Space Phys., **21**, 1567-1606.

Osborn, T.R., 1980. Estimates of the local rate of vertical diffusion from dissipation measurements. J. Phys. Oceanogra., **10**, 83-89.

Phillips, O.M., 1977. The Dynamics of the Upper Ocean, 2nd ed., Cambridge University Press, New York.

Saunders, P.M., 1987. Flow through Discovery Gap. J. Phys. Oceanogr., **17**, 631-643.

Thorpe, S.A., 1973. Turbulence in stably stratified fluids: A review of laboratory experiments. Bound. Layer Meteor., **5**, 95-119.

Wunsch, C. and S. Webb., 1979. The climatology of deep ocean internal waves. J. Phys. Oceanogr., **9**, 235-243.

BENTHIC BOUNDARY LAYERS ON SLOPES

S.A.THORPE
Department of Oceanography
The University
Southampton SO9 5NH U.K.

ABSTRACT

This note draws attention to the lack of information about the structure of the benthic boundary layer in regions of sloping topography. More observations are needed to establish, for example, the effect of internal waves on diapycnal mixing in such regions, if we are to be able to assess the local or ocean-basin-scale consequences of boundary mixing.

Observations on the western slope of the Porcupine Bank (S.W. of Ireland) at 3000m have detected inversions in potential density with overturning scales of some 30 m at heights above the slope of 100 m. The internal tides in the area are nearly 'self-resonant' on reflection from the bottom slope, and it is suggested that the inversions may be due to wave overturn or breaking induced by 'self-resonance'.

Necessary conditions for such 'self-resonance' at tidal frequences are common on the continental slopes. The nature of the outer boundary layer structure (that beyond the region dominated by bottom stress) in such regions of large internal waves may be essentially different from that above horizontal topography; the boundary layers on slopes belong to a novel class in which the mean production of turbulent kinetic energy may be dominated by internal wave breaking rather than by, for example, the bottom stress.

1 INTRODUCTION

Knowledge of the potential density, salinity, velocity and turbulent structure of the benthic boundary layer in the deep ocean is limited to a very few observations. None is so comprehensive that variations in the overall structure of, say, temperature can be related to simultaneous measurements of turbulence, and consequently our understanding of the mechanisms which support turbulence is limited.

Over nearly horizontal bottoms, for example the abyssal plains, layers of uniform temperature and salinity, 20-100 m in thickness, are observed adjacent to the boundary (Armi and Millard, 1976; Armi and D'Asaro, 1980; D'Asaro, 1982a), sometimes perturbed by the passage of thermal fronts (Armi and D'Asaro, 1980; Elliot and Thorpe, 1983; Thorpe 1983). The geothermal heat flux is generally small so that the Monin-Obukov length scale, L, usually exceeds the thickness of these uniform layers even

though the currents are weak, and a regime of forced convection exists
(see Wimbush,1970). The near-bed structure appears to be similar to that
in the constant stress layer of the atmosphere with a logarithmic velocity
profile (Wimbush and Munk, 1968) and a drag coefficient of about 1.9×10^{-3}
(Elliott, 1984). There is no concensus of agreement regarding the
appropriate scale thickness of the uniform layer. The commonly used Ekman
boundary layer formulation

$$h = \frac{0.4 \, u_*}{f} \qquad (1)$$

depending on the friction velocity u_* and Coriolis parameter, f, was found
to be typically only half, sometimes much less, than values observed
(D'Asaro, 1982a). In more stratified conditions, Weatherly and Martin
(1978) suggest the form

$$h = \frac{1.3 \, u_*}{f(1 + N/f)^{\frac{1}{4}}} \qquad (2)$$

where N is the buoyancy frequency, whilst Nabratov and Ozmidov (1987)
argue that

$$h = c \, \frac{u_*}{N} \qquad (3)$$

on basis of a relation between the Osmidov length scale and that of the
vertical scale of overturning, mixing, eddies. (The constant of
proportionality, c, appears to be of order 0.1). Data is insufficient to
establish any firm scaling, and it may be that the frequency of occurance
of fronts can play a role in determining the mean thickness. Provided
their vertical scale is much greater than the boundary layer thickness,
internal waves are little effected by the presence of the layer. They
moreover appear to play a minor role in the energetics of the boundary
layer which is mainly driven by low frequency motions (D'Asaro, 1982b).

Near sloping topography the observed thickness of near-bottom uniform
layers seems generally less than over the near-horizontal abyssal plains
(Armi & Millard, 1976; Thorpe, 1987b). This might reflect the somewhat
increased value of N in such shallower regions which would lower the
estimates of h in (2) and (3). Eriksen (1985) has however pointed out
that on sloping topography the modification of internal waves due to
reflection at the boundary may have important consequences. Eriksen
(1982) had found observed that near critical frequencies, $\sigma = \sigma_c$, where the
inclination of the wave group velocity to horizontal, ß (given by

$$\sin^2 \beta = (\sigma^2 - f^2)/N^2 - f^2) \qquad (4)$$

where σ is the wave frequency), matches the bottom slope, α, near-bottom moored observations show a spectral enhancement which decays roughly

exponentially from the ocean bottom with a scale of about 100 m. Reflection leads to an imbalance of the energy flux which must therefore be redistributed close to the boundary. Eriksen argues that a small percentage of the flux imbalance can, if used to mix the fluid, account for much of basin-averaged diapycnal diffusivity estimated to occur by diffusive-advective balance (e.g. Munk, 1966).

The processes by which the flux imbalance may be utilised to mix the fluid have yet to be fully described, but the possibility exists that the structure of the boundary layer itself is no longer determined as in (1) to (3) by the stress on the sea bed but, in addition, by the turbulence induced by the reflecting internal wave field. A purpose of this note is to draw attention to this idea, and to the singularities which occur in internal waves on reflection, not only at $\alpha=\beta$, but in conditions of 'self-resonance'. The effect of these singularities is to promote local regions of steep, or overturning, isopycnals and turbulent mixing some distance from the sea-bed. For a simple wave train the extent of the region in which such mixing occurs will depend on the vertical wavelength of the internal waves, on their steepness, on the slope of the bottom, α, and how well this is matched to the slope at which the waves are exactly self-resonant.

2 REFLECTION OF INTERNAL WAVES

The linear reflection of internal waves from smooth slopes has been studied by Phillips (1966) and Wunsch (1971; Wunsch also considered the non-linear consequences for driving Eulerian flows). Cacchione and Wunsch (1974) showed that regular arrays of vortices aligned with axes parallel to the slope contours are generated in laboratory experiments near critical slopes, $\alpha=\beta$. These vortices have not been studied further and it is not known whether they occur only at small Reynolds number (based, say, on the velocity scale induced by the waves and a boundary layer thickness $(\nu/\sigma)^{\frac{1}{2}}$ where ν is the kinematic viscosity and σ the wave frequency), or are a large Reynolds number phenomena which may be significant in the ocean.

Non-linear reflection results in singularities occurring at values of β other than α; the incident and reflected wave components can resonate (Thorpe, 1987a) locally producing high isopycnal gradients even when the

428

steepness of the incident wave is very small. This 'self-resonance' is likely to be more potent than other forms of weak wave interaction which occur in mid-water because the phase of the reflected wave is locked to that of the incident wave with which it interacts. The case $\alpha < \beta$ has been examined in detail. Second-order interaction occurs when the wavenumber $(2k, 2l, n_I + n_R)$ and frequency 2σ of a second-order wave component satisfy the first order (linear) dispersion relation, where (k, l, n_I) is the wave number of the incident wave of frequency σ, measured in axes up, along and normal to the sloping bottom, and n_R is the normal wavenumber component of the first order reflected wave. It may be shown that

$$n_I + n_R = \frac{k \sin 2\alpha}{\sin^2\alpha - \sin^2\beta} \tag{5}$$

and the dispersion relation for wavenumber (k, l, n) and frequency σ is

$$\sigma^2 = \frac{N^2\left[(k \cos\alpha - n \sin\alpha)^2 + l^2\right] + f^2(k \sin\alpha + n \cos\alpha)^2}{k^2 + l^2 + n^2} \tag{6}$$

Figure 1 shows the ratio l/k of along-slope to up-slope wavenumbers at which resonance is found as a function of α when $\sigma/f = 1.25$ and $f/N = 0.12, 0.16$ and 0.20. (This particular choice is explained later). These correspond to values of β of 5.1, 6.9 and 8.8 deg respectively. When l/k

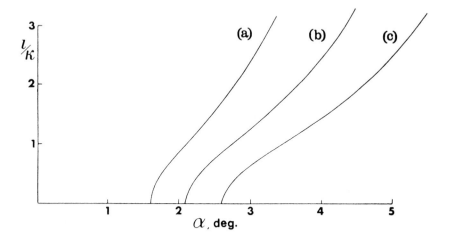

Fig.1 Conditions of self-resonance at second order of internal waves reflecting from a uniform slope of angle α. l/k is the ratio of the long-slope to up-slope wave numbers of the internal wave. The curves are for $\sigma = 1.4 \times 10^{-4}$ s^{-1} (M$_2$) and $f = 1.12 \times 10^{-4}$ s^{-1} (51 deg N) and (a) $f/N = 0.12$; (b) $f/N = 0.16$; (c) $f/N = 0.20$. The slope, α, for self-resonance increases with f/N and l/k, tending to β as l/k tends to infinity.

is small so that the internal wave propagates almost normal to the isobars of the slope, resonance occurs at values of α which are small and significantly less than ß. (As 1/k tends to infinity, α tends to ß). The width of the ß domain in which resonance significantly modifies the wave form has typically a spread of only 1 deg about the resonant angle (see Thorpe, 1987a, fig. 8c). Wave modification near the α=ß singularity is similarly confined to a narrow band of incident wave angles ß; the tuning is fine. In general, given a bottom slope and a randomly orientated field of internal waves, the effects of resonance is determined by both the directionality of the wave field and the distribution of energy with frequency.

3 OBSERVATIONS.

We have made observations of the structure of the boundary layer on the Continental Slope west of the Porcupine Bank at a water depth of about 3300 m using moored current meters, CTD profiles and a moored resistance thermometer array (Thorpe, 1987b). The bottom slope is about 2.2 deg in the main study area, rising from 0.4 deg at 4000m to a maximum of 10.6 deg at 2500 m. The local values of f and N are 1.12×10^{-4} s^{-1} and 7.1×10^{-4} s^{-1} respectively, so that $f/N = 0.16$. Although the temperature in the layer adjacent to the sea bed was very occasionally uniform to heights of about 30 m, the region above 10 m is usually stratified and any persistent 'mixed layer' is of extent smaller than 10 m, below the range of observations. The density gradient near the sea bed is strongly modulated by the M_2 tide ($\sigma/f = 1.25$, ß = 6.9 deg) which dominates the near bed motion. The baroclinic component of the tide travels upslope almost normal to the slope contours (1/k is small) with a downward phase propagation (upward group velocity) clearly apparent on the bottom 150 m. The vertical wavelength is 700-1000 m. No significant spectral peak is apparent at inertial frequencies or at a frequency of 1.15×10^{-4} s^{-1} of a wave with propagation angle ß which matches the bottom slope.

Although the conditions are well removed from the critical slope where α=ß, the M_2 internal wave is close to being self-resonant; the bottom slope is close to the values shown by the curve f/N=0.16 at small 1/k in figure 1. Overturns, remote from the boundary, are frequent, the largest observed being on scales of 30-40 m, 100 m off the bottom.

Precise estimates of the mean displacement scale, d, in the area are not available. The best available estimates are about 10 m which, given the empirical relationship

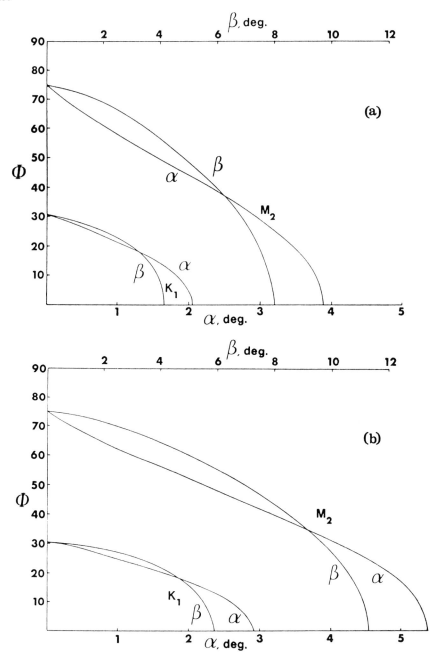

Fig.2   The bottom slope, α, as a function of the latitude, φ, (in deg.) at which self-resonance is found at second order for internal waves at M_2 and K_1 frequencies with $1/k = 0$ (propagation on-slope). Also shown is the slope, ß, of the internal wave group velocity. (a) $N = 7.1 \times 10^{-4}$ s^{-1} (0.41 cyc.hr^{-1}) (b) $N = 1\times 10^{-3}$ s^{-1} (0.57 cyc,hr^{-1}).

$$d = 1.25 \ \varepsilon^{\frac{1}{2}} \ N^{-3}/2 \qquad\qquad (7)$$

between the displacement scale and the Osmidov length scale (Dillon, 1982; Crawford, 1986), provides an estimate for the rate of dissipation of turbulent kinetic energy per unit mass, ε. The magnitude of ε implies that $\varepsilon/\nu N^2 \gg 1$, where ν is the kinematic viscosity, and that, within the lower 150 m of the water column, the turbulent energy generated by the overturns is comparable to, or exceeds, that produced by the working of the bottom stress (see Thorpe 1987b). It appears therefore that, whilst within a few metres of the sea bed turbulence is dominated by the bottom stress (conditions similar to those found over the abyssal plains), above this layer is a region of comparable turbulent production dominated by the presence, and possible self-interaction, of internal waves.

4 DISCUSSION

The spectral peak at M_2 tidal frequencies often dominates the high frequency energy spectrum near the continental slopes. Whilst this may derive mainly from the barotropic tides, there is also likely to be significant energy in the baroclinic tides derived from the direct interaction between waves and topography (Cox and Sandstrom, 1962; Bell, 1975; Baines, 1974; Princenberg and Rattray, 1975). The average slope of the sea bed in the depth range 1000-4000 m varies from about 0.7 to 2 deg. (Moore and Mark, 1986) and, except near turning latitudes, is poorly matched to the slope of the internal tide in this depth range. Other, lower frequency, waves may be affected by the critical slope ($\alpha=\beta$) effects, as Eriksen (1982) has found. Self-resonance can however, as we have seen, occur at small bottom slopes. Figure 2 shows the slope of the M_2 and K_1 tides, β, and the bottom slope, α, necessary for self-interaction, as functions of latitude, ϕ, for two typical values of N assuming that $1/k=0$. (Note that the upper scale is for β and the lower for α). Self-interaction at tidal frequencies can occur over a broad range of latitudes.

In practice the situation is more complex than we have suggested. The continental slope is not smooth but variable and irregular in both along and upslope directions. There remain several elements to be examined, such as the steadiness of the tides, the effect of varying N, and the effect of the turbulent layer induced by the bottom stress on wave reflection. More observations are planned to study the detailed structure of the boundary layer and to examine the important role played by internal waves in the dynamics of the benthic boundary layer on regions of sloping topography. The internal waves appear to play an important part in the diapycnal mixing in these regions which may thus belong in a special category of boundary layer types, akin to the breaker zone on the sea shore, where the production of turbulence is dominated by wave breaking rather than by the working of the stress on the seabed.

REFERENCES

Armi, L. and Millard, R.C. 1976 The bottom boundary layer of the deep ocean. J.Geophys. Res. 81, 4983-4990.

Armi, L. and D'Asaro, E. 1980 Flow structures of the benthic ocean. J.Geophys.Res. 85, 469-483.

Baines, P.G. 1974 The generation of internal tides over steep continental slopes. Phil.Trans. Roy.Soc.Lond. A 277, 27-58.

Bell, T.H. 1975 Topographically generated internal waves in the open ocean. J.Geophys.Res. 80, 320-327.

Cacchione, D. and Wunsch, C. 1974 Experimental study of internal waves over a slope. J.Fluid Mech. 66, 223-240.

Cox, C. and Sandstrom, H. 1962 Coupling of internal and surface waves in water of variable depth. J.Oceanogr. Soc. Jap. 20, 499-513.

Crawford, W.R. 1986. A comparison of length scales and decay times of turbulence in stably stratified flows. J.Phys.Oceanogr. 16, 1847-1854.

D'Asaro, E. 1982a Velocity structure of the benthic ocean. J.Phys. Oceanogr. 12, 313-322.

D'Asaro, E. 1982b Adsorption of internal waves by the benthic boundary layer. J.Phys. Oceanogr. 12, 323-336.

Dillon, T.M. 1982 Vertical overturns : a comparison of Thorpe and Osmidov length scales. J. Geophys.Res. 87, 9601-9613.

Elliot, A.J. 1984 Measurements of the turbulence in an abyssal boundary layer. J.Phys. Oceanogr. 14, 1779-1786.

Elliott, A.J. and Thorpe, S.A. 1984 Benthic observations in the Madeira Abyssal Plain. Oceanologica Acta 6, 463-466.

Eriksen, C.C. 1982 Observations of internal wave reflexion off sloping bottoms. J.Geophys.Res. 87, 525-538.

Eriksen, C.C. 1985 Implications of ocean bottom reflexion for internal wave spectra and mixing. J.Phys.Oceanogr. 15, 1145-1159.

Moore, J.S. and Mark, R.K. 1986 World slope map. EOS, Trans. Amer. Geophysical Union, 67, 1353-1362.

Munk, W.H. 1966 Abyssal recipes. Deep-Sea Res. 13, 707-730.

Nabratov, V.N. and Ozmidov, R.V. 1987 Investigation of the bottom boundary layer in the ocean. Oceanology, 27(1) 5-11.

Phillips, O.M. 1966 'The dynamics of the upper ocean', Cambridge University Press, 261pp.

Princeberg, S.J. and Rattray, M. 1975 Effects of continental slope and variable Brunt-Vaisala frequency on coastal generation of internal tides. Deep-Sea Res. 22, 1251-263.

Thorpe, S.A. 1983 Benthic observations on the Madeira Abyssal Plain : Fronts. J.Phys.Oceanog. 13, 1430-1440.

Thorpe, S.A. 1987a On the reflection of a train of finite amplitude internal waves from a uniform slope. J.Fluid Mech, 178, 279-302.

Thorpe, S.A. 1987b Current and temperature variability on the continental slope, Phil.Trans.Roy.Soc.Lond. A (in press)

Weatherley, G.L. and Martin, P.J. 1978 On the structure and dynamics of the oceanic boundary layer. J.Phys.Oceanogr. 8, 557-570.

Wimbush, M. 1970 Temperature gradient above the deep-sea floor. Nature, 227, 1041-1043.

Wimbush, M. and Munk, W., 1971 The benthic boundary layer. In 'The Sea', 4, 731-758 Ed. by Maxwell, John Wiley.

Wunsch, C. 1971 Note on some Reynolds stress effects on internal waves on slopes. Deep-Sea Res., 18, 583-591.

MIXING IN A THERMOHALINE STAIRCASE

R.W. SCHMITT
Woods Hole Oceanographic Institution, Woods Hole, MA 02543, USA

ABSTRACT

The field program C SALT (*Caribbean - Sheets And Layers Transects*) focused on the strong thermohaline staircase in the western tropical North Atlantic. Large- and small-scale surveys and temporal monitoring found an extensive and persistent sequence of subsurface mixed layers. Good evidence for salt finger activity is found in the thermal microstructure in the interfaces, plumes in the mixed layers, and large scale changes in the temperature and salinity of the layers. Dissipation rates of kinetic energy were lower than expected from the laboratory $\frac{4}{3}$ power law, but consistent with a "Stern number"-based flux law, in which the flux is inversely proportional to interface thickness. Observations of thermal microstructure also support a Stern number of order one. Both the dissipation measurements and the Stern number flux law yield estimates for the vertical salt diffusivity of order 1×10^{-4} m^2/s. Very thin interfaces found during an earlier survey suggest that higher diffusivities may transiently occur.

1 INTRODUCTION

The thermohaline staircase, in which a series of layers well mixed in temperature and salinity are separated by high gradient interfaces or sheets, is a striking form of oceanic fine-structure (Figure 1). It has long been associated with the double-diffusive instabilities of salt fingering and diffusive layering, since these processes form similar structures in the laboratory (Turner, 1973; Stern and Turner, 1969; Kelley, 1987). Double-diffusion at the interfaces releases the potential energy in the unstably distributed component; the net unstable buoyancy flux keeps the adjacent layers stirred. In the salt finger case, vertical flows in tall narrow cells, a few centimeters wide, exchange heat but not salt laterally. A staircase will have a higher flux than a non-steppy profile with fingers, because of the sharp gradients at the interfaces.

The parameter which determines the strength of double-diffusion is the density ratio, defined as:

$$R_\rho = \alpha T_z / \beta S_z \, ,$$

where $\alpha = -\frac{1}{\rho} \frac{\partial \rho}{\partial T}\big|_{S,p}$, $\beta = \frac{1}{\rho} \frac{\partial \rho}{\partial S}\big|_{T,p}$, T_z = vertical temperature gradient, S_z = vertical salt gradient, and p is the pressure. The salt fingering instability occurs when $1 < R_\rho < \frac{K_T}{K_S}$, where $\frac{K_T}{K_S}$, the ratio of the molecular heat and salt diffusivities, is about 100. The most active fingering occurs when $R_\rho \to 1.0$; the available energy in the salt field is proportional to $\frac{1}{(R_\rho-1)}$ (Schmitt and Georgi, 1982) and the growthrate increases as well (Schmitt, 1979a). Much of the ocean is favorable for salt fingers in the sense that $R_\rho > 1.0$, but the regions with $1.0 < R_\rho < 1.6$ are less common and more likely to show staircase profiles (Schmitt, 1981).

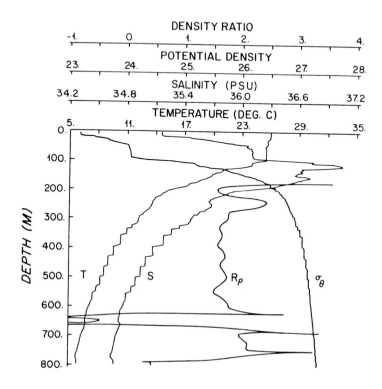

Figure 1. A thermohaline staircase from Station 49 during the spring cruise from the C-SALT survey area. Profiles of temperature, salinity, potential density (σ_θ) and density ratio (R_ρ) are shown. The density ratio was computed using least squares fits over a sliding 40 m vertical interval. The layers only appear when the density ratio is less than 1.7.

The oceanic staircases seem to be close analogs of laboratory steps; this view has been reinforced by direct observations of salt fingers using optical and towed conductivity methods in the steps beneath the Mediterranean water in the eastern North Atlantic (Williams, 1975; Magnell, 1976). Carrying the analogy even further, Stern and Turner (1969) and Lambert and Sturges (1976) used the laboratory flux laws to estimate the "effective vertical eddy diffusivity" due to the fingering interfaces. This calculation yields values of $\sim 5 \times 10^{-4}$ m^2/s which is significantly larger than usual thermocline values of less than 1×10^{-5} m^2/s (Gregg, 1987a). Such calculations must be considered an upper bound, since any number of oceanic processes may interfere with the fingers, but it suggests that the staircases are areas of enhanced vertical exchange. Schmitt (1981) has pointed out that a greater flux of salt in such low R_ρ regions can increase the density ratio and help to maintain the constant R_ρ shape of the mean T-S relation. Thus, there is a compelling need to quantify the rate of mixing by salt fingers, which may make a modest contribution to mixing even outside the staircase regions (Schmitt and Evans, 1978).

C-SALT (Schmitt, 1987) was designed to examine the large and small-scale structure of the staircase which is found east of Barbados in the tropical North Atlantic. A large area there has had many observations of staircases in the past (Boyd and Perkins, 1987). The strong salinity gradients of the region are maintained by the confluence of salinity maximum water (Worthington, 1976) originating at the surface of the subtropical North Atlantic in a region of high evaporation, and the low salinity Antarctic Intermediate Water, which comes via the South Atlantic (Wust, 1978). Between the salinity maximum at 150 m depth and the salinity minimum at about 700-800 m depth a sequence of well-mixed layers can be found (Figure 1). There are typically ten layers 5 to 30 m thick. The layers are separated by thin interfaces with temperature changes of up to 1.0°C and salinity changes of order 0.1 PSU (Practical Salinity Units or 10^{-3}). The staircase does not usually extend to the depth of either salinity extrema and is only found where the density ratio is less than 1.7. A set of recent papers reports on the preliminary findings from C-SALT (Schmitt et al., 1987; Marmorino et al., 1987; Lueck, 1987; Gregg and Sanford, 1987; Kunze et al., 1987).

Some of the results from C-SALT were consistent with prior conceptions of the structure of a salt finger staircase, but others were surprisingly different. In Section 2, the primary results are briefly reviewed. In Section 3, the laboratory flux laws are discussed and compared with the oceanic measurements, and estimates of the vertical diffusivity are made. Finally, results are summarized in Section 4.

2 THE C-SALT STAIRCASE

2.1 Survey Results

The C-SALT survey region (Figure 2) was occupied twice; during spring (March - April) and fall (November), 1985. A current meter mooring was deployed for the eight months between cruises. AXBTs were used to define the limits of the staircase, a grid of CTD stations was occupied during both seasons, and intensive fine- and microstructure work was performed from two ships in the fall. The region occupied by the steps was similar between the two periods (Schmitt et al., 1987) and similar to the composite of historical observations assembled by Boyd and Perkins (1987). The historical data and the persistence for eight months of at least three mixed layers within the aperture of the current meter array leads one to believe that the staircase is a permanent feature of the region. The layers were also found to be laterally coherent for great distances. Repeated lowerings of the CTD as the ship steamed slowly (a CTD tow-yo) indicated a minimum horizontal coherence scale of 30 km (Figure 3). Longer sections with XBTs extended the

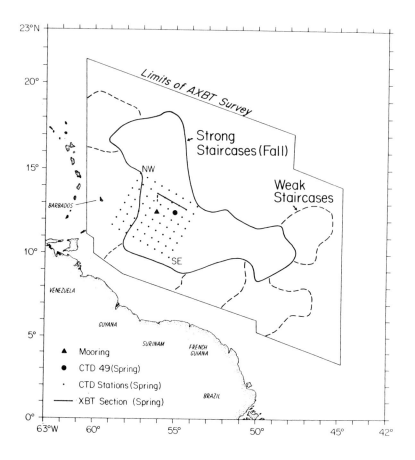

Figure 2. The regions surveyed by the ships and aircraft during the spring and fall of 1985 in C-SALT. The areas occupied by strong and weak staircases are indicated along with locations of the stations shown in the other figures and the stations occupied during the CTD surveys (•). The tow-yo is indicated by the short dashed line (- - -), at the northwest end of the XBT section (___).

coherence to over 200 km for some layers. However, significant temperature and salinity changes within the layers begin to become apparent over these distances and it is not always obvious how to relate individual layers in CTD stations separated by 55 km or more. Examination of the Temperature - Salinity characteristics of the layers, however, reveals a surprising pattern in which the layer T-S values are grouped in distinct lines (Figure 4). The lines allow one to relate layers in widely separated CTD stations and extend the horizontal correlation scale to 400-500 km. The trends are such that layers get warmer, saltier, and denser to the north and west.

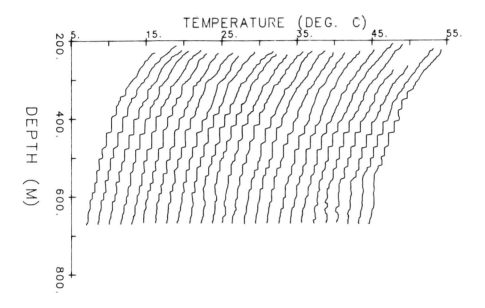

Figure 3. The offset temperature profiles from a CTD tow-yo. The left-most profile lies to the south of the others; the casts were taken as the ship steamed slowly to the north. The separation between casts is about 1.3 km; the total distance covered is 33 km.

This direct evidence for water mass conversion within the layers permits clear identification of the mixing processes involved. That is, the lines cut across locally referenced potential density surfaces with a well defined slope such that

$$\alpha T_x/\beta S_x \; = \; 0.85 \; (\pm \; 0.02) \text{ in the spring,} \qquad \alpha T_x/\beta S_x \; = \; 0.84 \; (\pm \; 0.03) \text{ in the fall} ,$$

when averaged over the ten principal layers. This horizontal density ratio for the layers is distinctly different from those resulting from isopycnal processes (1.0) or vertical processes (1.6, the value of R_ρ). It strongly supports the concept that salt fingers make a substantial

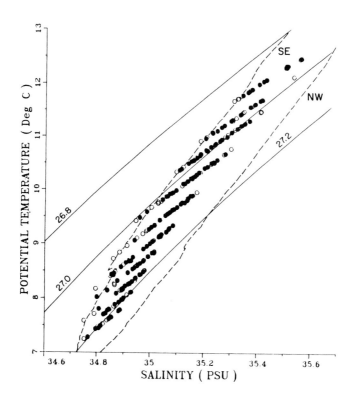

Figure 4. Potential Temperature - Salinity values of the layers from CTD stations during the spring cruise. The solid circles are from mixed layers more than 10 m thick, the open circles are from layers 5-10 m thick. The data from the fall cruise fell along nearly the same lines. Also shown are the potential Temperature - Salinity relationships obtained from the stations at the northwest and southeast corners of the survey area (- - - -). The trend of data from the layers crosses isopycnal surfaces (the 26.8, 27.0, and 27.2 potential density curves are shown), becoming warmer, saltier, and denser from southeast to northwest.

contribution to water mass conversion, since a vertically divergent finger flux would transport heat and salt with a flux ratio near this value, according to laboratory experiments
(Turner, 1967; Schmitt, 1979b; McDougall and Taylor, 1984) and theory (Schmitt, 1979a).
Indeed, the flux ratio must be less than one, or no staircase could be formed. Further
interpretation of this result is offered below. However, it does provide evidence that salt
fingering is the dominant process maintaining the C-SALT staircases.

2.2 Fine- and Microstructure Results

To examine the fine and microscales there were deployments of a towed thermistor chain,
a towed microstructure vehicle, a vertical microstructure profiler, and an optical shadowgraph instrument. These generated data which both reinforced and contradicted previous
notions about salt fingers. Some of the noteworthy results:

• Towed thermistor chain measurements revealed weak plumes within the layers and
narrow band thermal microstructure in the interfaces (Marmorino et al., 1987). The temperature gradient spectrum displays a peak at a wavelength of 5 cm. This is in excellent
agreement with the size predicted for the fastest-growing finger by the model of Schmitt
(1979a). This wavelength is given by:

$$L = 2\sqrt{2}\,\pi\,(g\alpha T_z/\nu K_T)^{-\frac{1}{4}}\,M^{-1}$$

where M is the non-dimensional wavenumber. For a 2 m-thick interface with a 0.5°C temperature contrast, a fastest-growing finger wavelength of 5.5 cm is predicted. The plumes
appear to have an order one aspect ratio; that is, the spacing between plumes is comparable to the layer thickness. Thus, these observations correspond nicely with the laboratory
picture of interfacial salt fingers driving convective overturning in the layers. However,
temperature inversions within the layers were also found, suggestive of lateral intrusions.

• Towed microstructure data also revealed narrow-band temperature structure but
found that it had low vertical coherence over several centimeters and low dissipation rates
of kinetic energy (Lueck, 1987). The low vertical coherence may indicate that the fingers
are tilted by shear. This not necessarily a flux-limiting feature as Lueck suggests,
since a well-defined vertical alignment is best observed in the laboratory in older, run-
down fingers. Newly formed fingers at low-density ratios in the laboratory tend to be less
structured, consistent with the broader bandwidth found in theory (Schmitt, 1979) and
recent numerical simulations (C. Shen, personal communication, 1987).

• Dropped profiler data also showed low dissipations and modest shears across the
interfaces (Gregg and Sanford, 1987; Gregg, 1987b).

• The shadowgraph images revealed that the interfaces were populated with thin laminae, strongly tilted to be nearly horizontal (Kunze et al., 1987). Occasional observations
of more isotropic structures, suggestive of turbulence, were also obtained. However, the
dominant structures in the interfaces were the horizontal laminae, which were found only

in salt finger-favorable conditions. These results are in surprising contrast to the predominantly vertical structure seen in other salt finger regions by Williams (1975) and Schmitt and Georgi (1982). A simple interpretation of the laminae as tilted salt fingers may be possible. However, Linden (1974) has shown that fingers will align with the shear to form vertical sheets. The laminae are also much smaller (0.5 cm) than the 5 cm scale structure seen with the towed and dropped instruments. The shadowgraph system emphasizes the smallest scales since it responds to the second derivative of index of refraction variations. At the 0.5 cm scale, salinity makes the primary contribution to the signal. The laminae could thus be the signature of unmixed salt anomalies being carried through the sheets, or they could be a manifestation of a secondary instability, perhaps in the form of corrugations on the walls of the sheets. This may correspond to the Holyer (1984) instability or the varicose structures observed in the experiments of Turner and Chen (1974). More laboratory work is clearly needed to understand the shadowgraph results.

Of these observations, the thermal microstructure in the interfaces, the plumes in the layers, the occurrence of the layers only at low density ratios, and the evidence for water-mass conversion seen in the large-scale changes in layer properties, provide a convincing case that the C-SALT staircase is maintained by salt fingering. Even though the tilted laminae differ from earlier observations, their appearance in finger-favorable regions only, and their dissimilarity from turbulence, supports a double-diffusive interpretation of the mixing in the staircase. The low levels of dissipation observed by Gregg and Sanford (1987) and Lueck (1987) raise questions about the intensity of the convection which are explored in the next section.

3 MIXING IN THE STAIRCASE

The vertical buoyancy flux due to salt transport by fingers at a laboratory interface has been parameterized as a function of the $\frac{4}{3}$ power of the salinity difference across the interface (Turner, 1967; Schmitt, 1979b; McDougall and Taylor, 1984). That is,

$$g\beta F_s = C (K_T)^{\frac{1}{3}} (g\beta \Delta S)^{\frac{4}{3}} \tag{1}$$

where βF_s is the vertical density flux due to salt, C is an empirical coefficient ranging from 0.05 (Schmitt, 1979) to 0.3 (McDougall and Taylor, 1984), g is the gravitational acceleration, K_T is the thermal diffusivity and ΔS is the salinity difference across the interface. The total vertical buoyancy flux is given by:

$$gF_\rho = g(\beta F_S - \alpha F_T) = g\beta Fs(1-\gamma) \tag{2}$$

where αF_T is the density flux due to heat and γ is the ratio of the heat and salt density fluxes. This ratio has been attributed values from 0.15 (Linden, 1973) to ~ 0.8 (Schmitt, 1979b). It can be broken into three parts:

$$\gamma = \gamma_{CV} + \gamma_{CN} + \gamma_{CA} \tag{3}$$

where γ_{CV} is the convective flux ratio due to the fingers alone, γ_{CN} is a correction due to thermal conduction across the interface (vertical salt diffusion being negligible for heat-salt fingers) and γ_{CA} is a correction due to asymmetric entrainment caused by the nonlinearity of the equation of state for sea water (the cabbeling effect, Schmitt (1979b), McDougall (1981)). Turner (1967), Schmitt (1979b) and McDougall and Taylor (1984) subtracted γ_{CN} from the total and reported values of 0.56, 0.72 ($\pm$ 0.06), and 0.5, respectively. The theoretical value of γ_{CV} for the fastest-growing finger at a density ratio of 1.6 is $\gamma_{CV} = 0.62$, according to the theory of Schmitt (1979a), but any flux ratio between 0.016 and 1.0 is possible for growing salt fingers at a density ratio of 1.6. The cabbeling effect should be negligible for the oceanic interfaces, but may have elevated the flux ratio reported by Schmitt (1979b).

The conductive correction is often ignored but it is easily estimated by using another relationship derived from the collective instability analysis of Stern (1969). This says that the non-dimensional group (the Stern number):

$$A = g\beta F_S(1 - \gamma)/\nu N^2 \tag{4}$$

is of order one. Stern's original analysis obtained a value of unity; subsequent work by Holyer (1981) indicates that a value of $\frac{1}{3}$ is the theoretical upper limit for stability. Laboratory measurements in the heat-salt system have yielded values of A from 1 (Schmitt, 1979b) to 4 (McDougall and Taylor, 1984). Kunze (1987) argues that this criterion for finger stability is equivalent to a finger Richardson number being order one. Whether or not the specific mechanism of collective instability applies, the empirically determined values of A allow it to be used as an additional flux law. That is:

$$g\beta F_s = A \ \nu N^2/(1-\gamma) = A\nu \ [(R_\rho-1)/(1-\gamma)] \ g\beta S_z \ . \tag{5}$$

Assuming that the conductive heat flux is porportional to $K_T \ \alpha \ T_z$, the conductive correction to the flux ratio is given by (McDougall and Taylor, 1984):

$$\gamma_{CN} = (K_T/\nu) \ (1 - \gamma_{CV}) \ R_\rho/A(R_\rho - 1) \ . \tag{6}$$

For the laboratory experiments of Schmitt (1979b) this correction is about 0.1, bringing the total flux ratio into rough agreement with the observed horizontal density ratio of 0.85. Alternative interpretations of this slightly high observed flux ratio may be developed by assuming that it results from a combination of fingering and vertical or isopycnal mixing. Also, models of lateral intrusions by McDougall (1985) and Posmentier (personal communication, 1986) show that intrusive cores may have $\alpha T_x/\beta S_x \simeq$ 0.8-0.9. However, the generally monotonic staircase profiles have only minor inversions, and this mechanism seems unlikely to play a primary role.

When combined with the $\frac{4}{3}$ power law, the Stern number relationship provides an estimate of the expected interface thickness. That is, the interface thickness, h, is given by:

$$h = [A(R_\rho - 1)/C(1 - \gamma)] \; \nu \; (g \; K_T)^{-\frac{1}{3}} \; (\beta \Delta S)^{-\frac{1}{3}} \; . \tag{7}$$

When this expression is evaluated for a salinity difference of 0.05 PSU, (typical of the C-SALT staircase) the value of h ranges from 3 to 70 cm for the range of possible values of A, C and γ (Table 1). This is smaller than the interface thicknesses of 1-10 m usually seen in the fall CTD profiles. However, not all interfaces were this thick. To explore this issue we must examine the unprocessed CTD data, since the normal one decibar pressure sorting will obscure any thinner interfaces. Figure 5 contains four conductivity profiles through

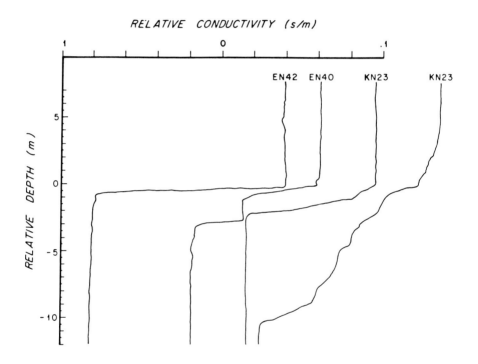

Figure 5. Detailed conductivity profiles from two spring CTD stations (Endeavor 40, 42) and a fall station (Knorr 23). The thin interfaces from the spring stations had $R_\rho = 1.4$, whereas the thicker interfaces from the fall had $R_\rho = 1.6$.

C-SALT interfaces. The conductivity channel is displayed as it has the fastest responding sensor. The profile from Endeavor station 42, from the spring survey, shows an interface thinner than 0.5 m, and an extremely strong conductivity jump (the temperature contrast exceeds 1.0°C!). This may be an instance where the finger interface was actually as thin as predicted by the laboratory flux laws. The profile from Endeavor station 40 shows a

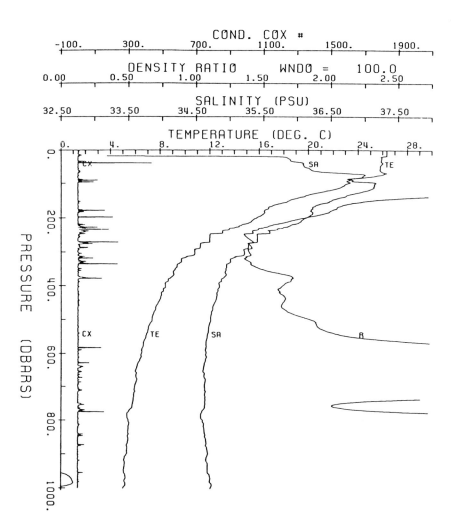

Figure 6.a. Temperature, salinity, density ratio and conductivity Cox number from a spring station (Endeavor station 42).

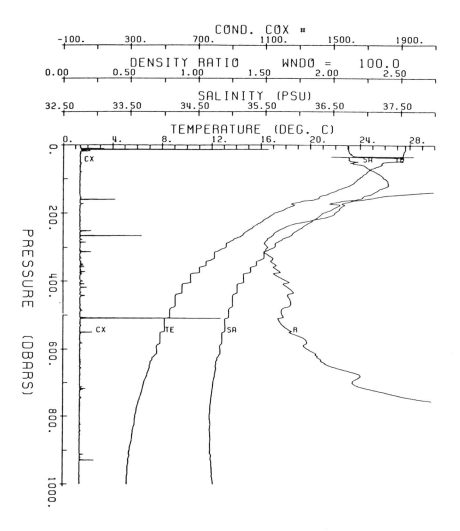

Figure 6.b. Temperature, salinity, density ratio and conductivity Cox number from a fall station (Knorr station 23).

double interface, similar to the ocean observations of Williams (1975) and the laboratory experiments of Linden (1978). Each of the interfaces may be sufficiently thin to satisfy the laboratory relationship, though the overall flux would be reduced by the decreased salt contrast. The profiles from Knorr station 23, from the fall survey, show 2 m and 12 m-thick interfaces which exceed the laboratory-based thickness prediction by 1-2 orders of magnitude. The examples chosen perhaps overemphasize the contrast between the spring and fall profiles, since Endeavor station 42 contains the sharpest interface observed, and there were instances of thick interfaces in the spring. However, it is unfortunate that the high resolution microstructure work was performed in rather thick interfaces which would have a much reduced flux if the Stern number, A, is assumed to be an order one constant. It is interesting to note that the sharp Endeavor interfaces had a density ratio of 1.4 whereas the more typical thick interfaces of the fall generally have an R_ρ near 1.6. They are also characterized by higher levels of microstructure, as indicated by the conductivity Cox number profiles of Figure 6a and 6b. These observations suggest that there is an evolution in the life of a staircase, since in the absence of advective renewal, the flux of salt by the fingers will increase R_ρ with time. It may be that a very low density ratio is required to initiate a staircase but that once formed the mixed layers can be maintained by the reduced fluxes which would result as R_ρ increased.

Since the interfaces are generally thicker than required by equation (7) and the dissipation measurements of Gregg and Sanford (1987) and Lueck (1987) suggest that the buoyancy flux is also lower than predicted by the $\frac{4}{3}$ power law, it is worth comparing the various flux laws with the dissipation measurements in some detail. The effective vertical salt diffusivity is also of interest since it will also apply to tracers, nutrients, etc. as well. In order to estimate a vertical diffusivity (K^s) from the dissipation measurements, we simply assume that the vertical buoyancy flux is equal to the dissipation rate of kinetic energy (ϵ):

$$g\beta Fs(1-\gamma) = \epsilon \ , \tag{8}$$

and that the mean vertical salt gradient is related to the stability (N^2) by:

$$g\beta S_z(R_\rho-1) = N^2 \ . \tag{9}$$

This yields a simple expression for the vertical salt diffusivity (Hamilton *et al.*, 1987):

$$K^s = [(R_\rho - 1)/(1 - \gamma)] \ \epsilon/N^2 = (1 \rightarrow 4) \ \epsilon/N^2 \ . \tag{10}$$

For the conditions observed in C-SALT, this leads to a multiplier for ϵ/N^2 of one to four depending on whether the flux ratio is 0.5 or 0.85. In contrast, the equivalent expression in a turbulent regime (Osborn, 1980) is:

$$K^\rho = [R_f/(1 - R_f)] \ \epsilon/N^2 \simeq (0.1 \longrightarrow 0.2) \ \epsilon/N^2 \tag{11}$$

where R_f is the flux Richardson number, usually taken to be 0.15. Thus, there is a difference of a factor of 5 to 40 between the vertical diffusivity expressions depending on whether the dissipation results from salt fingers or turbulence. This reflects the radically different energetics of the two mixing mechanisms.

In order to compare the flux laws with the dissipation measurements, specific evaluations are made for the C-SALT staircase during the fall microstructure work. The McDougall and Taylor flux law is evaluated with C = 0.16 and γ = 0.5 while the Schmitt flux law is evaluated with C = 0.078 and γ = 0.85, in order to explore the range of possible estimates. Similarly, the Stern number criterion is used as a flux law and applied with values of both 1 and 4. We choose an average ΔS of 0.05 PSU, resulting from a 0.35 PSU change over seven interfaces and a vertical distance of 190 m. The background salinity gradient is 1.8×10^{-3} PSU/m, the interface thickness is taken to be 2 m, $\nu = 1.3 \times 10^{-6}$ m^2/s, $K_T = 1.4 \times 10^{-7}$ m^2/s, $R_\rho = 1.6$, and $\beta = 0.76 \times 10^{-3}$/PSU. The results are shown in Table 1.

TABLE 1:
Comparison of Salt Finger Flux Estimates.

Method	Eq. (1)	Eq. (1)	Eq. (5)		Eq. (8)	
	McDougall & Taylor C = 0.16 γ = 0.5	Schmitt C = 0.078 γ = 0.85	Stern number A = 4	A = 1	Dissipation G & S	Lueck
Buoyancy Flux (10^{-10} W/kg)	110	16	5.6	1.4	1.9	5
Salt Eddy Diffusivity (10^{-4} m^2/s) γ = 0.5	16		0.83	0.21	0.28	0.74
γ = 0.85		8	2.8	0.71	0.94	2.5

The order of magnitude difference between the buoyancy fluxes calculated by the two $\frac{4}{3}$ power laws results from the differences in C and γ. However, both estimates are above the buoyancy flux as inferred from the dissipation measurements. Gregg and Sanford report average dissipations of 1.9×10^{-10} W/kg, and Lueck reports less than 5×10^{-10} W/kg. The flux estimated by the Stern number criterion is in reasonable agreement with these dissipation measurements. This flux law has been used by Gargett and Schmitt (1982) and is equivalent to that developed by Kunze (1987).

There is additional support in the microstructure data for the validity of equation (5).

This can be seen in the Cox number estimates for the interfaces. A relationship is easily derived between the Cox number and the Stern number by assuming that the primary balance in the heat equation is between the vertical advection of heat and horizontal diffusion (Gargett and Schmitt, 1982). That is:

$$w'\alpha\bar{T}_z = K_T \, \nabla^2 \alpha T' \ . \tag{12}$$

Multiplying by $\alpha T'$ and horizontally averaging we get:

$$\alpha F_T = \overline{w'\alpha T'} = K_T \, \overline{(\nabla \alpha T')}^2 / \alpha \bar{T}_z \ . \tag{13}$$

From the definition of A we can also write the heat flux as:

$$\alpha F_T = \gamma \beta F_S = [A\gamma(R_\rho - 1)/R_\rho(1 - \gamma)] \, \nu \alpha \bar{T}_z \ . \tag{14}$$

Combining the two yields an expression for the interfacial Cox number as a function of Prandtl number, Stern number, flux ratio and density ratio:

$$C_x = \overline{(\nabla \alpha T')}^2 / (\alpha \bar{T}_z)^2 = [A\gamma(R_\rho - 1)/R_\rho(1 - \gamma)] \, \nu / K_T \ . \tag{15}$$

The Prandtl number $(\frac{\nu}{K_T})$ is about 9-10 for the C-SALT interfaces. The other terms combine to yield factors of about 0.38 for $\gamma = 0.5$ and $A = 1$, to 8.2 for $\gamma = 0.85$ and $A = 4$. This gives a range of 3 to 80 for the predicted Cox numbers. The reported Cox numbers range from 8 (Marmorino et al., 1987) to 10-20 (Lueck, 1987) to 59 (Gregg and Sanford, 1987) in the C-SALT interfaces. Thus, it appears that intermediate to high values of both the Stern number and flux ratio will yield good agreement; having both A and γ low seems unlikely from the observed Cox numbers. This favors the salt diffusivity estimates of order 1×10^{-4} m^2/s in Table 1. Closer examination of the variation of Cox number with density ratio and Prandtl number in the microstructure data may further constrain the values of A and γ.

The reasonably high value of the effective salt diffusivity despite the low dissipations emphasizes a profound difference between double-diffusion and turbulence. That is, for salt fingers with a high flux ratio, the primary energy transfer is from the potential energy of the salt field into the potential energy of the temperature distribution, with relatively little being dissipated. For turbulence, the energy is drawn out of the shear field, most is dissipated and relatively little goes into potential energy. Thus, interpretation of oceanic dissipation measurements in terms of vertical diffusivities is not straightforward if both turbulence and double-diffusion are present. This suggests that alternative approaches to the assessment of mixing in the ocean, such as deliberate tracer release experiments (Ledwell and Watson, 1987), will be particularly valuable in both thermohaline staircases and the mixed turbulence-salt finger regime of the central waters.

An eddy diffusivity for salt of $\sim 1 \times 10^{-4}$ m^2/s is only a factor of 6 less than that estimated by Lambert and Sturges (1976), for a staircase not far from the C-SALT region. It is significantly larger than the value of less than 1×10^{-5} m^2/s thought to characterize the main thermocline in mid-gyre (Gregg, 1987a). Since there is evidence that the staircase was even more active in the spring, there is no reason to doubt that salt fingers are making a significant contribution to both the maintenance of the staircase and the observed water-mass conversions in the region.

4 CONCLUSIONS

The C-SALT field program provided the first intensive look at a salt fingering staircase system. The large scale surveys and monitoring showed that the staircase is widespread and persistent. Individual layers can be tracked for hundreds of kilometers and display distinct changes in temperature and salinity which are consistent with the expected flux ratio of salt fingers. Fine- and microstructure measurements reveal the narrow band signatures of salt fingers in the interfaces and convective plumes in the layers. The observed low dissipations indicate that the $\frac{4}{3}$ power law derived from laboratory measurements does not generally apply to these oceanic fingers. The fingering interfaces were also thicker than predicted from laboratory results. However, the observed Cox numbers in the interfaces do support the Stern number-based flux law, which also yields dissipations consistent with the observations. These dissipation measurements provide an estimate of the vertical eddy diffusivity for salt near 1×10^{-4} m^2/s for the fall observation period. The vertical diffusivity would have been higher for the thin interfaces found during the spring survey. For future estimates of the effects of salt fingers, it appears preferable to use a relationship dependent on the interface thickness (such as equation (5) or that given by Kunze, 1987) instead of the $\frac{4}{3}$ power law.

The C-SALT data clearly establish the importance of salt fingering for water-mass conversion in parts of the main thermocline. They also raise a number of questions about the phenomenology of salt fingers that should stimulate further laboratory work. The uncertainties in the values of the flux ratio and the Stern number are now significant issues to be resolved for the further interpretation of oceanic data. The interface thickness is observed to be highly variable. Is this related to internal wave activity? How do internal waves affect the fluxes? How do steady and time varying shears affect the fingers? What is the cause of the small-scale horizontal laminae observed with the shadowgraph?

On the observational side, the C-SALT data raise the issue of relating kinetic energy dissipation measurements to vertical diffusivity in a double-diffusive system. If some of the dissipation is caused by turbulence, how is it distinguished from that due to double diffusion so that the proper formula can be applied? Microstructure instruments which thoroughly document the finescale shear and density environment are clearly required. Also, it appears that the deliberate tracer release experiments proposed by Ledwell and Watson (1987) will be especially useful in both salt finger staircases and the mixed salt finger and turbulence regime of the central waters.

5 ACKNOWLEDGMENTS

Thanks to D. Kelley, E. Kunze, T. McDougall and J. Taylor for comments on the manuscript. This research was supported by the National Science Foundation grant No. OCE 84-09323. This is contribution number 6637 of the Woods Hole Oceanographic Institution.

6 REFERENCES

Boyd, J. D. and H. Perkins, 1987. Characteristics of thermohaline steps off the northeast coast of South America, July, 1983. Deep-Sea Res., 34(3), 337-364.

Gargett, A. E. and R. W. Schmitt, 1982. Observations of salt fingers in the Central Waters of the Eastern North Pacific. J. Geophys. Res., 87, 8017-8030.

Gregg, M. C., 1987a. Diapycnal mixing in the thermocline: A review. J. Geophys. Res., 92(C5), 5249-5286.

Gregg, M. C., 1987b. Mixing in the thermohaline staircase east of Barbados. Proceedings of the 19th International Liege Colloquium on Ocean Hydrodynamics, (this volume).

Gregg, M. C. and T. Sanford, 1987. Shear and turbulence in a thermohaline staircase. Deep-Sea Res., in press.

Hamilton, J. M., M. R. Lewis and B. R. Ruddick, 1987. Vertical fluxes of nitrate associated with salt fingers in the world's oceans. J. Geophys. Res., submitted.

Holyer, J. Y., 1981. On the collective instability of salt fingers. J. Fluid Mech., 110, 195-207.

Holyer, J. Y., 1984. The stability of long, steady two-dimensional salt fingers. J. Fluid Mech., 147, 169-185.

Kelley, D. E., 1987. Explaining effective diffusivities within diffusive oceanic staircases. Proceedings of the 19th International Liege Colloquium on Ocean Hydrodynamics (this volume).

Kunze, E., A. J. Williams III, and R. W. Schmitt, 1987. Optical microstructure in the thermohaline staircase east of Barbados. Deep-Sea Res., in press.

Kunze E., 1987. Limits on growing, finite length salt fingers: a Richardson number constraint. J. Mar. Res., 45, 533-556.

Lambert, R. B. and W. E. Sturges, 1976. A thermohaline staircase and vertical mixing in the thermocline. Deep-Sea Res., 24, 211-222.

Ledwell, J. R. and A. J. Watson, 1987. The use of deliberately injected tracers for the study of diapycnal mixing in the ocean. Proceedings of the 19th International Liege Colloquium on Ocean Hydrodynamics (this volume).

Linden, P. F., 1973. On the structure of salt fingers. Deep-Sea Res., 20, 325-340.

Linden, P. F., 1974. Salt fingers in a steady shear flow. Geophys. Fluid Dynamics, 6, 1-27.

Linden, P. F., 1978. The formation of banded salt finger structure. J. Geophys. Res., 83, 2902-2912.

Lueck, R., 1987. Microstructure measurements in a thermohaline staircase. Deep-Sea Res., in press.

Magnell, B., 1976. Salt fingers observed in the Mediterranean outflow region (34°N, 11°W) using a towed sensor. J. Phys. Oceanogr., 6, 511-523.

Marmorino, G. O., W. K. Brown and W. D Morris, 1987. Two-dimensional temperature structure in the C-SALT thermohaline staircase. Deep-Sea Res., in press.

McDougall, T. J., 1981. Double-diffusive convection with a nonlinear equation of state. Part II, Laboratory experiments and their interpretation. Prog. Oceanog., 10, 91-121.

McDougall, T. J. and J. R. Taylor, 1984. Flux measurements across a finger interface at low values of the stability ratio. J. Mar. Res., 42, 1-14.

McDougall, T. J., 1985. Double-diffusive interleaving. Part II: finite amplitude steady state interleaving. J. Phys. Oceanogr., 15, 1542-1556.

Osborn, T. R., 1980. Estimates of the local rate of vertical diffusion from dissipation measurements. J. Phys. Oceanogr., 10, 83-89.

Schmitt, R. W., 1979a. The growth rate of supercritical salt fingers. Deep-Sea Res., 26, 23-40.

Schmitt, R. W., 1979b. Flux measurements on salt fingers at an interface. J. Mar. Res., 37, 419-436.

Schmitt, R. W., 1981. Form of the temperature - salinity relationship in the Central Water: evidence for double-diffusive mixing. J. Phys. Oceanogr., 11(7), 1015-1026.

Schmitt, R. W., 1987. The Caribbean Sheets and Layers Transects (C-SALT) Program. EOS, Trans. Amer. Geophys. Union, 68,(5), 57-60.

Schmitt R. W. and D. L. Evans, 1978. An estimate of the vertical mixing due to salt fingers based on observations in the North Atlantic Central Water. J. Geophys. Res., 83, 2913-2919.

Schmitt, R. W. and D. T. Georgi, 1982. Finestructure and microstructure in the North Atlantic Current. J. Mar. Res., 40 (suppl.), 659-705.

Stern, M. E., 1969. The collective instability of salt fingers. J. Fluid Mech., 35, 209-218.

Stern M. E. and J. S. Turner, 1969. Salt fingers and convecting layers. Deep-Sea Res., 16, 497-511.

Turner, J. S., 1973. Buoyancy effects in fluids. Cambridge University Press, Cambridge, pp. 251-287.

Turner, J. S., 1967. Salt fingers across a density interface. Deep-Sea Res., 14, 599-611.

Williams, A. J., III, 1975. Images of ocean microstructure. Deep-Sea Res., 22, 811-829.

Worthington, L. V., 1976. On the North Atlantic Circulation. Johns Hopkins Oceanographic Studies No. 6, The Johns Hopkins University Press, Baltimore, 110 pp.

Wüst, G., 1978. The Stratosphere of the Atlantic Ocean. English translation, W. J. Emery, Editor. Amerind Pub. Co., New Delhi, 112 pp.

MIXING IN THE THERMOHALINE STAIRCASE EAST OF BARBADOS

M.C. GREGG
Applied Physics Laboratory and School of Oceanography, College of Ocean and Fishery Sciences, University of Washington, Seattle, WA 98105 (USA)

ABSTRACT

During the C-SALT measurements in November 1985 the average dissipation rate, ε, in the well-defined thermohaline staircase east of Barbados was only 2% of that expected. When driven by thermal convection, turbulent boundary layers in the atmosphere, ocean, and laboratory exhibit $\varepsilon \approx J_B$, where J_B is the buoyancy flux into the layer. Using observed ΔT and Δs across the steps in flux laws obtained for laboratory salt fingering experiments (McDougall and Taylor, 1984) gives J_B about 50 times the observed dissipation rates. The average Cox numbers across the staircase were slightly higher than those found at the same N in thermoclines lacking staircases. The dominant thermal microstructure consisted of nearly continuous temperature inversions 5–10 cm thick throughout the interfaces. They may be signatures of salt fingers strongly tilted from the vertical by interfacial shear; average velocity differences across the interfaces were usually 15–25 mm s^{-1}, much greater than estimated vertical velocities in salt fingers. In summary, the sharpness of the layers and the thermal microstructure in the interfaces are evidence of active mixing by strongly tilted salt fingers; the unexpectedly low dissipation rates, however, demonstrated that the turbulent buoyancy fluxes driving the mixing are apparently much weaker than expected. The horizontal structures found by other C-SALT investigators suggest that it may not be possible to understand the staircase simply by vertical processes and that it must be considered as a three-dimensional field.

1 INTRODUCTION

The role of salt fingering in the ocean remains a major puzzle. Although much of the thermocline is diffusively unstable to fingering, the distinctive staircases produced in laboratory experiments are found in only a few oceanic regimes. Schmitt and Evans (1978) argue that fingering occurs intermittently where internal wave strain locally increases the gradients, but reversals of the wave field disrupt the fingers before layers are formed. The finger growth rate depends on $R_\rho \equiv \alpha \overline{T_z}/\beta \overline{s_z}$ and is shorter than the local buoyancy period only when R_ρ is less than 2 (α is the thermal expansion coefficient; β is the coefficient of haline contraction). In most thermoclines $R_\rho \gtrsim 2$ and layers are not found. Prominent staircases have been found where $R_\rho < 2$. Under the Mediterranean outflow $R_\rho = 1.3$, in the Tyrrhenian Sea $R_\rho = 1.15$, and in the main thermocline east of Barbados $R_\rho = 1.6$. Finding staircases in low-R_ρ profiles, and nowhere else, provided the initial evidence for attributing their origin to double diffusion. Discovery of vertically banded shadowgraph images in the interfaces under the Mediterranean outflow (Williams, 1974) was taken as confirmation.

Understanding the dynamics of one staircase where salt fingering appears dominant is a first step in assessing the role of fingering in typical thermoclines. The C-SALT program examined the staircase east of Barbados. Although this field has the highest R_ρ, and possibly the weakest fingering, it is the most accessible because of its relatively shallow depth. The major objectives of C-SALT were to map the three-dimensional structure, observe temporal evolution, and measure mixing levels (Schmitt, 1987). Previously, no measurements of the viscous and thermal dissipation rates, ε and χ, had been reported from staircases.

Our measurements were made during November 1985, when the RV *Knorr* followed a drogued buoy for 5 days near 12°N, 56° 30′ W (well within the staircase field). Data taken from the *Knorr* with the Multi-Scale Profiler (MSP) are used to determine average levels of shear, stratification, and mixing. Then structures within the layers and interfaces are examined for clues about formation of the microstructure. Laboratory experiments provide the framework for interpreting the observations, but, as usual, the degree to which the experiments simulate the ocean is an issue. Before proceeding, some pertinent characteristics of the MSP are needed to understand the data and their limitations.

2 THE MULTI-SCALE PROFILER

The MSP was designed to relate microstructure directly to fine-scale shear by measuring both. Developed and operated jointly with Tom Sanford, the MSP avoids the ambiguities inherent in sensing shear and microstructure from different vehicles (Gregg and Sanford, 1980; Gregg et al., 1986). To provide a stable platform, it is 4.3 m long, 0.42 m in diameter, and weighs 295 kg in air (Fig. 1). Drag brushes at the top and droppable lead weights at the bottom are adjusted to yield a fall rate of 0.25 m s^{-1}. Typical drops during C-SALT went from the surface to 7 MPa (700 m), below the staircase.

A large suite of sensors measures finestructure and microstructure over scales from 1 cm to the profile length (Table 1). Multiple channels provide redundancy as well as overlapping spatial resolution. Voltage signals are digitized with a 16-bit analog-to-digital converter, and frequency signals are digitized with a hybrid period/frequency counter yielding high resolution with short gate times. Solid state memory records the data, which are subsequently read on deck without opening the instrument; this shortens the time between drops and protects the electronics.

Fine-scale velocity is sensed electromagnetically and acoustically. Short blades below the drag brushes cause a rotation about every 15 m, which is useful for monitoring low-frequency drift of both velocity channels. Two sets of electrodes measure voltages induced by seawater moving through the earth's magnetic field (Sanford et al., 1978). The electrode ports are mounted midway down the plastic tube encasing the aluminum pressure cylinder. Velocity is obtained directly, but has a depth independent, unknown offset. This does not affect the shear. A Neil Brown Instrument Systems (NBIS) acoustic current meter on the lower end cap measures changes in the travel time across two orthogonal 0.2-m-long paths, which are perpendicular to the vehicle axis. Because the instrument responds quickly to water motions over scales larger than the instrument,

Fig. 1. Deployment of the MSP.

the acoustic velocities are relative to the MSP. They are converted to water motion using the model of Evans et al. (1979). The long vertical wavelength structure absent from the uncorrected acoustic current meter data (left panel of Fig. 2) is often restored to about a centimeter per second, as evident in the right-hand panel. Noise of several centimeters per second in the EM data is larger than in previous measurements and results from the slow rotation rate; the rate had been decreased to minimize distortion of the dissipation-scale velocity measurements. Also apparent is the better resolution of wavelengths 10 m and smaller in the acoustic record. This record was therefore used for scientific analysis.

Fine-scale temperature and conductivity are measured with a set of Sea-Bird probes and with a Fastip thermistor and a 3-cm NBIS cell. The Sea-Bird temperature (Pederson, 1969) and conductivity (Pederson and Gregg, 1979) are very stable, allowing in-situ calibration of the other pair. The Sea-Bird sensors, however, give a coarser spatial resolution (Gregg and Hess, 1985) than do the thermistor (Gregg and Meagher, 1980) and NBIS cell (Gregg et al., 1982). Neither data set

TABLE 1

Characteristics of MSP data channels. Vertical resolutions give inherent sensor limitations; less resolution is obtained where weak mean gradients yield signals less than the noise.

Sensor	Sample Rate/Hz	Resolution/m
Finestructure		
EM velocity (2)	25	10
NBIS velocity	62.5	0.2
Sea-Bird temperature	25	0.2
Sea-Bird conductivity	25	0.3
NBIS conductivity	125	0.1
Thermometrics Fastip thermistor	125	0.01
Microstructure		
Airfoil shear (2)	125	0.01
Thermometrics Fastip thermistor	250	0.01
Nose two-needle conductivity	250	0.01
Strut-mounted two-needle conductivity (2)	250; 500	0.01
Vehicle Motion		
Paroscientific Digiquartz pressure	25	0.08
Develco flux-gate magnetometers (3)	25	N.A.
Sundstrand Q-flex accelerometers (3)	25	N.A.

was corrected for vertical resolution; hence, the Sea-Bird pair smoothes wavelengths smaller than about 1 m, and the other pair produces sharp spikes. Average gradients across each step and layer were determined by defining their boundaries to the nearest kilopascal (10 cm) and extrapolating salinity from the uniform level in the layers. To show the sharpness of the steps, plots were made with the NBIS data.

The microstructure of horizontal velocity is sensed with two airfoil probes (Osborn, 1974). Because previous measurements demonstrated horizontal isotropy, both probes are oriented to sense the same velocity component. This facilitates comparing the two ε profiles to detect plankton impacts, which are usually not simultaneous at both sensors. The probes were constructed following procedures developed by Neil Oakey and can resolve wavelengths of about 1 cm (Ninnis, 1984). The slow, stable descent produces few vibration spikes compared with lighter, faster-falling vehicles (Fig. 3). Consequently, the noise level is low, with minimum dissipation rates between 10^{-10} and 10^{-11} W kg^{-1}. Data processing is similar to that used for another instrument (Shay and Gregg, 1986), with

$$\varepsilon = 7.5\,\nu \int_{0.5\,\text{cpm}}^{k_c} \phi_S(k_z)dk_z \quad [\text{W kg}^{-1}], \qquad (1)$$

where ϕ_S is the shear spectrum, ν is the kinematic viscosity, and k_c is the wavenumber at which the spectrum begins to be affected by noise.

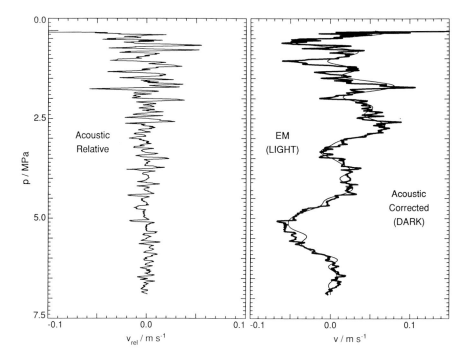

Fig. 2. Northward velocity, v, for MSP 70. The relative velocity from the acoustic current meter, on the left, shows pulses of several centimeters per second in the staircase, between 3 and 6 MPa. These occurred across the steps. After correction for vehicle motion, the velocity from the acoustic current meter overlays the large-scale structure of the data from the electrodes.

Scalar microstructure is measured with high-pass, high-gain circuits for temperature (Gregg et al., 1978) and for conductivity (Meagher et al., 1982), using a Fastip thermistor and a two-needle electrode. One two-needle conductivity probe is mounted on the lower end cap, and two others can be mounted on struts to examine signals along a helical trajectory. Noise of the temperature microstructure is low and repeatable (Fig. 4). The rate of diffusive dissipation of thermal fluctuations is

$$\chi_T = 6 \, \kappa_T \int\limits_{0.5 \text{ cpm}}^{k_c} \phi_{TG}(k_z) dk_z \quad [\text{K}^2 \text{s}^{-1}], \tag{2}$$

where ϕ_{TG} is the temperature gradient spectrum and k_c is the wavenumber at which the spectrum falls to the noise level.

Expressions (1) and (2) are both based on the assumption of isotropy. In (2) the effect of isotropy is included as a factor of 3 times the χ_T contribution due to only the observed vertical temperature gradient. In (1) isotropy appears as a factor of 3.75 times the sum of the two horizontal gradients, which have been found equal in previous measurements. Gargett et al. (1984) found that turbulence in a fjord was isotropic at dissipation scales when $\varepsilon \geq 200\nu N^2$.

458

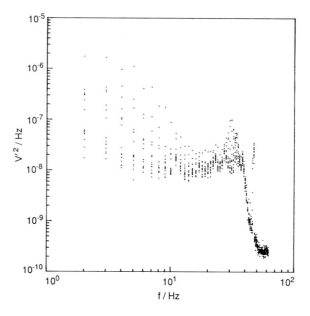

Fig. 3. Superimposed spectra from one of the airfoil channels, illustrating a wide range of signal levels. The largest signals, from well above the staircase, fall to the noise level by 15 Hz, which corresponds to 60 cpm (cycles per meter). A small vibration spike occurs near 30 Hz. Another near 50 Hz is greatly attenuated by anti-alias filters.

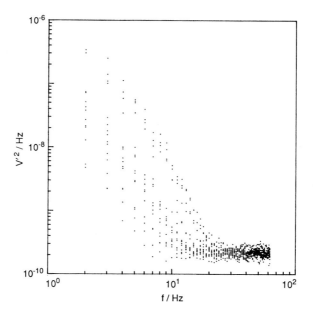

Fig. 4. Superimposed spectra from the microstructure temperature gradient circuit. All the signals were less than the noise level for frequencies greater than 25 Hz.

Three-axis accelerometers and magnetometers measure vehicle motion. The magnetometer output is routinely used to measure orientation in converting fine-scale velocities to earth-centered coordinates. Tilt and vibration are monitored with the accelerometers, but to date there has been no need to incorporate the data into the routine analysis.

3 SHEAR AND STRATIFICATION

The principal staircase was between 3 and 6 MPa (300 to 600 m) and usually had 9 or 10 well-developed layers. Overall, average thicknesses were 17 m for layers and 10 m for steps (Table 2). Typically, however, the layers were thicker (20–30 m) and the interfaces thinner (2–5 m) in the bottom half of the staircase. N^2 and $S^2 \equiv (\Delta u / \Delta z)^2 + (\Delta v / \Delta z)^2$ were computed with $\Delta z = 0.5$ m. Both have local maxima of 0.01 to 0.02 s^{-1} in the interfaces (Fig. 5). Shear typically drops to 0.001–0.002 s^{-1} in the layers. The time scale for large-scale motions to cross the layer is, at most, $S^{-1} \approx 20$ minutes, or about 1 hour for a complete overturn. The frequent density instabilities within the layers are plotted as $N = 0$.

Successive MSP profiles frequently showed slow, steady movement across steps (Gregg and Sanford, 1980), but little persistence across layers. Therefore, the average shear across the layers in Table 2 is simply a measure of the mean velocity difference and could represent the velocity of large-scale eddies, internal waves, or average shear. Gradient Richardson numbers, $Ri \equiv N^2/S^2$, formed from the average gradients across steps and layers, are high across the steps and nearly zero in the layers (Fig. 6). Statistics for all steps found $Ri > 5$ in 60% of the interfaces; only 10% had $Ri \leq 1$, and none had $Ri \leq 0.25$. Thus, on average, little mixing due to shear instability is expected across the steps, although low Ri regions may occur locally within the steps. Because $N \approx 0$ in the layers while $S \approx 0.001$, $Ri \approx 0$; i.e., the layers are more homogeneous in density than in velocity. A very low Richardson number due to $N \approx 0$, however, does not guarantee turbulence; shear is required for turbulent production.

TABLE 2

Average characteristics for the staircase between 3 and 6 MPa. The gradients are averages computed between the defined boundaries of each layer and step, e.g., Fig. 6. In the center of the staircase $\alpha = 1.7 \times 10^{-4}$ K^{-1}, and $\beta = 0.76$ (concentration units)$^{-1}$.

	ΔZ /m	S /s^{-1}	N /s^{-1}	T_z /K m^{-1}	ε /W kg^{-1}	χ_T /K^2 s^{-1}
Layers	17.2	8.4×10^{-4}	≈ 0	≈ 0	1.4×10^{-10}	6.7×10^{-9}
Steps	10.2	3.9×10^{-3}	6.7×10^{-3}	1.0×10^{-1}	4.9×10^{-10}	1.1×10^{-7}
Overall			3.5×10^{-3}	1.8×10^{-2}	1.9×10^{-10}	2.4×10^{-8}

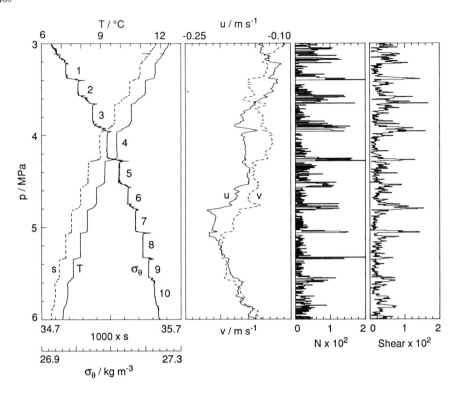

Fig. 5. Stability frequency, N, and shear, $(S^2)^{1/2}$, over successive 0.005-MPa intervals have maxima in the interfaces. Density derived from the Sea-Bird sensors was used for N. The coarser resolution underestimates N somewhat, but the error is less than that produced by spikes in the other data, used here for temperature, salinity, and density.

The MSP shear levels are similar to those obtained in the staircase under the Mediterranean outflow (Simpson et al., 1979) and in a staircase north of Haiti (Evans, 1981). Simpson et al., however, estimated $0.3 \le Ri \le 1$ in the interfaces, using nearly simultaneous CTD profiles for N^2. Evans found $Ri \approx 0$ in the layers and $30 \le Ri \le 100$ in the steps. The C-SALT Richardson numbers are much closer to Evans's results, although infrequent low values were found in our data.

4 DISSIPATION RATES

Viscous dissipation rates over successive 0.005-MPa intervals ranged from nearly 10^{-11} W kg^{-1} to slightly more than 10^{-8} W kg^{-1} (Fig. 7). Distinguishing noise from signal is difficult at the lower end of the dissipation range. Therefore, 10^{-10} W kg^{-1} was conservatively defined as the noise level and used as the lower bound of the plot.

Signals larger than 10^{-10} W kg^{-1} were found in most layers (Fig. 7), and it is these signals exceeding the noise that determine the layer average. Even so, averages for most layers were less

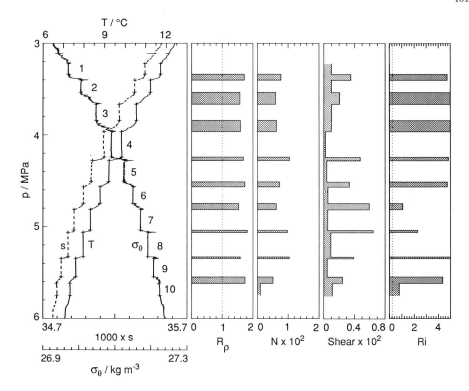

Fig. 6. Boundaries defining steps and layers are shown by + symbols on the temperature plot. N, S, and Ri are averages between those boundaries. Where probe resolution affected temperature and salinity, uniform levels within the layers were extrapolated to the boundaries. Owing to the weak stratification and to the small but significant shear, $Ri \approx 0$ in the layers. Although S is large in the steps, the very strong stratification produces high Richardson numbers.

than 10^{-10} W kg^{-1}. In Fig. 7, layers 5 and 6 were the exceptions. The overall layer average is 1.4×10^{-10} W kg^{-1} (Table 2).

Dissipation rates were larger in the interfaces, but the increase may have been partly caused by the response of the airfoil probes to both the strong mean shear and the large temperature gradients in the interfaces. In any event, the increase was too small to exceed the threshold for production of a turbulent buoyancy flux in a stratified profile. Van Atta and his colleagues report that turbulence produces a net buoyancy flux in decaying grid wakes only if $\varepsilon > (15\text{–}25) \, \nu N^2$ (Stillinger et al., 1983; Rohr et al., 1984). Gargett et al. (1984) found that spectral isotropy was assured if $\varepsilon > 200 \, \nu N^2$. As seen in Fig. 7, $\varepsilon < 25 \, \nu N^2$ in the steps. Therefore, assuming spectral isotropy is not warranted in the interfaces, those dissipation rates are upper bounds.

462

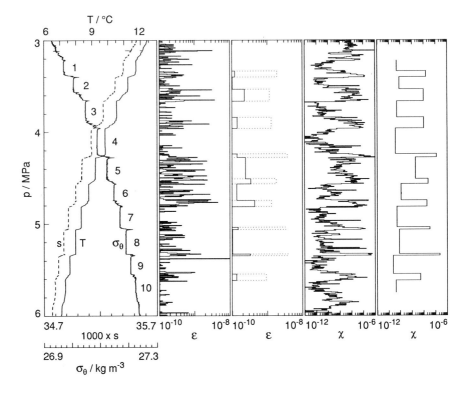

Fig. 7. Viscous and diffusive dissipation rates over 0.005-MPa intervals (second and fourth panels) and averaged over each step and layer (third and fifth panels) for MSP 71. Dotted lines in panel 3 are $\varepsilon_{tr} = 24.5\,\nu N^2$, the minimum dissipation rate for turbulence in stratified fluids to produce a buoyancy flux (Stillinger et al., 1983). Dissipation rates in all the steps are much less than ε_{tr}. Only layers 5 and 6 have average dissipation rates above the noise level.

Average dissipation rates across the staircase and within the interfaces were somewhat lower than those found at the same stratification in thermoclines lacking staircases (Fig. 8 and Table 2).

Salt fingering fluxes have been determined in laboratory experiments by observing temperature and salinity as functions of time in layers above and below a fingering interface. The interfacial fluxes of heat, J_q, and of salt, J_s, required to produce those changes are computed and parameterized as functions of Δs and R_ρ across the interface. We used polynomial fits to the McDougall and Taylor (1984) fluxes, which were obtained for $1.2 \le R_\rho \le 2.2$. (The polynomials were kindly supplied by T. McDougall as a private communication.) The buoyancy flux is

$$J_B = (g/\rho)\,[(\alpha/c_p)J_q - \beta J_s] \quad [\text{W m}^{-2}]\,, \tag{3}$$

where c_p is the specific heat. Both the heat and salt fluxes are negative, i.e., downward, but, because $\beta J_s > \alpha J_q/c_p$ for salt fingering, the buoyancy flux is positive. Averaged over steps 0 through 9, $J_B = (0.77 \pm 0.39) \times 10^{-8}$ W kg^{-1}, and $J_q = -43.9 \pm 21.1$ W m^{-2}.

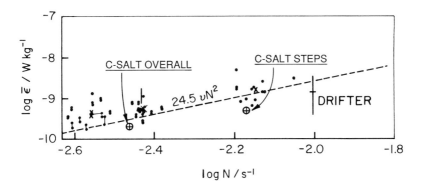

Fig. 8. Average viscous dissipation rates between 3 and 6 MPa compared with observations in thermoclines lacking staircases. The comparison values were taken from Gregg et al. (1986), which built on previous work by several authors. The C-SALT data lie below nearly all the others. The Drifter observations were made in a diffusively stable profile.

For steady homogeneous turbulence, the turbulent kinetic balance equation in a simple shear flow reduces to

$$\overline{u_1' u_3'} \frac{\partial \overline{u}_1}{\partial x_3} = J_B - \varepsilon . \tag{4}$$

Thus a positive buoyancy flux is a source of turbulent energy. For turbulent thermal convection in nearly steady state, with negligible mean shear and advection, large coherent eddies carry the buoyancy flux and are the energy-containing scales for the turbulent cascade. Therefore, the turbulent kinetic energy balance reduces to $J_B \approx \varepsilon$; turbulence generated by the release of potential energy is dissipated locally by viscosity. This balance has been confirmed in laboratory experiments (Willis and Deardorff, 1974), in the planetary boundary of the atmosphere (Kaimal et al., 1976), and in the surface layer of the ocean (Shay and Gregg, 1984). Although dissipation rates have not been reported from laboratory salt fingering experiments, Larson and Gregg (1983) argued that $J_B \approx \varepsilon$ should obtain when fingering is the dominant process producing turbulence. The thermal Rayleigh number, $R_a \equiv g \alpha \Delta T (\Delta z)^3 / \kappa_T v \approx 10^{10} \Delta T (\Delta z)^3$, is at least 10^{10} for the layers (the salinity Rayleigh number is $R_s \equiv R_\rho^{-1} R_a$), much larger than needed for the transition from steady to turbulent convection (Turner, 1973). Contrary to expectation, $\overline{\varepsilon}$ in the layers was only 2% of J_B predicted using the McDougall and Taylor (1984) flux formulation. Lueck (1987) found a similar result. Either the buoyancy flux was transmitted without producing turbulence or it was much smaller than given by McDougall and Taylor.

If the layers were fully turbulent their average thickness, $\Delta z = 17$ m, would be the energy-containing scale. Velocity differences across the layers were $\Delta u \approx S \Delta z \approx 0.014$ m s^{-1}. Taking Δu as the magnitude of the energy-containing velocities yields $\varepsilon \approx (\Delta u)^3 / \Delta z = 1.8 \times 10^{-7}$ W kg^{-1}.

This estimate, based on dimensional analysis, is usually accurate to within an order of magnitude (Tennekes and Lumley, 1972). Clearly, $\Delta z \approx 17$ m and $\Delta u \approx 0.014$ m s^{-1} are not the turbulent scales.

The other possible balance in (4) is turbulent production by the Reynolds stress and local dissipation. Using the average dissipation and shear yields a Reynolds stress of $\overline{u_1'u_3'} = \varepsilon/\partial\overline{u}_1/\partial x_3 \approx 10^{-7}$ m^2 s^{-1}. The corresponding velocity scale is $(u_1'u_3')^{\frac{1}{2}} \approx 0.4$ mm s^{-1}, much smaller than the observed velocity fluctuations. Shear production cannot be ignored.

Diffusive dissipation rates, χ_T, were well above the noise, with the largest values in the steps typically a factor of 100 greater than in the layers (Fig. 7 and Table 2). Using the average χ_T and the respective mean gradients, the Cox numbers, $C_T \equiv \chi_T/2\kappa_T T_z^2$, were 259 across the staircase and 59 across the steps, somewhat larger than found in "normal" thermoclines (Fig. 9). These are upper estimates, however, owing to the factor of 3 for isotropy in Eq. (2). In addition to the low value of $\varepsilon/\nu N^2$, shadowgraph images (Kunze et al., 1987) also reveal horizontal layering that is inconsistent with isotropy.

Osborn and Cox (1972) argue that the vertical turbulent heat flux is given by $J_q \equiv -\rho c_p \kappa_T C_T T_z$ when the turbulence is in a steady balance between local production and dissipation, with negligible advective transport. The average Cox numbers yield -2.6 W m^{-2} across the full staircase and -3.3 W m^{-2} just across the steps. Owing to the lack of isotropy, both estimates should be reduced by a factor of 3. By comparison, the McDougall and Taylor (1984) fluxes give -43.9 W m^{-2}.

How well the microstructure observations estimate the fluxes is uncertain, but the most likely comparisons are $J_B \approx \varepsilon$ in the layers and application of the Osborn-Cox model in the interfaces. The approximate balance of $J_B \approx \varepsilon$ has been observed in layers driven by the buoyancy flux of turbulent thermal convection, but there is no evidence that it should hold within fingering interfaces.

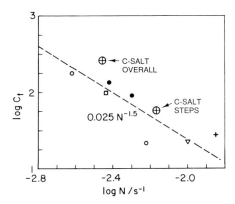

Fig. 9. Average Cox numbers across the steps and across the staircase compared with those from thermoclines lacking staircases (Gregg, 1987). Dissipation-scale isotropy was assumed in all cases. The step average lies close to the trend of the other data, the overall average slightly above.

Although not experimentally verified under any conditions, the Osborn-Cox model is formulated for thermal microstructure produced by vertical motion against the local mean temperature gradient. Convection is likely to carry microstructure into the layers from the interfaces, in contradiction to the assumptions. Using $J_B \approx \varepsilon$ in the layers gives a buoyancy flux through the layers of about 2% of that estimated from McDougall and Taylor (1984). Using the Osborn-Cox model in the interfaces, with no isotropy, yields a heat flux about 3% of the McDougall and Taylor estimate. The agreement is probably fortuitous, but both point to turbulent fluxes much lower than estimated from the only laboratory flux laws stated to be applicable for $R_\rho < 2$.

6 PATTERNS AND STRUCTURES

Some interfaces in the upper half of the staircase are resolved into thin well-mixed layers and subinterfaces. These are infrequent in the lower half of the staircase, where the simple interfaces range from 1 to 10 m thickness. The interfaces and subinterfaces contain nearly continuous temperature inversions. In one example (Fig. 10), Thorpe's (1977) resorting technique gives maximum vertical scales of 13 cm for the inversions; most are much less. Another example contains nearly regular structures with a spacing of 5 to 6 cm (Fig. 11).

Without salinity the static stability of the small inversions cannot be determined, but the low dissipation rates suggest that they are not associated with active turbulence. Owing to the strong mean shear across the interfaces and vertical velocities in salt fingers of about 0.1 mm s^{-1} (Kunze, 1987), the most likely interpretation of the small-scale inversions is that they are signatures of tilted salt fingers. Schmitt (personal communication, 1987) estimates finger diameters of 5.5 cm for these interfaces, in excellent agreement with the observed thickness.

The shadowgraph images (Kunze et al., 1987) are consistent with tilted fingers, but have a thickness of about 0.5 cm. By contrast Williams (1974) observed vertically banded shadowgraph images in the Mediterranean outflow staircase, and the thickness of the images agreed with estimated finger diameters. Why the C-SALT interfaces differ so much from those under the outflow is not understood.

Salt-stratified meter-scale temperature inversions within the layers are the other notable structures found in the data. The example in Fig. 12 has an inversion of 0.06°C with a salinity increase that produces a net increase in σ_θ. The homogeneous zones above and below the inversion are moving at several centimeters per second relative to the center of the inversion near 4.03 MPa. Thermistor chain tows found nearly vertical temperature fluctuations within layers. They were typically 0.005°C, much smaller than the inversion in Fig. 12 (Marmorino et al., 1987). Therefore, the relationship between the inversions and the plumes remains to be determined.

Plumes having $\theta' \approx 0.005$°C correspond to $\rho' \approx 3 \times 10^{-4}$ kg m^{-3} if they have the same θ-s relation as the mean profile. Because $J_B \equiv -(g/\rho)\overline{\rho' u_3'}$, a buoyancy flux of 10^{-10} W kg^{-1} requires velocities of only $u_3' \approx 0.1$ mm s^{-1}, much smaller than the observed fluctuations.

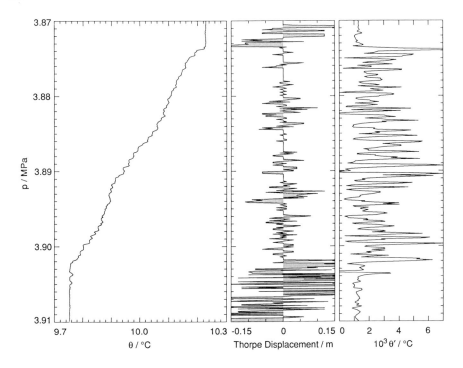

Fig. 10. Small temperature inversions occur throughout this 2.6-m-thick interface, which is representative of those observed during C-SALT. Thorpe (1977) displacements (middle panel) are less than 13 cm, and most are only a few centimeters. The panel on the right shows the temperature microstructure data. The electronics used with the microstructure channel have a relatively flat transfer function for wavelengths between 1 and 10 cm. Because the dominant variability is in this range, the data were scaled as temperature rather than as temperature gradient. The zero level is arbitrary.

7 SUMMARY

The principal findings are:

- On average, the layers were homogeneous in density, but were less well-mixed in velocity. Therefore, $N^2 \approx 0$ in the layers. Average shears of $\approx 10^{-3}$ s^{-1} and the weak stratification produced $Ri \approx 0$. Some layers contained salinity and temperature inversions that may be signatures of convective plumes.

- Average viscous dissipation rates were 1.4×10^{-10} W kg^{-1} in the layers. Consequently, the buoyancy flux in the layers due to turbulent convection must have been $\approx 10^{-10}$ W kg^{-1}, or about 2% of the flux predicted by applying McDougall and Taylor (1984) to observed ΔT and Δs across the steps.

- Only 10% of the steps had $Ri < 1$ and none had $Ri < 0.25$. Thus, although shear was high across the steps, shear instability should not have been an important factor in interfacial mixing.

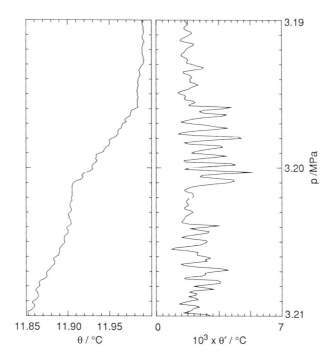

Fig. 11. Resolution of an interface in the upper part of the staircase, revealing a thin layer (3.201–3.204 MPa) and two high-gradient regions with continuous temperature inversions. Above the thin layer the inversions are nearly regular with vertical wavelengths of about 6 cm. The zero level of θ' is arbitrary.

- In spite of the low dissipation rates, thermal signatures consistent with tilted salt fingers occurred throughout the interfaces, in the form of continuous inversions having scales of about 5 cm. Removing the isotropy factor, the Osborn-Cox model gives $J_q = -1.1\ \mathrm{W\ m^{-2}}$, compared with $-43.9\ \mathrm{W\ m^{-2}}$ given by applying McDougall and Taylor (1984). Dissipation rates in the steps averaged $4.9 \times 10^{-10}\ \mathrm{W\ kg^{-1}}$. Even though this was computed from (1), assuming isotropy, it is much less than needed for isotropy using Gargett et al. (1984). Thus, the ε measurements provide further evidence for layering in the interfaces.

8 DISCUSSION

The large-scale C-SALT observations show sharply defined layers having great lateral coherence and a distinct slope across isopycnal surfaces (Schmitt, this volume), consistent with what we know of salt fingering. The low dissipation rates and the horizontally layered thermal structure of the interfaces, however, contradict the expected microstructure signatures.

After the microstructure measurements, Kunze (1987) noted that the $\Delta s^{4/3}$ flux law fitted to laboratory data is based on fingers extending completely through the interfaces. Although it was apparent from earlier CTD observations that the steps were several meters thick, compared with

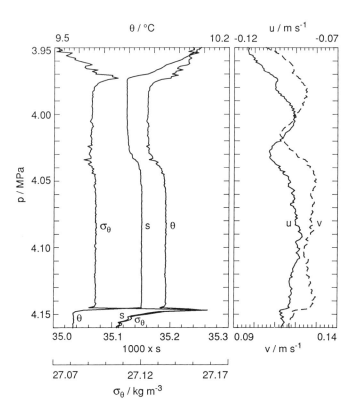

Fig. 12. Fine-scale salt-stabilized temperature inversion within a layer.

10–30 cm expected for fingers, this contradiction had not been considered. Assuming a Richardson number constraint on the fastest-growing fingers, Kunze (1987) obtained fluxes an order of magnitude smaller than those given by the $\Delta s^{4/3}$ law and thus much closer to the observed microstructure levels.

The structures observed in the interface have yet to be reconciled with the strong shear. Linden (1974) demonstrated that in the presence of shear the fingering instability forms two-dimensional sheets aligned in the direction of the mean shear. For linear shear in a unidirectional flow, the sheets are vertical and are aligned "downwind." The horizontally layered structures observed with MSP, with the horizontal tows (Marmorino et al., 1987; Lueck, 1987), and with shadowgraph profiles (Kunze et al., 1987) are not consistent with sheets. Other recent experiments (N. Larson, personal communication, 1987) demonstrate that both sheets and tilted fingers can occur in sheared flows. For the C-SALT interfaces, Kunze (1987) estimates fingering velocities of 0.1 mm s^{-1} and maximum finger lengths of 0.3 m. The observed interfacial shears would make such a finger nearly horizontal at the end of its trajectory, consistent with the C-SALT observations.

The combined C-SALT observations starkly demonstrate how little is known about salt fingering in the ocean. Resolution of the contradictions requires understanding the physics of salt fingering layers to a degree comparable with our knowledge of thermal convection. Direct measurements of fluxes, of dissipation rates, and of the fingers and convective cells are needed. Sheared regimes are necessary to simulate the ocean. This is an order of magnitude greater effort than made to date, but not more than is being devoted to stratified turbulence experiments.

If the oceanic fluxes are as low as presently estimated from the microstructure, it is possible that staircases develop only in locations of low mean flows, where the fluxes can act for long periods. The alternative is formation in confined localities where the fluxes are much larger, followed by weaker mixing, perhaps driven by the ambient internal waves, that can maintain the layers over a large area. For the C-SALT staircases, using floats to define the mean flow is an essential first step in resolving this issue. Until this accessible, well-defined staircase is understood, we are unlikely to determine the fluxes of salt fingering in the subtropical thermocline.

Acknowledgments. The MSP C-SALT work was conducted jointly with Tom Sanford on a project funded by the National Science Foundation, Grant OCE 8410741. Valuable comments were received from D. Kelley, E. Kunze, C. Gibson, J. Riley, E. D'Asaro, R. Schmitt, and G. Marmorino. School of Oceanography Contribution No. 1730.

9 REFERENCES

Evans, D. L., 1981. Velocity shear in a thermohaline staircase. Deep-Sea Res. 28: 1409–1415.

Evans, D. L., Rossby, H. T., Mork, M. and Gytre, T., 1979. YVETTE — a free-fall shear profiler. Deep-Sea Res., 26A: 703–718.

Gargett, A. E., Osborn, T. R. and Nasmyth, P. W., 1984. Local isotropy and the decay of turbulence in a stratified fluid. J. Fluid Mech., 144: 231–280.

Gregg, M. C., 1987. Diapycnal mixing in the thermocline: A review. J. Geophys. Res., 92: 5249–5286.

Gregg, M. C. and Hess, W. C., 1985. Dynamic response calibration of Sea-Bird temperature and conductivity probes. J. Atmos. Oceanic Tech., 2: 304–313.

Gregg, M. C. and Meagher, T. B., 1980. The dynamic response of glass rod thermistors. J. Geophys. Res., 85: 2779–2786.

Gregg , M. C. and Sanford, T. B., 1980. Signatures of mixing from the Bermuda Slope, the Sargasso Sea and the Gulf Stream. J. Phys. Oceanogr., 10: 105–127.

Gregg, M. C., and Sanford, T. B., 1987. Shear and turbulence in thermohaline staircases. Deep-Sea Res. (in press).

Gregg, M. C., Meagher, T. B., Pederson, A. M. and Aagaard, E. A., 1978. Low noise temperature microstructure measurements with thermistors. Deep-Sea Res., 25: 843–856.

Gregg, M. C., Schedvin, J. C., Hess, W. C. and Meagher, T. B., 1982. Dynamic response calibration of the Neil Brown conductivity cell. J. Phys. Oceanogr., 12: 720–742.

Gregg, M. C., D'Asaro, E. A., Shay, T. J. and Larson, N., 1986. Observations of persistent mixing and near-inertial internal waves. J. Phys. Oceanogr., 16: 856–885.

Kaimal, J. C., Wyngaard, J. C., Haugen, D. A., Cote, O. R., Izumi, Y., Caughey, S. J. and Readings, C. J., 1976. Turbulence structure in the convective boundary layer. J. Atmos. Sci., 33: 2152–2169.

Kunze, E., 1987. Limits on growing, finite-length, salt fingers: A Richardson number constraint. J. Mar. Res., 45: 533–556.

Kunze, E., Williams, A. J., III, and Schmitt, R. W., Jr., 1987. Optical microstructure in the thermohaline staircase east of Barbados. Deep-Sea Res. (in press).

Larson, N. G. and Gregg, M. C., 1983. Turbulent dissipation and shear in thermohaline intrusions. Nature, 306: 26–32.

Linden, P. F., 1974. Salt fingers in a steady shear flow. Geophys. Fluid Dyn., 6: 1–27.

Lueck, R. G., 1987. Microstructure measurements in a thermohaline staircase. Deep-Sea Res. (in press).

Marmorino, G. O., Brown, W. K. and Morris, W. D., 1987. Two-dimensional structure in the C-SALT thermohaline staircase. Deep-Sea Res. (in press).

McDougall, T. J. and Taylor, J. R., 1984. Flux measurements across a finger interface at low values of the stability ratio. J. Mar. Res., 42: 1–14.

Meagher, T. B., Pederson, A. M. and Gregg, M. C., 1982. A low-noise conductivity microstructure instrument. Oceans'82: Conference Record, Sept. 20–22, 1982, Marine Technology Society, Washington, D. C., 283–290.

Ninnis, R., 1984. The effects of spatial averaging on airfoil probe measurements of oceanic velocity microstructure. University of British Columbia, Vancouver (unpublished manuscript).

Osborn, T. R., 1974. Vertical profiling of velocity microstructure. J. Phys. Oceanogr., 4: 109–115.

Osborn, T. R. and Cox, C. S., 1972. Oceanic fine structure. Geophys. Fluid Dyn., 3: 321–345.

Pederson, A. M., 1969. An accurate low cost temperature sensor. Trans. Mar. Tech. Soc. Marine Temperature Measurements Symposium, 135–154.

Pederson, A. M. and Gregg, M. C., 1979. Development of a small in-situ conductivity instrument. IEEE J. Ocean. Eng., OE-4: 69-75.

Rohr, J. J., Itsweire, E. C. and Van Atta, C. W., 1984. Mixing efficiency in stably-stratified decaying turbulence. Geophys. Astrophys. Fluid Dyn., 29: 221–236.

Sanford, T. B. , Drever, R. G. and Dunlap, J. H., 1978. A velocity profiler based on the principles of geomagnetic induction. Deep-Sea Res., 25: 183–210.

Schmitt, R. W., Jr., 1987. The Caribbean Sheets and Layers Transects (C-SALT) program. EOS, Transactions, American Geophysical Union, 68: 57-60.

Schmitt, R. W., Jr., and D. L. Evans, 1978. An estimate of the vertical mixing due to salt fingers based on observations in the North Atlantic central water. J. Geophys. Res., 83: 2913–2919.

Shay, T. J. and Gregg, M. C., 1984. Turbulence in an oceanic convective mixed layer. Nature, 310: 282–285.

Simpson, J. H., Howe, M. R. and Morris, N. C. G., 1979. Velocity shear in the steps below the Mediterranean outflow. Deep-Sea Res., 26: 1381–1386.

Stillinger, D. C., Helland, K. N. and Van Atta, C. W., 1983. Experiments on the transition of homogeneous turbulence to internal waves in a stratified fluid. J. Fluid Mech., 131: 91–122.

Tennekes, H., and Lumley, J. L., 1972. A First Course in Turbulence. MIT Press, Boston, 300 pp.

Thorpe, S. A., 1977. Turbulence and mixing in a Scottish loch. Phil. Trans. Roy. Soc. London, 286: 125–181.

Turner, J. S., 1973. Buoyancy Effects in Fluids. Cambridge Univ. Press, Cambridge, 367 pp.

Williams, A. J. III, 1974. Salt fingers observed in the Mediterranean outflow. Science, 185: 941–943.

Willis, G. E. and Deardorff, J. W., 1974. Laboratory model of the unstable planetary boundary layer. J. Atmos. Sci., 31: 1297–1307.

LAYER THICKNESSES AND EFFECTIVE DIFFUSIVITIES IN "DIFFUSIVE" THERMOHALINE CONVECTION IN THE OCEAN

Konstantin N. FEDOROV
Institute of Oceanology, Acad. Sci. U.S.S.R, 117218 Moscow, U.S.S.R.

ABSTRACT

A parametrization is proposed to determine the characteristic convective layer thickness, H, of "diffusive"-type thermohaline staircases in the ocean and the related effective diffusivities for heat K_T and salt K_S in terms of the large-scale vertical temperature gradient $\frac{\partial T}{\partial z}$ and of the stability ratio $\bar{R}_\rho$ in the inversion layer. A critical Rayleigh number Ra_{cr}^* for an initial disturbance ΔT_0 on a smooth temperature profile is then defined in terms of the conductive interface thickness δ which corresponds to ΔT_0 at the beginning of a step formation. The results are compared with observations and with the earlier treatment of the same problem by Kelley (1984).

"Diffusive" thermohaline convection - one of the types of double-diffusive convection - may develop in hydrostatically stable oceanic regions where temperature and salinity increase downwards. Its basic principles are treated by Turner (1965, 1968, 1973), Huppert (1971), and Huppert and Linden (1979), while the earliest examples of convective steps in lakes and oceans are given by Hoare (1968), Neshiba et al. (1971) and Fedorov (1970, 1976). In this paper a parametrization is proposed to predict some characteristics of the step structure which could result from "diffusive" convection on the basis of the undisturbed background T,S-structure. The characteristics to be determined are :

1) the thickness, H, of convective layers before their possible merging;

2) the effective diffusivities for heat and salt, K_T and K_S, respectively;

3) a critical boundary-layer Rayleigh number, Ra_{cr}^*, which determines the critical value of an initial disturbance on a smooth temperature profile, ΔT_0, from which a series of convective steps could start growing.

The first attempt to solve a similar problem was undertaken by Kelley (1984). Using data from 15 pertinent observations of step-like structures in the ocean and some lakes, Kelley suggested a scheme of parametrization for the effective heat and salt fluxes in terms of large scale vertical gradients of T and S. For this purpose, a length scale $\left[\frac{k_T}{N}\right]^{\frac{1}{2}}$ was introduced for layer thicknesses on dimensional grounds. Then, the dependence of H on $\left[\frac{k_T}{N}\right]^{\frac{1}{2}}$ was found through an intermediate empirical relation between a Rayleigh number, Ra_H, and the stability parameter

$\overline{R}_\rho$. Here, $Ra_H = \dfrac{g\ \alpha\ \Delta T\ H^3}{\nu\ k_T}$, $\overline{R}_\rho = \dfrac{\beta\ \dfrac{\overline{\partial S}}{\partial z}}{\alpha\ \dfrac{\overline{\partial T}}{\partial z}}$, g is the acceleration due to gravity, T the tem-

perature, S the salinity, ρ the density of water, $\alpha = -\dfrac{1}{\rho}\ \dfrac{\partial \rho}{\partial T}$, $\beta = \dfrac{1}{\rho}\ \dfrac{\partial \rho}{\partial S}$, ΔT is the tempera-

ture step between the superimposed convective layers, ν the kinematic viscosity, k_T the molecu-lar conductivity, and $\overline{N}$ is the average Brunt-Vaisala frequency in the temperature inversion layer; z is directed downwards, and an overbar corresponds to a vertical average over several layer thicknesses.

An alternative to Kelley's attempt is proposed here, which does not imply any criticism of his approach. Indeed, both the following analysis – first presented, in Russian, by Fedorov (1986) – and the original study by Kelley (1984) are best viewed as exploratory. There is of course a motivation to seeking alternatives. Since the vertical heat flux F_T through the system of convective steps depends upon ΔT and $\overline{R}_\rho$ (Turner, 1973), i.e., upon $\dfrac{\overline{\partial T}}{\partial z}$, H and $\overline{R}_\rho$, it may

be more advantageous to base the relevant length scale also on $\dfrac{\overline{\partial T}}{\partial z}$ or on $\overline{N}_T = \left[g\ \alpha\ \dfrac{\overline{\partial T}}{\partial z}\right]^{\frac{1}{2}}$

instead of $\overline{N}$. The sought for parametrization, which is of the type $K_T = \phi\left[\overline{R}_\rho\right]$, may in this case take a simpler form. It may also be practical to apply the new length scale to a quantity more directly related to the heat flux F_T than the convective layer thickness H. In our opinion, such a quantity is the thickness δ of the high gradient interface between convective layers, taken at a time during its cyclic variations when the heat flux through the interface is still of a fully conductive nature. This approach may help to avoid the necessity of using empirical clo-sure of the scaling problem through Ra_H.

Let us therefore introduce a length scale

$$\delta_T = \left[\frac{k_T}{\overline{N}_T}\right]^{\frac{1}{2}} = \left[\frac{k_T^2}{g\ \alpha\ \dfrac{\overline{\partial T}}{\partial z}}\right]^{\frac{1}{4}} \tag{1}$$

and consider the thickness of the high gradient conductive interface, which is

$$\delta = f_0(\overline{R}_\rho)\ \delta_T \quad , \tag{2}$$

when scaled by (1). Here $f_0(\overline{R}_\rho)$ is a function to be determined.

It is assumed here that the conductive interface thickness attains its "natural" limit just prior to the occurrence of the cyclic convective instability of the two thin boundary layers adjacent to the highest gradient core. Hence, for the interface, Nu = 1. This assumption is based on the model of convection proposed by Howard (1966) and experimentally tested by Ginsburg and Fedorov (1978). The above assumption allows us to write :

$$- F_T = K_T \frac{\overline{\partial T}}{\partial z} = k_T \frac{\Delta T}{\delta} \quad , \tag{3}$$

where K_T, the effective diffusivity, is defined by the first equality in (3). Since $\Delta T = H \dfrac{\overline{\partial T}}{\partial z}$, one can obtain from (3) and (2) the following expression for K_T :

$$K_T = \frac{k_T H}{\delta} = \frac{k_T H}{f_0(\overline{R}_\rho) \, \delta_T} \quad . \tag{4}$$

According to Turner (1973),

$$-F_T = 0.085 \, f(\overline{R}_\rho) \, k_T \left[\frac{g \, \alpha}{k_T \, \nu} \right]^{\frac{1}{3}} \Delta T^{\frac{4}{3}} \quad , \tag{5}$$

where $f(\overline{R}_\rho)$ is the ratio between the heat flux through a liquid interface and the heat flux through a thin rigid conductive plate under the same temperature difference ΔT between the adjacent convective layers. One empirical expression for $f(\overline{R}_\rho)$ was obtained by Marmorino and Caldwell (1979) :

$$f(\overline{R}_\rho) = 0.101 \, \exp \left\{ 4.6 \, \exp \left[- 0.54 \, (\overline{R}_\rho - 1) \right] \right\} . \tag{6}$$

Since $K_T = \dfrac{- F_T}{\dfrac{\overline{\partial T}}{\partial z}}$ by definition, we may use (5) to obtain another expression for K_T :

$$K_T = 0.085 \, f(\overline{R}_\rho) \left[\frac{g \, \alpha}{k_T \, \nu} \right]^{\frac{1}{3}} H^{\frac{4}{3}} \left[\frac{\overline{\partial T}}{\partial z} \right]^{\frac{4}{3}} k_T \quad . \tag{7}$$

Equating the right hand sides of (4) and (7), we get :

$$H = \left[0.085 \, F(\overline{R}_\rho) \right]^{-3} Pr \, \delta_T \quad , \tag{8}$$

where a new function $F(\overline{R}_\rho) = f(\overline{R}_\rho) \, f_0(\overline{R}_\rho)$ has been introduced. Inserting into (8) the numerical values of H, Pr, and δ_T for the 15 cases reported by Kelley (1984), we find that

$$F\ (\overline{R}_\rho) \approx \text{const} = n \quad , \tag{9}$$

where $n = 3.67 \pm 0.26$.

Because of (9), the expression (8) for H takes the very simple form

$$H = 32.9\ \text{Pr}\ \delta_T \quad , \tag{10}$$

and, correspondingly, one can get from (10) and (2)

$$\frac{H}{\delta} \approx 9\ \text{Pr}\ f(\overline{R}_\rho) \quad . \tag{11}$$

From (11) and (6), it follows that the ratio $\dfrac{H}{\delta}$ is much greater when $\overline{R}_\rho \to 1$ than at larger values of $\overline{R}_\rho$. This may mean that with the intensification of turbulent convection in quasi-homogeneous layers (*i.e.,* when $\overline{R}_\rho$ is approaching 1) the conductive interfaces between them become thinner, which corresponds to the qualitative description of the interface behaviour in turbulent entrainment experiments by Rouse and Dodu (1955).

To the contrary, according to (10), H does not depend on $\overline{R}_\rho$. Therefore, (10) may be used to estimate H from the measured background values of $\dfrac{\partial T}{\partial z}$ only, while using the value of Pr for the observed temperature range. This may be very advantageous when measurements are made with a temperature probe only, say an XBT. Figure 1 shows a comparison of the thicknesses of convective layers calculated from (10), H_{calc}, with observed thicknesses, H_{obs}. This figure shows not only Kelley's 15 cases (circles) used to determine n, but also a number of independent observations (crosses) taken from other sources (Fedorov, 1976; Paka, 1984; Fedorov *et al.*, 1986; Perkin and Lewis, 1983). One of the new sources of data is the thermohaline staircase (Fig. 2) recently recorded in the Mediterranean Sea south of Peloponnesian Peninsula and described by Fedorov *et al.* (1986). The correlation coefficient between H_{calc} and H_{obs} is 0.99.

From (4) and (11), the following final expression for K_T may be obtained

$$K_T \approx 9\ \nu\ f(\overline{R}_\rho) \quad . \tag{12}$$

The coefficients K_S and K_ρ for effective salt and mass diffusivities can easily be expressed *via* K_T (see, *e.g.,* Huppert, 1971 or Fedorov, 1976), and the corresponding formulae found in Kelley (1984).

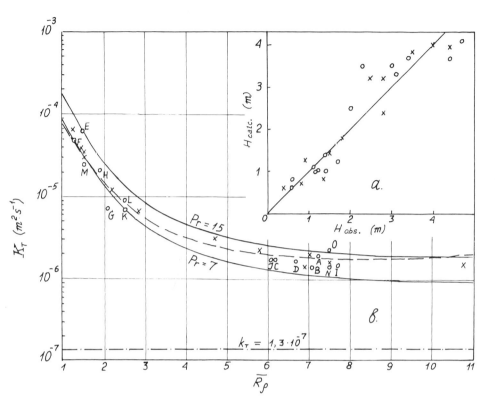

Fig. 1. a : Comparison of the calculated (H_{calc}) and the observed (H_{obs}) layer thicknesses. b : Dependence of the effective eddy diffusivity, K_T, on $\overline{R}_\rho$; the circles represent the 15 calibration data points taken from Kelley (1984) and the crosses correspond to 12 independent data points.

Figure 1b shows the dependence of K_T upon $\overline{R}_\rho$ for Pr = 7 and Pr = 15 which correspond to $\nu = 10^{-6}$ m^2 s^{-1} (T $\approx$ 22°C) and $\nu = 2 \cdot 10^{-6}$ m^2 s^{-1} (T $\approx - 1$°C) for sea water. One can see that the 15 calibration points (circles) corresponding to the data used by Kelley (1984) and the 12 independent points (crosses) are located between the two theoretical curves obtained by using the two values of ν in (12) (solid lines). Kelley's theoretical curve (dashed line) is found between those two extremes. Note that, in this figure and in figure 3, Kelley's data are annotated by the letters A to O.

The relevant Rayleigh number, which, in our opinion, relates to the convective step formation, should be expressed through the high gradient conductive interface thickness, δ, associated

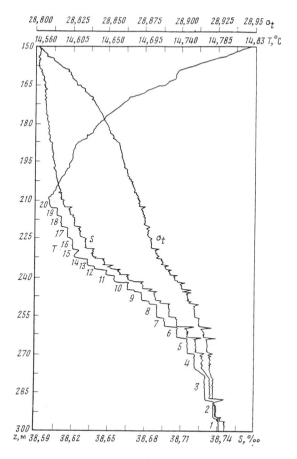

Fig. 2. Diffusive step-like structure in the Mediterranean Sea south of Greece: T is the temperature, S the salinity, and σ_t the density anomaly (for details see Fedorov *et al.*, 1986).

with the growing initial disturbance on the smooth temperature profile, ΔT_0:

$$Ra^* = \frac{g\,\alpha\,\Delta T_0\,\delta^3}{k_T\,\nu} \qquad . \tag{13}$$

This parameter will be critical (Ra^*_{cr}) when the value of δ substituted in (13) corresponds to Nu = 1 as in (2) and (3). The treatment here is again exactly as in Ginsburg and Fedorov (1978), where the physics of the boundary-layer Rayleigh number Ra^* is explained in detail in connection with the onset of convection in water cooled through its upper free surface. Taking into account (1), (2), and (9), and with $\Delta T_0 = \Delta T = H\,\dfrac{\overline{\partial T}}{\partial z}$, we obtain

$$Ra^*_{cr} = 1628\ /\ [f(\overline{R_\rho})]^3 \tag{14}$$

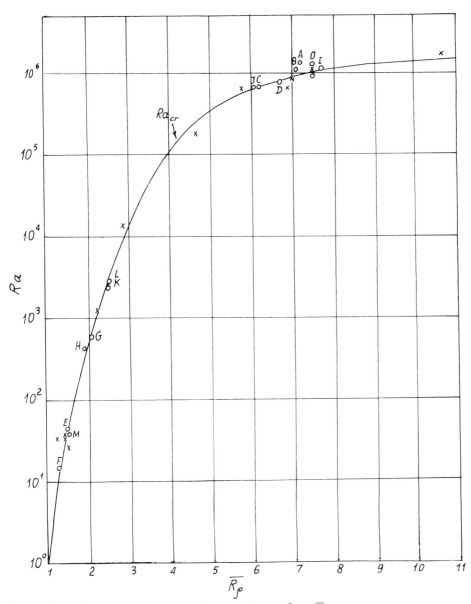

Fig. 3. Dependence of the interface Rayleigh number, Ra*, on $\overline{R}_\rho$. The solid line corresponds to critical values of the Rayleigh number (Ra^*_{cr}).

The dependence of Ra_{cr}^* on $\overline{R}_\rho$, according to (14), is shown on figure 3 (solid curve). The points correspond to values of Ra^* calculated from (13) with $\Delta T_0 = \Delta T$ taken from observational data and δ calculated from (2). It is of interest to note that the values of both Ra^* and Ra_{cr}^* vary in a very broad range from 10^0 to 10^6, increasing, as it should be expected, with increasing $\overline{R}_\rho$. For $3 \leq \overline{R}_\rho \leq 4$, the values of Ra_{cr} are of the same order of magnitude (10^4) as estimated by Turner (1968) for his experiments with a stratified fluid heated from below.

CONCLUSIONS

The results represented by (10), (12) and (14) call for additional comments. The estimates of H obtained from (10) relate only to layers which have not yet merged with the adjacent ones. Such layer merging is often the case in the lower portion of "diffusive" staircases, both in the laboratory (*e.g.*, Linden, 1976) and in the ocean (*e.g.*, Fedorov *et al.*, 1986). According to (10), H does not depend on $\overline{R}_\rho$ and may be estimated solely from $\dfrac{\partial T}{\partial z}$. This does not come as a surprise, since we have been aware for a long time that similar H values characterize the "diffusive" staircases observed in different ambient conditions, with $\overline{R}_\rho$ ranging from 1.5 to 7-8 (see *e.g.*, Table 1 in Kelley, 1984). In order to estimate K_T and Ra_{cr}^* from (12) and (14), one should know $\overline{R}_\rho$ and, consequently, one must have an estimate of $\dfrac{\partial S}{\partial z}$ to match the known $\dfrac{\partial T}{\partial z}$ value. However, according to (4), $\dfrac{K}{k_T} = \dfrac{H}{\delta}$, so that it may eventually be possible to estimate K_T from direct measurements of H and δ in the ocean. At present, however, conductive interfaces with typical δ of the order of few centimeters can not be resolved even with the best CTD-probes.

ACKNOWLEDGEMENTS

The author wishes to acknowledge helpful suggestions made by Trevor McDougall and Dan Kelley in the process of reviewing this paper.

REFERENCES

Fedorov, K.N., 1970. On the step-like structure of temperature inversions in the ocean. *Izv. Akad. Nauk SSSR, Fizika atmosfery i okeana*, v. 6, No. 11, 1178-1188 (in Russian).

Fedorov, K.N., 1976. *Tonkaya termohalinnaya struktura vod okeana (The Thermohaline Finestructure of the Ocean)*, 184 pp., Gidrometeoizdat, Leningrad (in Russian). English translation, 170 pp., 1978, Pergamon Marine Series, v. 2, Pergamon Press.

Fedorov, K.N., 1986. Layer thicknesses and effective diffusivities in "diffusive" thermohaline convection in the ocean. *Doklady Akad. Nauk SSSR*, v. 287, No. 5, 1230-1233 (in Russian).

Fedorov, K.N., V.T. Paka, G.R. Gamsakhurdia, M.V. Emelyanov, 1986. Analysis of a series of convective steps in a temperature inversion. *Izv. Akad. Nauk SSSR, Fizika atmosfery i okeana*, v. 22, No. 9, 969-978 (in Russian).

Ginsburg, A.I., K.N. Fedorov, 1978. On the critical boundary-layer Rayleigh number in water cooled through its upper free surface. *Izvestia Akad. Nauk SSSR, Fizika atmosfery i okeana*, v. 14, No. 4, 433-436 (in Russian).

Hoare, R.A., 1968. Thermohaline convection in Lake Vanda, Antarctica. *J. Geophys. Res.*, 73 (2), 607-612.

Howard, L.N., 1966. Convection at high Rayleigh number. *Proc. 11th Int. Congress of Appl. Mech., Munich, Germany*, 1109-1115.

Huppert, H.E., 1971. On the stability of a series of double-diffusive layers. *Deep-Sea Res.*, 18, No. 10, 1005-1022.

Huppert, H.E., and P.F. Linden, 1979. On heating a stable salinity gradient from below. *J. Fluid mech.*, 95 (3), 431-464.

Kelley, D., 1984. Effective diffusivities within oceanic thermohaline staircases. *J. Geophys. Res.*, 89, No.C6, 10,484-10,488.

Marmorino, G.O., and D.R. Caldwell, 1976. Heat and salt transport through a diffusive thermohaline interface. *Deep-Sea Res.*, 23, No. 1, 59-67.

Neshiba, S., V.T. Neal, and W. Denner, 1971. Temperature and conductivity measurements under Ice Island T-3. *J. Geophys. Res.*, 76 (33), 8107-8120.

Paka, V.T., 1984. Thermocline finestructure in the central part of the Crommwell Current. *Izv. Akad. Nauk SSSR, Fizika atmosfery i okeana*, 20, No. 1, 86-94 (in Russian).

Perkin, R.G., and E.L. Lewis, 1983. In "Frozen Sea Research Group", a section of the *1983 Annual Review of Activities,* Institute of Ocean Sciences, Patricia Bay, Canada, 34-35.

Rouse, H. and J. Dodu, 1955. Diffusion turbulente à travers une discontinuité de densité. *La Houille Blanche*, No. 4, 522-532.

Turner, J.S., 1965. The coupled turbulent transport of salt and heat across a sharp density interface. *Int. J. Heat and Mass Transfer*, 8, No. 5, 759-767.

Turner, J.S., 1968. The behaviour of a stable density gradient heated from below. *J. Fluid Mech.*, 33 (1), 183-200.

Turner, J.S., 1973. *Buoyancy Effects in Fluids*, 368 pp. Cambridge University Press, London, England.

Explaining Effective Diffusivities Within Diffusive Oceanic Staircases

Dan E. Kelley
Woods Hole Oceanographic Institution
Woods Hole, MA 02543 USA

Abstract

Before parameterizing double-diffusive fluxes as functions of large-scale vertical gradients of S and T, we must first parameterize the thickness H of the layers within the staircases. It has been shown (Kelley, 1984) that is possible for diffusive-case thermohaline staircases: oceanographic variability in H is collapsed onto a function of large-scale density ratio $R\rho$ when H is scaled by $(\kappa_T/<N>)^{1/2}$ (κ_T=molecular diffusivity; $<N^2>$=large-scale vertical gradient of buoyancy). Further confirmation of this scaling comes from additional data shown in this paper and in Fedorov's (1988) paper elsewhere in this volume.

Accordingly, we can find least-squares fits of the dependence of H on $<N>$ and $R\rho$. The scaling implies that the large-scale diffusivities $\{K_T, K_S\}$ are independent of $<N>$, so we can formulate them as functions of $R\rho$ (equations (3) and (4)) .

This allows the incorporation of double-diffusive fluxes in thermocline models. However, it is important to understand the scaling mechanistically before applying the diffusivities to cases not tested yet—e.g., the initiation of frontal interleaving. This paper aims at developing this understanding. I suggest that the scaling results from a balance between layer-merging and interface-splitting, processes which have opposite effects on H. A balance point is set by a critical value of a convection-based interfacial Richardson number, Ri. A relationship between Ri and H is developed. It shows that the critical value of Ri translates into the observed oceanographic relationship between $H/(\kappa_T/<N>)^{1/2}$ and $R\rho$. Thus, the mechanism quantitatively explains both the scaling and the large-scale diffusivities.

1   Significance Of Layer Thickness

1.1   The goal: incorporation of double-diffusion fluxes in large-scale models

Implications of the existence of double-diffusive staircases

Diffusive-case thermohaline staircases[1] are commonly found in the Arctic (Neshyba et al, 1971,1972; Padman & Dillon, 1987), the Antarctic (Middleton & Foster, 1980), salty lakes (Hoare, 1968; Newman, 1976), and at the boundaries between intrusions across thermohaline fronts (Perkin & Lewis, 1984).

The existence of well-formed thermohaline staircases in the presence of turbulent mixing, which disrupts such regular structures, indicates that the *double-diffusive fluxes must exceed tur-*

---

[1] *Terminology:* "Thermohaline staircases", systems comprising series of well-mixed horizontal layers separated by high-gradient interfaces, are called diffusive-case if both temperature (T) and salinity (S) increase with depth ($-z$) and finger-case if both S and T decrease with depth. This paper deals *only* with the diffusive-case.

bulent fluxes. This is because the timescale for a layer of thickness H to be disrupted by mixing[1] is H^2/K_T^{mix}, whereas the timescale for it to be reformed by double-diffusive fluxes is H^2/K_T^{dd}. Thus, a staircase can remain intact only if

$$K_T^{dd} > K_T^{mix};$$

that is, only if double-diffusion overwhelms mixing. This inequality demonstrates a common, but seldom stated, assumption well worth emphasizing: wherever well-formed staircases are found, double-diffusion must contribute significantly to vertical fluxes.

The prospect of parameterizing double-diffusive fluxes in terms of large-scale vertical gradients

Where can we expect to find staircases? How can K_T^{dd} be predicted *a priori*? In general, how can double-diffusion be incorporated into models resolving only larger scales? Such questions motivate most studies of double-diffusion. Since the focus of this Liege meeting is on connecting small-scale mixing with large-scale processes, these questions are particularly relevant here. We are making some progress towards the answers, but the path is seen increasingly as being non- monotonic.

Let me explain this gloomy comment by a short digression into the salt-finger case. Recent measurements in the C-SALT region, East of Barbados, call into question fundamental assumptions about finger convection [see Ray Schmitt's (1988) discussion in this volume]. It was hoped that C-SALT would supplement or overthrow the existing (sketchy) criteria for where finger staircases should occur. It did. It was also hoped that the shear and optical microstructure measurements would fit conveniently into the framework established by previous theoretical, laboratory and field studies. They did not. That's another story, better told by Gregg & Sanford (1987), Kunze *et al.* (1987) and Lueck (1987). For present purposes, the most disturbing suggestion of the C-SALT measurements is that the oft-used 4/3 flux laws, based on measurements in the laboratory, may not apply to the ocean. Until new flux laws [for example, those proposed by Kunze (1988)] are accepted, we cannot hope to incorporate salt-finger fluxes in large-scale models.

Luckily, this difficulty does not carry over to diffusive-case staircases. Oceanographic measurements confirm the laboratory-based 4/3 flux laws to within their factor-of-2 uncertainty (Gregg & Cox, 1972; Padman & Dillon, 1987). Accordingly, we may confidently apply the laboratory-based 4/3 flux laws to diffusive-case staircases in the ocean[2].

Starting from this framework, we need take only one more step before parameterizing diffusive-case double-diffusive fluxes in terms of large-scale gradients.

---

[1] The diffusivities (m^2 s^{-1}) for mixing and double-diffusion are here denoted K_T^{mix} and K_T^{dd}.

[2] A new theory suggests that the exponent in these laws is not 4/3, but instead lies between 5/4 and 4/3 (Kelley, 1986, 1988b). In spite of this development, the 4/3 law is used here for its familiarity. The difference in results when the two theories are used is minimal since each has empirical parameters which are adjusted to force the predictions to agree with measurements.

1.2   Why layer thickness is needed, and how to scale it

Why layer thickness is needed

The 4/3 flux laws (described in detail below) relate vertical fluxes of heat and salt to the temperature and salinity steps across interfaces (ΔT and ΔS). We can make some progress towards parameterizing fluxes in terms of large-scale ($< >$) quantities by noting that

$$(\Delta S, \Delta T) = H (<\partial S/\partial z>, <\partial T/\partial z>), \tag{1}$$

where H is the layer thickness defined as *the distance between the midpoints of adjacent inter-faces*[1].

We can parameterize the fluxes *only if* we can predict H in terms of large-scale properties of the water column.

How to scale layer thickness using dimensional analysis

In a previous paper (Kelley, 1984) I described a dimensional analysis which leads to a natural scaling for H. I'll review the analysis here for completeness.

We must find a length scale, H_0, with which to scale layer thickness H, thereby getting a dimensionless layer thickness which we will denote G. According to the usual method of dimensional analysis, H_0 must be expressed in terms of the parameters which describe the staircase. A list of relevant parameters, with notes about their possible contributions to the scaling, follows:

Large-scale shear:
    In general, we lack measurements of large-scale shear, so as a practical matter it cannot be used in a scaling for available data. It is likely that the shear varies over wide ranges in the available data set, ranging from small values in lakes, to larger values in oceans covered by ice, to still larger values in frontal intrusions in the ice-free mid-latitude ocean. Grouping the observations in these categories, it has been verified that large-scale shear does not enter into the scale (Kelley , 1984, 1986).

The double-diffusive buoyancy flux (or contributions to it from the salt or heat flux):
    These cannot be included in the scaling, because doing so leads to a circular result. The 4/3 flux laws state that flux is proportional to the 4/3 power of the temperature step across interfaces. Since this step is proportional to H, specifying H is equivalent to specifying buoyancy flux. In other words, scaling H by buoyancy flux is equivalent to scaling H by H itself: a circular scaling.

The Coriolis parameter:
    This cannot be involved, since the convective shears are large enough that friction outweighs rotation. A detailed demonstration of this was given by Kelley (1987a).

Molecular diffusivities:
    Since double-diffusion is based on molecular diffusion, the diffusivities for salt, heat and momentum—κ_S, κ_T, and ν—should appear in the scaling. Each of these diffusivities has dimensions *length²×time⁻¹*. From the set of three we must select one, the other two being taken into account through dimensionless ratios. According to dimensional analysis, any selection is equivalent. We may arbitrarily select κ_S, κ_T, or ν as the dimensional parameter. Let's select κ_T as the dimensional parameter,

---

[1]   This definition of H should be followed exactly when comparing new observations to those presented here. An alternative, equally reasonable, definition is the thickness of the well-mixed interior of the layer. This leads to biased estimate typically 20% lower.

allowing for the other two diffusivities *via* the Prandtl number $Pr = \nu/\kappa_T$ and the Lewis number $Le = \kappa_S/\kappa_T$. A rationale, or at least a mnemonic, for this choice is that diffusive-case convection is driven by the temperature distribution, so we might expect larger values of κ_T to lead to enhanced convection, to larger fluxes, and to thicker layers. Note, however, that this rationale is not necessary for, or even particularly relevant to, the dimensional analysis.

Large-scale Brunt-Vaisala parameter:
The vertical gradient of buoyancy smoothed over several layer thicknesses, $<N^2>$, should be included in the scaling. For a given buoyancy flux we might expect *a priori* to find thinner layers in regions with large $<N^2>$ than in regions with small $<N^2>$, because more work must be done to homogenize a layer of given thickness for large $<N^2>$. This suggests that $<N>$ be included in the scaling.
The individual components to $<N>$ made by the salinity and temperature fields also should be included, because their ratio, the 'density ratio',

$$<R\rho> = \frac{\beta<\partial S/\partial z>}{\alpha<\partial T/\partial z>}$$

is a major parameter determining the magnitude of the double-diffusive fluxes. Here, β and α are $\rho^{-1}\partial\rho/\partial S$ and $-\rho^{-1}\partial\rho/\partial T$ respectively.

Combining the elements retained from this list, κ_T and $<N>$, we get the scale

$$H_0 = \left(\frac{\kappa_T}{<N>}\right)^{1/2}.$$

Dimensional analysis thus suggests that the appropriate dimensionless measure of layer thickness is

$$G \equiv \frac{H}{(\kappa_T/<N>)^{1/2}}, \tag{2}$$

and that G should be a well-defined function of the other dimensionless parameters: $R\rho$, Pr, and Le.

To test assumptions of the dimensional analysis, then, one must test whether (2) collapses data onto a well-defined curve

$$G = G(R\rho, Pr, Le).$$

1.3 Testing the scaling with oceanographic data

Data selection
In Kelley (1984), equation (2) was tested using H, $<N>$, and $<R\rho>$ measured from published $\{S, T\}$ profiles. The data came from staircases in diverse locales: ice-covered saline lakes, ice-free (25 °C) saline lakes, polar oceans, interleaving fronts in polar regions, and interleaving fronts in temperate waters. This forms a cross-section of all known occurrences of diffusive-case staircases in the world ocean.

Analyzing published profiles has advantages and disadvantages compared to more detailed analysis based on new measurement programmes. The main *disadvantage* is crudeness.

The errors in extracting values of H, $<N>$, and $<R\rho>$ are large, typically ~20%. Furthermore, it's impractical to measure properties layer-by-layer; instead it's easier to formulate averages over 3-5 layers. The main *advantages* are speed of analysis and wide geographical coverage. Because the data represent all the regions where diffusive-case staircases have been observed, the results should be generally applicable.

Data analysis

I'll use $(G, R\rho)$ data from Kelley (1984), additional data described in Kelley (1986), and AIWEX Beaufort Sea data collected by Padman & Dillon (1987). The Kelley (1984) data have been re-analyzed from scratch to produce this figure. The difference between the data points in the two versions of the analysis confirms that the errors in both G and $R\rho$ are ~20%.

Figure 1[1] shows that scaling layer thickness by $(\kappa_T/<N>)^{1/2}$ collapses data onto a relatively tight function G of $R\rho$. Figure 1 does not display the variation of G with Le and Pr. Any such variation is lost in the noise (Kelley, 1986).

A verification that the scaling captures the variation of H with $<N>$ was given by Kelley (1984). The scaling successfully collapses oceanographic variability in H onto a tight function G of $R\rho$. Fedorov (1988) has further tested the scaling with independent data (see Figure 1 of his paper in this volume). With this addition, the data set is doubled. Fedorov's figures reproduce the Kelley (1984) data. This makes it easy to verify that the new points lie in the old data cloud: Fedorov's data further validates the scaling (2).

In order to get on with a discussion of Physics behind the scaling, only a few more words will be said about the data. Padman & Dillon (1987) tested (2) with measurements in the Beaufort Sea, calculating layer thickness data with a computer algorithm that identifies interfaces as regions of strong vertical gradients of T. This technique surpasses the graphical method used

1

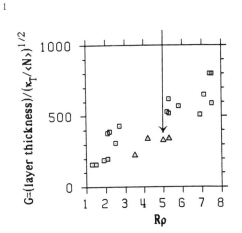

Figure 1:

G = layer thickness divided by $(\kappa_T/<N>)^{1/2}$, vs large-scale density ratio $R\rho$. Data sources are as listed in Kelley (1984, 1986), with the addition of the new Padman & Dillon (1987) Beaufort Sea measurements (triangles). The Kelley (1984) data were re-analyzed for this presentation, confirming that the error in estimating H, $<N>$ and $R\rho$ from published figures is ~20%.

The long vertical arrow requires separate discussion. It applies to the profile shown in Figure 3(b,c). The pointed end of the arrow lies on the value of G calculated using the layer thickness of the observed steps of Figure 3(b). The starting end of the arrow, however, is calculated using the thickness of the so-called "parent" steps inferred to precede the observed steps. This presumed parent profile is shown in Figure 3(c). Section 2.1 explains this inference in detail.

here, in terms of both accuracy and quantity of data. It represents the beginning of a second stage of testing the G scaling. To compare the two techniques, I've analyzed a profile in their paper using my graphical method, and shown the results as the triangles in Figure 1. The values of G fall below the main body of the data cloud. Even though this disagreement is small enough to be comparable to the scatter, it is systematic and therefore worth investigating. This will be done in section 2 below. In the meantime, it will only be noted that the scatter in Figure 1 is small enough (<30%) that the practical purpose behind the scaling has been accomplished. According to the 4/3 flux laws, the flux is approximately linearly related to G (see below), so *an empirical fit to G as a function of Rρ using the data of Figure 1 will yield flux predictions accurate to ~30%.*

1.4 Parameterizing effective diffusivities K_T and K_S

What K_T and K_S mean

It is common in physical oceanography to express fluxes in terms of large-scale gradients. This is done by formulating effective diffusivities, defined as the ratio of property flux to property gradient. This permits flux calculations from data that resolve property gradients but that don't resolve the physical parameters which actually govern the flux. It allows incorporation of the small-scale phenomenon in analytical and numerical models, and allows rough calculations based on low-resolution data. It is not, in general, physically justifiable: there are many cases in which fluxes simply are not related to gradients.

I'll present effective diffusivities here simply because they are needed before double-diffusion can be incorporated into any model of the thermocline mixing.

Calculating K_T and K_S from data

According to the 4/3 flux laws, summarized below, vertical fluxes of heat and salt can be calculated based on the salinity and temperature steps across interfaces, ΔT and ΔS. According to (1), these steps are determined in terms of H together with $<\partial S/\partial z>$ and $<\partial T/\partial z>$. The gradients are in turn determined by $<N>$ and $<R\rho>$. Consequently, the data set used to calculate the $(G,R\rho)$ points in Figure 1 is sufficient to calculate double-diffusive fluxes.

Dividing the fluxes by the gradients smoothed over several layer thicknesses, we get effective diffusivities.

How the G scaling implies that K_S and K_T can be parameterized as functions of Rρ

There is no *a priori* justification for formulating K_S and K_T as functions only of $R\rho$, Pr and Le, for K_S and K_T might well depend on $<N>$. However, the G scaling of (2) implies that K_S and K_T are *independent* of $<N>$. This is demonstrated as follows. According to (1), the 4/3 flux law:

$$flux \propto (\Delta S, \Delta T)^{4/3}$$

is equivalent to

$$flux \propto (H<N^2>)^{4/3},$$

where a function of $R\rho$ is dropped here and below since we are following only the $<N>$ dependence. Using the G scaling, (2), this can be rewritten

$$flux \propto [G<N>^{3/2}]^{4/3} \ .$$

Because G is a function of $R\rho$ but not of $<N>$, as verified in Figure 1 and Kelley (1984), this implies

$$flux \propto <N>^2.$$

Finally, we can substitute into the definition

$$effective \ diffusivity = \frac{flux}{gradient} \ ,$$

noting that

$$gradient \propto <N>^2,$$

to get

$$effective \ diffusivity \propto <N>^0;$$

that is, K_S and K_T are independent of $<N>$.

Thus, the success of G scaling in accounting for the dependence of H on $<N>$ directly implies that it is meaningful to formulate K_S and K_T as functions of $R\rho$.

Formulating K_T and K_S as functions of $R\rho$

Figure 2[1] shows K_T plotted as a function of $R\rho$, based on the data of Figure 1. K_T is here calculated directly from the data using the 4/3 flux laws. K_T is a fairly tight function of $R\rho$. It is exceeds the molecular diffusivity κ_T by 1 to 2 orders of magnitude. As $R\rho$ increases from 1.5 to 7.5, the value of K_T decreases from 10^{-5} m^2 s^{-1} to 10^{-6} m^2 s^{-1}, or from 100 to 10 times larger than κ_T.

In order to derive a convenient formula for application, a least-squares fit was made of $\log(K_T)$ to $\log(R\rho)$. The result, the curve shown in Figure 2, is

$$K_T = (1.3 \pm 0.3) \times 10^{-5} \ R\rho^{(-1.1\pm0.2)}, \quad m^2 \ s^{-1}. \tag{3}$$

The corresponding salt diffusivity is given by

$$K_S = K_T \frac{\gamma}{R\rho} \tag{4}$$

1

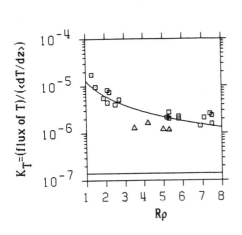

Figure 2:
Effective temperature diffusivity, K_T (in m^2 s^{-1}), calculated directly from the data as the ratio of the temperature flux to $<\partial T/\partial z>$. The curve is an empirical fit explained in the text (equation 3).

where the flux ratio γ is thought to decrease from ~1 at $R\rho = 1$ to ~0.15 at $R\rho = 2$ and remain constant at 0.15 for $R\rho > 2$ [according to laboratory experiments performed by Turner (1965) and analyzed by Huppert (1971)].

The standard error in the fit of $\log(K_T)$ used for (3) is 0.2. This translates into a factor-of-1.6 error in K_T. This is less than the factor-of-2 scatter in the flux data used to generate the empirical 4/3 flux laws, so *the scaling of layer thickness is accurate enough to allow us to parameterize large-scale effective diffusivities.*

A note of caution about the practical use of the {K_T, K_S} formulations

Having developed empirical diffusivity formulae, (3) and (4), for practical application, the remaining problem is one of *understanding*. Dimensional analysis cannot reveal what, if any, mechanism controls layer thickness. It would be prudent, if not necessary, to *understand* the mechanism before applying (3) and (4) universally. Several questions should be answered first. For example, before modelling cross-frontal intrusions driven by double-diffusive fluxes, we first need to know how rapidly staircases adjust to the scaling. Does the scaling hold under all conditions? What large-scale effects are generated by adjustment to the scaling?

The purpose of this paper is to outline a hypothetical mechanism which might control layer thickness. Without relying upon empirical adjustment, this mechanism quantitatively predicts the {G,$R\rho$} data distribution. Readers are fortunate to find another hypothetical mechanism in this volume, presented by Fedorov (1988). A third hypothesis will soon appear in a paper by Padman & Dillon (1988). Each of the three hypotheses is controversial; none has been tested adequately in either the laboratory or the ocean.

2   WHAT CONTROLS LAYER THICKNESS?

2.1   H is not set during layer formation

The hypothesis that H is set during staircase formation

An natural hypothesis is that the observed layer thickness in oceanic staircases is equal to the layer thickness established when the staircase was first formed. This hypothesis is used, in various ways, by Turner (1965), by Linden (1978), by Fedorov (1988), and by Padman & Dillon (1988). I claim it is fundamentally wrong. In showing why, I'll motivate my alternative hypothesis that initial layer size is irrelevant, that the equilibrium thickness is set by a balance between competing effects of layer-merging and interface-splitting .

Staircase creation in the laboratory

In a seminal laboratory experiment, Turner (1965) showed that bottom-heating a salt-stratified fluid causes the formation of a diffusive-case staircase. He outlined a theory which predicts accurately the thicknesses of the first two layers formed. Algebraic complexity prevented extension to staircases with more than 2 layers.

Huppert & Linden (1979) repeated the experiment with a taller tank, in which they could create longer staircases. To interpret their observations, they constructed a numerical model of the formation and evolution of staircases, based largely on Turner's theory. The model started with a smooth gradient heated from below. It was assumed, motivated by the laboratory observations, that a thermal boundary layer formed next to the heat source, and that this boundary layer grew progressively more Rayleigh-unstable until at some time it began to convect. Their model tracked the value of the boundary-layer Rayleigh number Ra_{bl}. When Ra_{bl} became critical, the model replaced the boundary layer with a well-mixed layer retaining the original heat and salt content. Subsequently, this convecting layer grew thicker, as background stratified fluid was entrained across the layer's upper interface. At the same time, a thermal boundary layer above this interface continued to develop, eventually to become Rayleigh unstable, leading to the formation of a second layer. In this way, the model tracked the formation of a sequence of many layers. The further evolution of the layers was based on layer-integrated conservation equations for S and T in the layers and 4/3 flux laws at the interfaces.

In order to obtain reasonable agreement between the model and the laboratory observations, Huppert & Linden (1979) had to add one more thing to the model. This was what they called 'layer-merging'—the coalescence of adjacent layers to form new, thicker, layers. In the experiments, layer-merging occurred mostly near the buoyancy source. They hypothesized that it occurred when convergent buoyancy flux caused adjacent layers to evolve to equal density. With the addition of layer-merging, the predictions of the model agreed very well with observations. Huppert & Linden's (1979) Figures 11, 12 and 13 show that the shape of the temperature profiles agreed to ~25% in terms of the temperature of a given layer, and the interface location agreed to ~10%. Their Figures 16 and 17 show that the general shape of the observed staircase—which takes into account both the creation of new layers at the top of the staircase and layer-merging in the interior—is predicted well by the model. Their Figures 18 and 19 show that the total height of the staircase was predicted to within ~20%. Their Figure 20 shows that the mean thickness of individual layers was predicted to within ~20%.

What these laboratory observations indicate about the distribution of H and G

This review of Huppert & Linden's (1979) experiments and model contributes to the present discussion of the distribution of layer thickness by telling us two distinct things.

The thickness of layers in thermohaline staircases is not equal to the thickness the layers had when they are first formed.
Because of layer-merging, layers are much thicker than they were at the time of their formation from the stratified fluid. This contradicts theories of layer-thickness control described by Fedorov (1988—in this volume), and by Padman & Dillon (1988).

Panel (a) of Figure 3[1] illustrates this point with the temperature profile predicted by a numerical model of a bottom-heated salinity gradient. This numerical model, based directly on the Huppert & Linden model, is described in detail in Kelley (1986). The simulation shown here mimics Huppert & Linden's (1979) Figure 1, except that here the evolution is followed over a time interval ten times longer. Layers near the bottom are much thicker than the layers above. As in the laboratory experiments, layer-merging has increased H near the buoyancy source.

The vertical variation of layer-thickness is monotonic, decreasing with distance from the buoyancy source .

This pattern, obvious from Figure 3(a), contrasts with oceanic profiles, which often have thin layers sandwiched between thicker layers. This has been noted since the first CTD observations of oceanic staircases, in both the finger case and the diffusive case. In one of the earliest discussions, Neshyba *et al.* (1971) called the thin layers they observed in a diffusive-case Arctic staircase "intermediate layers". For illustration, I've shown in Figure 3(b) a profile provided by Ron Perkin (IOS Patricia Bay, BC, Canada). Near the top and bottom of the profile, layer thickness is quite uniform, but near the centre there are three pairs of thin layers sandwiched between layers twice as thick. I'll call these thin layers "sandwich" layers.

Temperature profiles of an Arctic diffusive-case staircase, repeated at 4 minute intervals over 30 hours by Neshyba *et al.* (1972), showed that the sandwich layers sometimes appeared and then disappeared. Visually, it looks as though the interfaces split up, forming new layers. Kelley (1987b) has cautioned against interpreting this as temporal evolution, because the time-scale for layer-merging [estimated using the Huppert & Linden (1979) model] is much larger than the timescale of variability. The temperature evolution at a point in space is likely to be predominated by the advection of the spatially varying temperature field. Kelley (1987b) suggested that the length scales of the sandwich layers is probably O(100) m. The reason for this particular patchiness scale is presently unknown, and may remain so until we better understand how sandwich layers are formed.

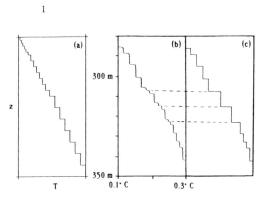

Figure 3:
(a) Temperature profile, in arbitrary linear units, from a numerical model of a bottom-heated salinity-gradient. Recently-formed layers near the top are much thinner than (merged) layers below.
(b) A LOREX Arctic profile (Ron Perkin, pers. comm.). Note the thick-thin-thick pattern of layer thickness in the middle of the profile. In the text these thin layers are called "sandwich" layers and the thick-thin-thick pattern is called the "sandwich" pattern.
(c) Parent profile presumed to have existed before the thin sandwich layers in (b) were formed.

2.2 Proposal: *H* is set by a balance between layer-merging and interface-splitting

Lack of oceanic observations

Neither the formation nor the later evolution of staircases has yet been observed in the ocean. Measurements made at a single location, as in the Neshyba *et al.* studies (1971, 1972) and the Padman & Dillon (1987) study, cannot distinguish between temporal evolution and advecting spatial variability (Kelley, 1987b). For the present, we must rely on laboratory experiments and numerical models for guidance.

Guidance from laboratory experiments and numerical models

When the layer-thickness in Huppert & Linden's (1979) experiments is scaled as in equation (2), the value of *G* is an order of magnitude smaller than in the ocean. The scaled layer thickness in their staircases increased with time, but never reached oceanic values over the course of their experiments. If the Huppert & Linden (1979) numerical model is a reliable predictor—as it is over the course of the experiments—we can expect layers to continually thicken by layer-merging, but *not* to form a sandwich pattern. This seems to imply that the sandwich layers were formed subsequent to the formation of the surrounding staircase. Such formation of layers in the interior of staircases was never observed by Huppert & Linden (1979).

The notion of "parent" profiles

Let us suppose that some process, to be discussed later, creates sandwich layers from interfaces. Since we know that interface migration will be negligible (Kelley, 1987b), we can extrapolate backward in time to guess at the shape of the profiles which existed prior to the creation of the sandwich layers. I'll call these hypothetical profiles the "parent" profiles. For example, the observed profile shown in Figure 3(b) leads to the guess at a parent profile shown in Figure 3(c).

Comparison of G in parent profiles and observed profiles

The long arrow in Figure 1 shows both the value of *G* calculated using the observed profile in Figure 3(b) and the value of *G* calculated using the presumed parent profile in Figure 3b. The parent value of *G* is large compared to the rest of the data cloud, while the observed value of *G* lies much closer to the data cloud.

This single example suggests that the creation of sandwich layers brings staircases closer to the observed oceanic $G=G(R\rho)$ relationship. The observed *G* falls somewhat below the rest of the data cloud, which largely represents profiles lacking the sandwich pattern. Similar results are found in other Arctic profiles (Kelley, 1986; Padman & Dillon, 1987). The evidence is very sketchy so far, but it seems to show that the sandwich pattern is associated with anomalously low values of *G*, while *G* calculated for the presumed parent layers is anomalously high.

It is too early to state this as more than casual observation. More work is needed before we can properly relate the value of *G* to the presence of sandwich layers, and before we can claim empirical understanding of the pattern of distribution of layer-thickness generally. This work will involve both new analysis of already collected data, and collection of new data.

Hoping that useful guidance for this work might come from speculation about how the sandwich layers are formed, I'll devote the rest of this paper to outlining a theory for the creation of the layers, and to explaining the relevance of this theory to the observed oceanic distribution of layer thickness.

Observations of formation of sandwich layers

The creation of sandwich layers has never been directly observed in the ocean[1]. Neither, for that matter, has the more fundamental process of *staircase* formation. This is the biggest problem oceanographers must face in trying to understand the nature of oceanic thermohaline staircases. For the time being, we must build our understanding on laboratory experience and numerical models designed to mimic laboratory staircases.

For *finger-case* staircases, there may be laboratory examples of the creation of sandwich layers, by a process we might call "interface-splitting". Linden (1978) set up 2-layer sugar-salt staircases in which the initial interface thickness was not of the equilibrium value. The interface thickness changed with time. In some cases, the interface broke down to form a third, thinner, convecting layer. In other cases it didn't. Given the exploratory nature of the experiments, Linden could not suggest a quantitative criterion for breakdown. He speculated that the criterion might be related to the interfacial Stern number but could not test the hypothesis, lacking flux measurements. The Stern number, a dimensionless ratio of the buoyancy flux to the interfacial density gradient, is thought to measure the so-called collective stability of the salt fingers (Stern, 1969; Holyer, 1981).

For *diffusive-case* staircases, it is unlikely that the Stern-number mechanism applies, because the physics of the interface is completely different. (Predominantly, there is no vertical motion within the core of the interface; the fluid there is quiescent.) There have been no laboratory observations of the creation of, or even the existence of, sandwich layers in diffusive-case staircases. This may be because the laboratory staircases have been too small, in scaled terms, to match oceanic staircases. In the most extensive experiments done to date, described by Huppert & Linden (1979), the layers [scaled with equation (2)] were ten times thinner than oceanic layers. Since the layer thickness tended to increase with time in their experiments, their staircases may have been too young to match oceanic staircases. It is possible that the process which creates sandwich layers does so only when the staircase has evolved long enough to have sufficiently thick layers.

Two hypotheses for the formation of sandwich layers

Lacking observational guidance, it is possible at present only to speculate about how the sandwich layers are created. Two general classes of mechanisms can be suggested: *The remote-formation hypothesis:* The layers might be created at a thermohaline front some horizontal distance away from the interior of the staircase, later to intrude laterally into the body of the staircase. *The local-formation hypothesis:* The layers might be formed in the interior of the staircase.

---

[1] The closest we've come is in the dye injection experiments done by Woods (1968) in the finger-case staircases of the Mediterranean Sea. These showed breaking Kelvin-Helmholtz waves on salt-finger interfaces, and so hint at the creation of sandwich layers.

Remote-formation hypothesis for sandwich layers

The layers could be created at regions with strong horizontal variations of S and T, perhaps at the sides of the staircase. This would be analogous to the creation of cross-frontal intrusions at density-compensated fronts (Ruddick & Turner, 1979; Toole & Georgi, 1981; Niino, 1986). The hypothesis here is that the intrusions later become the sandwich layers. This may explain the intrusive features seen by Marmorino *et al.* (1987) in their tow-thermister-chain measurements in the C-SALT region. However, it cannot in general explain the existence of sandwich layers, for it requires lateral variations which are not present in all staircases with sandwich layers. Examples of staircases containing sandwich layers which have little horizontal variation are those in bottom-heated lakes [see, for example, Newman (1976)].

Local-formation hypothesis for sandwich layers

There are two ways in which local changes to the staircase could account for the sandwich pattern. First, isolated layers of normal thickness could shrink because of inward *interface migration*, thus forming the sandwich pattern. Second, the sandwich layers could be created by a *transformation* of the interfaces.

The inward-migration mechanism is unlikely. It requires isolated layers to be selected for inward migration of their interfaces, adjacent layers being unaffected. According to the (weak) theoretical guidance we have to date, individual interfaces cannot be isolated for anomalous migration in this way; the distribution of migration speed should be smooth with depth (Kelley, 1987b).

The interface-transformation mechanism seems more likely, partly because we know of mechanisms by which density interfaces can break down. The most familiar example of interface breakdown is Kelvin-Helmholtz instability, in which a strongly sheared density interface develops waves which eventually break, or are otherwise disrupted, resulting in local mixing. The details of the transition to turbulence and of the subsequent mixing of the interface are not fully understood, and nor is the final state after mixing [compare Thorpe (1973) to Thorpe (1984); see also Thorpe (1987)]. Furthermore, laboratory observations so far have been restricted to fluids stratified by a single solute; when both S and T contribute to ρ, as in double-diffusive interfaces, the behaviour may be different.

The 'interface-splitting' hypothesis

In spite of these fundamental uncertainties, we can profit from constructing a speculative scenario—one which is not properly constrained by direct observation in the laboratory *or* the ocean—to account for the creation of new, relatively thin, layers from interfaces. I call this the 'interface-splitting' hypothesis:

Step 1: The interface is sheared strongly enough that the Richardson number

$$Ri = \frac{\Delta\rho\ gh}{\rho\ \Delta U} \tag{5}$$

falls below a critical value, perhaps 1/4 as theory and laboratory measurement suggest (Thorpe, 1973). Here $\Delta\rho$ and ΔU are the density step and velocity step across the interface of thickness h, and g is the gravitational acceleration.

Step 2: Interfacial waves grow on the interface and eventually either break or cause secondary billow instabilities (Thorpe, 1973, 1984, 1987).

Step 3: This is followed by local mixing of the interface. The interface thickens by entraining fluid from the layers. As this is happening, the Richardson number increases, being proportional to h (Thorpe, 1973). The details and the energetics of the mixing are not presently understood (Thorpe, 1984, 1987).

Step 4: The mixing ceases, perhaps when the Richardson number exceeds a critical of about 1 (Thorpe, 1973). The final result is a thickened patch of water in which density and velocity may form approximately linear profiles with depth (Thorpe, 1973 figure 14). In the double-diffusive case, we can expect that the double-diffusive boundary layers will be disrupted.

These four steps summarize the general conception of Kelvin-Helmoltz mixing of sharp density interfaces. Many elements are still not understood in their entirety: the critical value of the interfacial Richardson number; the relevance of the Reynolds number in determining whether mixing occurs; the mechanism of mixing, whether by breaking of the interfacial waves or by secondary convective instabilities; and the details of the final state after mixing (Thorpe, 1984, 1987). Worse still, from the present point of view, is that these experiments have *never* been done for a double-diffusively convecting fluid.

Presuming that the details can at some point be worked out sufficiently, let's continue with the scenario, now turning *entirely* to speculation for the last step:

Step 5: This leaves a thicker interface, whose double-diffusive boundary layers have been destroyed. As far as the surrounding layers are concerned, this stratified region looks just like the stratified region above a growing staircase. So what happens next is similar. The adjacent layers migrate inwards, entraining the fluid formerly in interface. Thermal boundary layers advance in front of the migrating edges of the layers. If the disrupted region is relatively thick, new layers will be formed because the boundary layers will have time to become Rayleigh unstable. If it is relatively thin, the boundary layers of the migrating edges will reach other, and a new interface will be created in place of the old one. The final state thus depends on the thickness of the disrupted interface. It can consist of either a new interface, or of one or more new layers.

It is important to note before going further that this scenario is wholly invented. The process has never been observed in the ocean or laboratory. This is not too surprising. The spatial-temporal resolution and coverage of the oceanographic observations are too limited to observe the process in action. In the laboratory experiments done to date the staircases have been too small to have interface-splitting (see below).

This scenario predicts that the interfacial Richardson number should be greater than 1/4 (otherwise, Kelvin-Helmoltz instability occurs, increasing the Richardson number). The next step is to formulate the Richardson number of the interface.

Why an interfacial Richardson number based on large-scale shear can't explain the G scaling

Denote the large-scale shear $<S>$. The shear, like the density gradient, is relatively low within the layers (Gregg & Sanford, 1987), so that the *shear*^{2} across the interface is intensified by the ratio $(H/h)^2$. But N^2 is intensified by only H/h. Thus, the interfacial Richardson number can be written

$$Ri_0 = \frac{h}{H} \frac{<N>^2}{<S>^2} ,$$

or

$$Ri_0 = \frac{h}{H} <Ri>,$$

where $<Ri>$ is the large-scale Richardson number.

It will soon be shown that h/H is determined by the details of the double-diffusion. On the other hand $<Ri>$ is set by the large-scale flow, which is independent of double-diffusion. For G to be an *intrinsic* property of the staircase, it cannot be determined by Ri_0, which is in turn determined by $<Ri>$, a large-scale property unaffected by the presence of absence of staircases.

Formulation of Richardson number based on convective shear

There is another Richardson number available: that based on the *convective shear* across the interface. In the rest of this section I'll roughly sketch a formulation of this Richardson number and its relationship to G. In the interests of brevity, I'll skip details which add only O(1) refinements to the formulation and which don't affect the functional form of the relationship between Ri and G. These details mostly relate to the convection model and to the nature of the temperature and velocity boundary layers comprising the interface. More complete discussion can be found in my thesis (Kelley, 1986) and will later appear in the available literature (Kelley, 1988b). Here, the presence of ~ rather than = is a code that such O(1) factors have been dropped.

Laboratory measurements of diffusive-case interfaces indicate that the opposing action of randomly oriented convection cells leads to a cross-interface **velocity jump** ΔU roughly equal to the convective velocity scale U_{conv}:

$$\Delta U \sim U_{conv} \tag{6}$$

(Kelley, 1986, 1988b)[1]. These laboratory measurements confirm to within a factor of 2 the theoretical prediction that the convection velocity is given by the expression

$$U_{conv} \sim \frac{\kappa_T}{H} Ra^{1/2}, \tag{7}$$

(Kelley, 1986, 1988b). Here

$$Ra \equiv \frac{g\alpha\Delta T H^3}{\nu\kappa_T} \tag{8}$$

is the layer Rayleigh number. An O(1) numerical factor has been left out of (7) for brevity of presentation here; the complete derivation can be found in Kelley (1986, 1988b). Equation (7) comes from equating the rate at which energy is supplied by the convective buoyancy flux with

---

[1] Since the Kelley (1986, 1988b) references are not readily accessible, interested readers should examine Figure 3 of Kelley (1987a) for an overview of this laboratory result.

the rate at which energy is removed by viscous dissipation in boundary layers near the interfaces. Part of the reason to trust in the accuracy of (7) is that application to single-component convection yields accurate predictions of the dependence of convective velocity on the Rayleigh number, without empirical adjustment of any of the parameters of the problem (Kelley, 1986, 1988b). It is not possible to extend the theory to the salt-finger case because the convection near the finger interfaces is not well understood.

The **density jump** across the interface is given by

$$\frac{\Delta\rho}{\rho_0} = \alpha \, \Delta T \, (R\rho - 1), \tag{9 a}$$

which can be rewritten

$$\frac{\Delta\rho}{\rho_0} = (R\rho - 1) \, Ra \, Pr \, \frac{\kappa_T^2}{gH^3} \tag{9 b}$$

using the definitions of Ra and Pr.

Now the **interfacial Richardson number**, defined by (5), can be written approximately as

$$Ri_{conv} \approx Pr \, (R\rho - 1) \frac{h}{H} . \tag{10}$$

using (6), (7) and (9b). This approximation is good to the same order as (7), or to about a factor of 2.

Equation (10) will be closed in terms of known quantities if we can express the **thickness** h **of the interface** in terms of the thickness H of the layer. This can be done using the Nusselt number, Nu, a dimensionless measure of the temperature flux defined as

$$Nu \equiv (temperature \, flux) \frac{H}{\kappa_T \Delta T} .$$

Measurements in the laboratory (Marmorino & Caldwell, 1976) and in the ocean (Gregg & Cox, 1972) confirm the theoretical prediction that

$$\frac{H}{h} = Nu. \tag{11}$$

where the Nusselt number is calculated using the 4/3 flux laws

$$Nu = C \, Ra^{1/3}. \tag{12}$$

Here $C = C(R\rho)$ is an empirical factor found to be

$$C = 0.0086 \exp \left(\frac{4.6}{\exp [\, 0.54 \, (R\rho - 1)]} \right) \tag{13}$$

in experiments done by Marmorino and Caldwell (1976). Laboratory experiments done by others have yielded values of C which differ by a factor of 2 (see a review in Kelley, 1986), so the uncertainty in the 4/3 flux laws is a factor of 2.

Since, by definition of G, we can write the Rayleigh number as

$$Ra \equiv \frac{G^4}{Pr \, (R\rho - 1)} , \tag{14}$$

we can now write the convection Richardson number as:

$$\boxed{Ri_{conv} \sim \frac{Pr^{4/3} \, (R\rho - 1)^{4/3}}{C \, G^{4/3}}} \tag{15}$$

This is a closed expression for Ri_{conv} in terms of G and $R\rho$. It suffers a factor of 2 uncertainty due to uncertainties in the 4/3 flux laws. As stated at the first of the section, it also contains $O(1)$ factors which account for details of boundary-layer structure I'm concealing here for brevity.

Nature of relationship between G and Ri_{conv}

The inverse relation between Ri_{conv} and G given in (15) implies that Kelvin-Helmholtz instability should occur preferentially between *thick* layers. This is in accord with the pattern seen in oceanographic observations, suggested by the example in Figure 3: sandwich layers are usually found in parts of staircases where otherwise layer thickness would otherwise be larger than typical.

Calculating Ri_{conv} for oceanic data

Translating the $(G, R\rho)$ data in Figure 1 into $(Ri_{conv}, R\rho)$, and filling in $O(1)$ factors left out here for brevity [given in Kelley (1986, 1988a and 1988b)], we get Figure 4[1].

The main features of Figure 4 are:
- Ri_{conv} depends only weakly on $R\rho$; the $(R\rho-1)$ term in (15) varies by a factor of 20 over the range of the data, but this variation is cancelled by the variation of C and G.
  COMMENT: Since the variation of C over this range is not known to better than a factor of 2, we can reasonably say that the $G(R\rho)$ variation makes Ri_{conv} *independent of R\rho*.
- Ri_{conv} is of order 0.1 to 1.
  COMMENT: The lower limit is reminiscent of the 1/4 limit for Kelvin-Helmholtz instability. The upper limit is about the same as the Richardson number for sheared interfaces which have undergone Kelvin-Helmholtz mixing (Thorpe, 1973). This suggests that interfaces in oceanic diffusive-case staircase are on the verge of Kelvin-Helmholtz instability set up by the convective shear.
- The parent layers—that is, the layers *presumed* to have existed prior to the creation of the sandwich layers—had an anomalously low values of Ri_{conv}. In contrast, the *observed* layers have an anomalously high values.

1

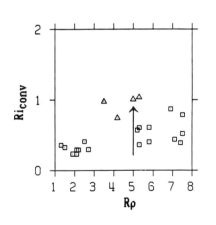

Figure 4:
Convection-based Richardson numbers calculated using (15) on the data in Figure 1. The arrow refers to the observed and presumed parent profiles of Figure 3b and 3c. The pointed end of the arrow is at the value of Ri_{conv} calculated using actual profile. The starting end of the arrow is at Ri_{conv} calculated using the presumed thickness of parent layers.

COMMENT: According to the speculative interface-splitting scenario, this can be interpreted as follows: the parent layers were so thick, and the convection within them so vigorous, that the interfaces became Kelvin-Helmoltz unstable. Then the interfaces were transformed (somehow) into new layers, the result being a reduction in average layer thickness, a decrease in shear, and an increase in Ri_{conv}.

A layer-thickness control mechanism consistent with the Ri_{conv} distribution

This distribution leads to a speculative hypothesis for the control of layer thickness in the ocean: H is set by a balance between two competing tendencies:

- Flux convergences cause **layer-merging**, *increasing H* and reducing Ri_{conv}.

- When Ri_{conv} falls below ~0.1, **interface-splitting** creates new layers, increasing H and increasing Ri_{conv} to O(1).

This implies that Ri_{conv} should be restricted to lie in the range O(0.1) to O(1). This agrees well with observations (Figure 4). Using (15) to translate this critical range in Ri_{conv} into a corresponding function $G=G(R\rho)$, we recover the observed oceanic distribution of layer thickness.

Thus, the prediction of the Ri_{conv} range, founded on theory and observation for sheared density interfaces, yields a reasonable prediction of the oceanographic distribution of layer thickness, providing a physical explanation of the G scaling of equation (2).

Comment on Ri_{conv} in laboratory experiments done to date

The Huppert & Linden (1979) experiments had values of G an order of magnitude smaller than the oceanographic values. This translates to Ri_{conv} values ten times larger than in the ocean. Since the oceanographic values of Ri_{conv} appear to be marginally critical with respect to Kelvin-Helmholtz instability, this seems to imply that the laboratory staircases were much too small to be susceptible to interface-splitting. This may explain the lack of observation of interface-splitting.

3   SUMMARY AND CONCLUSIONS

New data (Figure 1) further verify the Kelley (1984) layer thickness scaling

$$G = \frac{H}{(\kappa_T/\langle N\rangle)^{1/2}} = \text{function of } (R\rho), \tag{2}$$

allowing the construction of effective diffusivities for heat and salt fluxes in terms of the large-scale density ratio $R\rho$: $K_T = K_T(R\rho)$ is shown in Figure 2. Equations (3) and (4) give K_T and K_S as functions of $R\rho$. Independent data presented by Fedorov (1988) elsewhere in this volume further verify both the scaling and the effective diffusivities.

As noted in the original paper about layer thickness in the ocean (Kelley, 1984), it is not clear how to explain the observed $G=G(R\rho)$ relationship. However, we have some new clues:

- The laboratory measurements and the results of a numerical model described by Huppert & Linden (1979) suggest that layer-merging greatly thickens layers. The layer thickness we observe is not equal to the layer thickness at the time the staircase was being formed. This contradicts Fedorov's (1988) proposal that the functional relationship $G = G(R\rho)$ can be explained in terms of initial layer thickness.

- Numerical models based on the laboratory experiments suggest that layer-merging might continue without limit, continually increasing G.
- In laboratory staircases and in numerical simulations, layer thickness is monotonic with depth (Figure 3a).
- Oceanic staircases sometimes have thin "sandwich" layers separated between thick layers (Figure 3b). Interpretation of laboratory and numerical results suggests that the thin "sandwich" layers must have been formed after the surrounding layers were formed.
- Extrapolating backwards in time, we can guess at the state of staircases before the sandwich layers were formed. The layers in these hypothetical staircases are anomalously thick compared to the observed $G(R\rho)$ distribution.

This suggests that the *creation of sandwich layers* constrains the layer thickness distribution $G=G(R\rho)$. A speculative scenario for the creation of sandwich layers was sketched. In it, sandwich layers are created following Kelvin-Helmholtz disruption of interfaces being sheared by convection in the adjoining layers.

To test this scenario, an interfacial convective Richardson number Ri_{conv} was formulated. (This can be done for the diffusive-case, but not for the finger-case.) Translating the data from $(G,R\rho)$ space (Figure 1) into $(Ri_{conv},R\rho)$ space (Figure 4) we found that:

(i) Ri_{conv} is limited to a small range (0.1 to 1).

(ii) Ri_{conv} is nearly independent of $R\rho$.

(iii) The largest values of Ri_{conv} are found in regions of layer sandwiching.

(iv) The sandwich pattern appears to occur in staircases which previously had anomalously large values of G.

Points (i) and (ii) suggest that Ri_{conv} is constrained to lie within a Kelvin-Helmholtz critical range. Points (iii) and (iv) suggest that the Ri_{conv} constraint—in particular, the reduction of G by the creation of new layers after Kelvin-Helmholtz mixing—is responsible for the observed $G = G(R\rho)$ relationship.

The suggested explanation of the observed $G=G(R\rho)$ relationship is that layer thickness is set by a competition between the tendency of layer-merging to increase H (decreasing Ri_{conv}, which is inversely proportional to H) and the tendency of interface-splitting to decrease H (increasing Ri_{conv}). The balance point is determined by the latter: new sandwich layers are created by interface-splitting, which occurs when the surrounding layers get so thick that Ri_{conv} falls below O(0.1). After splitting, Ri_{conv} rises to O(1), as in Thorpe's (1973) experiments with a singly-stratified fluid. Thus the equilibrium state has $0.1 < Ri_{conv} < 1$.

When the range $0.1 < Ri_{conv} < 1$ *is translated from* $(Ri_{conv},R\rho)$ *space into* $(G,R\rho)$ *space, the magnitude of G and its variation with Rρ are explained quantitatively*

This explanation is hypothetical. It is based on reasonable inferences from observations, but major components are not tested by observations. How may the explanation be tested?

LABORATORY TESTS: The first step will be to set up laboratory staircases with sufficiently thick layers. There must be at least 4 layers, so that the interior layers will be bounded by natural double-diffusive interfaces. De-stabilizing buoyancy fluxes should be applied at the ends of the staircase, perhaps in such a way that there is no net buoyancy gain by the staircase as a whole. Will layer-merging occur? What is the the value of the interfacial Richardson number? Are the interfaces disrupted when the layers exceed a certain size? Do new layers grow in place of the disrupted interfaces? Does background shear disrupt the process, or is it necessary as a catalyst?

OCEANOGRAPHIC TESTS: Observing layer-merging and interface-splitting in the ocean will be difficult. Many techniques should be used: dye release; towed thermistor chains; profiling and towed microstructure instruments; mid-depth buoys (measuring shear, density and optical microstructure); and profiling optical shadowgraphs. Only optical shadowgraphs are likely to yield unequivocal evidence of either layer-merging or interface-splitting. Even if we were to be satisfied to measure the gross effects on staircase structure, using conventional S and T measurements, the logistics will be hard. The main difficulty is the wide range of time and space scales. The layer-merging and interface-splitting events are expected to be brief, lasting only $N^{-1} \sim 10$ minutes. Rapid profiler cycling would be required to see both the event and the local effect on the staircase. The observation period will have to be quite long, because the time between events will be $>> H^2/K_T \sim 10$ days ($H \sim 1$ m; $K_T \sim 10^{-6}$ m^2 s^{-1}). To avoid confusing advective and temporal effects, it will be necessary to map the horizontal variation around tracked portions of staircases. This will require the fine spatial resolution of thermistor chains. For example, the horizontal lengthscales are estimated to be O(100) m for Arctic staircases (Kelley, 1987b). This brings up a final point: the present view has ignored horizontal variability altogether. We might well expect layer-merging and interface-splitting to occur in patches. How this confounds theory and observation is yet to be determined.

4   ACKNOWLEDGEMENTS

This paper stems from work done in my thesis at Dalhousie University, where I was funded by the Izaak Walton Killam foundation and by the Natural Sciences and Engineering Research Council of Canada and supervised by Barry Ruddick and Chris Garrett. I thank Barry and Chris for help and encouragement during the work, and Chris for asking me to attend this Liege meeting. This paper was improved greatly as a result of insightful comments from Mike Gregg and Eric Kunze. I thank Konstantine Fedorov for comments on the paper and for interest in the work; if you found this paper interesting, you should flip the pages of this book to find his paper. Vicki Cullen at the graphics department here at WHOI was kind enough to allow me to use her LaserWriter freely, resulting in this camera-ready copy.

5   REFERENCES

Fedorov, K. N., 1988. Layer thickness and effective diffusivities under the "diffusive type" thermohaline convection in the ocean. In this volume.

Gregg, M. C. and C. S. Cox, 1972. The vertical microstructure of temperature and salinity. *Deep-Sea Res.*, 19: 355-376.

Gregg, M. C. and T. B. Sanford, 1987. Shear and turbulence in thermohaline staircases. *Deep-Sea Res.*, 34(10): 1689-1696.

Hoare, R. A., 1968. Thermohaline convection in Lake Vanda, Antarctica. *J. Geophys. Res.*, 73: 607-612.

Holyer, J. Y., 1981. On the collective instability of salt fingers. *J. Fluid Mech.*, 110: 195-207.

Huppert, H. E., 1971. On the stability of a series of double-diffusive layers. *Deep-Sea Res.*, 18: 1005-1021.

Huppert, H. E. and P. F. Linden, 1979. On heating a stable salinity gradient from below. *J. Fluid Mech.*, 95: 431–464.

Kelley, D. E., 1984. Effective diffusivities within ocean thermohaline staircases. *J. Geophysical Res.*, 89: 10484–10488.

Kelley, D. E., 1986. *Oceanic thermohaline staircases*. PhD dissertation, Dalhousie University, Halifax, Nova Scotia.

Kelley, D., 1987a. The influence of planetary rotation on oceanic double-diffusive fluxes. *J. Mar. Res.*, 45: 829-841.

Kelley, D., 1987b. Interface Migration in Thermohaline Staircases. *J. Physical Oceanogr.*, 17: 1633-1639.

Kelley, D., 1988a. A possible mechanism for the control of layer thickness in oceanic diffusive staircases. (In preparation)

Kelley, D., 1988b. New flux laws for single-component convection and double-diffusion. (In preparation)

Kunze, E., 1987. Limits on growing, finite-length salt fingers: A Richardson number constraint. *J. Mar. Res.*, 45: 533-556.

Kunze, E., A. J. Williams, and R. W. Schmitt, 1987. Optical microstructure in the thermohaline staircase east of Barbados. *Deep-Sea Res.*, 34(10): 1697-1704.

Linden, P. F., 1978. The formation and destruction of fine-structure by double-diffusive processes. *Deep-Sea Res.*, 23: 895-908.

Lueck, R. G., 1987. Microstructure measurements in a thermohaline staircase. *Deep-Sea Res.*, 34(10: 1677-1688.

Marmorino, G. O. and D. R. Caldwell, 1976. Heat and salt transport through a diffusive thermohaline interface. *Deep-Sea Res.*, 23: 59-67.

Marmorino G. O., W. K. Brown and W. D. Morris, 1987. Two-dimensional temperature structure in the C-SALT thermohaline staircase. *Deep-Sea Res.*, 34(10):1667-1676.

Middleton, J. H., and T. D. Foster, 1980. Fine structure measurements in a temperature-compensated halocline. *J. Geophysical Res.*, 85:1107-1122.

Neshyba, S., V. T. Neal and W. W. Denner, 1971. Temperature and conductivity measurements under ice-island T-3. *J. Geophys. Res.*, 76: 8107-8120.

Neshyba, S., V. T. Neal and W. W. Denner, 1972. Spectra of internal waves: *In-situ* measurements in a multiple-layered structure. J. Physical Oceanogr., 2: 91-95.

Newman, F. C., 1976. Temperature steps in Lake Kivu: a bottom heated saline lake. *J. Physical Oceanogr.*, 6:157-163.

Niino, H., 1986. A linear stability theory of double-diffusive horizontal intrusions in a temperature-salinity front. *J. Fluid Mech.*, 171: 71-100.

Padman, L. and T. M. Dillon, 1987. Vertical fluxes through the Beaufort sea thermohaline staircase. *J. Geophysical Res.*, 92: 10799-10806.

Padman, L. and T. M. Dillon, 1988. Modelling of oceanic diffusive staircases. In preparation.

Perkin, R. G. and E. L. Lewis, 1984. Mixing in the West Spitsbergen Current. *J. Physical Oceanogr.*, 14: 1315-1325.

Ruddick, B. R. and J. S. Turner, 1979. The vertical length scale of double-diffusive intrusions. *Deep-Sea Res.*, 26A: 903-913.

Schmitt, R. W., 1988. Mixing in a thermohaline staircase. In this volume.

Stern, M. E., 1969. Collective instability of salt fingers. *J. Fluid Mech.*, 35: 209-218.

Thorpe, S. A., 1973. Experiments on instability and turbulence in a stratified shear flow. *J. Fluid Mech.*, 61: 731-751.

Thorpe, S. A., 1984. The transition from Kelvin-Helmholtz instability to turbulence. In *Internal gravity waves and small-scale turbulence*, ed. P. Müller and R. Pujalet. Hawaii Institute of Geophysics Special Publication.

Thorpe, S. A., 1987. Transitional phenomena and the development of turbulence in stratified fluids: a review. *J. Geophysical Res.*, 92: 5231-5248.

Toole, J. M. and D. T. Georgi, 1981. On the dynamics and effects of double-diffusively driven intrusions. *Prog. Oceanog.*, 10: 123-145.

Turner, J. S., 1965. The behaviour of a stable salinity gradient heated from below. *J. Fluid Mech.*, 33: 183-200.

Turner, J. S., 1965. The coupled turbulent transports of salt and heat across a sharp density interface. *Int. J. Heat Mass Transfer*, 8: 759-767.

Woods, J. D., 1968. Wave-induced shear instability in the summer thermocline. *J. Fluid Mech.*, 32: 791-800.

NUMERICAL EXPERIMENTS ON THERMOHALINE CONVECTIVE MOTIONS ACROSS INTERFACES OF INTRUSIONS

S.A.PIACSEK
SACLANT Research Centre, Via S. Bartolomeo 400,
19026 La Spezia, Italy

N.H. BRUMMELL
Department of Mathematics, Imperial College, London, SW7 2BZ

B.E.MCDONALD
Naval Ocean Research and Development Activity, NSTL Station,
MS 39529, USA

ABSTRACT

Thermohaline convection was studied via 2-D numerical models in several 2-layer and 3-layer situations. In particular, double-diffusive and weakly unstable gravitational motions were studied that arise at the interfaces between layers of an oceanic interleaving system, such as are found commonly in and near oceanic fronts. Both the flow details and the horizontally-averaged fluxes of heat and salt were investigated in 'run-down' situations. In general the results show that diffusive fluxes can be of the same magnitude as fingering fluxes, and last much longer because the interface keeps its identity longer. In the experiments performed here the fingers break up into 'blobs' and lose their identity rapidly, implying that to maintain a fingering interface some flow shear associated with a larger scale convection or advective flow is needed to supply new mean-property gradients. An interesting sequence of convective motions is observed when a diffusive interface is unstable initially at a subcritical Rayleigh number. In this case the initial perturbations (of the gravitational type) die out and vigorous diffusive motions set in, that eventually thicken the unstable region so that gravitational overturning once more sets in. To enable the integrations to reach longer time scales, a new type of numerical advective scheme was applied that could handle the advection of the salt field accurately even in the presence of marginally resolving grids.

1. INTRODUCTION:

It has been well established from oceanographic measurements and laboratory simulations that oceanic fronts, with their associated strong horizontal gradients of T and S, are subject to instabilities of the interleaving kind, in which a series of alternating warm and cold horizontal layers start intruding into the cold and warm sides of the front, respectively. The properties of these intrusions have been summarised by Fedorov (1978) and

Ruddick and Turner (1979). In particular, the vertical thicknesses of these layers have been found to fall into three ranges: 10-12m, 25-30m, and 100-120m, with horizontal extents from 1 to 10 km and life times of days to weeks. These length and time scales, and the gradients of T and S that occur along the intrusions and across the interfaces between them, are a function of their position within the front and the nature of the front. Different situations arise within fronts that are close to being density compensated (e.g. the subtropical front), and fronts that are accompanied by strong geostrophic motions (e.g. the North Wall of the Kuroshio). The intrusions themselves have been found so far mostly to be density compensated.

Observations on interleaving and intrusions have been mainly carried out in connection with frontal studies. Tabata (1970) has investigated the structure of the North Wall of the Kuroshio, and Schmitt and Georgi (1982) that of the North Atlantic Current; Horne (1978), Posmentier and Houghton (1978), and Voorhis et al. (1976) have examined the fine structure in the shelf/slope water front off Nova Scotia, the New York Bight and New England, respectively.

Quadfasel and Schott (1982) have studied the thermohaline structure off the Somali Coast; and Joyce et al (1978) and Toole (1981) have examined the intrusion characteristics of the Antarctic Polar Front.

It has been suspected from the beginning, and later verified by careful analysis of oceanographic data as well as by laboratory simulations and theoretical studies, that the interfaces between the intrusions are the seat of vigorous double-diffusive motions. In typical situations, one finds vertical gradients of temperature (T) and salinity (S) on both sides of the front, so that the actual T,S jumps that will appear across the intrusive interfaces, and the corresponding thermohaline motions that can become active across them, will depend on both the horizontal (frontal) gradients and the stratifications on two sides of front. It has been deduced from some of the observations that the interfaces form an alternating series of fingering and diffusive regions. Quadfasel and Schott (1982) have examined the stability properties of the lower and upper interfaces of intrusions and found that, for a cold intrusion, they correspond to a fingering upper interface and a diffusive lower interface, in agreement with the theoretical predictions of Turner (1978). Toole (1981) and Joyce et al (1978) have used flux considerations to infer double diffusive activity, and Posmentier and Houghton (1978) have used the T-S relations to come to the same conclusions. The horizontal and vertical motions and accompanying T/S fluxes that are found in these intrusions have important implications on water mass transformations, and the general dia- and isopycnal mixing processes in the ocean. These implications have been investigated through data analysis by Joyce et al

(1978) and Horne (1978), and through theoretical studies by Stern (1967), Joyce (1977), McDougall (1985) and Posmentier and Kirwan (1985).

Depending on the specific horizontal and vertical gradients on the two sides of the front, we may encounter stable interfaces with either fingering-type or diffusive type double-diffusive convection acting on them, or even unstable interfaces where (at least initially) the Rayleigh number is found to be subcritical. The latter will also be subject to double-diffusive convection at values of Rρ typically between .6 and 1.0, until such time that thermal diffusion and diffusive motions thicken the unstable layer and eventual gravitational overturning occurs.

In order to get a better understanding of the mechanisms that drive these intrusions, Turner (1978) and Ruddick and Turner (1979) have performed some laboratory experiments in which they deduced the spreading velocity, vertical fluxes and the vertical thickness of intrusive layers in a sugar-salt system, and made extrapolations to the ocean. More recently, Ruddick (1984) investigated the life of a 'cold' intrusion in a 3-layer run-down sugar salt system, and found that the fluxes associated with the diffusive motions can be important at certain times in the life of the intrusion, particularly when they change the density of the middle layer with respect to the upper layer, so that the fingers run into a convectively unstable situation.

The present numerical experiments aim to supplement the laboratory and theoretical studies by studying via direct simulation on the molecular scale the double-diffusive motions, which through their flux transports through the bounding interfaces control the evolution of an interleaving layer, such as its horizontal spreading, buoyant rise and property 'run-down'.

The numerical model intrusions will have the following limitations:

(a) small vertical extent (<25cm);

(b) 2-D quiescent fluid (no background turbulence);

(c) no large-scale shear present in any of the layers;

(d) all end up in a 'run-down' state.

In many ways, the numerical experiments follow the laboratory experiments of Ruddick (1984), and attempt to shed more information on the molecular-scale fluxes, particularly those associated with the diffusive interface.

Most up-to-date information on fluxes through diffusive interfaces is derived from sugar-salt systems rather than heat-salt systems. On the other hand, the fluxes through fingering interfaces have been thoroughly investigated in heat-salt systems [Schmitt (1979), McDougall and Taylor (1984)].

Most of the numerical experiments had density ratios of Rρ = 1.5 associated with them, with some experiments having one interface set at

Rρ=2.4. In these experiments we have tried to determine the temporal behaviour of the fluxes and the property content of the various layers, in order to understand the penetration of the layers by the double diffusive motions and the exchange of properties across the interfaces. So as to better resolve the details of these motions, some single interface experiments were also performed to (a) double the grid resolution, or (b) to double the total simulation volume.

2. THE NUMERICAL APPROACH

It has been found in a previous numerical simulation of salt fingers [Piacsek and Toomre,1980] that extending the calculations beyond a certain number of time steps and total simulation time results in the solution blowing up. In that study the critical simulation time was of the order of 90 seconds, which took about 2500 time steps at .04 sec each. In particular, the salinity maxima and minima have exceeded the values specified on the boundaries, which in that case were the largest (smallest) possible values allowed by physical and mathematical considerations. These difficulties were traced to the inability of ordinary finite difference approximations of the advective terms to handle the sharp gradients encountered in the system. One must remember that not only is the molecular diffusion of salt very low ($\sim 10^{-5}$ cm^2/s), but that the blobs of salt that separate from the tips of the fingers and race ahead cannot, by the virtue of their rapid motions, exchange any appreciable salinity content with their surroundings, in fact constituting one component of a diabatic, immiscible two-fluid system.

A numerical grid resolution of the so-called 'diffusive sublayer', which behaves as $(Ra_s)^{-1/3}$ with Ra_s being the Rayleigh number based on salt, is at best marginally possible in quasi-stationary layers, such as a diffusive interface, or slower-moving parts of the fingering system.

In order to overcome these difficulties in a practical way, we have adapted a set of numerical schemes originally developed for shock waves in plasmas [Boris and Book,1971; Zalesak,1979]. The contribution of the present work has been to adapt these so-called FCT (Flux-Corrected-Transport) algorithms to an incompressible fluid on the staggered grid of Piacsek and Toomre (1980).

In every other respect, the numerical method follows that of Piacsek and Toomre (1980), except for an improved version of the solver used to invert the Laplace's equation relating the stream function to the vorticity, adapting a recent solver technique based on FFT algorithms for vector computers [Moore (1985)].

One more remark must be made about the model concerning the stretching of the vertical coordinate. As in the 1980 study, the first stretching employed was a simple geometric progression, i.e. $DZ(j) = DZ(j-1)*FMULT$, with FMULT being a constant of the order of 1.03–1.06. In the current studies we have also used a stretching of the form $DZ(z) = c.z^2.exp(-z^2/a^2)$. In all experiments the minimum DZ was located in the centre of the diffusive interface, with a value of $DZ = .015$ cm. It was found that in general, the fingers are not very sensitive to its exact value. The highest values of DZ encountered were of order .08 cm, and these proved to be troublesome. The general conclusion we came to after all the runs with different stretchings that for the parameter values employed (see the discussion Section 3), the maximum DZ should not exceed .06 cm, or about 4 times the minimum value. In locations where DZ exceeded this value the flow developed unnatural distortions with horizontal elongations (trying to preserve mass and not being able to proceed correctly into the large grid spacing region). The horizontal grid spacing was constant for all experiments at $dx = .02$ cm.

3. DISCUSSION OF THE NUMERICAL RESULTS

In all the experiments carried out in this study, the horizontal boundary conditions applied were periodic, and the top and bottom boundaries were set to be insulating against both heat and salt, guaranteeing the so-called 'run-down' state of the system after sufficiently long times. The perturbation applied to an interface was always of the form

$$T,S\ (x,z) = (A_i,B_i).cos(4\pi x/L).exp[-(z-z_i)^2/d_i^2]$$

where typically the A_i-s and the B_i-s were chosen to be .001 times the respective jumps ΔT_i , ΔS_i across interface i, and d was set to $(3–5)xDZ$ (DZ being the minimum vertical grid spacing located at the centre of the respective interface). In three layer situations care was taken to balance the A_i-s and B_i-s, so that the perturbation energy imparted to each interface was the same. In all cases except one (the one being that of small vertical extent), the choice of a two-wave perturbation made the diffusive motions unhappy, and after a while they elongated their horizontal dimensions to a one-wave mode of convection. This transition was always accompanied by a large increase in the energy of the motions. The fingers were happy with the prescribed perturbation wavelength in each case. No attempt was made in any of the experiments to insert 'fastest growing' modes into the system, because this could have led to very different wavelengths at the different interfaces, for which the limited computational volume did not allow.

3.1   Experiment 1: Interaction of diffusive motions with weak gravitational

508

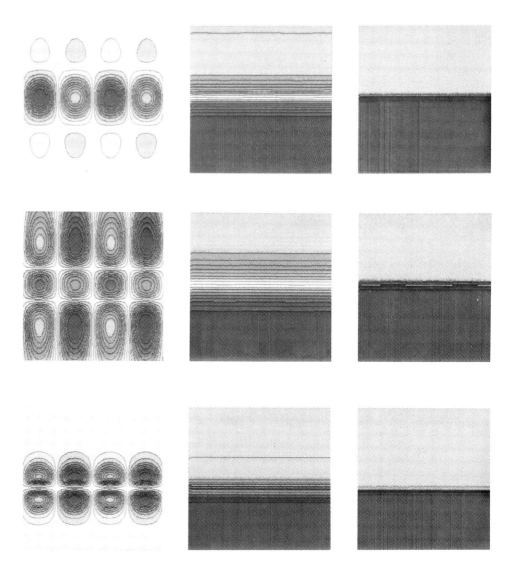

Fig. 1. Layout of 2-layer and 3-layer representation of interleaving systems. The top and bottom boundaries are insulating in each case, and the grid is stretched vertically to create a sufficiently large vertical layer thickness. Typically the maximum DZ is about 4-5 times the value in the diffusive interface and is located near the boundaries, as well as at the centre of the middle layer in a three-layer system.

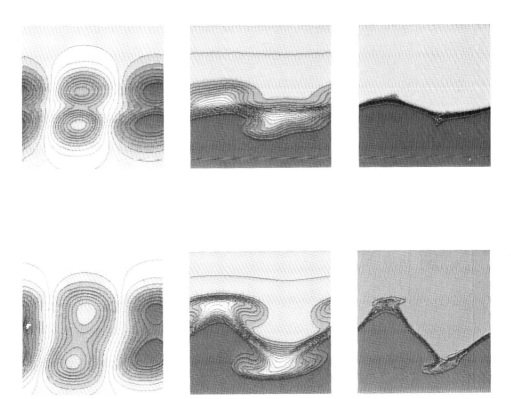

Fig. 2. Late time evolution of experiment 1: growth of gravitational (Rayleigh–Taylor) instability. The layout is that of Fig. 1, with successive rows representing snapshots at t=96,104 and 112 seconds, respectively. Corresponding to the interface deformation, the stream lines coalesce from symmetric vortex pairs on both sides of the interface to a single vortex centred on the interface, albeit with two maxima at this time.

convection. This experiment was carried out for a 3-layer system, with fingers at the upper and diffusive motions at the lower interface. However, the density ratio of the diffusive interface was set to $R\rho = .74$, and the motions at the diffusive interface dominated. Figs. 1 and 2 represent the early and late time evolution of the system, respectively. The perturbation as given by 2.1 gives rise to anti-symmetric motions in the sense that under an upwelling in the upper layer there is also an upwelling in the lower layer. Negative and positive contours of the streamlines are represented by dark and like contours and shadings, respectively. This perturbation is of the 'gravitational kind' and imparts a vertical velocity to the interface. Nevertheless, we see in Fig. 1 that this perturbation dies out quickly and double-diffusive motions, which do not have streamlines crossing the interface but are confined to their respective layers, set in vigorously. Fig. 3 illustrates the situation at late times, after thermal diffusion and advection by the diffusive motions have widened the unstable region for the Rayleigh number to go critical.

3.2 Experiment 2: Evolution of a strong fingering region in a 3-layer system. This experiment was designed to test the accuracy of the numerical scheme in treating the salt fingers, and to study their interaction with the middle layer and the diffusive interface. It is clear that the very thin interface is distorted without appreciable widening throughout the evolution of the system; even the separated 'blobs' remain segments of the (stretched) interface enclosing regions of lower (higher) salinity (Fig. 3a). After about 40 seconds the density of the middle layer approaches that of the lower one (Fig.3b). Because the salinity of the bottom layer was only slightly less than that of the top layer, the blobs broke up on hitting the interface; in cases where it is substantially less, the blobs keep going through, reminiscent of the "China Syndrome" until they reach a region of salinity equal to theirs. A surrounding cooler temperature field only slows them down, because they can gain and lose heat relatively fast and are only left with their salt anomaly.

3.3 Experiment 3: Wave-length selection and interface distortion of diffusive motions. This 2-layer experiment had a finer resolution in the vertical than either 1 or 2, and the layer thicknesses were 5 cm on either side of the interface, double that available to the diffusive motions in experiment 1 or 2. As mentioned before, the motions are happy at the prescribed perturbation wavelength only until t = 120 seconds or so; after that a strong nonlinear interaction in horizontal wavenumbers occurs which is reflected in the irregular shape of the convection cells (Fig.4). At the same time, the interface, which up to now has been perfectly smooth, thin, flat and

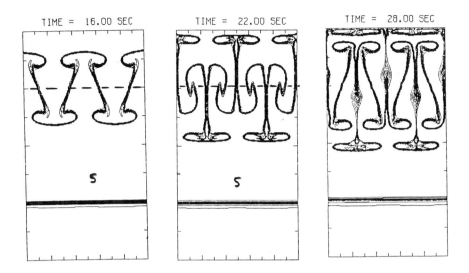

Fig. 3a. Time evolution of fingers in a 3-layer system (experiment 2). The motions at the lower, diffusive interface are too weak and do not disturb the interface. Note the uniform distortion of the fingering interface with a constant width as if it were a string; this should be so since in the elapsed time the salt could not diffuse to any discernible distance.

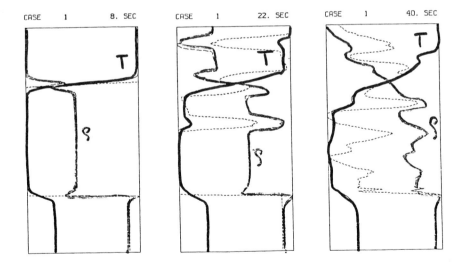

Fig. 3b. Time evolution of the mean T,S, and ς profiles in experiment 2. The diffusive interface remains sharp; the fingering interface is destroyed and the salinity of the upper layer is gradually added to the middle layer (since the upper and lower boundaries are insulating against both heat and salt.

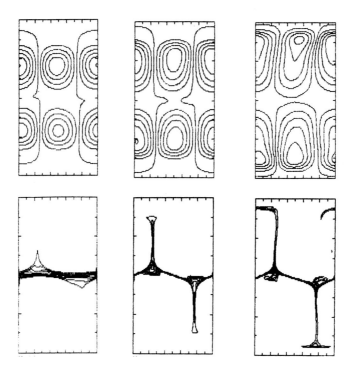

Fig. 4. Late-time evolution of experiment 3. The top row represents contours of streamlines and the lower one those of salinity. From left to right the columns represent successive snapshots at t=184, 200 and 216 seconds, respectively. Note the deformation in the interface and the thin sheets of salt carried upward by the convective motions, with a tendency to wrap them around the cells.

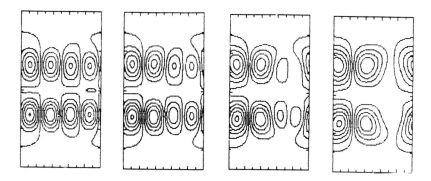

Fig. 5. Time sequence of wave-length transition in a diffusive system (experiment 3). From left to right, the frames represent snapshots of streamlines at t=104, 120, 136 and 162 seconds, respectively. The transition is from higher to lower wave number and is accompanied by a strong intensification of the convection energy (see Fig. 4).

undisturbed, undergoes some deformations and thin sheets of salinity are beginning to be pulled away form the interface, to be eventually wound around the circulating cells (Fig.4).

3.4 Experiment 4: Interaction of fingering and diffusive motions. This was a 3-layer experiment in which the middle layer had twice the thickness of the upper and lower layers, to allow for the penetration of motions from both interfaces. The fingering motions grew much faster initially, and at the beginning of the time sequence illustrated in Fig. 6, at t=72 sec, still have more energy.

But by the time the merging takes place, around t = 94 sec or so, the diffusive motions dominate and the fingering motions become totally disorganised.

Table 1 contains a summary of the density ratios applied at the interfaces of the various experiments, and the accompanying flux ratios. That of the fingers is consistent with previous measurements; for the diffusive cases the available information is scanty to comparison for the exact conditions encountered in these experiments.

4. Conclusions:

We can summarise our results in the following main conclusions:

(a) Diffusive motions are easier to model numerically than fingers, and can be maintained almost indefinitely in the computational simulations;

(b) Diffusive motions are much more sensitive to the horizontal wavelength than fingering motions;

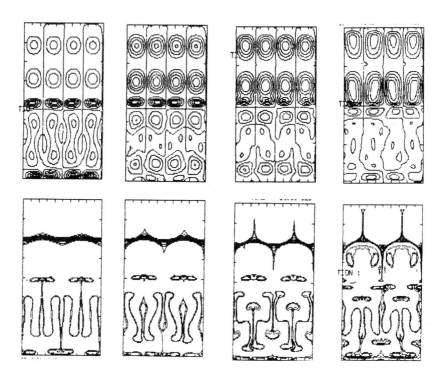

Fig. 6. Time evolution of a 3-layer system (experiment 4). The contour denotations are as in Fig. 6. From left to right, the successive columns depict snapshots at t=72,78,84, and 96 seconds, respectively. Initially the kinetic energy of the fingers is much larger, with the circulation concentrated around the separating 'blobs' of salt. Between t = 78 and 94 seconds the upper fingering vortices blend with the lower diffusive cells and lose their identity; at about this time the diffusive motions get much stronger, and the fingers overall seem to lose their structure.

(c) 2-D fingers in a quiescent fluid destroy an interface and change the properties of adjoining layers much more quickly than diffusive motions;

(d) Marginal and weakly unstable interfaces can be accelerated toward gravitational instability by double-diffusive motions;

(e) Fluxes of both types of motions across the interface can be comparable for certain values of R_ρ and at certain evolution times.

TABLE 1

INTERFACE STABILITY RATIOS AND FLUX RATIOS

Experiment	1	2	3	4
R_F	5.6	1.5	–	1.5
X_F	.50	.58	–	.54
R_D	1.60	2.4	1.5	1.5
X_D	.50	.51	.76	.62

where the flux ratio X_F is defined as F_H/F_S, with F_H and F_S being the total heat and salt transports (advective and conductive), respectively. X_D is the diffusive analog of X_F and is defined as F_S/F_H.

REFERENCES

Boris,J.P. and Book.D.L., 1973. I. SHASTA, A Fluid Transport Algorithm That Works. J. Comp. Phys. 11, 38.

Fedorov,K.N., 1978. The Thermohaline Fine Structure of the Ocean. Pergamon Press, Oxford, 169 pp.

Horne,E.P.W., 1978. Interleaving at the subsurface fronts in the slope water off Nova Scotia. J. Geophys. Res. 83, 3659.

Huppert,H.E., 1971. On the stability of a series of double diffusive layers. Deep-Sea Res. 18, 1005.

Joyce,T.M., 1977. A Note on the Lateral Mixing of Water Masses. J. Phys. Ocean. 7, 626.

Joyce,T.M.,Zenk,W. and Toole,J., 1978. The anatomy of the Antarctic Polar Front in the Drake Passage. J. Geophys. Res. 83, 6093.

Nagata,Y., 1970. Detailed temperature cross-section of the cold-water belt along the northern edge of the Kuroshio. J. Mar. Res. 28, 1.

McDougall,T.J. and Taylor,J.R., 1984. Flux measurements across a finger interface at low values of the stability ratio. J. Mar. Res. 42,1.

McDougall,T.J., 1985. Double-Diffusive Interleaving. Part II: Finite Amplitude Steady State Interleaving. J. Phys. Ocean. 15, 1542.

516

Moore,D.R., 1985. Efficient Explicit Real FFTs for Rapid Elliptic Solvers. Naval Ocean Research and Development Activity Technical Rep., 34 pp.

Piacsek,S.A. and Toomre,J., 1980. Nonlinear Evolution and Structure of Salt Fingers. In: J.C.J. Nihoul (Editor), Marine Turbulence. Elsevier, Amsterdam/Oxford/New York, 193–219.

Posmentier,E.S. and Houghton,R.W., 1978. Fine Structure Instabilities Induced by Double Diffusion in the Shelf/Slope Water Front. J. Geophys. Res. 83, 5135.

Posmentier,E.S. and Hibbard,C.B., 1982. The Role of Tilt in Double Diffusive Interleaving. J. Geophys. Res. 87, 518.

Posmentier,E.S. and Kirwan,A.D., 1985. The role of double diffusive interleaving in mesoscale dynamics: An hypothesis. J. Mar. Res. 43, 541.

Quadfasel,D.R. and Schott,F., 1982. Water-mass distributions at intermediate layers off the Somali Coast during the onset of the Southwest Monsoon. J. Phys. Ocean. 12, 1358.

Ruddick,B., 1984. The life of a thermohaline intrusion. J. Mar. Res. 42, 831.

Ruddick,B. and Turner,J.S., 1979. The vertical length scale of double-diffusive intrusions. Deep-Sea Res. 26A, 903.

Schmitt,R.W., 1979. Flux measurements on salt fingers at an interface. J. Mar. Res. 37, 419.

Schmitt,R.W. and Georgi,D.T., 1982. Finestructure and microstructure in the North Atlantic Current. J. Mar. Res. 40 (Suppl.), 659.

Stern,M.E., 1967. Lateral mixing of water masses. Deep-Sea Research 14, 747.

Stommel,H. and Fedorov,K.N., 1967. Small scale structure in temperature and salinity. Tellus 19, 306.

Toole,J.M., 1981. Intrusion characteristics in the Antarctic Polar Front. J. Phys. Ocean. 11, 780.

Toole,J.M. and Georgi,D.T., 1981. On the dynamics and effects of double-diffusively driven intrusions. Prog. in Ocean. 10, 123.

Turner,J.S., 1978. Double-diffusive intrusions into a density gradient. J. Geophys. Res. 83, 2887.

Voorhis,A.D.,Webb,D.C. and Millard,R.C., 1976. Current structure and mixing in the shelf/slope water front south of New England. J. Geophys. Res. 81, 3695.

Zalesak,S.T.,1979. Fully Multidimensional Flux-Corrected Transport Algorithms for Fluids. J. Comp. Phys. 31, 335.

REYNOLDS NUMBER EFFECTS ON TURBULENCE IN THE PRESENCE OF STABLE STRATIFICATION

A.E. GARGETT
Institute of Ocean Sciences, P.O. Box 6000, 9860 West Saanich Road, Sidney, B.C., (Canada).

ABSTRACT

Scalings derived for turbulence in the presence of stable stratification are highly dependent upon assumptions about the magnitude of relevant non-dimensional parameters and the degree of anisotropy exhibited by the largest overturning eddies. In particular, two determinate scalings derived for buoyancy-affected turbulence (turbulence existing at near unity value of the appropriate Froude number Fr) differ radically with magnitude of Re_w, the appropriate Reynolds number. This scale analysis shows that the parameter $(\varepsilon/\nu N^2)$, increasingly used to characterize turbulence "state", does not have a unique interpretation in terms of the ability of turbulence to effect a buoyancy flux. Moreover, strongly anisotropic (low Re_w) turbulence may cause differential transfer of heat and salt, even where mean gradients are stable to familiar double-diffusive effects.

1 INTRODUCTION

When attempting to interpret observations of turbulence in stably stratified fluids, one yearns for a scaling with the range and power of Kolmogoroff's (1941) scaling for high Reynolds number homogeneous isotropic turbulence, a scaling in terms of length and time scales which are functions of only two parameters; ε, the rate of dissipation of turbulent kinetic energy (TKE) per unit mass, and ν, the kinematic viscosity of the fluid. Unfortunately, such dreams are doomed to failure by (at least) two major differences between turbulence in stratified and unstratified fluids.

First, addition of a third parameter describing the degree of background stratification (normally taken as the Brunt-Vaisala frequency $N \equiv (g\rho_o^{-1}\partial\bar{\rho}/z)^{-1}$, where $\partial\bar{\rho}/\partial z$ is the (time-) mean stratification and ρ_o a reference density) leads immediately to both a diversity of possible scales which may be defined and to an increase in the number of non-dimensional parameters which must be specified. Thus we have moved from a system specified by a single non-dimensional number, the Reynolds number, and a single length scale $L_k = (\nu^3/\varepsilon)^{1/4}$, to a system in which we must also know the range of a Froude number, and choose among L_k, $L_b \equiv (\varepsilon/N^3)^{1/2}$, or perhaps $\delta \equiv (\nu/N)^{1/2}$.

Secondly, it is commonly believed that turbulence in the presence of stable stratification may exhibit significant anisotropy of the largest energy-containing scales. The argument is as follows. Buoyancy forces act directly only on the vertical component of momentum; any inefficiency of pressure forces in redistributing this momentum defect among all three momentum components will result in anisotropy of the form $w < u$ and hence $h < \ell$ where (u,w) are horizontal and vertical velocity scales and (ℓ,h) are horizontal and vertical length scales characteristic of the largest eddies. If the largest scale eddies can indeed be substantially anisotropic under certain conditions, then it is necessary to determine appropriate scales for two additional variables, namely $w \neq u$ and $h \neq \ell$.

In face of the additional complexities introduced by the presence of stratification, determinate scalings can only be derived by adding additional constraints. These may be in the form of restricting the range of various non-dimensional parameters, adding measurable parameters as external constraints, specifying the degree of anisotropy, etc., variety of choice which contributes to a wealth of possible solutions. It is thus essential that the assumptions or constraints underlying any proposed scaling be clearly outlined if it is to prove useful in understanding observations of stratified turbulence.

2 TWO DETERMINATE SCALINGS

As examples, I will briefly review two determinate scalings derived by Gargett (1988). These were originally derived merely as two determinate, internally consistent scalings, associated with clear choices of constraints. In fact, they both appear to describe appropriate (given the constraints) sets of observational data, hence may have significant implications for understanding both oceanic and laboratory observations (as discussed further in Sections 3 and 4).

The equations describing the motion of a non-rotating stably stratified fluid, under the Boussinesq approximation and the assumption of zero time-mean flow, are scaled allowing for possible anisotropy in the sense $w < u$, $h < \ell$. The full set of equations and assumptions are detailed in Gargett (1988): here I will examine only the vertical momentum equation (VME), since it is at the heart of the differences between stratified and unstratified scalings. Defining u_i as the component of velocity in direction x_i where $i = 3$ is vertical (positive upwards), p' as pressure (scale magnitude p) and ρ' (scale magnitude ρ) as the deviation of density from the local time-mean value $\rho(z)$, the terms in the VME

$$\frac{\partial u_3}{\partial t} + u_1 \frac{\partial u_3}{\partial x_1} + u_2 \frac{\partial u_3}{\partial x_2} + u_3 \frac{\partial u_3}{\partial x_3} = -\rho_0^{-1} \frac{\partial p'}{\partial x_3} + \nu \left(\frac{\partial^2 u_3}{\partial x_1^2} + \frac{\partial^2 u_3}{\partial x_2^2} + \frac{\partial^2 u_3}{\partial x_3^2} \right) + g\rho_0^{-1} \rho' \tag{1}$$

scale as $\qquad \dfrac{w^2}{h} \qquad\qquad \dfrac{\rho_0^{-1} p}{h} \qquad\qquad \dfrac{\nu w}{h^2} \qquad\qquad hN^2 \tag{2}$

Derivation of the scale magnitudes in (2) uses an assumption of horizontal homogeneity ($u_1 \sim u_2 \sim u$) as well as the scale relations $u/\ell \sim w/h$ from the continuity equation and $\rho' \sim h\bar{\rho}z$ from the density equation: see Cargett (1980) for complete details.

Examining first the ratio of the scale magnitudes of nonlinear and buoyancy terms in the VME, we find that

$$\frac{VN}{VB} \sim \left(\frac{w}{Nh} \right)^2 \sim \left(\frac{u}{N\ell} \right)^2 \sim \left(\frac{T_w}{T_e} \right)^2$$

is the (square) ratio of an internal wave period $T_w \propto N^{-1}$ to an "eddy" overturning time scale $T_e \propto u/\ell \sim w/h$. Thus an appropriate Froude number for this system is defined as

$$Fr \equiv \frac{u}{N\ell} \quad,$$

since if $Fr \gg 1$ buoyancy forces are unimportant and the turbulent motion is unaffected by stratification, while if $Fr \ll 1$ nonlinear forces are unimportant relative to buoyant restoring forces and the force balance reverts to that commonly associated with linear internal waves. Neither of these regions of Fr space are associated with the type of motion we are attempting to describe, namely motions in which neither nonlinear nor buoyancy forces are negligible: this must be the regime of $Fr \sim 0(1)$.

One must also be concerned with the relative magnitude of nonlinear and viscous forces: for homogeneous isotropic turbulence their ratio is the classical Reynolds number. Examining the ratio VN/VV from (2), it is clear that the appropriate form of the Reynolds number for turbulence which may be anisotropic is

$$Re_w \equiv \frac{wh}{\nu}$$

(the ratio HN/HV of the same terms in the horizontal momentum equation is also Re_w, hence this is a unique definition of a Reynolds number). Note that Re_w is formed from the smaller of the possible velocity and length scales, hence it is possible for Re_w to be small even while $Re_u \equiv u\ell/\nu$, formed from the horizontal velocity and length scales, remains large.

TABLE 1

Scaling relations and assumptions used to derive determinate scalings of
Table 2. For details, see Gargett (1987).

	High Re_w : $Re_w \gg 1$	Low Re_w : $Re_w \sim 1$
(1)	$\dfrac{u}{\ell} \sim \dfrac{w}{h}$	
(2)	$\rho \sim h\bar{\rho}_z$	
(3)	$\left(\dfrac{u}{N\ell}\right)^2 \sim 1$	
(4)	$\dfrac{u^3}{\ell} \sim \varepsilon$	
(5)	$u \sim w$	$\dfrac{wh}{\nu} \sim 1$
(6)	$p \sim \rho_0^{-1} u^2$	$p \sim \rho_0^{-1} w^2$ or $p \sim 0$

As is the case for non-stratified turbulence, the character of the motion
depends strongly upon Re_w, and Gargett (1988) derives determinate scalings for
two extremes of Re_w, a high $Re_w \gg 1$ case and a low $Re_w \sim 1$ case. Table 1 is a
summary of the scale relations and/or assumptions which apply to each
situation. The first four scale relations are common to both situations:
(1) comes from the continuity equation, (2) from the density equation, (3) is
the condition Fr $\sim$ 0(1) discussed above and (4) comes from the turbulent
kinetic energy equation under the assumption that

$\eta \simeq (Nh/u)^2 \leq 1$,

where $\eta \equiv \gamma/\rho_0\varepsilon$, the ratio of the rate γ at which available potential
energy is dissipated relative to the rate $\rho_0\varepsilon$ at which turbulent kinetic
energy is dissipated, is what McEwan (1983) has called the mixing efficiency.
In both cases relation (5) is an externally imposed assumption, while (6) is
a resulting pressure scaling, derived from the horizontal momentum equation
in the high Re_w case, and from the VME in the low Re_w case. Note that the
difference between the pressure scalings is indicative of the fundamental
physical difference between the two cases. In the low Re_w case, the

magnitude of the pressure force is (at most) proportional to the smallest dynamic pressure, hence the relative inability of the pressure forces to restore isotropy by redistributing momentum between components.

These relationships immediately lead to the scale variables listed in Table 2. The high Re_w scaling is familiar, the scales for length L_b and velocity u_b having been in frequent use in scaling experimental results over the past decade: it is associated with $\eta \sim O(1)$. The low Re_w scaling, which may have $\eta \ll 1$, retains the scales u_b and L_b for the horizontal variables, but has vertical scales which are those previously found to characterize the final (linear) viscous decay of velocity fluctuations (Gibson, 1980; Pearson and Linden, 1983).

TABLE 2

Determinate velocity and length scales for turbulence which is affected by stable stratification. (Fr $\sim$ 1)

High Re_w : $Re_w \gg 1$, $\eta \sim 1$		Low Re_w : $Re_w \sim 1$, $\eta \lesssim 1$	
$u \sim \left(\dfrac{\varepsilon}{N}\right)^{1/2} \equiv u_b$	$w \sim u$	$u \sim u_b$	$w \sim (\nu N)^{1/2}$
$\ell \sim \left(\dfrac{\varepsilon}{N^3}\right)^{1/2} \equiv L_b$	$h \sim \ell$	$\ell \sim L_b$	$h \sim \left(\dfrac{\nu}{N}\right)^{1/2} \equiv \delta$

The major conclusions which may be drawn from this analysis are the following. First, while the basic effect of buoyancy is to set an upper limit of $O(N^{-1})$ for the period of the largest overturning eddies, fundamental differences in morphology (ℓ/h) and effect (η) of these eddies are determined by the Reynolds number, appropriately defined. The profound significance of the Reynolds number in the description of turbulence modified by the presence of stable stratification has not been fully appreciated. Second, since low Re_w turbulence may be highly anisotropic, spectral scalings may be path-dependent. Third, the parameter ($\varepsilon/\nu N^2$), presently widely used to estimate the "state" of turbulence, does not have a universal interpretation in terms of the ability of turbulence to effect a buoyancy flux: using Table 2 it is easy to show that if $Re_w \sim 1$, ($\varepsilon/\nu N^2$) $\sim \eta^{-1}$, but if $Re_w \gg 1$, ($\varepsilon/\nu N^2$) $\sim Re_w$. Put another way, if all one knows about a particular turbulent field is that ($\varepsilon/\nu N^2$) is large, one does not know whether to interpret this as evidence of a large Re_w, hence $\eta \sim 1$, or as a large η^{-1}, hence $\eta \ll 1$. The difference of course has significant repercussions.

3 OBSERVATIONAL EXAMPLES

Although these scalings were originally derived merely as internally consistent determinate solutions to the scaling problem (not the only solutions), Gargett (1988) has shown that the Knight Inlet observations of Gargett et al. (1984) are consistent with the high Re_w case, while the laboratory grid-stirring experiments of Stillinger et al. (1983) and Itsweire et al. (1986) exhibit many characteristics of the low Re_w regime. The clearest indication of this latter claim is that measured values of η are found to be inversely proportional to $(\varepsilon/\nu N^2)$, i.e.: this is a case where large values of the parameter $(\varepsilon/\nu N^2)$ do not imply large Re_w but small η, hence very inefficient conversion of kinetic to potential energy.

As a further example of the very different perspective offered by consideration of Re_w and degree of anisotropy of the large scale eddies, let us reexamine the "hydrodynamic phase diagrams" proposed by Gibson (1986, 1987) as a means of classifying turbulence into "active" and "fossil" states and various mixtures thereof. Figure 1(a) (redrawn from Fig. 8 of Gibson, 1987) has been interpreted as showing the evolution from "active" to "fossil" states of turbulence generated by a grid in a salt-stratified laboratory flow facility (Itsweire et al., 1986). The arrow at upper right indicates direction of increasing distance from the grid. The abscissa of this diagram involves a constant times the square root of the parameter $(\varepsilon/\nu N^2)$. The ordinate is another constant times the ratio of $L_b \equiv (\varepsilon/N^3)^{1/2}$ to the Ellison (1957) length scale $L_E \equiv \left(\overline{(\rho')^2}\right)^{1/2}/\bar{\rho}z$, which has been shown to be a reasonable approximation to the vertical scale h of overturning density surfaces (Itsweire, 1984). The left-hand scale is that of Gibson (1987) (who uses the notation L_R instead of L_b). On the right-hand scale, I have defined an anisotropy ratio $A \equiv \ell/h$, as the ratio of horizontal to vertical eddy length scales. Since $h \sim L_E$ and since both high and low Re_w scaling result in $\ell \sim L_b$, A is just some constant b times L_b/L_e: the scales at the right indicate the location of isotropy (i.e.: log (ℓ/h = 1) = 0) for various choices of b. Viewed in terms of A, the laboratory experiments are observed to originate as anistropic eddies (A $\gg$ 1), and evolve towards isotropy (A $\to$ constant in the lower left hand corner; A $\to$ 1 if b has a value of $\sim$2). Moreover, the assumption that this turbulence exists at low Re_w leads to the prediction that $A \equiv \ell/h \sim (\varepsilon/\nu N^2)^{1/2}$, hence that the experiment should evolve, as observed, along a line of slope +1 in this particular log/log representation. In this interpretation, the laboratory grid turbulence is never high Reynolds number homogeneous isotropic turbulence: it both originates and evolves as low Re_w turbulence.

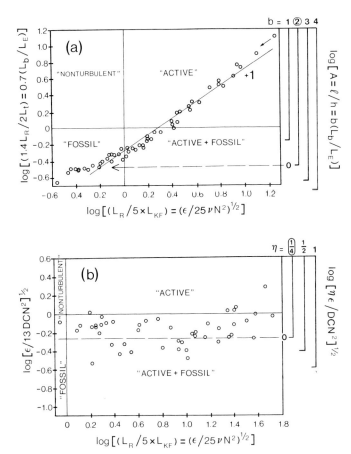

Fig. 1. (a) "Hydrodynamic phase diagram", after Fig. 8 of Gibson (1987) for
the laboratory grid turbulence experiments of Itsweire et al. (1986). In the
context of the scaling arguments reviewed in Sec. 2 of this paper, the
ordinate is re-interpreted as an anisotropy ratio $A \equiv \ell/h$ of horizontal to
vertical length scales of the largest eddies of the turbulent field. The
arrow gives the direction of evolution of the field (downstream of the
generating grid), and the line of slope +1 is the predicted course of
experiment evolution, based upon the assumption that this laboratory
turbulence exists at low Re_w (for details see text Sec. 3)

(b) An alternative form of "hydrodynamic phase diagram", after Fig. 9 of
Gibson (1987), for the lake observations of Dillon (1982). In the context of
the scaling arguments in Sec. 2, the behaviour of this data is consistent with
the assumption that the turbulence exists at $Re_w \gg 1$ (for further details see
text, Sec. 3)

Figure 1(b) shows another form of "hydrodynamic phase diagram" devised by
Gibson (1980) for cases (such as the lake measurements of Dillon (1982) shown
here) in which h and ℓ are not measured directly and only the microscale
variables ε and Cox number C are available. Again the abscissa is
proportional to $(\varepsilon/\nu N^2)^{1/2}$ while the ordinate involves the square root of the

ratio (ε/DCN^2) multiplied by some constant, for which Gibson (1980) proposes $(1/13)^{\frac{1}{2}}$. Now from the definition of C and the equation for conservation of available potential energy, it is easily seen that $\eta = \gamma/\rho_0\varepsilon \sim DCN^2$, so that turbulence existing at constant η will exhibit a constant value of ($\eta \varepsilon/DCN^2$) (Dillon, 1984). If we assume that the lake observations are characterized by high Re_w, then the scalings derived from this case predict that $\eta \sim O(1)$, hence that (ε/DCN^2) should vary little with variation in ($\varepsilon/\nu N^2$) $\propto Re_w$: this seems a reasonable description of the behaviour of the data shown in Fig. 1(b). From the scales shown at the right, a value of $\eta \sim \frac{1}{4}$ would appear to best characterize the lake date (the large observed scatter is to be expected in a quantity such as η which is the ratio of two non-normally distributed observed variables, ε and C). Similar diagrams given by Gibson (1987; his figures 11 and 12) for two oceanic data sets reported by Dillon (1982) also suggest states of high Re_w and "constant" η, with a tendency for η to increase with wind speed: Dillon's high wind oceanic data is scattered around log ($\varepsilon/13DCN^2$)$^{\frac{1}{2}}$ = -0.6, associated with $\eta \sim 1$ (right-hand scale of Fig. 1(b)). From the point of view of this paper, such turbulence is perfectly "active", in the sense that $Re_w \gg 1$.

Thus, most available field observations (the surface layer measurements of Dillon (1982) which have already been examined and the Knight Inlet observations of Gargett et al. (1984)) appear to have the character of high Re_w turbulence. The only exceptions are the very near-surface lake estimates of Imberger and Boashash (1986), which roughly approximate the low Re_w behaviour of the laboratory experiments.

The major difference between the present interpretations of "hydrodynamic phase diagrams" and that of Gibson (1987) lies in the estimated ability of a particular turbulent field to transport mass in the vertical. The conclusions drawn on the basis of the scale analysis of Sec. 1 are that the laboratory turbulence exists at low Re_w and consequently is extremely inefficient at transporting mass, while the oceanic observations suggest turbulence which is high Re_w and reasonably efficient at mass transport. These are almost exactly opposite in character to conclusions which have been drawn from the Gibson fossil turbulence scenario (Gibson, 1980). Since perhaps the major goal of microscale oceanic observations is to quantify diapycnal mass fluxes, it is obvious that these differences matter, and that the issues must be resolved. To do so, it is essential to measure characteristics of the turbulent field which will allow independent determination of the Reynolds number $Re_w = wh/\nu$ and the degree of anisotropy $A = \ell/h$ of the largest overturning eddies. We may no longer comfortably measure just those parameters (ε and N) which we are presently most easily able to measure in the ocean and still feel that we can draw unambigious conclusions as to the magnitude of the associated buoyancy flux.

4 Differential Transport of T and S

If the largest scales of turbulent eddies proved to be substantially anisotropic in some regions of the ocean, there is (in addition to the problem discussed above) the unsettling possibility that such regions might be characterized by differential diapycnal transfer rates for heat (or temperature T) and salt S, even where the mean T and S gradients are stable with respect to classical double-diffusive instabilities.

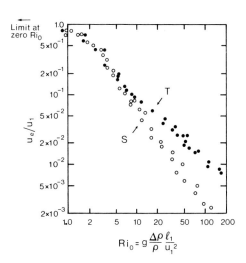

Fig. 2. The grid-stirring experiments of Turner (1968), here scaled with the horizontal turbulent length (ℓ_1) and velocity (u_1) scales subsequently reported by Thompson and Turner (1975), show that at large values of a Richardson number Ri_o, the entrainment rate (u_e) of properties across a stable density interface between two homogeneous layers depends upon whether the stratification is produced by temperature T or salt S. Such differential transport of T and S may be important in the ocean if the large turbulent eddies prove significantly anisotropic under certain conditions (for details see Sec. 4).

The origin of this possibility lies in grid-stirring experiments, originally carried out by Turner (1968) and later quantified by Thompson (1969), in which turbulence generated by shaking a grid caused transport across a density interface produced by either temperature or salt differences. Figure 2, redrawn from Turner (1973), shows that the interfacial flux of stratifying property (quantified by the entrainment velocity u_e) depends upon which property is being used to provide the density difference between the (homogeneous) upper and lower layers. In Figure 2, the interfacial Richardson number Ri_o has been formed using the horizontal integral length scale ℓ_1 and rms horizontal velocity u_1 reported by Thompson and Turner (1975). These parameters were measured at the depth of the interface but in the absence of the interface, necessary because density fluctuations would have contaminated the hot-film measurement technique used to determine u_1, hence ℓ_1. This diagram has never elicited much alarm from oceanographers, because of a general belief that oceanic turbulence exists at a (suitably defined) turbulence Richardson number near unity, hence within the region of Figure 2 where the T and S transport values have converged.

Two factors should perhaps increase the sense of unease experienced in viewing Figure 2. First, Altman (1987) has demonstrated that essentially the same results are found when T and S are present together: i.e., when both T and S contribute to $\Delta\rho$ (since the physical mechanisms leading to differential transport are not well understood at present, it was necessary to check experimentally that no unforeseen mechanism would arise to equalize the transports if both T and S were present together). Secondly, the Richardson number Ri_o calculated from the properties (u_1, ℓ_1) of turbulent eddies unaffected by the interface stratification may considerably overestimate an "effective" Richardson number Ri_e formed from properties of the turbulent eddies when they are in contact with the interface, the only position in which they are able to effect density tranport across the interface.

As a simple illustration of the possible magnitude of this effect, consider an individual turbulent eddy as a deformable entity which conserves mass (volume) and angular momentum while flattening during its initial contact with the density interface. Evidence from recent laboratory experiments by Hannoun et al. (1987) shows that eddies near the interface have larger horizontal velocities and length scales than eddies in the mixed layers well away from the interface. Thus if $u_1 \sim w_1$, $\ell_1 \sim h_1$ are the initial dimensions of the eddy, $u > u_1$, $\ell > \ell_1 > h$ are its "final" dimensions (see Figure 3 for schematic). Conserving mass $h_1 \ell_1^2 \sim h \ell^2$ and angular momentum $uh \sim u_1 h_1$ results in

$$Ri_e = \frac{g\Delta\rho}{\rho}\frac{\ell}{u^2} \sim \left(\frac{\ell_1}{\ell}\right)\left(\frac{h}{h_1}\right)Ri_o << Ri_o$$

Thus an "effective" Richardson number of $0(1)$ may correspond to a much larger value of Ri_o in Figure 2, and we have the possibility that differential transport may exist in the ocean as well as the laboratory.

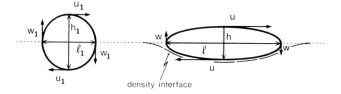

Fig. 3. Schematic of changes in morphology experienced as an initially isotropic eddy (left) comes into contact with a stable density interface (right).

This possibility is associated with various disturbing implications. It implies that the easiest laboratory experiments to do (those in which stratification is produced by salinity, eliminating the need for elaborate thermal insulation) might never tell us everything we need to know in order to parameterize mixing in the ocean. It implies that salt fingers in the presence of such turbulence can be characterized by any value of the ratio $R_f \equiv \alpha F_T/\beta F_s$ of the density flux due to heat αF_T relative to that due to salt βF_s, depending upon the relative magnitude of the vertical velocity associated with the turbulence (w) to that associated with the salt fingers (w_s): if $w \ll w_s$, $R_f < 1$, while if $w \gg w_s$, $R_f > 1$. (The laboratory measurements of Altman (1987) alone indicate that the condition of $\alpha F_T \sim \beta F_s$ (w ~ w_s), taken by Linden (1971) as evidence of the dominance of turbulence over salt fingering, is actually a condition for the approximate equality of contributions due to one process with $\alpha F_T > \beta F_s$ (low Re_w anisotropic turbulence) and another with $\alpha F_T < \beta F_s$ (salt fingering).) More generally, the existence of differential transport of heat and salt even in double-diffusive stable regimes would have significant consequences for areas such as water mass modification, estimation of vertical transport of biologically active nutrients and proper interpretation of results of purposeful tracer experiments, to mention just a few.

REFERENCES

Altman, D.B., 1987. Differential property transport due to incomplete mixing in a stratified fluid, (in preparation).

Dillon, T.M., 1982. Vertical overturns: A comparison of Thorpe and Ozmidov length scales. J. Geophys. Res., 87, 9601-9613.

Dillon, T.M., 1984. The energetics of overturning structures: implications for the theory of fossil turbulence. J. Phys. Oceanogr, 14, 541-549.

Ellison, T.H., 1957. Turbulent transport of heat and momentum from an infinite rough plate. J. Fluid Mech., 2, 456-466.

Gargett, A.E., 1988. Scaling of turbulence in the presence of stable stratification. J. Geophys. Res., accepted.

Gargett, A.E., Osborn T.R. and Nasmyth P.W., 1984. Local isotropy and the decay of turbulence in a stratified fluid. J. Fluid Mech, 144, 231-280.

Gibson, C.H., 1987. Fossil turbulence and intermittency in sampling oceanic mixing processes. J. Geophys. Res., 2 (C5), 5389-5404.

Gibson, C.H., 1986. Internal waves, fossil turbulence, and composite ocean microstructure spectra. J. Fluid Mech., 168, 89-117.

Gibson, C.H., 1980. Fossil temperature, salinity and vorticity turbulence in the ocean. Marine Turbulence, ed. J.C.J. Nihoul, Elsevier Series, 28, 221-258.

Hannoun, I.A., Fernando, H.J.S. and List, E.J., 1986. Turbulence structure near a sharp density interface. J. Fluid Mech., submitted.

Imberger, J. and Boashash, B., 1986. Application of Wigner-Ville distribution to temperature gradient microstructure: a new technique to study small scale variations. J. Phys. Oceanogr., 16(12), 1997-2012.

Itsweire, E.C., 1984. Measurements of vertical overturns in a stably stratified turbulent flow. Phys Fluids, 27, 764-766.

Itsweire, E.C., Helland, K.N., and Van Atta, C.W., 1986. The evolution of grid-generated turbulence in a stably stratified fluid. J. Fluid Mech., 162, 299-338.

Kolmogoroff, A.N., 1941. The local structure of turbulence in an incompressible viscous fluid for very large Reynolds number. C.R. Acad. Sci., USSR, 30, 301-305.

Linden, P.F., 1971. Salt fingers in the presence of grid-generated turbulence. J. Fluid Mech., 49, 611-624.

McEwan, A.D. (1983). Internal mixing in stratified fluids. J. Fluid Mech., 128, 59-80.

Pearson, H.J. and Linden, P.F., 1983. The final stage of decay of turbulence in stably stratified fluid. J. Fluid Mech., 134, 195-203

Stillinger, D.C., Helland, K.N. and Van Atta, C.W., 1983. Experiments on the transition of homogeneous turbulence to internal waves in a stratified fluid. J. Fluid Mech., 131, 91-122.

Thompson, S.M., 1969. Turbulent interfaces generated by an oscillating grid in a stably stratified fluid. PhD Thesis, University of Cambridge. U.K. 184 pp.

Thompson, S.M. and Turner, J.S., 1975. Mixing across an interface due to turbulence generated by an oscillating grid. J. Fluid Mech., 67, 349-368.

Turner, J.S., 1968. The influence of molecular diffusivity on turbulent entrainment across a density interface. J. Fluid Mech., 33, 639-656.

Turner, J.S., 1975. Buoyancy Effects in Fluids. Cambridge University Press, U.K. 367 pp.

COMMENT ON "REYNOLDS NUMBER EFFECTS ON TURBULENCE IN THE PRESENCE OF STABLE STRATIFICATION"

C. H. GIBSON
Departments of Applied Mechanics and Engineering Sciences and Scripps Institution of Oceanography, University of California, San Diego La Jolla, CA 92093, USA.

The paper by Gargett (1988, this volume) is based on the ideas that turbulence can be highly anisotropic and that the Ozmidov length scale represents the largest horizontal eddy size at all stages of development of turbulence in a stably stratified fluid. Neither of these ideas is correct, and consequently the proposed classification scheme for stratified turbulence is invalid. The "strongly anisotropic" turbulence at small "appropriate" Reynolds numbers simply doesn't exist.

The grid turbulence measurements of Stillinger et al. (1983) and Itsweire et al. (1986) are described as examples of such strongly anisotropic turbulence using the hydrodynamic phase diagram of Gibson (1987) in Figure 1a. The left ordinate is the turbulence activity coefficient $A_T = (\varepsilon / \varepsilon_o)^{1/2}$ proposed by Gibson (1980) as a means of classifying microstructure according to its hydrodynamic state: if $A_T > 1$, the microstructure is actively turbulent; if $A_T < 1$, the microstructure is fossil turbulence at the largest scales. A_T represents the ratio of the microstructure Froude number to the critical Froude number at the buoyancy-inertial force transition point, termed the "point of fossilization" by Gibson (1980). The abscissa $(\varepsilon / 25 \, v \, N^2)^{1/2}$ of Figure 1a represents the ratio of the microstructure Reynolds number to the critical Reynolds number at the buoyant-inertial-viscous transition. When this value is less than one the microstructure is fossil turbulence at all scales. For references and a more complete description of the fossil turbulence paradigm and microstructure classification scheme, see Gibson (1988, this volume).

Gargett has reinterpreted A_T as an anisotropy ratio $A = l/h$, where l is the largest horizontal eddy size and h is the largest vertical eddy size. She assumes that l is the Ozmidov scale and h is the Ellison, or overturn, scale. However, the Ozmidov scale is an upper bound for l (and h) and is not a measure of turbulence eddy sizes as assumed by Gargett. The stratified grid turbulence data of Itsweire et al. (1986) shown in Figure 1a progresses from about 10 mesh lengths in the active turbulence quadrant (upper right) to the point of fossilization where the points cross the (horizontal) critical Froude number line. The microstructure is then mixed active and fossil turbulence as its Reynolds number decreases to the buoyant-inertial-viscous value at the vertical line (the turbulent eddy size at this point is proportional to the fossil Kolmogoroff scale $L_{KF} = (v / N)^{1/2}$) and it leaves the test section as completely fossil turbulence.

At no point is the microstructure strongly anisotropic, even after buoyancy forces begin to suppress the turbulence, since the resulting saturated internal wave motions are nearly isotropic. The isotropy of grid turbulence has been measured in many ways by many authors, including Itsweire *et al.* (1986). Transverse and streamwise spectra are within 10% of the isotropic ratio at all scales, as shown by Schedvin *et al.* (1974, Fig. 9).

According to the present paper the "laboratory experiments are observed to originate as anisotropic eddies (A >> 1) and evolve toward isotropy". Perhaps the best counter example to this interpretation is the case of (isotropic) unstratified turbulence where the Ozmidov scale and, consequently, the anisotropy ratio A go to infinity at all distances downstream. The Gargett conclusions that the ratio $\varepsilon / (v N^2)$ does not have a unique interpretation in terms of the ability of turbulence to effect a buoyancy flux and that "strongly anisotropic turbulence" (fossil turbulence) may cause differential transfer of heat and salt are interesting, and probably correct but for different reasons.

Additional reference

Schedvin, J., G. Stegen and C. Gibson, 1974. Universal similarity at high grid Reynolds numbers. J. Fluid Mech., vol. 65, pp. 561-579.

REPLY

A. E. GARGETT
Institute of Ocean Sciences, P.O Box 6000, 9860 West Saanich Road, Sydney, B.C., Canada.

The paper entitled "Reynolds number effects on turbulence in the presence of stable stratification" (Gargett, this volume) uses a scaling analysis to conclude that, whether the largest eddies are isotropic or not, L_b is an appropriate horizontal length scale l for turbulence characterized by $Fr \sim 1$ and by either $Re_w \sim 1$ or $Re_w \gg 1$, both ends of the range expected of a vertical Reynolds number. It is not claimed that this result can be applied if the conditions underlying the discussion are not met : thus the counter-example of unstratified turbulence ($N = 0$, hence $Fr \to \infty$) offered by Gibson is not relevant.

What range of parameters is occupied by the grid turbulence experiments of Stillinger et al. (1983) and Itsweire et al. (1986; hereafter IHV)? Since measured overturning scales rarely lie outside the velocity dissipation range, and then only near the grid, Re_w is arguably not large. However the important parameter to estimate is $Fr \equiv u/Nl$. I shall use reported values of rms fluctuating velocity u' to estimate u (leading to an upper bound on Fr, since u' may contain wave as well as turbulence velocity variance, IHV) and for l consider either L_b or the grid mesh length M (preferred by C. Van Atta, pers. comm.). For the three stratifications $\overline{N} = (0.90, 0.56, 0.27 \text{ rad s}^{-1})$ of experiment R36 of IHV, one obtains upper bound estimates of either $Fr (L_b) \leq (0.58, 0.59, 0.46)$ or $Fr (M) \leq (0.46, 0.85, 1.76)$ at $x/M = 5$. Either set of values suggests that Fr is indeed near 1 immediately behind the grid, hence the turbulence cannot be imagined to evolve, even initially, free from the influence of buoyancy. Thus there seem to be some grounds for exploring the assumption that the experiments evolve in the $Fr \sim 1$, $Re_w \sim 1$ parameter range.

Now while this parameter range *may* be characterized by anisotropy of the largest overturning scales, it need not be, hence the question of isotropy is separate from the question of parameter range. First, in the context of results from stratified flows in the UCSD facility, Gibson's claim that the isotropy of grid turbulence has been demonstrated is misleading. IHV do not report investigations of the degree to which isotropic inter-relationships are satisfied by their various measured spectra, and the fact that Schedvin et al. (1974) demonstrate isotropy in a non-stratified ($Fr \to \infty$) system with a grid mesh Reynolds number $Re_m \equiv \overline{U}M/\nu$ of 2–4×10^5 seems of dubious relevance to the question of isotropy in a strongly stratified ($Fr \sim 1$) system run at a maximum Re_m of 9500. Concerning the isotropy of the stratified experiments, it seems generally agreed that the Ellison scale L_E provides a representative vertical overturning scale h, at least in the initial stages of turbulence development (Itsweire, 1984). Using h = L_E and taking for l the minimum of L_b and M yields *lower bound* estimates of (5.1, 5.4, 4.9) ≤ l/h at x/M = 5 for the runs of experiment R36, suggesting that initial anisotropy of the largest overturning eddies is likely.

Thus using an assumption of initially anisotropic eddy structure evolving at low Re_w and Fr ~ 1, the paper under discussion provides an interpretation of the UCSD laboratory experiments which, although different from the previous interpretation, appears consistent with observed features of these experiments, as reported. Differentiating between these views will require more careful definition and measurement of the morphology of the largest overturning structures. I have suggested (Gargett, 1988) a particular measurement (of the spectrum of vertical velocity as a function of vertical wavenumber) which should be most different under the two scenarios, and I await this measurement with interest.

AN OVERVIEW OF THE 19TH INTERNATIONAL LIEGE COLLOQUIUM ON
OCEAN HYDRODYNAMICS; "SMALL-SCALE TURBULENCE AND MIXING IN THE OCEAN"

S.A.THORPE
Department of Oceanography
The University,
Southampton SO9 5NH, U.K.

(This, with minor editing, is the text of a 'Conclusions and Perspectives'
talk presented at the end of the Colloquium. The author wishes it to be
known that these are his own, biased views, not necessarily those of
other members of the SCOR Working Group 69 which was responsible for the
scientific organisation of the meeting. In this, at least, they are free
from blame. Except for the appended list, reference is made to papers
presented at the meeting, most or all of which are contained in the
Proceedings).

It has been an exciting and exhausting week!

I must begin by expressing some regrets. As a member of the
Scientific Programme Committee I am sorry we could not fit more talks
into the programme. As it is, there has perhaps been too little time for
discussion. I am disappointed that Professor Fedorov was the only
person to come from the U.S.S.R. An international meeting on small-scale
turbulence in the ocean is not complete without adequate representation of
the fine work done in the U.S.S.R. in this field. I ask Professor Fedorov
to convey our regrets to his colleagues who are not with us.

At the beginning of the meeting Konstantin Fedorov suggested that
we should be able, as one of the objectives of the meeting, to judge
progress in the subject since the last meeting 8 years ago. I was not
present at that meeting so I can only guess that perhaps the main areas of
progress which we have heard about are:-

(1) the use of purposeful tracers (pioneered by Ledwell with Broecker and
Watson),

(2) the continued use of natural tracers (Sarmiento),

(3) the impressive extension of the careful and wonderful microstructure studies, especially in particular regions of the World Ocean, the Equatorial Undercurrent, the Strait of Gibraltar, in salt finger layers of the Caribbean and in Meddies (contributions from Peters, Gregg, Wesson, Oakey and Sandford),

(4) the use of acoustic methods to supplement microstructure or fine structure observations (e.g. Sandstrom, and Wesson and Gregg),

(5) the increased emphasis on applying ideas of mixing to problems of biological application (several speakers, including Stigebrandt),

(6) the work on understanding the importance of diapycnal and isopycnal diffusion, rather than horizontal and vertical diffusion, especially in the way the mixing processes are represented in large-scale models. Here I include also McDougall's careful work on neutral surfaces.

(7) the rather exciting developments in ideas about vortical modes or of the zoo of Intrathermocline Eddies (Muller, and verbal discussion by Fedorov),

(8) the work of which we have just heard on double diffusive processes, especially the C-SALT experiment (Schmitt),

(9) the beginnings of studies of the benthic boundary layer on slopes (Pingree, Garrett and Gilbert).

We have not had reported here some other very important observations which have been made in, for example, the upper ocean boundary layer (and I have in mind Shay and Gregg's, 1984, work which has established similarity with the atmospheric boundary layer in convective conditions, and advances in the study of surface wave breaking; for a review see Thorpe, 1985) and those of fine-structure in the Arctic using various methods, including acoustics. This should dispel a feeling of complacency that we have done a comprehensive job in covering knowledge of small scale ocean turbulence. The reason is of course that we have focussed here on the relationships between small-scale mixing and large-scale features, and on the related problem of parameterising

small-scale mixing in models of larger scale flows. This was the real
purpose and objective of the meeting, so let me try to view the highlights
from the perspective of this objective.

The subject was lucidly set out by early speakers. You will
recall especially Dirk Olbers vividly describing how ill-conditioned were
the early solutions used by Munk (1966) to estimate vertical diffusivity,
and the apparent sensitivity of model estimates of meridional heat flux to
variation in the choice of diapycnal diffusivity (although we were
reminded by Peter Rhines to be careful from reading too much into this
due to changes in boundary conditions!).

We heard too of the sensitivity of near-surface and shallow sea
circulation models to the choice of K_v and the present uncertainty about
how to account for surface waves. There are new ideas here about what to
do (for example, the use of transilient turbulence by Eric Kraus and
others) but these have not yet reached a stage at which observations of
turbulence can be of value in aiding parameterisation. Even
representation of Langmuir circulation - that embarrassingly obvious
processes of which we speculate so much but really know so little, - seems
not to have been incorporated in the parameterisation, and this is only
one of the coherent structures in the turbulent upper ocean boundary layer
flow which might be so included. Here is an area for further observation
and modelling.

On a larger scale, problems of air-sea exchange and ventilation
became very apparent. We are now conscious of the need to carefully
include seasonal (even diurnal) variability in circulation models, and
here the development of further tracer techniques is important and
valuable. I feel a certain disappointment in what has been learnt so far
from the natural or anthropogenic tracers, and there is considerable
potential for more tracer studies especially using tracers, like Radium
228 (about which Dr Sarmiento told us), which have localised sources.
There are other anthropogenic tracers (Tritium and the Freons) which will
yet provide much more information in the future.

Salt is still a natural tracer 'par excellence'. We had a really
exciting 'Morning of the Meddies'. At last the oceanographer's answer to
the astrophysicist! 'The cosmology of the Meddy galaxy, creation to
destruction' - including single double and triple ....etc helical

instrusions. 'Big Salty Blinies' – a moment to relish! But wait a
minute. We were not being persuaded of the importance of meddies in
transporting salt generally, although they may be important in some areas.
Rather they belong to the class of vortical modes, whose implication for
tracer distributions, streakiness, and iso- and dia-pycnal mixing require
further consideration. At last the need to consider the relation of iso-
and diapycnal mixing to which Peter Rhines drew attention become more
vivid and urgent. Part of the story seems to be as shown in figure 1 –
and I have plagiarised the ideas of Peter Rhines and Peter Muller. One
example is suggested in figure 2.

But does this relationship of iso- and diapycnal mixing –
supposing it can be supported and quantified – help our ability to predict
dispersion or diffusion of T, S, or other tracers in the ocean? Is it
worth further study? I vote yes! I think we need to continue to work at
reducing the degree of parameterisation in models if we are really to make
progress in understanding how different tracers are diffused. This was
one of Ann Gargett's worries. Are K_T, K_S etc. different? Surely it is
likely that they are, but where and when? This is presently beyond our
range of knowledge. More laboratory experiments, together with
imaginative use of models, may help guide progress.

The intermittency of turbulence in the ocean, a factor to which
Carl Gibson drew attention, is one of our greatest problems when it comes
to estimating diapycnal diffusion. Basically it goes back to the fact
that the ocean is full of transitional phenomena, of which there are a
great variety. Described in Thorpe[1] (1987) are examples of the stages, or
bifurcations, through which Kelvin-Helmholtz instability must develop
before a chaotic or turbulent flow is produced. Much of the ocean lies in
a multi-parameter regime in which a variety of processes are present
including, but not only, K-H I, each in different stages of transition.
Unfortunately (or challengingly, depending on your degree of pessimism or
optimism) the fluxes of momentum and heat and tracers are very sensitive
to position in parameter space, and their prediction is therefore subject
to considerable uncertainty. If we are lucky we can devise experimental
methods to integrate these fluxes – for example by purposeful tracer
experiments as Jim Ledwell has proposed, – but we must be careful in the

----

[1]This is in a special issue of J.Geophys. Res. devoted to the proceedings
of an IUTAM Conference on Ocean Mixing which contains several important
papers on the subject, especially laboratory studies (which were hardly
represented at the Colloquium).

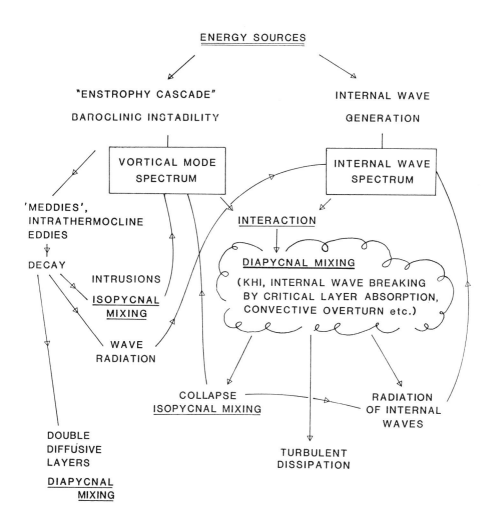

FIG. 1   PROCESSES OF ISO– AND DIA–PYCNAL MIXING.

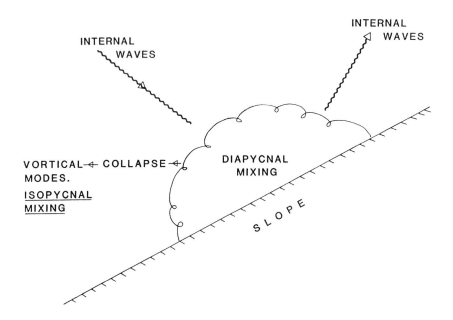

FIG. 2 AN EXAMPLE OF RELATED DIA— AND ISO—PYCNAL MIXING.

experiments to establish what ocean it is we are integrating (is it typical, or not?) and this means much more extensive studies of the environmental conditions, please.

I must refer to one or two of what I found to be additional highlights.

1.   Neil Oakey's attempt to relate turbulence measures such as mixing efficiency to environmental parameters, here the Turner angle. One would like to see other relationships tried. I am reminded too of the variation of diffusion coefficients with Ri which Hartmut Peters and Mike Gregg showed us. These are important advances in our search for appropriate parameterisation for use in numerical models. How this all relates to Ann Gargett's spectral parameterisation or Carl Gibson's fossils is not yet clear to me. Perhaps when the Proceedings are published.....?

2.   Trevor McDougall's work on neutral surfaces, especially the large
slope differences between neutral and potential density surfaces which
might lead to vertical velocity differences in order 10^{-6} ms^{-1}. Never
again shall we rest easy on our potential surfaces!

3.   The computer simulations shown as movie films, Greg Holloway's and
especially Anders Stigebrandt's. These help to give considerable insight
into interacting physical processes.

4.   The rapid responses found in diurnal cycling in the Equatorial
Undercurrent observed by Hartmut Peters and Mike Gregg. (Karen Thomas,
1987, at IOS, Wormley, U.K. has reported a nocturnal cooling of 30mK
reaching to 145m in 3 hrs W. of Portugal in the Spring time). It is almost
as if the day-time layer reaches quasi-dynamic equilibrium which is
suddenly upset when the surface heat flux changes sign! I am reminded of
the ideas of marginal stability of turbulent flows, ideas going back to
Willem Malkus.

There were other points of excitement. In Table 1 are my
selection of 'Quotes of the Week' and, since I am now at a University, I
have naturally offered marks (out of 10).

Finally let me speculate on the highlights of the next Liège
meeting on Ocean Turbulence (in 1995?).

1.   We shall see considerable success in the use of natural tracers
and inverse modelling, probably with World Ocean eddy-resolving models
running.

2.   We shall see the impact of some international, and
multi-disciplinary, programmes such as the World Ocean Circulation
Experiment and the (presently styled) Joint Global Ocean Flux Study which
will lend further stimulus to the study of biologically and chemically
related problems of mixing.

3.   We shall have a better idea of the life-story of vortical modes,
the 'cosmology' I referred to earlier.

4.   One of the joys of coming to an international meeting, apart from
meeting old, and making new, friends, is that of learning new acronyms.

540

QUOTES OF THE WEEK.   Marks/10
'We are discovering more and more ways in which the ocean   9
can dissipate energy'.   (P.Rhines in his talk. This is
something which makes the subject so fascinating and challenging)

'....100 km scale processes being controlled by centimetre   9
scale things'. (N. Oakey about microstructure. This too
is part of the fascination – the subtle relation of scales).

'....rebellious working group...' (K.Fedovov, about SCOR   6
Working Group 69. 'Vive la révolution')

'....I offer a bottle of wine.....' (E. Kraus in introducing   20
one of his talks in which he asked the audience to spot an error
he had found. A most subtle way to keep audience attention:
I offer 20 out of 10 in the hope of reward).

'Does it matter to the models that there is a depth   10
variation in K?' (A. Gargett in a discussion. Getting to
the heart of the matter. Full marks!).

'We will all be dissipated'. (P.Rhines, in a discussion   9
relating to energy cascade. No doubt!)

'Anything that wriggles is turbulence' (denied by C. Gibson,   0
in his talk on fossil turbulence).

'It refuses to move' (Projectionist having difficulties
in R.Pingree's talk)   -10

'Let us suppose that eddies are a set of deformable   0
billiard balls.' (A. Gargett's assumption in a model to
explain different diffusive rates for T and S. No Ann,
you won't win the World Snooker Championship that way!)

'There is nothing to beat a really good equation'
(T.McDougall, on seeing a computer simulation movie).   0

'It's the same story but it looks better in colour'
(S. Piacsek, on introducing coloured slides of salt fingers.
A late entry, unmarked).

TABLE 1

Here are some of this week's: VKE, HRCP, NSPV, AMP, LPCP, EKE, BERTHA,
IPV, C-SALT. At the next meeting there are bound to be more!

May I, on your behalf, and that of the Scientific Programme
Committee, express our thanks to all speakers and presenters of posters,
but particularly to the invited speakers, for taking the trouble to
explain their work so carefully, and especially our gratitude to
Professor Jacques Nihoul and his Secretariat for hosting the meeting. It
has been a most pleasant and scientifically stimulating occasion.

REFERENCES

Munk, W.H. 1966 Abyssal recipes. Deep-Sea Res. 13, 707-730.

Shay, T.M. & Gregg, M.C. 1984 Turbulence in an oceanic convective mixed
    layer. Nature,. 310, 282-285 (see also 311, 84).

Thomas, K.J.H. 1987 A Lagrangian study of the diurnal heating of the
    upper ocean.   Ph.D. Dissertation, University of Southampton.
    Unpublished.

Thorpe, S.A. 1985. Small-scale processes in the upper ocean boundary
    layer. Nature, 318, 519-522.

Thorpe, S.A. 1987 Transitional phenomena and the development of
    turbulence in stratified fluids: a review.
    J.Geophys. Res. 92, 5231-5248.